Advances in Intelligent Systems and Computing

Volume 872

The series "Advances in Intelligent Systems and Computing" contains publications on theory, applications, and design methods of Intelligent Systems and Intelligent Computing. Virtually all disciplines such as engineering, natural sciences, computer and information science, ICT, economics, business, e-commerce, environment, healthcare, life science are covered. The list of topics spans all the areas of modern intelligent systems and computing such as: computational intelligence, soft computing including neural networks, fuzzy systems, evolutionary computing and the fusion of these paradigms, social intelligence, ambient intelligence, computational neuroscience, artificial life, virtual worlds and society, cognitive science and systems, Perception and Vision, DNA and immune based systems, self-organizing and adaptive systems, e-Learning and teaching, human-centered and human-centric computing, recommender systems, intelligent control, robotics and mechatronics including human-machine teaming, knowledge-based paradigms, learning paradigms, machine ethics, intelligent data analysis, knowledge management, intelligent agents, intelligent decision making and support, intelligent network security, trust management, interactive entertainment, Web intelligence and multimedia.

The publications within "Advances in Intelligent Systems and Computing" are primarily proceedings of important conferences, symposia and congresses. They cover significant recent developments in the field, both of a foundational and applicable character. An important characteristic feature of the series is the short publication time and world-wide distribution. This permits a rapid and broad dissemination of research results.

More information about this series at http://www.springer.com/series/11156

Bing-Yuan Cao · Yu-Bin Zhong
Editors

Fuzzy Sets and Operations Research

Editors
Bing-Yuan Cao
Foshan University
Foshan, Guangdong, China

and

Guangzhou Vocational
College of Science and Technology
Guangzhou, China

Yu-Bin Zhong
School of Mathematics
and Information Science
Guangzhou University
Guangzhou, China

ISSN 2194-5357 ISSN 2194-5365 (electronic)
Advances in Intelligent Systems and Computing
ISBN 978-3-030-02776-6 ISBN 978-3-030-02777-3 (eBook)
https://doi.org/10.1007/978-3-030-02777-3

Library of Congress Control Number: 2018959138

This Springer imprint is published by the registered company Springer Nature Switzerland AG
The registered company address is: Gewerbestrasse 11, 6330 Cham, Switzerland

Preface

The Ninth Academic Conference of Fuzzy Information and Engineering Branch, China Operations Research Society (ACFIEBORSC'2017) on July 13–18, 2017, in Hu He Hao Te; and The Second Representative Conference and the Third Academic Conference of the Operational Research Society of Guangdong Province (RCORCGD'2017) on December 29–30, 2017, in Shenzhen, were held separately, both in China.

The monograph is published by *Advances in Intelligent and Soft Computing* (AISC), Springer, ISSN: 1867-5662.

We have received more than 100 submissions from both conferences this year. After each paper of them undergoes a rigorous review process, 35 papers from the conference were accepted because only high-quality papers are included in the book, which are divided into six main parts:

Part 1 focuses on "Fuzzy Measure and Integral."
Part 2 themes on "Fuzzy Topology and Algebras."
Part 3 topics discussed on "Classification and Recognition."
Part 4 has ideas circling around "Control and Fuzziness."
Part 5 has dissertations on "Extension of Fuzzy Set and System."
Part 6 subjects on "OR."
Part 7 focuses on "Others."

We appreciate sponsors from Fuzzy Information and Engineering Branch of ORSC; Operations Research Society of Guangdong Province; China Guangdong, and undertakers from Shenzhen University, Guangdong, China.

Our heartfelt thanks to Liaoning University of Engineering and Technology; Hong Kong and Macao Operations Research Society; Foshan University, China; Room Dispenser Live Rooms co., China; China Charity Magazine; HKU Alumni Association; Yunchuang Daye (Shenzhen) Media Co., Ltd.; Shenzhen Institute of Excellence Innovation; Wong Wang Health Consulting Management (Shenzhen) Co., Ltd.; China Education and Research Foundation for co-sponsorships.

We are showing gratitude to Academician Zhang Jing-zhong of the Chinese Academy of Sciences; Prof. Wang Peizhuang, a pioneer of fuzzy mathematics in China; President Hao Zhi-feng of Foshan University; and Prof. N. Hadi of Mazandaran University from Iran to participate and to make reports.

We are grateful to the Editorial Committee and reviewers for their selfless dedication. We wish to express our heartfelt appreciation to all the authors and participants for their great contributions that made these conferences possible and all the hard work worthwhile.

Finally, we thank the publisher, Springer, for publishing the AISC (Note: Our series of conference proceedings by Springer), and by China Science and Education Press (Hong Kong).

March 2018

Bing-Yuan Cao
Yu-Bin Zhong

Organization

Conference General Chair

Bing-yuan Cao, China

Honorary Chair

Lotfi A. Zadeh, USA

Co-chair

Zhi-feng Hao, China

Steering Committee

J. C. Bezdek, USA
Z. Bien, Korea
D. Dubois, France
Gui-rong Guo, China
M. M. Gupta, Canada
Xin-gui He, China
Abraham Kandel, Hungary
J. Kacprzyk, Poland
E. Mamdani, UK
R. P. Nikhil, India
M. Sugeno, Japan
Hao Wang, China
P. Z. Wang, USA
Witold Pedrycz, Canada
Jing-zhong Zhang, China
H. J. Zimmermann, Germany

Program Committee of ACFIEBORSC'2017

Chair

Zeng-liang Liu, China

Secretary

Yu-bin Zhong, China

Program Committee of RCORCGD'2017

Chair

Bing-yuan Cao

Secretary

Xue-hai Fan, China

Members

R. Ameri, Iran
K. Asai, Japan
Shi-zhong Bai, China
J. P. Barthelemy, France
Rajabali Borzooei, Iran
Tian-you Chai, China
Guo-qing Chen, China
Mian-yun Chen, China
Shui-li Chen, China
Ovanes Chorayan, Russia
Sen-lin Cheng, China
He-pu Deng, Australia
Ali Ebrahimnejad, Iran
M. Fedrizzi, Italy
Jia-li Feng, China
Yin-jun Feng, China
Kai-zhong Guo, China
Si-cong Guo, China
Ming-hu Ha, China
Li-yan Han, China
Cheng-ming Hu, USA
Bao-qing Hu, China
Zhe-xue Huang, Australia
Hiroshi Inoue, Japan
Li-min Jia, China
X. Q. Jin, Macau
Guy Jumarie, Canada
Jim Keller, USA
E. E. Kerre, Belgium
K. H. Kim, USA
N. Kuroki, Japan
D. Lakov, Bulgaria
Tsu-Tian Lee, Taiwan China
Dong-hui Li, China
Hong-xing Li, China
Jun Li, China
Tai-fu Li, China
Yu-cheng Li, China
T. Y. Lin, USA
Zhong-fei Li, China

Bao-ding Liu, China
Zhi-qiang Liu, Hongkong
Ming Ma, China
Sheng-quan Ma, China
D. Dutta Majumder, India
Hong-hai Mi, China
M. Mizumoto, Japan
Zhi-wen Mo, China
J. Motiwalla, Singapore
M. Mukaidono, Japan
J. Mustonen, Finland
Shohachiro Nakanishi, Japan
Michael Ng, Hongkong
Jin-ping Ou, China
Hadi Nasseri, Iran
H. Prade, France
D. A. Ralescu, USA
E. Sanchez, France
Victor V. Senkevich, Russia
Qiang Shen, UK
Kai-quan Shi, China
Jie Song, China
Zhen-ming Song, China
Lan Su, China
Enric Trillas, Spain
P. Veeramani, India
José Luis Verdegay, Spain
Guo-qing Wang, China
Xi-zhao Wang, China
Xue-ping Wang, China
Zhen-yuan Wang, USA
Fu-yi Wei, China
Xin-jiang Wei, China
Ber-lin Wu, Taiwan China
Yun-dong Wu, China
Ci-yuan Xiao, China
Yang Xu, China
Ze-shui Xu, China
R. R. Yager, USA
T. Yamakawa, Japan
Liang-zhong Yi, China
Xue-hai Yuan, China
Qiang Zhang, China
Cheng-yi Zhang, China
Yun-jie Zhang, China
Shao-hui Zeng, Hongkong
Ya-lin Zheng, China
Kai-qi Zou, China

Contents

Fuzzy Measure and Integral

Trapezoidal Intuitionistic Approximations of Intuitionistic Fuzzy Numbers Preserving the Width

Shu-yang Li[1,2(✉)] and Hong-xing Li[1]

[1] School of Control Science and Engineering, Dalian University of Technology, Dalian 116024, China
lishuyang0515@163.com
[2] Department of Basic Science, Dalian Naval Academy, Dalian 116018, China

Abstract. The problem is discussed about approximating a sequence of intuitionistic fuzzy numbers or aggregating the sequence of intuitionistic fuzzy numbers and then approximating the output of aggregation with condition of preserving the width. An interesting conclusion is obtained.

Keywords: Intuitionistic fuzzy numbers
Trapezoidal intuitionistic approximation · Width

1 Introduction

Intuitionistic fuzzy sets have been used widely in various fields, since they can better depicit the vague information. The numbers acts as the main carriers for conveying information, and thus intuitionistic fuzzy numbers play an important role among intuitionistic fuzzy sets. The simple and regular membership functions and non-membership functions will produce effortless calculation and they are much more direct and comprehensible. So some researchers have shown great interest in approximation of intuitionistic fuzzy numbers.

Ban [1] gave the nearest interval approximations of intuitionistic fuzzy numbers. Then Ban [2] discussed trapezoidal approximations which were expressed by value, width, width and weighted expected value. Ban and Coroianu [3] presented a new approach to obtain trapezoidal approximations of intuitionistic fuzzy numbers. When approximation methods of intuitionistic fuzzy numbers were investigated, more and more researchers [4–6] have paid attention to aggregation of intuitionistic fuzzy information. One basic question should be discussed. Should the approximation be done before aggregation or after aggregation? The problem is discussed about approximating the given intuitionistic fuzzy numbers or aggregating the given intuitionistic fuzzy numbers and then approximating the output of aggregation with condition of preserving the width. The results of the two methods are compared in this paper.

This paper is organised in the following manner. Some important definitions and propositions of intuitionistic fuzzy numbers are given in Sect. 2. Main results on the

B.-Y. Cao and Y.-B. Zhong (Eds.): ICFIE 2017, AISC 872, pp. 3–10, 2019.
https://doi.org/10.1007/978-3-030-02777-3_1

approximation with restrictions on the width are discussed in Sect. 3. Finally, conclusions are given in Sect. 4.

2 Preliminaries

Definition 2.1 [7]. Let X be a given nonempty set. An intuitionistic fuzzy set in X is an object A given by $A = \{\langle x, \mu_A(x), \nu_A(x)\rangle : x \in X\}$, where $\mu_A : X \to [0,1]$ and $\nu_A : X \to [0,1]$ satisfy the condition $0 \le \mu_A(x) + \nu_A(x) \le 1$, for every $x \in X$.

Definition 2.2 [8]. An intuitionistic fuzzy set $A = \{\langle x, \mu_A(x), \nu_A(x)\rangle : x \in X\}$ such that μ_A and $1 - \nu_A$, where $(1-\nu_A)(x) = 1 - \nu_A(x)$, $\forall x \in R$, are fuzzy numbers is called an intuitionistic fuzzy number.

Let $A = \langle \mu_A, \nu_A\rangle$ be intuitionistic fuzzy number, where $\mu_A = \mu_A(t_1, t_2, t_3, t_4)$ and $1 - \nu_A = (1-\nu_A)(s_1,\ s_2, s_3,\ s_4)$. Let us adopt the following notation [9]:

$$l = \frac{t_1+t_2}{2}, u = \frac{t_3+t_4}{2}, x = t_2 - t_1, y = t_4 - t_3,$$
$$l' = \frac{s_1+s_2}{2}, u' = \frac{s_3+s_4}{2}, x' = s_2 - s_1, y' = s_4 - s_3.$$

Thus according to the above notations, $\mu_A(t_1, t_2, t_3, t_4) \equiv \mu_A(l, u, x, y)$, $(1-\nu_A)(s_1, s_2, \ldots s_3, s_4) \equiv (1-\nu_A)(l', u', x', y')$,

$$(\mu_A)_\alpha = \left[l + x\left(\alpha - \frac{1}{2}\right), u - y\left(\alpha - \frac{1}{2}\right)\right], \tag{1}$$

$$(1-\nu_A)_\alpha = \left[l' + x'\left(\alpha - \frac{1}{2}\right), u' - y'\left(\alpha - \frac{1}{2}\right)\right]. \tag{2}$$

Proposition 2.1 [10]. Let $A = \langle \mu_A, \nu_A\rangle$ *and* $B = \langle \mu_B, \nu_B\rangle$ be intuitionistic fuzzy numbers, where $\mu_A = \mu_A(l_1, u_1, x_1, y_1)$, $1 - \nu_A = (1-\nu_A)(l'_1, u'_1, x'_1, y'_1)$, $\mu_B = \mu_B(l_2, u_2, x_2, y_2)$, $1 - \nu_B = (1-\nu_B)(l'_2, u'_2, x'_2, y'_2)$. Then

$$\begin{aligned} d^2(A,B) = &\frac{1}{2}(l_1 - l_2)^2 + \frac{1}{2}(u_1 - u_2)^2 + \frac{1}{24}(x_1 - x_2)^2 + \frac{1}{24}(y_1 - y_2)^2 \\ &+ \frac{1}{2}(l'_1 - l'_2)^2 + \frac{1}{2}(u'_1 - u'_2)^2 + \frac{1}{24}(x'_1 - x'_2)^2 + \frac{1}{24}(y'_1 - y'_2)^2. \end{aligned}$$

Proposition 2.2 [10]. Let $A = \langle \mu_A, \nu_A\rangle$ be intuitionistic fuzzy number, where $\mu_A = \mu_A(l, u, x, y)$, $1 - \nu_A = (1-\nu_A)(l', u', x', y')$. Then $A = \langle \mu_A, \nu_A\rangle$ is a trapezoidal intuitionistic fuzzy number iff $x, y, x, y' \ge 0, x + y \le 2(u - l)$ *and* $x' + y' \le 2(u' - l')$.

Proposition 2.3 [10]. Let $A = \langle \mu_A, \nu_A \rangle$ be intuitionistic fuzzy number and $T_e(A) = \langle \mu_{T_e}, \nu_{T_e} \rangle$ be extended trapezoidal intuitionistic approximation of A. Then

$$d^2(A,X) = d^2(A, T_e(A)) + d^2(T_e(A), X)$$

holds for every extended trapezoidal intuitionistic fuzzy number $X = \langle \mu_X, \nu_X \rangle$.

Proposition 2.4 [11]. The width of intuitionistic fuzzy number $A = \langle \mu_A, \nu_A \rangle$ is defined as

$$W(A) = \frac{1}{2}\int_0^1 \left((\mu_A)_U(\alpha) - (\mu_A)_L(\alpha)\right)d\alpha + \frac{1}{2}\int_0^1 \left((1-\nu_A)_U(\alpha) - (1-\nu_A)_L(\alpha)\right)d\alpha.$$

Substituting into Proposition 2.4 by Eqs. (1) and (2), we get the following result.

Proposition 2.5. Let $A = \langle \mu_A, \nu_A \rangle$ be intuitionistic fuzzy number, where $\mu_A = \mu_A(l,u,x,y)$, $1-\nu_A = (1-\nu_A)(l',u',x',y')$, then the width of A is given by

$$W(A) = \frac{1}{2}(u - l + u' - l').$$

From Propositions 2.3, 2.4 and Eqs. (12) and (19) in [10], we can proof the following conclusion.

Proposition 2.6. Let $T_e(A)$ be extended trapezoidal intuitionistic approximation of intuitionistic fuzzy number A. Then $W(A) = W(T_e(A))$.

In the following section, the Karush-Kuhn-Tucker theorem will be used to prove the main results of this paper.

Theorem 2.1 [12]. Let $f, g_1, \cdots, g_m : R^n \to R$ be convex and differentiable functions. Then $\bar{x}$ solves the convex programming problem

$\min f(x)$

满足$g_i(x) \le h_i, i \in \{1, \cdots, m\}$

if and only if there exists $\xi_i, i \in \{1, \cdots, m\}$, such that

$$\begin{aligned}
&(\text{i})\nabla f(\bar{x}) + \sum_{i=1}^{m} \xi_i \nabla g_i(\bar{x}) = 0;\\
&(\text{ii}) g_i(\bar{x}) - h_i \le 0;\\
&(\text{iii}) \xi_i \ge 0;\\
&(\text{iv}) \xi_i(h_i - g_i(\bar{x})) = 0.
\end{aligned}$$

3 Trapezoidal Intuitionistic Approximation Preserving the Width

3.1 Approximation of Intuitionistic Fuzzy Numbers

The width of intuitionistic fuzzy numbers $A_1, A_2, \cdots, A_n$ can be defined as [8]:

$$W(A_1, A_2, \cdots, A_n) = \frac{1}{n}(W(A_1) + \cdots + W(A_n)). \tag{3}$$

By Proposition 2.6, we have

$$W(A_1, A_2, \cdots, A_n) = \frac{1}{2n}\sum_{i=1}^{n}\left(u_i^e - l_i^e + u_i'^e - l_i'^e\right) \tag{4}$$

Now we want to approximate intuitionistic fuzzy numbers $A_1, A_2, \cdots, A_n$. $T_{(A_1,\cdots,A_n)} = \left\langle \mu_{T_{(A_1,\cdots,A_n)}}, \nu_{T_{(A_1,\cdots,A_n)}} \right\rangle$ is called trapezoidal intuitionistic approximation preserving the width of $A_1, A_2, \cdots, A_n$ if $\min_{B \in IF^T(\mathrm{R})} d^2((A_1, A_2, \cdots, A_n), B) = d^2\left((A_1, A_2, \cdots, A_n), T_{(A_1,\cdots,A_n)}\right)$ under the condition

$$W(A_1, A_2, \cdots, A_n) = W\left(T_{(A_1,\cdots,A_n)}\right), \tag{5}$$

where $\mu_{T_{(A_1,\cdots,A_n)}} = \mu_{T_{(A_1,\cdots,A_n)}}(l, u, x, y)$, $1 - \nu_{T_{(A_1,\cdots,A_n)}} = \left(1 - \nu_{T_{(A_1,\cdots,A_n)}}\right)(l', u', x', y')$. Taking into account Proposition 2.3, to minimize $d^2\left(A_i, T_{(A_1,\cdots,A_n)}\right)$ is as to minimizing $d^2\left(T_e(A_i), T_{(A_1,\cdots,A_n)}\right)$. $T_e(A_i) = \langle \mu_{T_i}, \nu_{T_i} \rangle$ is weighted extended trapezoidal intuitionistic approximation of A_i, where $\mu_{T_i} = \mu_{T_i}\left(l_i^e, u_i^e, \delta_i^e, \sigma_i^e\right), 1 - \nu_{T_i} = (1 - \nu_{T_i})\left(l_i'^e, u_i'^e, \delta_i'^e, \sigma_i'^e\right)$. By Proposition 2.1, we have

$$\begin{aligned}\sum_{i=1}^{n} d^2\left(T_e(A_i), T_{(A_1,\cdots,A_n)}\right) = &\frac{1}{2}\sum_{i=1}^{n}\left(l - l_i^e\right)^2 + \frac{1}{2}\sum_{i=1}^{n}\left(u - u_i^e\right)^2 + \frac{1}{24}\sum_{i=1}^{n}\left(x - x_i^e\right)^2 + \frac{1}{24}\sum_{i=1}^{n}\left(y - y_i^e\right)^2 \\ &+ \frac{1}{2}\sum_{i=1}^{n}\left(l' - l_i'^e\right)^2 + \frac{1}{2}\sum_{i=1}^{n}\left(u' - u_i'^e\right)^2 + \frac{1}{24}\sum_{i=1}^{n}\left(x' - x_i'^e\right)^2 + \frac{1}{24}\sum_{i=1}^{n}\left(y' - y_i'^e\right)^2.\end{aligned} \tag{6}$$

From Eqs. (3) and (5), we get

$$u - l + u' - l' = \frac{1}{n}\sum_{i=1}^{n}\left(u_i^e - l_i^e + u_i'^e - l_i'^e\right) \tag{7}$$

Let $k_1 = l - u + \frac{1}{n}\sum_{i=1}^{n} u_i^e - \frac{1}{n}\sum_{i=1}^{n} l_i^e + \frac{1}{n}\sum_{i=1}^{n} u_i'^e - \frac{1}{n}\sum_{i=1}^{n} l_i'^e$, $h_1 = l' - u_i'^e + k_1$. Substituting into Eq. (6) by Eq. (7), we get

$$\sum_{i=1}^{n} d^2(T_e(A_i), T_{(A_1,\cdots,A_n)}) = \frac{1}{2}\sum_{i=1}^{n}(l - l_i^e)^2 + \frac{1}{2}\sum_{i=1}^{n}(u - u_i^e)^2 + \frac{1}{24}\sum_{i=1}^{n}(x - x_i^e)^2 + \frac{1}{24}\sum_{i=1}^{n}(y - y_i^e)^2 + \frac{1}{2}\sum_{i=1}^{n}(l' - l_i'^e)^2 + \frac{1}{2}\sum_{i=1}^{n}h_1^2 + \frac{1}{24}\sum_{i=1}^{n}(x' - x_i'^e)^2 + \frac{1}{24}\sum_{i=1}^{n}(y' - y_i'^e)^2. \tag{8}$$

So minimum of Eq. (8) can be reduced to the following problem.

Problem 3.1. Find $(l, u, x, y, l', x', y') \in \Omega_1$, which minimizes Eq. (8) under the condition Eq. (7), where

$$\Omega_1 = \{(l, u, x, y, l', x', y') \in R^7 | x, y, x', y' \geq 0, x + y \leq 2(u - l), x' + y' \leq 2k_1\}.$$

According to Theorem 2.1, $T_{(A_1,\cdots,A_n)} = \left\langle \mu_{T_{(A_1,\cdots,A_n)}}, \nu_{T_{(A_1,\cdots,A_n)}} \right\rangle$ is a solution of Problem 3.1 if and only if there exist $\vartheta_1, \vartheta_2, \vartheta_3, \vartheta_4, \vartheta_5, \vartheta_6$ such that the following system of equations holds:

$$\begin{aligned} &A_1 + \sum_{i=1}^{6}\vartheta_i B_{1i} = 0,\\ &C_1 = 0,\\ &F_1 \geq 0, \end{aligned} \tag{9}$$

where

$A_1 = \left(\sum_{i=1}^{n}(l - l_i^e) + \sum_{i=1}^{n}h_1, \sum_{i=1}^{n}(u - u_i^e) - \sum_{i=1}^{n}h_1, \frac{1}{12}\sum_{i=1}^{n}(x - x_i^e), \frac{1}{12}\sum_{i=1}^{n}(y - y_i^e),\right.$ $\left.\sum_{i=1}^{n}(l' - l_i'^e) + \sum_{i=1}^{n}h_1, \frac{1}{12}\sum_{i=1}^{n}(x' - x_i'^e), \frac{1}{12}\sum_{i=1}^{n}(y' - y_i'^e)\right)^T$, $B_{11} = (0, 0, -1, 0, 0, 0, 0)^T$, $B_{12} = (0, 0, 0, -1, 0, 0, 0)^T$, $B_{13} = (0, 0, 0, 0, 0, -1, 0)^T$, $B_{14} = (0, 0, 0, 0, 0, 0, -1)^T$, $B_{15} = (2, -2, 1, 1, 0, 0, 0)^T$, $B_{16} = (-2, 2, 0, 0, 0, 1, 1)^T$, $C_1 = (\vartheta_1 x, \vartheta_2 y, \vartheta_3 x', \vartheta_4 y', \vartheta_5 (x + y - 2u + 2l), \vartheta_6(x' + y' - 2k_1))^T$, $F_1 = diag\{x, y, x', y', 2(u - l) - x - y, 2k_1 - x' - y', \vartheta_1, \vartheta_2, \vartheta_3, \vartheta_4, \vartheta_5, \vartheta_6\}$.

3.2 Approximation of an Aggregation

Now we want to approximate the intuitionistic fuzzy number $\bar{A} = \frac{1}{n}(A_1 + A_2 + \cdots + A_n)$ under the condition

$$W(\bar{A}) = W(T_{\bar{A}}) \tag{10}$$

$T_{\bar{A}} = \left\langle \mu_{T_{\bar{A}}}, \nu_{T_{\bar{A}}} \right\rangle$ is called trapezoidal intuitionistic approximation preserving the width of $\bar{A}$ if $\min_{B \in IF^T(R)} d^2(\bar{A}, B) = d^2(\bar{A}, T_{\bar{A}})$, where $\mu_{T_{\bar{A}}} = \mu_{T_A}(\bar{l}, \bar{u}, \bar{x}, \bar{y})$, $1 - \nu_{T_{\bar{A}}} = (1 - \nu_{T_{\bar{A}}})(\bar{l}', \bar{u}', \bar{x}', \bar{y}')$, with respect to Eq. (34). By Propositions 2.5 and 2.6, we have

$$W(\bar{A}) = \frac{1}{2}\left(\frac{1}{n}\sum_{i=1}^{n} u_i^e - \frac{1}{n}\sum_{i=1}^{n} l_i^e + \frac{1}{n}\sum_{i=1}^{n} u_i'^e - \frac{1}{n}\sum_{i=1}^{n} l_i'^e\right). \tag{11}$$

From Eqs. (10) and (11), we get

$$\bar{u} - \bar{l} + \bar{u}' - \bar{l}' = \frac{1}{n}\sum_{i=1}^{n} u_i^e - \frac{1}{n}\sum_{i=1}^{n} l_i^e + \frac{1}{n}\sum_{i=1}^{n} u_i'^e - \frac{1}{n}\sum_{i=1}^{n} l_i'^e. \tag{12}$$

Taking into account Proposition 2.3, to minimize $d^2(\bar{A}, T_{\bar{A}})$ is as to minimizing $d^2(T_e(\bar{A}), T_{\bar{A}})$. $T_e(\bar{A}) = \langle \mu_{\bar{T}_e}, v_{\bar{T}_e} \rangle$ is weighted extended trapezoidal intuitionistic approximation of the intuitionistic fuzzy number $\bar{A}$, where $\mu_{\bar{T}_e} = \mu_{\bar{T}_e}\left(\frac{1}{n}\sum_{i=1}^{n} l_i^e, \frac{1}{n}\sum_{i=1}^{n} u_i^e, \frac{1}{n}\sum_{i=1}^{n} x_i^e, \frac{1}{n}\sum_{i=1}^{n} y_i^e\right)$, $1 - v_{\bar{T}_e} = (1 - v_{\bar{T}_e})\left(\frac{1}{n}\sum_{i=1}^{n} l_i'^e, \frac{1}{n}\sum_{i=1}^{n} u_i'^e, \frac{1}{n}\sum_{i=1}^{n} x_i'^e, \frac{1}{n}\sum_{i=1}^{n} y_i'^e\right)$.

From Proposition 2.1, we have

$$\begin{aligned} d^2(T_e(\bar{A}), T_{\bar{A}}) = & \frac{1}{2}\left(\bar{l} - \frac{1}{n}\sum_{i=1}^{n} l_i^e\right)^2 + \frac{1}{2}\left(\bar{u} - \frac{1}{n}\sum_{i=1}^{n} u_i^e\right)^2 + \frac{1}{24}\left(\bar{x} - \frac{1}{n}\sum_{i=1}^{n} x_i^e\right)^2 \\ & + \frac{1}{24}\left(\bar{y} - \frac{1}{n}\sum_{i=1}^{n} y_i^e\right)^2 + \frac{1}{2}\left(\bar{l}' - \frac{1}{n}\sum_{i=1}^{n} l_i'^e\right)^2 + \frac{1}{2}\left(\bar{u}' - \frac{1}{n}\sum_{i=1}^{n} u_i'^e\right)^2 \\ & + \frac{1}{24}\left(\bar{x}' - \frac{1}{n}\sum_{i=1}^{n} x_i'^e\right)^2 + \frac{1}{24}\left(\bar{y}' - \frac{1}{n}\sum_{i=1}^{n} y_i'^e\right)^2. \end{aligned} \tag{13}$$

Let $k_2 = \bar{l} - \bar{u} + \frac{1}{n}\sum_{i=1}^{n} u_i^e - \frac{1}{n}\sum_{i=1}^{n} l_i^e + \frac{1}{n}\sum_{i=1}^{n} u_i'^e - \frac{1}{n}\sum_{i=1}^{n} l_i'^e, h_2 = \bar{l}' - \frac{1}{n}\sum_{i=1}^{n} u_i'^e + k_2$.
Substituting into Eq. (13) by Eq. (12), we get

$$\begin{aligned} d^2(T_e(\bar{A}), T_{\bar{A}}) = & \frac{1}{2}\left(\bar{l} - \frac{1}{n}\sum_{i=1}^{n} l_i^e\right)^2 + \frac{1}{2}\left(\bar{u} - \frac{1}{n}\sum_{i=1}^{n} u_i^e\right)^2 + \frac{1}{24}\left(\bar{x} - \frac{1}{n}\sum_{i=1}^{n} x_i^e\right)^2 \\ & + \frac{1}{24}\left(\bar{y} - \frac{1}{n}\sum_{i=1}^{n} y_i^e\right)^2 + \frac{1}{2}\left(\bar{l}' - \frac{1}{n}\sum_{i=1}^{n} l_i'^e\right)^2 + \frac{1}{2}h_2^2 \\ & + \frac{1}{24}\left(\bar{x}' - \frac{1}{n}\sum_{i=1}^{n} x_i'^e\right)^2 + \frac{1}{24}\left(\bar{y}' - \frac{1}{n}\sum_{i=1}^{n} y_i'^e\right)^2. \end{aligned} \tag{14}$$

So minimum of Eq. (14) can be reduced to the following problem.

Problem 3.2. Find $(\bar{l},\bar{u},\bar{x},\bar{y},\bar{l}',\bar{x}',\bar{y}')\in\Omega_2$, which minimizes Eq. (14) under the condition Eq. (12), where

$$\Omega_2=\{(\bar{l},\bar{u},\bar{x},\bar{y},\bar{l}',\bar{x}',\bar{y}')\in R^7|\bar{x},\bar{y},\bar{x}',\bar{y}'\geq 0,\bar{x}+\bar{y}\leq 2(\bar{u}-\bar{l}),\bar{x}'+\bar{y}'\leq 2k_2\}.$$

According to Theorem 2.1, $T_{\bar{A}}=\left\langle \mu_{T_{\bar{A}}},\nu_{T_{\bar{A}}}\right\rangle\in IF^T(R)$ is a solution of Problem 3.2 if and only if there exist $\tau_1,\tau_2,\tau_3,\tau_4,\tau_5,\tau_6$ such that the following system of equations holds:

$$\begin{aligned}&A_2+\sum_{i=1}^{6}\tau_i B_{2i}=0,\\&C_2=0,\\&F_2\geq 0,\end{aligned}\tag{15}$$

where
$A_2=\left\{\bar{l}-\frac{1}{n}\sum_{i=1}^{n}l_i^e+h_2,\bar{u}-\frac{1}{n}\sum_{i=1}^{n}u_i^e-h_2,\frac{1}{12}\left(\bar{x}-\frac{1}{n}\sum_{i=1}^{n}x_i^e\right),\frac{1}{12}\left(\bar{y}-\frac{1}{n}\sum_{i=1}^{n}y_i^e\right),\bar{l}'-\frac{1}{n}\sum_{i=1}^{n}l_i'^e+h_2,\frac{1}{12}\left(\bar{x}'-\frac{1}{n}\sum_{i=1}^{n}x_i'^e\right),\frac{1}{12}\left(\bar{y}'-\frac{1}{n}\sum_{i=1}^{n}y_i'^e\right)\right\}^T$, $B_{21}=(0,0,-1,0,0,0,0)^T$, $B_{22}=(0,0,0,-1,0,0,0)^T$, $B_{23}=(0,0,0,0,0,-1,0)^T$, $B_{24}=(0,0,0,0,0,0,-1)^T$, $B_{25}=(2,-2,1,1,0,0,0)^T$, $B_{26}=(-2,2,0,0,0,1,1)^T$, $C_2=(\tau_1\bar{x},\tau_2\bar{y},\tau_3\bar{x}',\tau_4\bar{y}',\tau_5(\bar{x}+\bar{y}-2\bar{u}+2\bar{l}),\tau_6(\bar{x}'+\bar{y}'-2k_2))^T$, $F_2=diag\{\bar{x},\bar{y},\bar{x}',\bar{y}',2(\bar{u}-\bar{l})-\bar{x}-\bar{y},2k_2-\bar{x}'-\bar{y}',\tau_1,\tau_2,\tau_3,\tau_4,\tau_5,\tau_6\}$.

If we compare (7) with (12) and (9) with (15), the bellow result can be obtained immediately.

Theorem 3.1. The trapezoidal intuitionistic approximation preserving the width of $A_1,A_2,\cdots,A_n$ is the trapezoidal intuitionistic approximation preserving the width of $\bar{A}$.

Proof. We can get Eqs. (15) substituting $\vartheta_1,\vartheta_2,\vartheta_3,\vartheta_4,\vartheta_5,\vartheta_6$ in Eqs. (9) by $n\tau_1,n\tau_2,n\tau_3,n\tau_4,n\tau_5,n\tau_6$. Since the width introduced in Proposition 2.4 is linear, from (3) we obtain.

$$W(A_1+\cdots+A_n)=\frac{1}{n}(W(A_1)+\cdots+W(A_n))=W\left(\frac{1}{n}(A_1+\cdots+A_n)\right)=W(\bar{A}).$$

4 Conclusion

We obtain a conclusion the approximations preserving the width are the same in two cases if we choose the average as the aggregation operator. The conclusion in Sect. 3 can be probably extended to case of approximation under other additional constraints.

Acknowledgements. Thanks to the support by Scientific Research Development Fund of Dalian Naval Academy and National Natural Science Foundation of China (No.61374118).

References

1. Ban, A.I.: Nearest interval approximation of an intuitionistic fuzzy number. In: Reusch, B. (ed.) Computational Intelligence, Theory and Applications, pp. 229–240. Springer, Heidelberg (2006)
2. Ban, A.I.: Trapezoidal approximations of intuitionistic fuzzy numbers expressed by value, ambiguity, width and weighted expected value. Notes Intuit. Fuzzy Sets **14**(1), 38–47 (2008)
3. Ban, A.I., Coroianu, L.C.: A method to obtain trapezoidal approximations of intuitionistic fuzzy numbers from trapezoidal approximations of fuzzy numbers. Notes Intuit. Fuzzy Sets **15**(1), 13–25 (2009)
4. Wan, S.P., Dong, J.Y.: Power gemetric operators of trapezoidal intuitionistic fuzzy numbers and application to multi-attribute group decision making. Appl. Soft Comput. **29**(1), 153–168 (2015)
5. Xu, J., Wan, S.P., Dong, J.Y.: Aggregating decision information into Atanassov's intuitionistic fuzzy numbers for heterogeneous multi-attribute group decision making. Appl. Soft Comput. **41**(C), 331–351 (2016)
6. Chen, S.M., Chang, C.H.: Fuzzy multiattribute decision making based on transformation techniques of intuitionistic fuzzy values and intuitionistic fuzzy geometric averaging operators. Inf. Sci. **352**(1), 133–149 (2016)
7. Atanassov, K.T.: Intuitionistic fuzzy sets. Fuzzy Sets Syst. **20**(1), 87–96 (1986)
8. Grzegorzewski, P.: Intuitionistic fuzzy numbers-principles, metrics and ranking. In: Atanassov, K.T., Hryniewicz, O., Kacprzyk, J. (eds.) Soft Computing Foundations and Theoretical Aspects, pp. 235–249. Academic House Exit, Warszawa (2004)
9. Yeh, C.T.: Trapezoidal and triangular approximations preserving the expected interval. Fuzzy Sets Syst. **159**(11), 1345–1353 (2008)
10. Li, S.Y., Li, H.X.: An approximation method of intuitionistic fuzzy numbers. J. Intell. Fuzzy Syst. **32**(6), 4343–4355 (2017)
11. Ban, A.I., Lucian, C.C.: Approximate solutions preserving parameters of intuitionistic fuzzy linear systems. Notes Intuit. Fuzzy Sets **17**(1), 58–70 (2011)
12. Rockafellar, R.T.: Convex Analysis. Princeton University Press, Princeton (1970)

New Definition of the Definite Integral of Fuzzy Valued Function Linearly Generated by Structural Elements

Tian-jun Shu(✉) and Zhi-wen Mo

College of Mathematics and Software Science, Sichuan Normal University, Chengdu 610066, China
605519161@qq.com, mozhiwen@263.net

Abstract. A kind of fuzzy distance gives a new definition of the definite integral of fuzzy valued function linearly generated by structural element. Then the new definition is used to study the basic properties of the definite integral of fuzzy valued function linearly generated by structural elements defined on the interval $[a, b]$. They are Newton Leibniz formula, addition together with multiplication, interval additivity, boundedness, local protection and the first mean value theorem for integrals. After that, By studing some theorems within Darboux sum definition of fuzzy valued function linearly generated by structural elements, The first integrable condition and the second integrable condition of fuzzy valued functions for linear generation of structural elements on $[a, b]$ was given. Meanwhile, investigate the integrability of fuzzy valued function linearly generated by structural elements which is continuous or bounded on [a, b].

Keywords: Structural element · Fuzzy valued function
Fuzzy distance · Definite integral · Integrable condition

1 Introduction

Since Zadeh put forward the fuzzy set theory in 1965, after fifty years of development, the fuzzy system theory and its application result have bee remarkable and perfect. The notion of fuzzy valued function for linear generation of structural elements is proposed by Guo in literature [10]. For the definite integral of the fuzzy valued function linearly generated by structural elements, it has the different form of expression because of the different form of fuzzy distance. In this paper, the definite integral of fuzzy valued function linearly generated by structural elements is defined by this fuzzy distance which was given in literature [11]. Then the related properties about the definite integral of fuzzy valued functions linearly generated by structural elements on the interval $[a, b]$ are investigated. Simultaneously, the integrable condition of fuzzy valued function linearly generated by structural elements on $[a, b]$ is given, also discuss the related applications.

B.-Y. Cao and Y.-B. Zhong (Eds.): ICFIE 2017, AISC 872, pp. 11–22, 2019.
https://doi.org/10.1007/978-3-030-02777-3_2

2 Brief Introduction of the Fuzzy Valued Function of Linear Construction Theory

Definition 2.1 [9]. E is the fuzzy structural element over the field R of real numbers. If its membership function E(x)($x \in R$) has following:

(1) E(0) = 1, and E(1+0) = E(–1–0) = 0.
(2) If $x \in [-1, 0)$, then E(x) is increasingly monotonic function being right continuous, and if $x \in (0, 1]$, then E(x) is decreasing monotonic function being left continuous.
(3) If $x \in (-\infty, -1) \cup (1, +\infty)$, then E(x) = 0.

Obviously, E is a regularly convex fuzzy set over the field R, and it is a boundedly closed fuzzy number.

Definition 2.2 [10]. $\widetilde{A}$ is finite fuzzy number. If there is a fuzzy structural element E, and finite real number $a \in R$, r$\in R^+$, such that $\widetilde{A} = a + rE$ ($r \to 0^+$), then $\widetilde{A}$ is said to be a fuzzy number linearly generated by E. All fuzzy numbers linearly generated by E is denoted as the symbol $\varepsilon(E)$, and write $\varepsilon(E) = \{\widetilde{A} \mid \widetilde{A} = a + rE, \forall\, a \in R,\, r \in R^+\}$.

All in this paper, there must be $\widetilde{A} \in \varepsilon(E)$, on account of the decomposition theorem of fuzzy set, $\widetilde{A} = \bigcup_{\lambda\in[0,1]} \lambda A_\lambda = \bigcup_{\lambda\in[0,1]} \lambda[a + rE_\lambda^-, a + rE_\lambda^+]$.

Definition 2.3 [11]. If X and Y are two real number sets. $\widetilde{N}(f)$ is a set consisted of all fuzzy numbers on Y, and $\widetilde{f}$ is a mapping from X to $\widetilde{N}(f)$. In other words, for arbitrary x $\in$ X, there exist only $\widetilde{y} \in \widetilde{N}(f)$ with it correspondence. It is recorded as $\widetilde{y} = \widetilde{f}(x)$, then $\widetilde{f}(x)$ is said to be a fuzzy value function defined on X. If E is a regular fuzzy structural element on $\widetilde{N}(f)$, such that $\widetilde{f}(x) = h(x) + \omega(x)E$ (x$\in$ X, h(x) and ω(x) are bounded function on X, even $\omega(x)$¿0), then $\widetilde{f}(x)$ is said to be a fuzzy valued function linearly generated by E. We use the symbol $\widetilde{N}(E_f)$ to denote all of fuzzy valued function linearly generated by E, and write $\widetilde{N}(E_f) = \{\widetilde{f}(x) \mid \widetilde{f}(x) = h(x) + \omega(x)E,\, \forall\, x \in X,\, \omega(x) > 0\}$.

All in this paper, there must be $\widetilde{f}(x) \in \widetilde{N}(E_f)$, on account of the decomposition theorem of fuzzy set, $\widetilde{f}(x) = \bigcup_{\lambda\in[0,1]} \lambda \widetilde{f}_\lambda(x) = \bigcup_{\lambda\in[0,1]} (h(x) + \omega(x)E_\lambda) = \bigcup_{\lambda\in[0,1]} [h(x) + \omega(x)e_\lambda^-, h(x) + \omega(x)e_\lambda^+]$.

Definition 2.4 [12]. $\widetilde{a}$ and $\widetilde{b}$ are arbitrary fuzzy numbers. The fuzzy distance of $\widetilde{a}$ and $\widetilde{b}$ is defined as $\widetilde{d}(\widetilde{a}, \widetilde{b}) = \bigcup_{\lambda\in[0,1]} \lambda[\sup_{\lambda\le\mu\le 1} |\widetilde{a}_\mu^- - \widetilde{b}_\mu^-|, \sup_{0\le\lambda\le\mu} (|\widetilde{a}_\mu^- - \widetilde{b}_\mu^-| \vee |\widetilde{a}_\mu^+ - \widetilde{b}_\mu^+|)]$.

If $\widetilde{a}$, $\widetilde{b}$ and $\widetilde{c}$ are the fuzzy numbers, then the distance obviously follows:

(1) $\widetilde{d}(\widetilde{a}, \widetilde{b}) \ge 0$, $\widetilde{a} = \widetilde{b}$ if and only if $\widetilde{d}(\widetilde{a}, \widetilde{b}) = 0$.
(2) $\widetilde{d}(\widetilde{a}, \widetilde{b}) = \widetilde{d}(\widetilde{b}, \widetilde{a})$.
(3) $\widetilde{d}(\widetilde{a}, \widetilde{b}) \le \widetilde{d}(\widetilde{a}, \widetilde{c}) + \widetilde{d}(\widetilde{c}, \widetilde{b})$.

3 A New Definition of the Definite Integral of $\widetilde{f}(x)$ belonged to $\widetilde{N}(E_f)$

Definition 3.1. For the closed interval $[a, b]$, there are n-1 points, it follows $a = x_0 < x_1 < x_2 < \cdots < x_{n-1} < x_n = b$. Sub $[a, b]$ into n intervals as $\Delta_i = [x_{i-1}, x_i]$, $i = 1, 2 \cdots n$. These closed intervals constitute a segmentation on [a, b], which is denoted as $T = \{\Delta_1, \Delta_2 \cdots \Delta_n\}$. The length of the interval Δ_i is $\Delta x_i = x_i - x_{i-1}$. We have $\|T\| = \max\limits_{1 \le i \le n} \{\Delta x_i\}$ as the modulus of segmentation T. Arbitrarily pick point $\xi_i \in \Delta_i$, $i = 1, 2 \ldots n$, there the formula $\sum\limits_{i=1}^{n} \widetilde{f}(\xi_i) \cdot \Delta x_i$ is named an integral sum of $\widetilde{f}(x)$ on [a, b].

Definition 3.2. Let the definition domain of $\widetilde{f}(x)$ be on the interval $[a, b]$. If $\forall$ $\varepsilon > 0$, there exists positive δ, related to segmentation T on [a, b] and the set of points $\{\xi_i\}(\xi_i \in \Delta_i)$ selected in it. Whenever $\|T\| < \delta$, such that $\widetilde{d}(\sum\limits_{i=1}^{n} \widetilde{f}(\xi_i) \cdot \Delta x_i, \widetilde{A}) < \varepsilon$, then $\widetilde{f}(x)$ is fuzzy integrable on [a, b], and $\widetilde{A}$ is the fuzzy definite integral of $\widetilde{f}(x)$ on [a, b]. It is recorded as $\widetilde{A} = \int_a^b \widetilde{f}(x) d_x$.

3.1 The Properties of Definite Integral of $\widetilde{f}(x)$ belonged to $\widetilde{N}(E_f)$

Definition 3.1.1. ($\varepsilon - \delta$ definition of the limit of $\widetilde{f}(x)$) Let the definition domain of $\widetilde{f}(x)$ to $U^0(x_0; \delta')$ $\widetilde{A} \in \varepsilon(E)$. If $\forall$ $\varepsilon > 0$, there exists positive $\delta(< \delta')$, whenever $0 <| x - x_0 |< \delta$, such that $d(\widetilde{f}(x), \widetilde{A}) < \varepsilon$. We have $\widetilde{A}$ is the limit of $\widetilde{f}(x)$ when x tends to x_0, and write $\lim\limits_{x \to x_0} \widetilde{f}(x) = \widetilde{A}$.

Definition 3.1.2. Let $\widetilde{f}(x)$ be defined on $U^o(x_0)$. If the limit of $\lim\limits_{x \to x_0} \frac{\widetilde{f}(x) - \widetilde{f}(x_0)}{x - x_0}$ is present, then it is said that $\widetilde{f}(x)$ is derivable at point x_0, and the limit is said to be the derivative of $\widetilde{f}(x)$ at the point x_0.

Definition 3.1.3. Let $\widetilde{f}(x)$ and $\widetilde{F}(x)$ be defined on the interval I. If $\widetilde{F}'(x) = \widetilde{f}(x)$, $x \in I$, then $\widetilde{F}(x)$ is named the primitive function of $\widetilde{f}(x)$ on the interval I.

Theorem 3.1.1. (Newton Leibniz theorem) If $\widetilde{f}(x)$ is continuous on the interval [a, b], and there is the primitive function $\widetilde{F}(x)$, that is $\widetilde{F}'(x) = \widetilde{f}(x), x \in [a, b]$, then $\widetilde{f}(x)$ is named fuzzy integrable on [a, b], and $\int_a^b \widetilde{f}(x) d_x = \widetilde{F}(b) - \widetilde{F}(a)$.

Proof. For any segmentation $T = \{a = x_0, x_1, x_2 \ldots x_{n-1}, x_n = b\}$ on [a, b], and the set of points $\{\xi_i\}(\xi_i \in \Delta_i)$ selected in it. $\widetilde{F}(x)$ satisfies Lagrange mean value theorem. There exist separately $\eta \in (x_{i-1}, x_i)$, $i = 1, 2 \ldots n$, such that $\widetilde{F}(b) - \widetilde{F}(a) = \sum\limits_{i=1}^{n} [\widetilde{F}(x_i) - \widetilde{F}(x_{i-1})] = \sum\limits_{i=1}^{n} \widetilde{F}'(\eta_i) \cdot \Delta_i = \sum\limits_{i=1}^{n} \widetilde{f}(\eta_i) \cdot \Delta_i$. Because

$\widetilde{f}(x)$ is continuous in the interval [a, b], it is uniformly continuous. For any $\varepsilon > 0$, there exists $\delta_1 > 0$, whenever $x', x'' \in [a, b]$ and $|x' - x''| < \delta$. Such that $\widetilde{d}(\widetilde{f}(x'), \widetilde{f}(x'')) < \frac{\varepsilon}{b-a}$. In this case, when $\triangle x_i \leq ||T|| < \delta$, take $\eta_i \in [x_{i-1}, x_i]$, so that $|\xi_i - \eta_i| < \delta$, we have $\widetilde{d}(\sum_{i=1}^{n} \widetilde{f}(\eta_i) \cdot \Delta_{x_i}, \widetilde{F}(b) - \widetilde{F}(a)) = \widetilde{d}(\sum_{i=1}^{n} \widetilde{f}(\eta_i) \cdot \Delta_{x_i}, \sum_{i=1}^{n} \widetilde{f}(\xi_i) \cdot \Delta x_i) = \sum_{i=1}^{n} \Delta x_i \widetilde{d}(\widetilde{f}(\eta_i), \widetilde{f}(\eta_i)) < \sum_{i=1}^{n} \Delta_{x_i} \cdot \frac{\varepsilon}{b-a} = \varepsilon$. This theorem is proved.

Theorem 3.1.2. If $\widetilde{f}(x)$ is integrable function on the interval [a, b], k is nonnegative constant, then $k\widetilde{f}(x)$ is also integrable function on [a, b], and $k \int_a^b \widetilde{f}(x) d_x = \int_a^b k \widetilde{f}(x) d_x$.

Proof. When k = 0, the conclusion is obvious. When $k \neq 0$, let $\widetilde{A} = \int_a^b \widetilde{f}(x) d_x$. Defined by fuzzy definite integral, for any $\varepsilon > 0$, there exists $\delta > 0$, whenever $\|T\| < \delta$, such that $\widetilde{d}(\sum_{i=1}^{n} \widetilde{f}(\xi_i) \cdot \Delta x_i, \widetilde{A}) < \frac{\varepsilon}{k}$. Hence $k\widetilde{d}(\sum_{i=1}^{n} \widetilde{f}(\xi_i) \cdot \Delta x_i, \widetilde{A}) < \varepsilon$. We have $\widetilde{d}(\sum_{i=1}^{n} k\widetilde{f}(\xi_i) \cdot \Delta x_i, k\widetilde{A}) < \varepsilon$. That has been proved $k\widetilde{f}(x)$ is also integrable function on [a, b], and $k \int_a^b \widetilde{f}(x) d_x = k\widetilde{A} = \int_a^b k\widetilde{f}(x) d_x$.

Theorem 3.1.3. If $\widetilde{f_1}(x)$ and $\widetilde{f_2}(x)$ are integrable functions on the interval [a, b], then $\widetilde{f_1}(x) + \widetilde{f_2}(x)$ is also integrable function on [a, b], and $\int_a^b \widetilde{f_1}(x) + \widetilde{f_2}(x) d_x = \int_a^b \widetilde{f_1}(x) d_x + \int_a^b \widetilde{f_2}(x) d_x$.

Proof. Owing to $\widetilde{f_1}(x)$ and $\widetilde{f_2}(x)$ are integrable functions on the interval [a, b]. Let $\widetilde{A_1} = \int_a^b \widetilde{f_1}(x) d_x$, $\widetilde{A_2} = \int_a^b \widetilde{f_2}(x) d_x$. For each $\varepsilon > 0$, there exist positive numbers δ_1 and δ_2. Whenever $\|T_1\| < \delta_1$, such that $\widetilde{d}(\sum_{i=1}^{n} \widetilde{f_1}(\xi_i) \cdot \Delta x_i, \widetilde{A_1}) = \bigcup_{\lambda \in [0,1]} \lambda \sum_{i=1}^{n} [\sup_{\lambda \leq \mu \leq 1} |h_1(\xi_i) \cdot \Delta x_i + \omega_1(\xi_i) \cdot \Delta x_i E_\mu^- - a_1 - r_1 E_\mu^-|, \sup_{0 \leq \lambda \leq \mu} (|h_1(\xi_i) \cdot \Delta x_i + \omega_1(\xi_i) \cdot \Delta x_i E_\mu^- - a_1 - r_1 E_\mu^-| \vee |h_1(\xi_i) \cdot \Delta x_i + \omega_1(\xi_i) \cdot \Delta x_i E_\mu^+ - a_1 - r_1 E_\mu^+|)] < \frac{\varepsilon}{2}$. Whenever $\|T_2\| < \delta_2$, such that $\widetilde{d}(\sum_{i=1}^{n} \widetilde{f_2}(\xi_i) \cdot \Delta x_i, \widetilde{A_2}) = \bigcup_{\lambda \in [0,1]} \lambda \sum_{i=1}^{n} [\sup_{\lambda \leq \mu \leq 1} |h_2(\xi_i) \cdot \Delta x_i + \omega_2(\xi_i) \cdot \Delta x_i E_\mu^- - a_2 - r_2 E_\mu^-|, \sup_{0 \leq \lambda \leq \mu} (|h_2(\xi_i) \cdot \Delta x_i + \omega_2(\xi_i) \cdot \Delta x_i E_\mu^- - a_2 - r_2 E_\mu^-| \vee |h_2(\xi_i) \cdot \Delta x_i + \omega_2(\xi_i) \cdot \Delta x_i E_\mu^+ - a_2 - r_2 E_\mu^+|)] < \frac{\varepsilon}{2}$. Take $\delta = \min\{\delta_1, \delta_2\}$, whenever $\|T\| < \delta$, we have $\widetilde{d}(\sum_{i=1}^{n} \widetilde{f_1}(\xi_i) \cdot \Delta x_i + \sum_{i=1}^{n} \widetilde{f_2}(\xi_i) \cdot \Delta x_i, \widetilde{A_1} + \widetilde{A_2}) = \bigcup_{\lambda \in [0,1]} \lambda \sum_{i=1}^{n} [\sup_{\lambda \leq \mu \leq 1} |(h_1(\xi_i) \cdot \Delta x_i + \omega_1(\xi_i) \cdot \Delta x_i E_\mu^- + h_2(\xi_i) \cdot \Delta x_i + \omega_2(\xi_i) \cdot \Delta x_i E_\mu^-) - (a_1 + r_1 E_\mu^- + a_2 + r_2 E_\mu^-)|, \sup_{0 \leq \lambda \leq \mu} (|(h_1(\xi_i) \cdot \Delta x_i + \omega_1(\xi_i) \cdot \Delta x_i E_\mu^- + h_2(\xi_i) \cdot \Delta x_i + \omega_2(\xi_i) \cdot \Delta x_i E_\mu^-) - (a_1 + r_1 E_\mu^- + a_2 + r_2 E_\mu^-)| \vee |(h_1(\xi_i) \cdot \Delta x_i + \omega_1(\xi_i) \cdot \Delta x_i E_\mu^+ + h_2(\xi_i) \cdot \Delta x_i + \omega_2(\xi_i) \cdot \Delta x_i E_\mu^+) - (a_1 + r_1 E_\mu^+ + a_2 + r_2 E_\mu^+)|)] = \bigcup_{\lambda \in [0,1]} \lambda \sum_{i=1}^{n} [\sup_{\lambda \leq \mu \leq 1} |(h_1(\xi_i) \cdot \Delta x_i + \omega_1(\xi_i) \cdot \Delta x_i E_\mu^- -$

$a_1 - r_1E_\mu^-) + (h_2(\xi_i)\cdot\Delta x_i + \omega_2(\xi_i)\cdot\Delta x_iE_\mu^- - a_2 - r_2E_\mu^-)|, \sup_{0\le\lambda\le\mu} (|(h_1(\xi_i)\cdot\Delta x_i + \omega_1(\xi_i)\cdot\Delta x_iE_\mu^- - a_1 - r_1E_\mu^-) + (h_2(\xi_i)\cdot\Delta x_i + \omega_2(\xi_i)\cdot\Delta x_iE_\mu^- - a_2 - r_2E_\mu^-)| \vee |(h_1(\xi_i)\cdot \Delta x_i + \omega_1(\xi_i)\cdot\Delta x_iE_\lambda^+ - a_1 - r_1E_\lambda^+) + (h_2(\xi_i)\cdot\Delta x_i + \omega_2(\xi_i)\cdot\Delta x_iE_\lambda^+ - a_2 - r_2E_\lambda^+)|)] \le \bigcup_{\lambda\in[0,1]} \lambda \sum_{i=1}^{n}[\sup_{\lambda\le\mu\le1} |h_1(\xi_i)\cdot\Delta x_i + \omega_1(\xi_i)\cdot\Delta x_iE_\mu^- - a_1 - r_1E_\mu^-| + |h_2(\xi_i)\cdot\Delta x_i + \omega_2(\xi_i)\cdot\Delta x_i\cdot E_\mu^- - a_2 - r_2E_\mu^-|, \sup_{0\le\lambda\le\mu} (|h_1(\xi_i)\cdot\Delta x_i + \omega_1(\xi_i)\cdot\Delta x_iE_\mu^- - a_1 - r_1E_\mu^-| + |h_2(\xi_i)\cdot\Delta x_i + \omega_2(\xi_i)\cdot\Delta x_iE_\mu^- - a_2 - r_2E_\mu^-| \vee |h_1(\xi_i)\cdot \Delta x_i + \omega_1(\xi_i)\cdot\Delta x_iE_\mu^+ - a_1 - r_1E_\mu^+| + |h_2(\xi_i)\cdot\Delta x_i + \omega_2(\xi_i)\cdot\Delta x_iE_\mu^+ - a_2 - r_2E_\mu^+|)] = \bigcup_{\lambda\in[0,1]} \lambda \sum_{i=1}^{n}[\sup_{\lambda\le\mu\le1} |h_1(\xi_i)\cdot\Delta x_i + \omega_1(\xi_i)\cdot\Delta x_iE_\mu^- - a_1 - r_1E_\mu^-|, \sup_{0\le\lambda\le\mu} (|h_1(\xi_i)\cdot\Delta x_i + \omega_1(\xi_i)\cdot\Delta x_iE_\mu^- - a_1 - r_1E_\mu^-| \vee |h_1(\xi_i)\cdot\Delta x_i + \omega_1(\xi_i)\cdot\Delta x_iE_\mu^+ - a_1 - r_1E_\mu^+|)] + \bigcup_{\lambda\in[0,1]} \lambda \sum_{i=1}^{n}[\sup_{\lambda\le\mu\le1} |h_2(\xi_i)\cdot\Delta x_i + \omega_2(\xi_i)\cdot\Delta x_iE_\mu^- - a_2 - r_2E_\mu^-|, \sup_{0\le\lambda\le\mu} (|h_2(\xi_i)\cdot\Delta x_i + \omega_2(\xi_i)\cdot\Delta x_iE_\mu^- - a_2 - r_2E_\mu^-| \vee |h_2(\xi_i)\cdot\Delta x_i + \omega_2(\xi_i)\cdot\Delta x_iE_\mu^+ - a_2 - r_2E_\mu^+|)] < \frac{\varepsilon}{2} + \frac{\varepsilon}{2} = \varepsilon$. Tt has been proved that $k\widetilde{f}(x)$ is also integrable function on [a, b], and $k\int_a^b \widetilde{f}(x)d_x = \int_a^b k\widetilde{f}(x)d_x$.

Corollary 3.1.1. If $\widetilde{f_1}(x)$ and $\widetilde{f_2}(x)$ are integrable functions on the interval [a, b], then $\widetilde{f_1}(x)-\widetilde{f_2}(x)$ is also integrable function on [a, b], and $\int_a^b \widetilde{f_1}(x)-\widetilde{f_2}(x)d_x = \int_a^b \widetilde{f_1}(x)d_x - \int_a^b \widetilde{f_2}(x)d_x$.

Theorem 3.1.4. (Interval additivity of fuzzy integral) $\widetilde{f}(x)$ is integrable on the interval [a, b] if and only if for any $c \in [a, b]$, $\widetilde{f}(x)$ is integrable functions on the interval [a, c] and [c, d], also$\int_a^b \widetilde{f}(x)d_x = \int_a^c \widetilde{f}(x)d_x + \int_c^b \widetilde{f}(x)d_x$.

Proof. Sufficiency. Because $\widetilde{f}(x)$ is integrable functions on the interval [a, c] and [c, d]. For each $\varepsilon > 0$. The segmentations of [a, c] and [c, d] are T' and T'' respectively. Whenever $\|T_1\| < \delta_1$, such that $\widetilde{d}(\sum_{i=1}^{n} \widetilde{f}(\xi_i') \cdot \Delta x_i, \widetilde{A}_1) < \frac{\varepsilon}{2}$; and whenever $\|T_2\| < \delta_2$, such that $\widetilde{d}(\sum_{i=1}^{n} \widetilde{f}(\xi_i'') \cdot \Delta x_i, \widetilde{A_2}) < \frac{\varepsilon}{2}$. Let T=$T' + T''$, it is a segmentation of $[a, b]$. Take $\delta = \min\{\delta_1, \delta_2\}$, whenever $\|T\| < \delta$. According to the proof of theorem 3.3, we have $\widetilde{d}(\sum_{i=1}^{n} \widetilde{f}(\xi_i') \cdot \Delta x_i + \sum_{i=1}^{n} \widetilde{f}(\xi_i'') \cdot \Delta x_i, \widetilde{A_1} + \widetilde{A_2}) \le \widetilde{d}(\sum_{i=1}^{n} \widetilde{f}(\xi_i') \cdot \Delta x_i, \widetilde{A}_1) + \widetilde{d}(\sum_{i=1}^{n} \widetilde{f}(\xi_i'') \cdot \Delta x_i, \widetilde{A_2}) < \frac{\varepsilon}{2} + \frac{\varepsilon}{2} = \varepsilon$. It has been proved that $\widetilde{f}(x)$ is integrable function on the interval [a, b].

Necessity. Because $\widetilde{f}(x)$ is integrable function on the interval [a, b], there is a segmentation T on [a, b]. For each ε>0, whenever $\|T\|$<δ, such that $\widetilde{d}(\sum_{i=1}^{n} \widetilde{f}(\xi_i)\cdot\Delta x_i, \widetilde{A}) < \varepsilon$. Add a point c on T to get a new segmentation T^*, and $\widetilde{d}(\sum_{i=1}^{n} \widetilde{f}(\xi_i^*)\cdot\Delta x_i^*, \widetilde{A}^*) \le \widetilde{d}(\sum_{i=1}^{n} \widetilde{f}(\xi_i)\cdot\Delta x_i, \widetilde{A}) < \varepsilon$. T^* exists in parts of $[a, c]$

and $[c, b]$, they are denoted as T' and T''. Whenever $\|T'\|<\delta$, such that $\widetilde{d}(\sum_{i=1}^{n} \widetilde{f}(\xi_i')\cdot \Delta x_i', \widetilde{A}') \leq \widetilde{d}(\sum_{i=1}^{n} \widetilde{f}(\xi_i^*)\cdot \Delta_{x_i}, \widetilde{A}) < \varepsilon$; and whenever $\|T''\|<\delta$, such that $\widetilde{d}(\sum_{i=1}^{n} \widetilde{f}(\xi_i)''\cdot \Delta x_i'', \widetilde{A}'') \leq \widetilde{d}(\sum_{i=1}^{n} \widetilde{f}(\xi_i^*)\cdot \Delta x_i, \widetilde{A}) < \varepsilon$. Then $\widetilde{f}(x)$ is integrable function on the interval [a, b].

The segmentation on [a, b] is T, the constant point c is one of the points. The segmentation T^* exists on parts of $[a, c]$ and $[c, b]$, Because $\sum_{i=1}^{n} \widetilde{f}(\xi_i)\cdot \Delta x_i = \sum_{i=1}^{n} \widetilde{f}(\xi_i')\cdot \Delta x_i' + \sum_{i=1}^{n} \widetilde{f}(\xi_i'')\cdot \Delta x_i''$. Whenever $\|T\|<\delta$, such that $\|T'\|<\delta$ and $\|T''\|<\delta$. Namely, $\int_a^b \widetilde{f}(x)d_x = \int_a^c \widetilde{f}(x)d_x + \int_c^b \widetilde{f}(x)d_x$. This theorem is fully proved.

Theorem 3.1.5. (Boundedness) If $\widetilde{f}(x)$ is integrable functions on the interval [a, b], then $\widetilde{f}(x)$ is bounded on the interval [a, b].

Proof. proof by contradiction. If $\widetilde{f}(x)$ is unbounded on the interval [a, b], then for any segmentation T on [a, b], there must be an short interval Δ_k belonged to T. $\widetilde{f}(x)$ is unbounded on Δ_k, ε_i is taken from the interval Δ_i, $i \neq k$, and it is recorded as $\widetilde{G} = |\sum_{i\neq k} \widetilde{f}(\xi_i)\cdot \Delta x_i|$. For any large positive fuzzy number $\widetilde{M}$, because $\widetilde{f}(x)$ is unbounded on Δ_k. There exists $\xi_k \subset \Delta_k$, such that $|\widetilde{f}(\xi_k)| > \frac{\widetilde{G}+\widetilde{M}}{\Delta x_k}$. Namely $|\sum_{i=1}^{n} \widetilde{f}(\varepsilon_i)\cdot \Delta x_i| \geq |\widetilde{f}(\varepsilon_k)\cdot \Delta_k| - |\sum_{i\neq k} \widetilde{f}(\varepsilon_i)\cdot x_i| > \frac{\widetilde{G}+\widetilde{M}}{\Delta x_k}\cdot \Delta x_k \text{-} \widetilde{G} = \widetilde{M}$. Therefore. no matter how small the $||T||$ is, only the point set ε_i is selected by the above method, such that the absolute value of integral sum is greater than any positive number given in advance. That is inconsistent with integration of $\widetilde{f}(x)$ on the interval $[a, b]$.

Theorem 3.1.6. (Local protection) If $\widetilde{f}(x)$ and $\widetilde{g}(x)$ are integrable functions on the interval $[a, b]$, and $\widetilde{f}(x) \leq \widetilde{g}(x)$, then $\int_a^b \widetilde{f}(x)d_x \leq \int_a^b \widetilde{g}(x)d_x$.

Proof. Let $\widetilde{F}(x) = \widetilde{g}(x) - \widetilde{f}(x)$, $x \in [a, b]$. Because $\widetilde{f}(x) \leq \widetilde{g}(x)$. $\widetilde{F}(x) \geq 0$. Then any integral sum of $\widetilde{F}(x)$ is nonnegative on the interval $[a, b]$. By the definition of fuzzy definite integral, $\int_a^b \widetilde{F}(x)d_x \geq 0$. Hence $\int_a^b \widetilde{F}(x)d_x = \int_a^b \widetilde{g}(x)d_x$-$\int_a^b \widetilde{f}(x)d_x \geq 0$. That is $\int_a^b \widetilde{f}(x)d_x \leq \int_a^b \widetilde{g}(x)d_x$.

Theorem 3.1.7. (The first mean value theorem for integrals) If $\widetilde{f}(x)$ is continuous on the interval $[a, b]$, there is at least one point $\xi \in [a, b]$, then $\int_a^b \widetilde{f}(x)d_x = \widetilde{f}(\xi)(b-a)$.

Proof. Because $\widetilde{f}(x)$ is continuous on the interval $[a, b]$, there exist maximum value $\widetilde{M}$ and minimum value $\widetilde{m}$. Such that $\widetilde{m} \leq \widetilde{f}(x) \leq \widetilde{M}$, $x \in [a, b]$. According to the properties of fuzzy integral inequality, $(b-a)\widetilde{m} \leq \int_a^b \widetilde{f}(x)d_x \leq (b-a)\widetilde{M}$.

It is said that $\widetilde{m} \le \frac{\int_a^b \widetilde{f}(x)d_x}{b-a} \le \widetilde{M}$. Because of the boundary value of continuous fuzzy valued function, there must be at least one point $\xi \in [a,b]$ made $\widetilde{f}(\xi) = \frac{\int_a^b \widetilde{f}(x)d_x}{b-a}$. We have $\int_a^b \widetilde{f}(x)d_x = \int_a^b \widetilde{f}(\xi)(b-a)$.

Theorem 3.1.8. If $\widetilde{f}(x)$ is integrable function on the interval [a, b], then $|\widetilde{f}(x)|$ is also integrable on the interval [a, b], and $| \int_a^b \widetilde{f}(x)d_x| \le \int_a^b \widetilde{|f(x)|}d_x$.

Proof. Because $\widetilde{f}(x)$ is integrable function on the interval [a, b], Let $\widetilde{A} = \int_a^b \widetilde{f}(x)d_x$. According to the definition of fuzzy definite integral, for any positive ε, there exists $\delta > 0$, related to segmentation T on [a, b], whenever $\|T\| < \delta$, hence $\widetilde{d}(\widetilde{s}(T), \widetilde{S}(T)) < \varepsilon$. By Definition 2.4, $\widetilde{d}(|\widetilde{s}(T)|, |\widetilde{S}(T)|) \le \widetilde{d}(\widetilde{s}(T), \widetilde{S}(T))$. Thereby that has been proved $|\widetilde{f}(x)|$ on the interval [a, b]. According to the nature of inequality $-|\widetilde{f}(x)| \le \widetilde{f}(x) \le |\widetilde{f}(x)|$, Applied Theorem 3.1.6. It is proved that $| \int_a^b \widetilde{f}(x)d_x| \le \int_a^b \widetilde{|f(x)|}d_x$.

3.2 Integrability Conditions of Definite Integral of $\widetilde{f}(x)$ Belonged to $\widetilde{N}(E_f)$

Definition 3.2.1. (Darboux sum definition of $\widetilde{f}(x)$) Let arbitrary segmentation on the interval [a, b] be $T = \{\Delta_i | i = 1, 2...n\}$. Because $\widetilde{f}(x)$ is bounded on the interval $[a,b]$, there are the supremum and infimum on Δ_i, they are respectively recorded as: $\widetilde{M_i} = \sup\limits_{x\in\Delta_i} \widetilde{f}(x)$, $\widetilde{m_i} = \inf\limits_{x\in\Delta_i} \widetilde{f}(x)$, $i = 1, 2...n$. Further on, $\widetilde{S}(T) = \sum\limits_{i=1}^{n} \widetilde{M_i} \cdot \Delta x_i$, $\widetilde{s}(T) = \sum\limits_{i=1}^{n} \widetilde{m_i} \cdot \Delta x_i$. They are respectively named Darboux upper sum and Darboux lower sum of $\widetilde{f}(x)$ with regard to segmentation T on [a, b], and they are collectively referred to as Darboux sum. For any $\xi_i \in \Delta_i$, $i = 1, 2...n$, it is obvious that $\widetilde{m}(b-a) \le \widetilde{s}(T) \le \sum\limits_{i=1}^{n} \widetilde{f}(\xi_i) \cdot \Delta x_i \le \widetilde{S}(T) \le \widetilde{M}(b-a)$.

Theorem 3.2.1. Let $\widetilde{f}(x)$ be bounded on the interval $[a,b]$. For the same segmentation $T = \{\Delta_i | i = 1, 2...n\}$ on the interval [a, b], the Darboux upper sum is the supremum of all fuzzy definite integral, and the Darboux lower sum is the infimum of all fuzzy definite integral. Namely, $\widetilde{S}(T) = \sup\limits_{\xi_i} \sum\limits_{i=1}^{n} \widetilde{f}(\xi_i) \cdot \Delta x_i$, $\widetilde{s}(T) = \inf\limits_{\xi_i} \sum\limits_{i=1}^{n} \widetilde{f}(\xi_i) \cdot \Delta x_i$, $\xi_i \in \Delta_i$.

Proof. By Definition 3.2.1, Darboux upper sum and Darboux lower sum are the upper bound and lower bound of all integral sum of $\widetilde{f}(x)$ bounded on [a, b]. Now we prove further that they are the least upper bound and the largest lower bound of all integral sum. For each $\varepsilon > 0$, On each Δx_i, since $\widetilde{M_i}$ is the supremum of $\widetilde{f}(x)$, the optional fetch $\xi_i \in \Delta_i$ makes the $\widetilde{f}(\xi_i) \cdot \Delta x_i > \widetilde{M_i} - \frac{\varepsilon}{b-a}$. We have $\sum\limits_{i=1}^{n} \widetilde{f}(\xi_i) \cdot \Delta x_i > \sum\limits_{i=1}^{n} (\widetilde{M_i} - \frac{\varepsilon}{b-a}) \cdot \Delta x_i = \sum\limits_{i=1}^{n} \widetilde{M_i} \cdot \Delta x_i - \frac{\varepsilon}{b-a} \cdot \sum\limits_{i=1}^{n} \Delta x_i = \widetilde{S}(T) - \varepsilon$.

Hence $\widetilde{S}(T)$ is the supremum of all integral sum. Analogously, it can be proved that the Darboux lower sum is the infimum of all fuzzy definite integral.

Theorem 3.2.2. Let $\widetilde{f}(x)$ be bounded on the interval $[a, b]$. For the segmentation $T = \{\Delta_i | i = 1, 2...n\}$ on the interval [a, b], if the segmentation T is added with P new points of division, the resulting segmentation is T_p, then $\widetilde{S}(T) \geq \widetilde{S}(T') \geq \widetilde{S}(T) - p(\widetilde{M} - \widetilde{m})||T||, \widetilde{s}(T) \geq \widetilde{s}(T') \geq \widetilde{s}(T) - p(\widetilde{M} - \widetilde{m})||T||$.

Proof. Here we prove the first inequality. The same can be proved the second inequality. Add p new points to the segmentation T at the same time, or add them to the segmentation T one by one. The segmentation T_1 can also be obtained equally. Now the case of p = 1 is proved firstly. One new point is added to T, which must fall within a interval Δ_k on T, and divide the interval Δ_k into two intervals. However, other interval $\Delta_i (i \neq k)$ on T is still the small interval on T_1. The supremums of $\widetilde{f}(x)$ on Δ_k' and Δ_k'' are respectively recorded as $\widetilde{M_k'}$, $\widetilde{M_k''}$. Even compare the Darboux upper sum $\widetilde{S}(T)$ and $\widetilde{S}(T_1)$, the difference between them is that only $\widetilde{M_k} \cdot \Delta x_k$ in $\widetilde{S}(T)$ is replaced by $\widetilde{M_k'} \cdot \Delta x_k'$ and $\widetilde{M_k''} \cdot \Delta x_k''$ in $\widetilde{S}(T_1)$. We have $\widetilde{S}(T) - \widetilde{S}(T_1) = \widetilde{M_k} \cdot \Delta x_k - (\widetilde{M_k'} \cdot \Delta x_k' + \widetilde{M_k''} \cdot \Delta x_k'') = \widetilde{M_k} \cdot (\Delta x_k' + \Delta x_k'' + (\widetilde{M_k'} \cdot \Delta x_k' + \widetilde{M_k''} \cdot \Delta x_k'') = (\widetilde{M_k} - \widetilde{M_k'}) \cdot \Delta x_k' + (\widetilde{M_k} - \widetilde{M_k''}) \cdot \Delta x_k''$, and because of $\widetilde{m} \leq \widetilde{M'}_k (or \widetilde{M_k''}) \leq \widetilde{M_k} \leq \widetilde{M}$. There is $0 \leq \widetilde{S}(T) - \widetilde{S}(T_1) \leq (\widetilde{M} - \widetilde{m}) \cdot \Delta x_k' + (\widetilde{M} - \widetilde{m}) \cdot \Delta x_k'' = (\widetilde{M} - \widetilde{m}) \cdot x \Delta_k \leq (\widetilde{M} - \widetilde{m})||T||$. So that when p = 1, the inequality $\widetilde{S}(T) \leq \widetilde{S}(T') \leq \widetilde{S}(T) - (\widetilde{M} - \widetilde{m}) \sum_{i=0}^{p-1} ||T_i||$ is established. For T_i, a point is added to get T_{i+1}. Similarly, we have $0 \leq \widetilde{S}(T_i) - \widetilde{S}(T_{i+1}) \leq (\widetilde{M} - \widetilde{m})||T_i||$, $(i = 0, 1, 2, ..., p-1. T_0 = T)$. By adding these inequalities corresponded to each i, it is obtained that $0 \leq \widetilde{S}(T) - \widetilde{S}(T') \leq (\widetilde{M} - \widetilde{m}) \sum_{i=0}^{p-1} ||T_i|| \leq p(\widetilde{M} - \widetilde{m})||T||$. In other words, the first inequality is proved fully.

Corollary 3.2.1. If T' and T'' are arbitrary two segmentations on [a, b], let $T = T' + T''$, it means the segmentation obtained by T' and T'' with all points (Repeated points are taken only once), then $\widetilde{S}(T) \leq \widetilde{S}(T')$, $\widetilde{s}(T) \geq \widetilde{s}(T')$, $\widetilde{S}(T) \leq \widetilde{S}(T'')$, $\widetilde{s}(T) \geq \widetilde{s}(T'')$.

Theorem 3.2.3. If T' and T'' are arbitrary two segmentations on [a, b], then $\widetilde{s}(T') \geq \widetilde{S}(T'')$.

Proof. Let segmentation $T = T' + T''$. By Theorem 3.2.1 and Corollary 3.2.1, there is $\widetilde{s}(T') \leq \widetilde{S}(T) \leq \widetilde{S}(T) \leq \widetilde{S}(T'')$.

Corollary 3.2.2. For arbitrary segmentation T on the interval [a, b], all Darboux upper sum is the supremum of existence, and all Darboux lower sum is the infimum of existence. They are respectively recoreded as $\widetilde{S} = \inf_T \widetilde{S}(T)$, $\widetilde{s} = \sup_T \widetilde{S}(T)$. Also we name $\widetilde{S}$ is the upper fuzzy integral of $\widetilde{f}(x)$ on [a, b], $\widetilde{S}$ is the lower fuzzy integral of $\widetilde{f}(x)$ on [a, b]. There must be $\widetilde{m}(b-a) \leq \widetilde{s} \leq \widetilde{S} \leq \widetilde{M}(b-a)$.

Theorem 3.2.4. (Darboux theorem) The upper fuzzy integral and the lower fuzzy integral are respectively the limit of Darboux upper sum and Darboux lower sum when $||T|| \to$ O. Namely $\lim\limits_{||T||\to 0} \widetilde{S}(T) = \widetilde{S}$, $\lim\limits_{||T||\to 0} \widetilde{s}(T) = \widetilde{s}$.

Proof. For any $\varepsilon > 0$, by the definition of $\widetilde{S}$, there exists a segmentation T', hence $\widetilde{S}(T') < \widetilde{S} + \frac{\varepsilon}{2}$. Assume that T consists of p points. For arbitrary segmentation T, $T + T'$ is more than p points than T. Owing to Theorem 3.2.2 and Corollary 3.2.1, $\widetilde{S} - p(\widetilde{M} - \widetilde{m})||T|| \leq \widetilde{S}(T + T') \leq \widetilde{S}(T')$, we have $\widetilde{S} \leq \widetilde{S}(T') + p(\widetilde{M} - \widetilde{m})||T||$. Therefore, whenever $(\widetilde{M} - \widetilde{m}) < \frac{\varepsilon}{2p}$, such that $\widetilde{S}(T) < \widetilde{S}(T') + \frac{\varepsilon}{2}$. Hence $\widetilde{S} \leq \widetilde{T} < \widetilde{S} + \varepsilon$. It is proved that $\lim\limits_{||T||\to 0} \widetilde{S}(T) = \widetilde{S}$. $\lim\limits_{||T||\to 0} \widetilde{s}(T) = \widetilde{s}$ can be proved similarly.

Theorem 3.2.5. (The first condition of integration) $\widetilde{f}(x)$ is integrable on the interval [a, b] if and only if the upper fuzzy integral and the lower fuzzy integral of $\widetilde{f}(x)$ are equal, that is $\widetilde{S} = \widetilde{s}$.

Proof. Sufficiency. Assume $\widetilde{S} = \widetilde{s} = \widetilde{A}$. By Darboux theorem, $\lim\limits_{||T||\to 0} \widetilde{S}(T) = \lim\limits_{||T||\to 0} \widetilde{s}(T)\widetilde{S} = \widetilde{A}$. According to Definition 3.2.1, for any $\varepsilon > 0$, there exists $\delta > 0$, whenever $0 < ||T|| < \delta$, such that $\widetilde{A} - \varepsilon < \widetilde{s}(T) \leq \sum\limits_{i=1}^{n} \widetilde{f}(\xi_i) \cdot \Delta x_i \leq \widetilde{S}(T) < \widetilde{A} + \varepsilon$. Hence $\widetilde{d}(\sum\limits_{i=1}^{n} \widetilde{f}(\xi_i) \cdot \Delta x_i, \widetilde{A}) < \varepsilon$. We have $\widetilde{f}(x)$ is integrable on the interval [a, b], and $\int_a^b \widetilde{f}(x) d_x = \widetilde{A}$.

Necessity. On account of integrability of $\widetilde{f}(x)$ defined on [a, b], take $\widetilde{A} = \int_a^b \widetilde{f}(x) d_x$. For each $\varepsilon > 0$, there exists $\delta > 0$, whenever $||T|| < \delta$, hence $\widetilde{d}(\sum\limits_{i=1}^{n} \widetilde{f}(\xi_i) \cdot \Delta x_i, \widetilde{A}) < \varepsilon$. By Theorem 3.2.1, $\widetilde{S}(T) = \sup\limits_{\xi_i} \sum\limits_{i=1}^{n} \widetilde{f}(\xi_i) \cdot \Delta x_i$; $\widetilde{s}(T) = \inf\limits_{\xi_i} \sum\limits_{i=1}^{n} \widetilde{f}(\xi_i) \cdot \Delta x_i$, $\xi_i \in \Delta_i$, even whenever $||T|| < \delta$, we have $\widetilde{d}(\widetilde{S}(T), \widetilde{A}) \leq \varepsilon$, $\widetilde{d}(\widetilde{s}(T), \widetilde{A}) \leq \varepsilon$. By Theorem 3.2.4, when $||T|| \to$ O, there must be $\lim\limits_{||T||\to 0} \widetilde{S}(T) = \widetilde{A}$, $\lim\limits_{||T||\to 0} \widetilde{s}(T) = \widetilde{A}$. According to Definition 2.4, There must be $\widetilde{S} = \widetilde{s} = \widetilde{A}$.

Theorem 3.2.6. (The second condition of integration) $\widetilde{f}(x)$ is integrable on the interval [a, b] if and only if for arbitrary positive number ε, there exist segmentation T on the interval [a, b]. Whenever $\|T\| < \delta$, such that$\widetilde{d}(\widetilde{s}(T), \widetilde{S}(T)) < \varepsilon$.

Proof. Sufficiency. Because $\widetilde{d}(\widetilde{s}(T), \widetilde{S}(T)) < \varepsilon$. , there is $\widetilde{s}(T) \leq \widetilde{s} \leq \widetilde{S} \leq \widetilde{S}(T)$, hence $0 \leq \widetilde{d}(\widetilde{S}, \widetilde{s}) \leq \widetilde{d}(\widetilde{S}(T) - \widetilde{s}(T)) < \varepsilon$. Because of the arbitrariness of ε, there is $\widetilde{S} = \widetilde{s}$. According to Theorem 3.2.5, we have $\widetilde{f}(x)$ is integrable on the interval [a, b].

Necessity. Assume $\widetilde{f}(x)$ is integrable on the interval [a, b]. Owing to Theorem 3.2.5, it can be obtained that $\lim\limits_{||T||\to 0} \widetilde{S}(T) = \lim\limits_{||T||\to 0} \widetilde{s}(T)\widetilde{S} = \widetilde{A}$. By Definition 3.1.1, for any positive ε, there exists $\delta > 0$, whenever $\|T\|<\delta$, such that $\widetilde{d}(\widetilde{S}(T), \widetilde{A}) < \frac{\varepsilon}{2}$, $\widetilde{d}(\widetilde{s}(T), \widetilde{A}) < \frac{\varepsilon}{2}$. In view of Definition 2.4, we have $\widetilde{d}(\widetilde{s}(T), \widetilde{S}(T)) < \widetilde{d}(\widetilde{s}(T), \widetilde{A}) + \widetilde{d}(\widetilde{A}, \widetilde{S}(T)) < \frac{\varepsilon}{2} + \frac{\varepsilon}{2} = \varepsilon$.

Theorem 3.2.7. If $\widetilde{f}(x)$ is continuous on the interval [a, b], then $\widetilde{f}(x)$ is integrable on the interval [a, b].

Proof. Because $\widetilde{f}(x)$ is continuous on the interval [a, b], uniformly continuous on $[a, b]$. In other words, fou each $\varepsilon > 0$, there exists $\delta > 0$, as well x', $x''\in[a, b]$, whenever $|x' - x''| < \delta$, such that $\widetilde{d}(\widetilde{f}(x'), \widetilde{f}(x'')) < \frac{\varepsilon}{b-a}$. Only the segmentation of [a, b] satisfies $||T|| < \delta$. On any interval Δ_i belonged to T, there is $\sup\limits_{x',x''\in\Delta_i} \widetilde{d}(\widetilde{f}(x'), \widetilde{f}(x'')) < \frac{\varepsilon}{b-a}$. Hence $\widetilde{d}(\widetilde{s}(T), \widetilde{S}(T))=\widetilde{d}(\sum\limits_{i=1}^{n} \widetilde{M_i} \cdot \Delta x_i, \sum\limits_{i=1}^{n} \widetilde{m_i} \cdot \Delta x_i) = \widetilde{d}(\sup\limits_{x\in\Delta_i} \widetilde{f}(x), \inf\limits_{x\in\Delta_i} \widetilde{f}(x)) \leq \sup\limits_{x',x''\in\Delta_i} \widetilde{d}(\widetilde{f}(x'), \widetilde{f}(x''))\cdot \sum\limits_{i=1}^{n} \Delta x_i < \frac{\varepsilon}{b-a}\cdot(b-a) = \varepsilon$. By Theorem 3.1.1, we have $\widetilde{f}(x)$ is integrable on the interval [a, b].

Theorem 3.2.8. If $\widetilde{f}(x)$ is a bounded function with finite discontinuous points on the interval [a, b], then $\widetilde{f}(x)$ is integrable on [a, b].

Proof. Without loss of generality. It is only proved that $\widetilde{f}(x)$ has only one break point on [a, b]. And assume that the breakpoint is the endpoint b. Assume that the upper and lower bounds of $\widetilde{f}(x)$ on [a, b] are respectively $\widetilde{M}$, $\widetilde{m}$. For any $\varepsilon > 0$, take $\delta' > 0$ and $\frac{\varepsilon}{2(b-a)} < \widetilde{d}(\widetilde{M}, \widetilde{m}) < \frac{\varepsilon}{2\delta'}$. Assume that the upper and lower bounds of $\widetilde{f}(x)$ on $\Delta' = [b-\delta', b]$ are individually $\widetilde{M'}$, $\widetilde{m'}$. By Theorem 4.1, There is a split $T' = \Delta_1, \Delta_2, ...\Delta_{n-1}$ for $[b - \delta', b]$, hence $\widetilde{d}(\widetilde{s}(T'), \widetilde{S}(T')) < \frac{\varepsilon}{2}$. Let $\Delta_n = \Delta'$, so $T = \{\Delta_1, \Delta_2...\Delta_{n-1}, \Delta_n\}$ is segmentation on [a, b]. For any $\varepsilon > 0$, there exists $\delta(\delta < \delta')$, whenever $||T|| < \delta$, such that $\widetilde{d}(\widetilde{s}(T), \widetilde{S}(T)) = \widetilde{d}(\widetilde{s}(T'), \widetilde{S}(T'))+\widetilde{d}(\widetilde{M'}, \widetilde{m'}) \cdot \delta' < \frac{\varepsilon}{2} + \frac{\varepsilon}{2\delta'} \cdot \delta' = \varepsilon$.

Theorem 3.2.9. If $\widetilde{f}(x)$ is a monotonic function on the interval $[a, b]$, then $\widetilde{f}(x)$ is integrable on [a, b].

Proof. Let $\widetilde{f}(x)$ be a monotonically increasing function, and $\widetilde{f}(a) < \widetilde{f}(b)$. (If $\widetilde{f}(a) = \widetilde{f}(b)$, even $\widetilde{f}(x)$ is function of constant fuzzy value, it is integrable obviously). For arbitrary segmentation T on [a, b], because $\widetilde{f}(x)$ is monotonically increasing. On arbitrary interval $\Delta_i = [x_{i-1}, x_i], i = 1, 2, ..., n$, which is belonged to T. There must be Supremum of $\widetilde{f}(x)$: $\widetilde{M_i} = \widetilde{f}(x_i)$, and infimum of $\widetilde{f}(x)$: $\widetilde{m_i} = \widetilde{f}(x_{i-1})$. Hence $\widetilde{d}(\widetilde{s}(T), \widetilde{S}(T)) \leq \sum\limits_{i=1}^{n} \widetilde{d}(\widetilde{f}(x_i), \widetilde{f}(x_{i-1})||T|| = \widetilde{d}(\widetilde{f}(b), \widetilde{f}(a))||T||$. Therefore, for any $\varepsilon > 0$, only $\widetilde{f}(b) - \widetilde{f}(a) < \frac{\varepsilon}{||T||}$. There exists $\delta > 0$, whenever $||T|| < \delta$, such that $\widetilde{d}(\widetilde{s}(T), \widetilde{S}(T)) \leq \varepsilon$. We have $\widetilde{f}(x)$ is integrable function on [a, b].

4 Conclusion

By meaning of a fuzzy distance, the definite integral of fuzzy valued function linearly generated by structural elements is defined. The application of this definition studies the basic properties and integrable condition of definite integral of fuzzy valued function linearly generated by structural elements in the paper.

Acknowledgements. Thanks to the support by National Natural Science Foundation of China (Grant No. 11671284) and Doctoral Fund of colleges and Universities (20135134110003).

References

1. An-Gui, L.I., Jin, H.W., Zhang, Z.H., Hua, W.B.: A new definition of fuzzy limit. J. Liaoning Tech. Univ. **23**(6), 845–847 (2004)
2. Zhang, G.: Fuzzy continuous function and it's properties. Fuzzy Sets Syst. **43**(2), 159–171 (1991)
3. Burgin, M.: Neoclassical analysis: fuzzy continuity and convergence. Fuzzy Sets Syst. **75**(2), 291–299 (1995)
4. Panigrahi, M., Panda, G., Nanda, S.: Convex fuzzy mapping with differentiability and its application in fuzzy optimization. Eur. J. Oper. Res. **185**(1), 47–62 (2008)
5. Kaleva, O.: Fuzzy differential equation. Fuzzy Sets Syst. **24**(3), 301–317 (1987)
6. Guo, S.Z., Liu, X.H., Han, J.: Extremal problem of fuzzy-valued function. Syst. Eng. -Theory Pract. **34**(3), 738–745 (2014)
7. Bi, S.J., Zhang, X.D.: Convergence and continuity of fuzzy-valued functions. J. Heilongjiang Commer. Coll. **18**(3), 330–333 (2002)
8. Jang, L.C., Kim, T.K., Jeon, J.D., Kim, W.J.: On Choquet integrals of measurable fuzzy number-valued functions. Bull. Korean Math. Soc. **41**(1), 95–107 (2004)
9. Guo, S.Z.: Brief introduction to fuzzy-valued function analysis based on the fuzzy structured element method(I). Math. Prac. Theory **21**(5), 87–93 (2002)
10. Guo, S.Z.: Brief introduction of fuzzy-valued function analytics base on fuzzy structured element method(II). Math. Prac. Theory **38**(2), 73–79 (2008)
11. Yin, F., Wang, P.F.: A new definition of fuzzy-valued function limit(continuity) and its derivability. J. North Univ. China 32(6), 662-666 (2011)
12. Guo, S.Z.: Principle of Fuzzy Mathematical Analysis Based on Structural Element Theory, pp. 97–113. Northeastern University Press, ShenYang (2004)
13. Tran, L., Duckstein, L.: Comparison of fuzzy numbers using a fuzzy distance measure. Fuzzy Sets Syst. **130**(3), 331–341 (2002)
14. Goetschel, R., Voxman, W.: Elementary fuzzy calculus. Fuzzy Sets Syst. **19**(1), 82–86 (2005)
15. Puri, M.L., Ralescu, D.A.: Differentials on fuzzy functions. J. Math. Anal. Appl. **91**(2), 552–558 (1983)
16. Burgin, M.: Theory of fuzzy limits. Fuzzy Sets Syst. **115**(115), 433–443 (2000)
17. Guo, S.Z.: Commonly express method of fuzzy-valued function based on structured element. Fuzzy Syst. Math. 19(1), 82-86(2005)
18. Shu-Juan, B.I.: The convergence of fuzzy value function defined in fuzzy number. J. Harbin Univ. Sci. Technol. **15**(2), 76–78 (2010)

19. Yang, L.B., Gao, Y.Y., Lin, W.X.: The Principle and Application of Fuzzy Mathematics, pp. 117–125. South China University of Technology Press, GuangDong (2001)
20. Guo, X.M., Zhang, X.: Limit of fuzzy valued function and its properties. J. ChuZhou Univ. **15**(5), 21–23 (2013)
21. Liu, H.L., Feng, R.P.: New fuzzy distance definition of fuzzy number. Fuzzy Syst. Math. **19**(2), 106–109 (2005)
22. Department of mathematics, East China Normal University: On Mathematical Analysis 3rd edn. pp. 23-55. Higher Education Press, Beijing (1981)
23. Department of mathematics, Tongji University: Higher Mathematics Book, 6th edn. pp. 23-38. Higher Education Press, Beijing (2007)

An Information Quantity-Based Uncertainty Measure to Incomplete Numerical Systems

Xin Guo[1(✉)], Yingzhuo Xiang[2], and Lan Shu[1]

[1] School of Mathematical Sciences, University of Electronic Science and Technology of China, Chengdu 611731, China
201521100502@std.uestc.edu.cn

[2] National Key Laboratory of Science and Technology on Blind Signal Processing, Chengdu 610041, China

Abstract. The uncertainty measure plays an important role in the analysis of data. At present, many uncertain measures for incomplete information systems or incomplete decision systems have been developed. However, these measures are mainly aimed at discrete valued information systems, but they are not suitable for real valued data sets. In this paper, we mainly study the uncertainty measurement method of incomplete numerical information systems. By introducing neighborhood tolerant rough sets model, each concept has a neighborhood tolerant subset called neighborhood tolerance granule. Neighborhood tolerance information quantity uncertainty measure is proposed. We then prove that it satisfies non-negativity and monotonicity, giving maximum and minimum values. On this basis, the concept of neighborhood-tolerance joint quantity and neighborhood-tolerance condition quantity is proposed, and the relation between them and neighborhood tolerance information is discussed. Theoretical analysis and experimental results show that, in incomplete numerical information systems, the uncertainty measures we propose are performed better than the neighborhood-tolerance approximation accuracy measure in some case.

Keywords: Neighborhood-tolerance rough set · Uncertainty measure
Feature selection · Neighborhood-tolerance information quantity

1 Introduction

The information system contains complete and incomplete. The main characteristic of incomplete information systems is the lack of information. An incomplete information system, if it can distinguish conditional attributes and decision attributes, can be called incomplete decision system. The concept in incomplete decision system is also called incomplete data. In addition, incomplete numerical data implies which has numerical characteristics. More importantly, tolerance relations and tolerance rough sets are tools for dealing with missing values. The main advantage of rough set theory in dealing with incomplete decision table is that it does not change the original system and can effectively eliminate unnecessary knowledge [1, 3, 4, 6–10].

In the literature [2], Chen et al. considering the classical rough set model cannot handle continuous complete data, the neighborhood rough set model is put forward.

B.-Y. Cao and Y.-B. Zhong (Eds.): ICFIE 2017, AISC 872, pp. 23–29, 2019.
https://doi.org/10.1007/978-3-030-02777-3_3

From the Angle of neighborhood granularity, each target has a neighborhood. The information quantity based uncertainty measurement was proposed, and they proved the validity of it. However, there is no one has been considered the uncertainty measurement in incomplete numerical information system from the Angle of neighborhood granularity. In this paper, considering the particularity of incomplete numerical information system, we put forward the neighborhood tolerance relation. Then, based on the neighborhood tolerance relation, the neighborhood tolerance rough set model is constructed. Consequently, we put forward an uncertainty measurement based on neighborhood-tolerance information quantity. Experimental results show this method is outperform the neighborhood-tolerance approximation accuracy measure.

The rest of this paper is structured as follows: In Sect. 2, we put forward a new method of uncertainty measurement in numerical incomplete information system: neighborhood-tolerance information quantity is proposed. Then we put forward the concept of neighborhood-tolerance joint quantity and the neighborhood-tolerance condition quantity, naturally we prove some properties of them. In Sect. 3, we use an instance of Mammographic data set in the UCI repository of machine learning database to verify the validity of the proposed uncertainty measurement method. Finally, we make a conclusion and give the future research direction in Sect. 4.

2 Uncertainty Measures in Incomplete Numerical Information System

The information quantity was first put forward by Hu et al. in literature [5]. It is used to measure the uncertainty of a fuzzy attribute set or a fuzzy equivalence relation. According to literature [5], Chen et al. proposed the information quantity-based uncertainty measures for neighborhood rough sets [2]. Which is used to measure the uncertainty of complete information systems. In what follows, we extend the information quantity-based uncertainty measures [2] to neighborhood tolerance environment. It is applied for evaluating a neighborhood tolerance attribute set or a neighborhood tolerance relation. Furthermore, we present its various extensional forms.

2.1 Information Quantity-Based on Neighborhood Tolerance Relation

Definition 1. Let $IIS = (U, A)$ be an incomplete information system, $B \subseteq A$, and $U/NT_B^{\delta} = \{NT_B^{\delta}(x_1), NT_B^{\delta}(x_2), \cdots, NT_B^{\delta}(x_{|U|})\}$. The neighborhood-tolerance information quantity of B is defined as:

$$Q_{\delta}(B) = -\frac{1}{|U|\log|U|}\sum_{i=1}^{|U|}\log\frac{|NT_B^{\delta}(x_i)|}{|U|}$$

Proposition 1. Suppose $IIS = (U, C \cup D)$ be an incomplete decision system. If $B_1 \subseteq B_2 \subseteq C$, then $0 \leq Q_{\delta}(B_1) \leq Q_{\delta}(B_2) \leq 1$.

Proof. Since $B_1 \subseteq B_2 \subseteq C$, we obtain $\forall x_i \in U, NT^{\delta}_{B_1}(x_i) \supseteq NT^{\delta}_{B_2}(x_i)$.

Hence, $U \supseteq NT^{\delta}_{B_1}(x_i) \supseteq NT^{\delta}_{B_2}(x_i) \supseteq \{x_i\}$. Thus, $1 \geq \frac{|NT^{\delta}_{B_1}(x_i)|}{|U|} \geq \frac{|NT^{\delta}_{B_2}(x_i)|}{|U|} \geq \frac{1}{|U|}$,

Then, $0 \geq \log \frac{|NT^{\delta}_{B_1}(x_i)|}{|U|} \geq \log \frac{|NT^{\delta}_{B_2}(x_i)|}{|U|} \geq \log \frac{1}{|U|}$.

Hence, $0 \leq -\frac{1}{|U|}\sum_{i=1}^{|U|} \log \frac{|NT^{\delta}_{B_1}(x_i)|}{|U|} \leq -\frac{1}{|U|}\sum_{i=1}^{|U|} \log \frac{|NT^{\delta}_{B_2}(x_i)|}{|U|} \leq \log|U|$.

Thus, $0 \leq -\frac{1}{|U|\log|U|}\sum_{i=1}^{|U|} \log \frac{|NT^{\delta}_{B_1}(x_i)|}{|U|} \leq -\frac{1}{|U|\log|U|}\sum_{i=1}^{|U|} \log \frac{|NT^{\delta}_{B_2}(x_i)|}{|U|} \leq 1$.

Therefore, $0 \leq Q_{\delta}(B_1) \leq Q_{\delta}(B_2) \leq 1$.

Proposition 2. Suppose $IIS = (U, C \cup D)$ be an incomplete decision system, $B \subseteq A$. If $0 \leq \delta_1 \leq \delta_2 \leq 1$, then $0 \leq Q_{\delta_2}(B) \leq Q_{\delta_1}(B) \leq 1$.

The proof progress is seminary to proposition 1, so I won't repeat it.

Definition 2. Let $IDS = (U, C, D)$ be an incomplete decision table, $B \subseteq C$, $U/NT^{\delta}_{B} = \{NT^{\delta}_{B}(x_1), NT^{\delta}_{B}(x_2), \cdots, NT^{\delta}_{B}(x_{|U|})\}$, let $U/NT^{\delta}_{D} = \{NT^{\delta}_{D}(x_1), NT^{\delta}_{D}(x_2), \cdots, NT^{\delta}_{D}(x_{|U|})\}$, for any $B \subseteq C$, the neighborhood-tolerance joint quantity of D and B is defined as:

$$Q_{\delta}(BD) = -\frac{1}{|U|\log|U|}\sum_{i=1}^{|U|} \log \frac{|NT^{\delta}_{B}(x_i) \cap NT^{\delta}_{D}(x_i)|}{|U|}$$

Proposition 3. Suppose $IIS = (U, C \cup D)$ be an incomplete decision system. If $B_1 \subseteq B_2 \subseteq C$, then $Q_{\delta}(B_1 D) \leq Q_{\delta}(B_2 D)$.

Definition 3. Let $IDS = (U, C, D)$ be an incomplete decision table, $B \subseteq C$, $U/NT^{\delta}_{B} = \{NT^{\delta}_{B}(x_1), NT^{\delta}_{B}(x_2), \cdots, NT^{\delta}_{B}(x_{|U|})\}$, let $U/NT^{\delta}_{D} = \{NT^{\delta}_{D}(x_1), NT^{\delta}_{D}(x_2), \cdots, NT^{\delta}_{D}(x_{|U|})\}$, for any $B \subseteq C$, the neighborhood-tolerance conditional quantity of B to D is defined as:

$$Q_{\delta}(B|D) = -\frac{1}{|U|\log|U|}\sum_{i=1}^{|U|} \log \frac{|NT^{\delta}_{B}(x_i) \cap NT^{\delta}_{D}(x_i)|}{|NT^{\delta}_{D}(x_i)|}$$

Proposition 4. $Q_{\delta}(B|D) = Q_{\delta}(BD) - Q_{\delta}(D)$.

Definition 4. Let $IDS = (U, C, D)$ be an incomplete decision table, $B \subseteq C$. The attribute set B is a relative reduction of IDS if B satisfies:

(1) $Q_{\delta}(B|D) = Q_{\delta}(C|D)$;
(2) $\forall a \in B, Q_{\delta}(B - a|D) < Q_{\delta}(B|D)$.

Definition 5. Suppose $IIS = (U, C \cup D)$ be an incomplete decision system, $B \subseteq C$ and $U/D = \{D_1, D_2, \cdots, D_l\}$ be equivalence classes generated by a decision attribute D on the universe U and a conditional attribute subset $B \subseteq C$. The approximation accuracy of

U/D related to B based on the neighborhood-tolerance information quantity in an incomplete information system is defined as follows:

$$q\alpha_B^\delta(U/D) = \alpha_B^\delta(U/D) * Q_\delta(B)$$

Proposition 5. Suppose $IIS = (U, C \cup D)$ be an incomplete decision system. If $B_1 \subseteq B_2 \subseteq C$, then $q\alpha_{B_1}^\delta(U/D) \leq q\alpha_{B_2}^\delta(U/D)$.

3 Experimental Analyses About the Uncertainty Measure

In this section, a series of experiments are designed to test the effectiveness of the neighborhood-tolerance information quantity based uncertainty measurement to the data sets with numerical attributes. We apply the following definition to normalize the data set.

Definition 6. Let the object set X be a set of nonnegative real numbers, for $x_{\max}$ and $x_{\min}$ are respectively the maximal and minimal values of X, the normalized object set X' is defined as:

$$X' = \left\{ y \middle| y = \frac{x - x_{\min}}{x_{\max} - x_{\min}}, x \in X \right\}$$

In order to test the effectiveness of our proposed uncertainty measurement, we construct some experiments on the Mammographic data set. It can be downloaded from UCI repository of machine learning database. Here, we randomly select partial of the data from Mammographic data set as experimental data in Table 1. Firstly, we normalize the data to a range 0–1 which is shown in Table 2. Then, we use the Euclidean distance to granulate the data and suppose the neighborhood parameter $\delta = 0.35$. The attribute subset B varies by the sequences $\{a_1\}$, $\{a_1, a_2\}$, $\{a_1, a_2, a_3\}$, $\{a_1, a_2, a_3, a_4\}$ and $\{a_1, a_2, a_3, a_4, a_5\}$. We can calculate the neighborhood-tolerance information quantities first. The experimental result is shown in Fig. 1. It can be seen from Fig. 1 that the values of the information quantities are increasing as the cardinality of condition attribute subset becomes bigger. As we know, the uncertainty is decreasing with the attribute subset increasing. The experimental results show that the measure our proposed vary inversely with the uncertainty of a numerical incomplete information system. These result show that our proposed measure is valid for evaluating the uncertainty of a numerical incomplete information system.

Figure 2 shows the values of the quantity measure of the data sets in Table 1 with respective to all attributes. The X-axis represents the threshold value δ, which step is 0.05 from 0 to 1. The Y-axis represents the value of the measure. It can be seen that the neighborhood-tolerance information quantity is monotonically decreasing with the increase of the value of δ. These results verify the proposition 4.

The measuring results of the quantity-based approximation accuracy method are shown in Fig. 3. It can be seen from Fig. 3 that both the values of the approximation accuracy and the quantity-based approximation accuracy measure are increasing with

Table 1. The experimental data of the Mammographic mass data set from UCI machine learning repository.

U	a_1	a_2	a_3	a_4	a_5	d	U	a_1	a_2	a_3	a_4	a_5	d
x_1	5	67	3	5	3	1	x_{16}	4	54	1	1	3	0
x_2	4	43	1	1	?	1	x_{17}	3	52	3	4	3	0
x_3	5	58	4	5	3	1	x_{18}	4	59	2	1	3	1
x_4	4	28	1	1	3	0	x_{19}	4	54	1	1	3	1
x_5	5	74	1	5	?	1	x_{20}	4	40	1	?	?	0
x_6	4	65	1	?	3	0	x_{21}	?	66	?	?	1	1
x_7	4	70	?	?	3	0	x_{22}	5	56	4	3	1	1
x_8	5	42	1	?	3	0	x_{23}	4	43	1	?	?	0
x_9	5	57	1	5	3	1	x_{24}	5	42	4	4	3	1
x_{10}	5	60	?	5	1	1	x_{25}	4	59	2	4	3	1
x_{11}	5	76	1	4	3	1	x_{26}	5	75	4	5	3	1
x_{12}	3	42	2	1	3	1	x_{27}	2	66	1	1	?	0
x_{13}	4	64	1	?	3	0	x_{28}	5	63	3	?	3	0
x_{14}	4	36	3	1	2	0	x_{29}	5	45	4	5	3	1
x_{15}	4	60	2	1	2	0	x_{30}	5	55	4	4	3	0

Table 2. The normalized experimental data set.

U	a_1	a_2	a_3	a_4	a_5	d	U	a_1	a_2	a_3	a_4	a_5	d
x_1	1	0.8125	0.6667	1	1	1	x_{16}	0.6667	0.5417	0	0	1	0
x_2	0.6667	0.3125	0	0	?	1	x_{17}	0.3333	0.5	0.6667	0.75	1	0
x_3	1	0.625	1	1	1	1	x_{18}	0.6667	0.6458	0.3333	0	1	1
x_4	0.6667	0	0	0	1	0	x_{19}	0.6667	0.5417	0	0	1	1
x_5	1	0.9583	0	1	?	1	x_{20}	0.6667	0.25	0	?	?	0
x_6	0.6667	0.7708	0	?	1	0	x_{21}	?	0.7917	?	?	0	1
x_7	0.6667	0.875	?	?	1	0	x_{22}	1	0.5833	1	0.5	0	1
x_8	1	0.2917	0	?	1	0	x_{23}	0.6667	0.3215	0	?	?	0
x_9	1	0.6042	0	1	1	1	x_{24}	1	0.2917	1	0.75	1	1
x_{10}	1	0.6667	?	1	0	1	x_{25}	0.6667	0.6458	0.3333	0.75	1	1
x_{11}	1	1	0	0.75	1	1	x_{26}	1	0.9792	1	1	1	1
x_{12}	0.3333	0.2917	0.3333	0	1	1	x_{27}	0	0.7917	0	0	?	0
x_{13}	0.6667	0.75	0	?	1	0	x_{28}	1	0.7292	0.6667	?	1	0
x_{14}	0.6667	0.1667	0.6667	0	0.5	0	x_{29}	1	0.3542	1	1	1	1
x_{15}	0.6667	0.6667	0.3333	0	0.5	0	x_{30}	1	0.5625	1	0.75	1	0

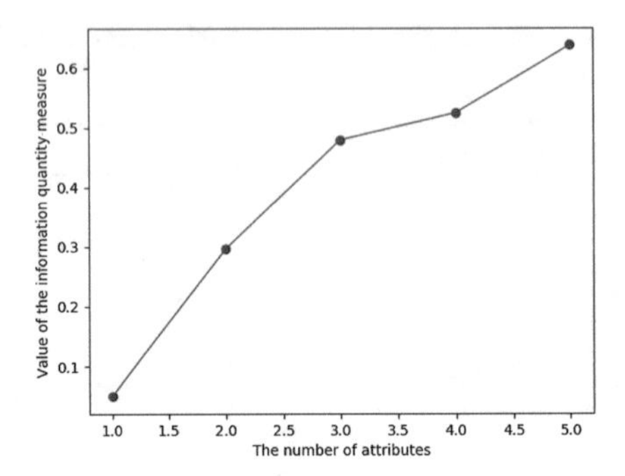

Fig. 1.

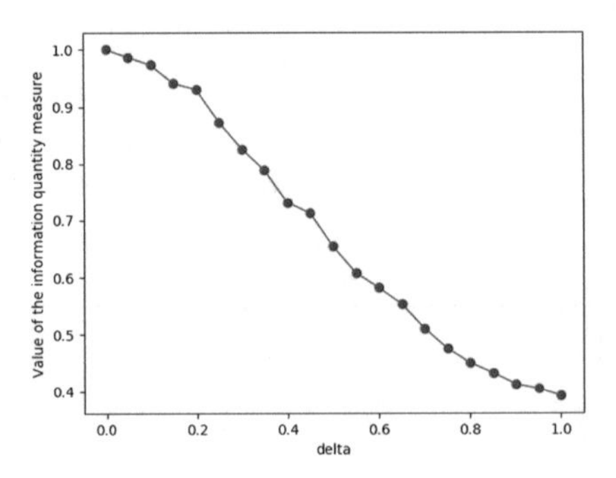

Fig. 2.

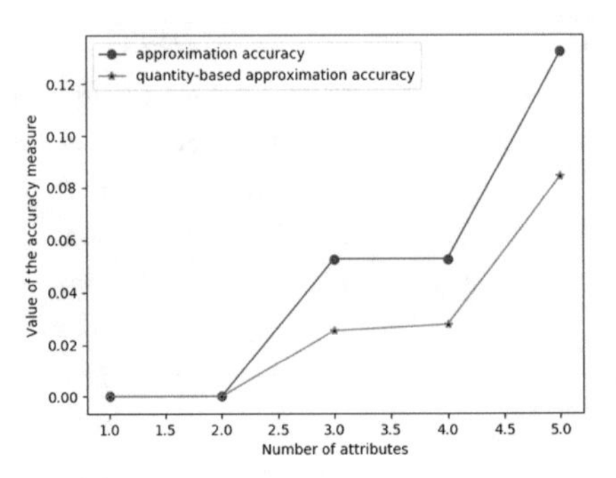

Fig. 3.

the number of selected attributes becoming bigger. However, there is no change as the number of attributes increases from $|B| = 3$ to $|B| = 4$. By contrast, the quantity-based approximation accuracy measure can discern them clearly. That means the quantity-based approximation accuracy measure is valid for evaluating the uncertainty of a numerical incomplete decision system and is more powerful for evaluating the uncertainty in some cases.

4 Conclusion

This article is mainly aimed at the uncertainty metrics of incomplete numeric information system. Considering the classical uncertainty measures are mainly designed for discrete-valued information system, but not suitable to real-valued data sets. we introduced neighborhood tolerance relation for incomplete numerical information system. Then, the neighborhood tolerance rough set model is constructed. From the Angle of neighborhood granularity, the concept of neighborhood-tolerance information quantity is proposed. Experiments show that our proposed uncertainty measure is reasonable for both incomplete numerical information system and incomplete numerical decision system. However, neighborhood tolerance relation processing incomplete information may be too loose. In future studies, we can improve the tolerance relation to deal with incomplete information system.

Acknowledgments. Thanks to the support by National Natural Science Foundation of China (G0501100110671030).

References

1. Hu, Q., Yu, D., Xie, Z.: Information-preserving hybrid data reduction based on fuzzy-rough techniques. Pattern Recogn. Lett. **27**(5), 414–423 (2006)
2. Chen, Y., Xue, Y., Ma, Y., et al.: Measures of uncertainty for neighborhood rough sets. Knowl.-Based Syst. **120**, 226–235 (2017)
3. Zhao, H., Qin, K.J.: Mixed feature selection in incomplete decision table. Knowl.-Based Syst. **57**, 181–190 (2014)
4. Qian, Y., Liang, J.C.: Combination entropy and combination granulation in incomplete information system. In: International Conference on Rough Sets and Knowledge Technology, pp. 184–190 . Springer, Heidelberg (2006)
5. Zheng, T., Zhu, L.J.: Uncertainty measures of neighborhood system-based rough sets. Knowl.-Based Syst. **86**, 57–65 (2015)
6. Pawlak, Z.J.: Rough sets. Int. J. Parallel Prog. **11**(5), 341–356 (1982)
7. Chen, Y., Wu, K., Chen, X., et al.: An entropy-based uncertainty measurement approach in neighborhood systems. Inf. Sci. **279**, 239–250 (2014)
8. Dai, J., Hu, H., Zheng, G., et al.: Attribute reduction in interval-valued information systems based on information entropies. Front. Inf. Technol. Electron. Eng. **17**, 919–928 (2016)
9. Yager, R.R.J.: Uncertainty modeling using fuzzy measures. Knowl.-Based Syst. **92**, 1–8 (2016)
10. Yao, Y., Deng, X.J.: Quantitative rough sets based on subsethood measures. Inf. Sci. **267**, 306–322 (2014)

An Inexact Reasoning Method Based on Sugeno Integral

Jun Shen[1(✉)], De-jun Peng[1], and Sheng-quan Ma[2]

[1] School of Mathematics and Statistics, Hainan Normal University, HaiKou 571158, China
407519785@qq.com

[2] School of Information Science and Technology, Hainan Normal University, HaiKou 571158, China

Abstract. Fuzzy rules, which are important tools in the fuzzy reasoning, have been widely used in expert systems to represent fuzzy and uncertain concepts. Using the fuzzy set as an interior representation can make the reasoning accord with the man's thought better. To represent the fuzziness and uncertainty more effectively and make the result more reasonable, some knowledge representation parameters such as threshold value, certain factor, local weight and global weight are introduced. When the interaction exists among fuzzy proposition in a fuzzy rule set, the parameter weight is displayed by the fuzzy measure. The paper proposes an inexact reasoning method based on matrix transformation, the interaction exists among fuzzy propositions and the confidence level of the rules are all considered. The method is mainly used in the incomplete inductive reasoning.

Keywords: Fuzzy measure · Sugeno integral · Matrix transformation

1 Introduction

Fuzzy Production Rules (FPRs) are widely used in expert system to represent fuzzy and uncertain concepts. FPRs are usually presented in the form of a fuzzy IF-THEN rule in which both the antecedent and the consequent are fuzzy concepts denoted by fuzzy sets. Some knowledge parameters such as certainty factor, local weight, global weight and threshold valued had been incorporated into the FPRs. For example, IF A AND B THEN C (CF, GW, λ), where A, B, C are all fuzzy set, CF is the certainty factor of the rule, GW denotes the global weight, λ is the threshold of the rule.

In [1], He presented the concepts of weighted fuzzy logics, the truth value of a conjunction of propositions is a weighted sum of the truth value of each proposition, where the truth values of the propositions and weights of the propositions are represented by crisp real values. In [2], Chen proposed an algorithm which allowed the truth values of the conditions appearing in the antecedent portions of the rules to be represented by trapezoidal fuzzy number. In [3, 4], Wuqiong proposed a weighted inexact reasoning algorithm based on matrix transformations. Given the fuzzy truth values of some conditions, the algorithm can performance the weighted fuzzy reasoning to evaluate the fuzzy truth value of the conditions automatically. When the interaction

B.-Y. Cao and Y.-B. Zhong (Eds.): ICFIE 2017, AISC 872, pp. 30–37, 2019.
https://doi.org/10.1007/978-3-030-02777-3_4

exists among fuzzy proposition in a fuzzy rule set, the parameter weight is displayed by the fuzzy measure. In this paper, an inexact reasoning method based on Sugeno integral is given, the interaction of the propositions and the confidence level of the rule are all considered. The method is mainly used in the inductive reasoning.

2 Preliminaries

2.1 Fuzzy Measure and Fuzzy Integral

Definition 1. Let $X = \{x_1, x_2, \cdots x_n\}$, $P(X)$ is the power set of X, $\mu : P(X) \to R$ is called fuzzy measure [5]. If the following conditions hold:

(1) $\mu(\emptyset) = 0$,
(2) IF $A \subset X$, then $\mu(A) \geq 0$,
(3) IF $A \subset X$, $B \subset X$, $A \subset B$, then $\mu(A) \leq \mu(B)$.

Definition 2. Let $X = \{x_1, x_2, \cdots, x_n\}$, μ be a fuzzy measure or a non-additive set function defined on the power set of X, f be a function from X to [0, 1]. The Sugeno integral [5] of f with respect to μ is defined by

$$(S)\int_X f d\mu = \bigvee_{i=1}^{n} [f(x_i) \wedge \mu(A_i)]$$

where, we assume without loss of generality that $0 = f(x_0) \leq f(x_1) \leq \cdots \leq f(x_n)$ and $A_i = \{x_i, x_{i+1}, \cdots, x_n\}$. If no confusion on arises, we can omit the X and in short denote the Sugeno integral by $(S)\int f d\mu$.

2.2 Fuzzy Production Rules

Fuzzy production rules are widely used in expert systems to represent fuzzy, imprecise, ambiguous and vague concepts [6]. FPRs are usually presented in the form of fuzzy IF-THEN rule. With the increasing complex of today's knowledge based system, some knowledge parameters such as certainty factor, local weight, global weight and threshold valued had been incorporated into the FPRs.

In this paper, the fuzzy rules are used for knowledge representation. Let R be a set of fuzzy rules, $R = \{R_1, R_2, \cdots, R_n\}$. The general formulation of a fuzzy rule $R_i \in R$, is as follows:

$$R_i : \text{IF } d_j \text{ THEN } d_k\ (CF, GW, \lambda),$$

where d_j and d_k are the conditions described by propositions, the truth values of the condition d_j and d_k are described by real values in [0, 1]; CF is a certainty factor indicating the degree of the belief of the rule R_i; GW denotes the global weight assigned to R_i and λ is the threshold value of R_i.

If a fuzzy rules contains either AND or OR connectors, then it is called a compound fuzzy product rule. The compound fuzzy product rules can be divided into the following rule types.

Type 1: IF d_{j1} AND d_{j2} AND $\cdots$ AND d_{jn} THEN d_k (CF, LW, GW, λ).

This type of rule can be equivalently be reduced into the following rule:

$$\text{IF}\, d_j\, \text{THEN}\, d_k\, (CF),$$

where, $d_j = d_{j1}$ AND d_{j2} AND $\cdots$ AND d_{jn}.

Type 2: IF d_{j1} OR d_{j2} OR $\cdots$ d_{jn} THEN d_k (CF, GW, λ).

This type can equivalently be decomposed into the following rules:

$$R_1 : \text{IF}\, d_{j1}\, \text{THEN}\, d_k\, (CF, GW, \lambda),$$

$$R_2 : \text{IF}\, d_{j2}\, \text{THEN}\, d_k\, (CF, GW, \lambda),$$

$$R_n : \text{IF}\, d_{jn}\, \text{THEN}\, d_k\, (CF, GW, \lambda).$$

3 A Fuzzy Reasoning Method Based on Sugeno Integral

We can use a rule matrix to represent the rules of a rule based system. The concepts of rule matrix are derived from [3, 4]. A rule matrix T associated with m conditions is an $m \times m$ rule matrix, $T(k, j) = CF$ indicating that there is a rule IF d_j THEN d_k (CF) in the knowledge base; $T(k, j) = 0$ indicates that the relation between the condition d_j and d_k is unknown; the diagonals of T are one by reflexivity.

For example, if the knowledge base of a rule-based system contains the following rules:

$$R_1 : \text{IF}\ d_1\, \text{THEN}\, d_2\, (CF = 0.3),$$

$$R_2 : \text{IF}\, d_2\, \text{THEN}\, d_3\, (CF = 0.7),$$

$$R_3 : \text{IF}\, d_3\, \text{THEN}\, d_4\, (CF = 0.6).$$

Then these rules can be represented by a rule matrix T as follows

$$T = \begin{bmatrix} 1 & 0 & 0 & 0 \\ 0.3 & 1 & 0 & 0 \\ 0 & 0.7 & 1 & 0 \\ 0 & 0 & 0.6 & 1 \end{bmatrix}$$

We can also use a matrix R to represent the truth values of the conditions, where $R[i] = r_i$ indicates that the truth value of the condition d_i is r_i, and $1 \leq i \leq m$. The truth values of the conditions d_1, d_2, d_4 are 0.3, 0.4, 0.5, respectively. $R[i] = 0$ indicates that the truth value of d_i is unknown. Then the truth values of the conditions can be represented by a truth value matrix R shown as follows,

$$R = \begin{bmatrix} 0.3 & 0.4 & 0 & 0.5 \end{bmatrix}^T.$$

If a production rule contains AND connectors, then it is a compound fuzzy production rule, for example

$$\text{IF } d_{j1} \text{ AND } d_{j2} \text{ AND} \cdots \text{AND } d_{jn} \text{ THEN } d_k \ (CF, \mu),$$

where CF is a certainty factor, $CF \in [0, 1]$, μ is a fuzzy measure based on $\{d_{j1}, \cdots, d_{jn}\}$, $d_{ji} \in [0, 1]$, $1 \leq i \leq n$. Assume that the truth values of the conditions d_{j1}, d_{j2}, … and d_{jn} are r_{j1}, r_{j2}, …, d_{jn}, respectively. The truth value of d_k can be evaluated as follows

$$r_k = CF \cdot (S) \int r_{ji} d\mu.$$

If there are two rules leading to the condition d_k,

$$R_1 : \text{IF } d_{j1} \text{ AND } d_{j2} \text{ AND} \cdots \text{AND } d_{jn} \text{ THEN } d_k \ (CF_1, \mu_1),$$

$$R_2 : \text{IF } d_{l1} \text{ AND } d_{l2} \text{ AND } \cdots \text{ AND } d_{lm} \text{ THEN } d_k \ (CF_2, \mu_2),$$

where, $CF_1 \in [0, 1]$, $CF_2 \in [0, 1]$, μ_1 is a fuzzy measure based on $\{d_{j1}, \cdots, d_{jn}\}$, $d_{ji} \in [0, 1]$, $1 \leq i \leq n$. μ_2 be a fuzzy measure based on $\{d_{l1}, \cdots, d_{lm}\}$, $d_{ji} \in [0, 1]$, $1 \leq i \leq m$. Assume that the truth values of the conditions d_{j1}, d_{j2}, … and d_{jn} are r_{j1}, r_{j2}, …, and r_{jn} respectively. The truth values of the conditions d_{l1}, d_{l2}, …, and d_{lm} are r_{l1}, r_{l2}, …, and r_{ln}. The truth value of d_k can be evaluated as follows

$$r_k = \max\left\{ CF_1 \cdot (S) \int r_{ji} d_{\mu_1} \ , \ CF_2 \cdot (S) \int r_{li} d_{\mu_2} \right\}.$$

If the rules in the knowledge base of a rule based system are defined by m conditions and there are k compound condition conditions appearing in the antecedent portion of the rules, an $(m+k) \times (m+k)$ rule matrix T and $(m+k) \times 1$ truth value matrix R will be written. Assume that the rule matrix T and the truth value matrix R have the following forms.

$$T = \begin{bmatrix} f_{11} & f_{12} & \cdots & f_{1(k+m)} \\ f_{21} & f_{22} & \cdots & f_{2(k+m)} \\ \vdots & \vdots & \vdots & \vdots \\ f_{(k+m)1} & f_{(k+m)2} & \cdots & f_{(k+m)(k+m)} \end{bmatrix}, R = \begin{bmatrix} r_1 \\ r_2 \\ \vdots \\ r_{k+m} \end{bmatrix}.$$

The transformation operation $R' = T * R$ is defined as follow

$$R' = T * R = \begin{bmatrix} \max\left(1 \times r_1, f_{12} \times r_2, \cdots, f_{1(m+k)} \times r_{m+k}\right) \\ \max\left(f_{21} \times r_1, 1 \times r_2, \cdots, f_{2(m+k)} \times r_{m+k}\right) \\ \vdots \\ \max\left(f_{(m+k)1} \times r_1, f_{(m+k)2} \times r_2, \cdots, 1 \times r_{m+k}\right) \end{bmatrix}.$$

Given the truth values of some conditions, we can get the truth values of other conditions according to the following method.

Fuzzy Reasoning Method

Step 1. Generate the rule matrix T.

Step 2. Evaluate the fuzzy truth value of k compound conditions. If

$$d_q = d_j \,\text{AND}\, d_s \,\text{AND} \cdots \text{AND}\, d_t,$$

then

$$R[q] = (S)\int R[\;]d\mu,\ (q = m+1,\ m+2, \cdots,\ m+k),$$

The $(m+k) \times 1$ truth value matrix be $R = [R[1], R[2], \cdots, R[m], R[m+1], \cdots, R[m+k]]^T$.

Step 3. Performing the transformation operations $R' = T * R$. If $R' = R$, go to Step 4, if $R' \neq R$, let $R[i] = R'[i]$, $i = 1, 2, \cdots, m$, go to Step 2.

Step 4. Let $R = R'$, put out R.

In the following, we use an example to illustrate the inexact reasoning process.

***Example* 3.1.** Assume that the knowledge base of a rule-based system contains the following rules:

$$R_1:\ \text{IF}\, d_1 \,\text{AND}\, d_2 \,\text{THEN}\, d_4\ (CF = 0.7, \mu),$$

$$R_2:\ \text{IF}\, d_3 \,\text{THEN}\, d_2\ (CF = 0.6),$$

$$R_3:\ \text{IF}\, d_1 \,\text{OR}\, d_3 \,\text{THEN}\, d_5\ (CF = 0.5),$$

and assume that d_6 is a compound condition, where $d_6 = d_1$ AND d_2. Based on the previous discussion, we can see that the rule R_1 can equivalently be reduced to

$$R_1:\ \text{IF}\, d_6 \,\text{THEN}\, d_4\ (CF = 0.7).$$

R_3 can equivalently be decomposed into the following rules:

$$R_{3a}: \text{IF}\, d_1\, \text{THEN}\, d_5\, (CF = 0.5),$$

$$R_{3b}: \text{IF}\, d_3\, \text{THEN}\, d_5\, (CF = 0.5).$$

Assume that the truth values of the conditions of d_1 and d_3 are 0.8 and 0.6, respectively, thus, initially, we let $R[1] = 0.8$, $R[2] = 0$, $R[3] = 0.6$, $R[4] = 0$, $R[5] = 0$. μ is a fuzzy measure based $\{d_1, d_2\}$, where $\mu(\{d_1\}) = 0.3$, $\mu(\{d_2\}) = 0.3$, $\mu(\{d_1, d_2\}) = 1$.

Step 1: Construct 6×6 rule matrix

$$T = \begin{bmatrix} 1 & 0 & 0 & 0 & 0 & 0 \\ 0 & 1 & 0.6 & 0 & 0 & 0 \\ 0 & 0 & 1 & 0 & 0 & 0 \\ 0 & 0 & 0 & 1 & 0 & 0.7 \\ 0.5 & 0 & 0.5 & 0 & 1 & 0 \\ 0 & 0 & 0 & 0 & 0 & 1 \end{bmatrix},$$

Step 2: Initially, we let $R[1] = 0.8$, $R[2] = 0$, $R[3] = 0.6$, $R[4] = 0$, $R[5] = 0$, the truth value of condition d_6 can be computed by

$$R[6] == CF \cdot (S) \int R[\,]d\mu = 0.7 \cdot [(0 \wedge 1) \vee (0.8 \wedge 0.3)] = 0.21.$$

The truth value matrix is

$$R = [0.8 \quad 0 \quad 0.6 \quad 0 \quad 0 \quad 0.21]^T.$$

Step 3: After performing the transformation operations $R' = T * R$, we can get the following result

$$R' = T * R = \begin{bmatrix} 1 & 0 & 0 & 0 & 0 & 0 \\ 0 & 1 & 0.6 & 0 & 0 & 0 \\ 0 & 0 & 1 & 0 & 0 & 0 \\ 0 & 0 & 0 & 1 & 0 & 0.7 \\ 0.5 & 0 & 0.5 & 0 & 1 & 0 \\ 0 & 0 & 0 & 0 & 0 & 1 \end{bmatrix} * \begin{bmatrix} 0.8 \\ 0 \\ 0.6 \\ 0 \\ 0 \\ 0.21 \end{bmatrix} = \begin{bmatrix} 0.8 \\ 0.36 \\ 0.6 \\ 0.147 \\ 0.4 \\ 0.21 \end{bmatrix}.$$

Because $R' = R$, we go to step 2.

Step 2: Let $R[1] = 0.8$, $R[2] = 0.36$, $R[3] = 0.6$, $R[4] = 0.147$, $R[5] = 0.4$, the truth value of condition d_6 can be computed by

$$R[6] == CF \cdot (S) \int R[\,]d\mu = 0.7 \cdot [(0.36 \wedge 1) \vee (0.8 \wedge 0.3)] = 0.252.$$

We get that the truth value matrix

$$R = \begin{bmatrix} 0.8 & 0.36 & 0.6 & 0.147 & 0.4 & 0.252 \end{bmatrix}^T.$$

Step 3: After performing the transformation operations $R' = T * R$, we can get the following result

$$R' = \begin{bmatrix} 0.8 & 0.36 & 0.6 & 0.147 & 0.4 & 0.252 \end{bmatrix}^T.$$

Because $R' = R$, we go to step 4.

Step 4: The system has completed the fuzzy reasoning process. The truth values of the conditions d_1, d_2, d_3, d_4, d_5 and d_6 are 0.8, 0.36, 0.6, 0.147, 0.4 and 0.252, respectively.

In the following, we analyze the finiteness of the proposed method based on [7, 8]. In order to analyze the finiteness problem, the concept of reachability tree will be used, where nodes z_i in the reachability tree is associated with a temporal content of the truth value matrix R[8] a new node z_{i+1} in the tree is generated from z_i by performing a transformation operation $R' = T * R$ and then a new temporal content of R is derived from R'. The proposed method constructs a reachability tree, where each node in the tree has only one direct successor [8]. An important property of reachability tree is that it is finite. The proof of the property requires the following lemma.

Lemma: In any infinite directed tree in which each node has only a finite number of direct successors, there is an infinite path leading from the root node.

Proof: See [7, p. 97].

Theorem: The reachability tree constructed by the algorithm is finite.

Proof: The proof is the same as the one shown in [8]. Assume that there exists an infinite reachability tree. Because each node in the reachability tree constructed by the algorithm has only one direct successor, by Lemma, there is an infinite number of nodes in the reachability tree. Because there are p rules which are defined by m conditions and because there are k compound conditions appears more than once in the implication chain formed by these rules, the maximum number of implications in the chain is limited by p, there p is the number of rules in the knowledge. Thus, the maximum number of transformation operation $R' = T * R$ needed to be performed by the algorithm is $p + 1$, wherever $R' = R$, the algorithm terminates. Therefore, there are at most $p + 1$ nodes in the reachability tree.

4 Conclusion

When the interaction exists among fuzzy propositions in a fuzzy rule set, the parameter weight is displaced by the fuzzy measure, an inexact reasoning method based on Sugeno integral is proposed. This method is mainly used in the incomplete inductive reasoning.

Acknowledgements. Thanks to the supported by the International Science and Technology Cooperation Foundation of China (Grant No. 2012DFA11270) by Natural Science Foundation of Hainan (Grant No. 117123).

Recommender. Thanks to Professor Sheng-quan Ma's recommendation of Hainan Normal University in China.

References

1. He, X: Weighted fuzzy logics and its applications. In: Proceedings of COMPSAC 1988, pp. 485–489 (1988)
2. Chen, S.M.: A weighted fuzzy reasoning algorithm for medical diagnosis. Decis. Support Syst. **11**(1), 37–43 (1994)
3. Lu, Z.D., Hu, H.P., Li, F.: A weighted inexact reasoning algorithm. Acta Math. Appl. Sin. **14** (1), 16–25 (1996)
4. Wu, Q.: A weighted inexact reasoning algorithm. Control Theory Appl. **26**(2), 1–3 (2007)
5. Wang, X.Z.: Fuzzy Measure and Fuzzy Integral and Its Application in the Classification Techniques, pp. 12–35. Science Press, Beijing (2008)
6. Wang, X.Z., Chen, A.X., Feng, H.M.: Upper integral network with extreme learning mechanism. Neurocomputing **74**(16), 2520–2525 (2011)
7. Peterson, J.L.: Petri Nets. Theory and the Modeling of Systems. Prentice-Hall, Englewood Cliffs (1981)
8. Chen, S.M.: A fuzzy reasoning approach for rule-based systems based on fuzzy logics. IEEE Trans. Syst. Man, Cybern.-Part B: Cybern. **26**(5), 769–778 (1996)

Fuzzy Topology and Algebras

Properties of Fuzzy Filters in EQ-algebras

Hai Xie(✉)

School of Science, Guilin University of Technology,
Guilin 541004, China
xiehai126@126.com

Abstract. In this paper, the concept of fuzzy filter of an EQ-algebra is firstly redefined. Further, the definition of fuzzy prime filter of an EQ-algebra is proposed. At last, the connections between fuzzy filters and fuzzy prime filters of an EQ-algebra are discussed.

Keywords: EQ-algebras · Fuzzy prefilters · Fuzzy filters
Fuzzy prime filters

1 Introduction

EQ-algebra is a special algebra and its preliminary version was presented for the first time by Novák in [10]. Novák and De Baets made formal elaboration about EQ-algebras in [11]. EQ-algebras are intended to become algebras of truth values for fuzzy type theory (FTT) [12] where the main connective is a fuzzy equality. An EQ-algebra has three basic binary operations (meet, multiplication and a fuzzy equality) and a top element. By the commutativity of multiplication, EQ-algebras can be divided into two classes: *commutative* EQ-algebras [10,11] and *non-commutative* EQ-algebras [5,6]. From the point of view of logic, the biggest difference between EQ-algebras and residuated lattices lies in the construction way of implication. The implication in EQ-algebras is derived from the fuzzy equality and it is not a residuation with respect to multiplication. Since every residuated lattice can be seen as an EQ-algebra but not vice versa, EQ-algebra is a generalization of residuated lattice in a certain sense. Using the connective Δ, an EQ-algebra can be extended to get an EQ_{Δ}-algebra [1]. Moreover, Dyba and Novák [1,4] presented EQ_{Δ}-logic on the basis of EQ_{Δ}-algebra. The above EQ-logics belong to the propositional EQ-logics. The definition and basic properties of first-order EQ-logic were introduced by Dyba and Novák [2]. Further, Dyba et al. [3] introduced non-commutative first-order EQ-logics. First-order EQ-logic is the predicate version of EQ-logic.

Filter theory plays an important role in studying various logical algebras. From logical point of view, filters correspond to sets of provable formulae. In [11], Novák introduced the definition of EQ-filter (i.e. filter) of an EQ-algebra. Liu and Zhang [7] introduced the notions of positive implicative prefilters (filters) and implicative prefilters (filters) in EQ-algebras and studied their basic

B.-Y. Cao and Y.-B. Zhong (Eds.): ICFIE 2017, AISC 872, pp. 41–54, 2019.
https://doi.org/10.1007/978-3-030-02777-3_5

characterizations. Mohtashamnia ans Torkzadeh [9] introduced and characterized a prefilter generated by a nonempty subset of an EQ-algebra. The notion of fuzzy EQ-filter (i.e. fuzzy filter) of an EQ-algebra was first introduced by Ma and Hu [8]. Ma and Hu carried out a preliminary research on fuzzy filters of EQ-algebras. But in this definition, the membership degree of top element **1** is required to equal 1. This condition is overly strict. We think that this restriction can be relevantly relaxed and the membership degree of top element **1** can be less than or equal to 1. In [11], Novák and De Baets pointed out that the filter theory of EQ-algebras is rather subtle and still requires more investigation. In this paper, we continue the study of fuzzy filters of EQ-algebras. Firstly, we propose the definition of fuzzy prefilter of an EQ-algebra. Next, the concept of fuzzy filter of an EQ-algebra is redefined. Finally, we introduce the notion of fuzzy prime filter of an EQ-algebra and discuss the connections between fuzzy filters and fuzzy prime filters of an EQ-algebra.

This paper is organized as follows: In Sect. 2, we mainly review the basic definitions of EQ-algebras and their important properties. In Sect. 3, we introduce the notion of fuzzy prefilter of an EQ-algebra and redefined the definition of fuzzy filter of an EQ-algebra, and study their basic properties. Meantime, we propose the concept of fuzzy prime filter of an EQ-algebra. Finally, the main results are summarized in Sect. 4.

2 Preliminaries

In this section, we first review the notion and some basic properties of an EQ-algebra, which will be used in the following sections.

Definition 2.1 ([11]). A (commutative) EQ-algebra $\mathcal{E}$ is an algebra of type (2, 2, 2, 0), i.e. $\mathcal{E} = \langle E, \wedge, \otimes, \sim, \mathbf{1}\rangle$, where for all $a, b, c, d \in E$,

(E1) $\langle E, \wedge, \mathbf{1}\rangle$ is a commutative idempotent monoid (i.e. $\wedge$-semilattice with top element $\mathbf{1}$),
(E2) $\langle E, \otimes, \mathbf{1}\rangle$ is a commutative monoid and $\otimes$ is isotone w.r.t. $\leq$ (with $a \leq b$ defined as $a \wedge b = a$),
(E3) $a \sim a = \mathbf{1}$, (Reflexivity Axiom)
(E4) $((a \wedge b) \sim c) \otimes (d \sim a) \leq c \sim (d \wedge b)$, (Substitution Axiom)
(E5) $(a \sim b) \otimes (c \sim d) \leq (a \sim c) \otimes (b \sim d)$, (Congruence Axiom)
(E6) $(a \wedge b \wedge c) \sim a \leq (a \wedge b) \sim a$, (Monotonicity Axiom)
(E7) $(a \wedge b) \sim a \leq (a \wedge b \wedge c) \sim (a \wedge c)$, (Monotonicity Axiom)
(E8) $a \otimes b \leq a \sim b$. (Boundedness Axiom)

In [11], three operations $\wedge$, $\otimes$, $\sim$ are called, respectively, meet (infimum), multiplication and fuzzy equality.

For all $a, b \in E$, two derived operations [11] are defined as follows:

$$a \to b := (a \wedge b) \sim a. \tag{1}$$

$$\widetilde{a} := a \sim \mathbf{1}. \tag{2}$$

El-Zekey et al. [5,6] thought that the multiplication $\otimes$ needs not be *commutative.* In this paper, we always assume that EQ-algebra is *commutative.*

Theorem 2.2 ([11]). Let $\mathcal{E}$ be an EQ-algebra. For all $a, b \in E$ such that $a \leq b$ it holds that

(i) $a \to b = \mathbf{1}$;
(ii) $a \sim b = b \to a$;
(iii) $c \to a \leq c \to b$ and $b \to c \leq a \to c$.

Theorem 2.3 ([11]). Let $\mathcal{E}$ be an EQ-algebra. For all $a, b, c, d, a', b', c', d' \in E$, it holds that

(i) $a \otimes b \leq a$, $a \otimes b \leq a \wedge b$, $c \otimes (a \wedge b) \leq (c \otimes a) \wedge (c \otimes b)$;
(ii) $a \sim b \leq a \to b$, $a \to a = \mathbf{1}$;
(iii) $(a \to b) \otimes (b \to a) \leq a \sim b$;
(iv) $a = b$ implies $a \sim b = \mathbf{1}$;
(v) $a \leq \widetilde{a}$ and $\widetilde{\mathbf{1}} = \mathbf{1}$;
(vi) $\widetilde{a} = \mathbf{1} \to a$ and $a \to \mathbf{1} = \mathbf{1}$;
(vii) $a \otimes (a \sim b) \leq \widetilde{b}$;
(viii) $\widetilde{a} \otimes \widetilde{b} \leq a \sim b$;
(ix) $b \leq \widetilde{b} \leq a \to b$;
(x) $((a \wedge b) \sim (c \wedge d)) \otimes (a \sim a') \otimes (b \sim b') \otimes (c \sim c') \otimes (d \sim c') \leq (a' \wedge b') \sim (c' \wedge d')$.

Two binary operations $\leftrightarrow$ and $\overset{\circ}{\leftrightarrow}$ [11] on E are defined by

$$\begin{aligned} a \leftrightarrow b &= (a \to b) \wedge (b \to a), \\ a \overset{\circ}{\leftrightarrow} b &= (a \to b) \otimes (b \to a). \end{aligned} \tag{3}$$

Theorem 2.4 ([11]). Let $\mathcal{E}$ be an EQ-algebra. For all $a, b \in E$ it holds that

(i) $(a \wedge b) \leftrightarrow a = (a \wedge b) \overset{\circ}{\leftrightarrow} a = a \to b$;
(ii) $a \overset{\circ}{\leftrightarrow} b \leq a \sim b \leq a \leftrightarrow b$;
(iii) If $\mathcal{E}$ is linearly ordered, then $a \overset{\circ}{\leftrightarrow} b = a \sim b = a \leftrightarrow b$.

Theorem 2.5 ([6]). The following properties hold in all EQ-algebras:

(i) $(a \sim b) \otimes (c \sim d) \leq (a \wedge c) \sim (b \wedge d)$;
(ii) $a \sim d \leq ((a \wedge b) \sim c) \sim ((d \wedge b) \sim c)$;
(iii) $a \sim d \leq (a \sim c) \sim (d \sim c)$;
(iv) $a \sim d \leq (b \to a) \sim (b \to d)$;
(v) $a \to d \leq (b \to a) \to (b \to d)$;
(vi) $b \to a \leq (a \to d) \to (b \to d)$.

Definition 2.6 ([11]). Let $\mathcal{E}$ be an EQ-algebra. We say that it is

(i) *good* if for all $a \in E$, $a \sim \mathbf{1} = a$,
(ii) a *lattice-ordered* EQ-algebra if it has a lattice reduct,
(iii) a *lattice* EQ-algebra (ℓEQ-algebra) if it is a *lattice-ordered* EQ-algebra in which the following substitution axiom holds, for all $a, b, c, d \in E$: $((a \vee b) \sim c) \otimes (d \sim a) \leq (d \vee b) \sim c$.

3 Fuzzy Filters and Fuzzy Prime Filters of EQ-algebras and Their Properties

3.1 Filters of EQ-algebras

Definition 3.1 ([11]). Let $\mathcal{E}$ be an EQ-algebra. A subset F of E is called a filter (or an EQ-filter) of $\mathcal{E}$, if for all $a, b, c \in E$ it holds that

(i) $\mathbf{1} \in F$,
(ii) If $a, b \in F$, then $a \otimes b \in F$,
(iii) If $a, a \to b \in F$, then $b \in F$,
(iv) If $a \to b \in F$, then $a \otimes c \to b \otimes c \in F$.

In Definition 3.1, if F only satisfies items (i) and (iii), then F is called a prefilter of $\mathcal{E}$.

Theorem 3.2. Let F_1, F_2 be two filters of an EQ-algebra $\mathcal{E}$. Then $F_1 \cap F_2$ is a filter of $\mathcal{E}$.

Proof. It is trivial. □

Let F be a filter of an EQ-algebra $\mathcal{E}$. In [11], an equivalence relation on E can be defined by

$$a \approx_F b \quad \Leftrightarrow \quad a \sim b \in F. \tag{4}$$

Theorem 3.3. Let F_1, F_2 be two filters of an EQ-algebra $\mathcal{E}$ and $F_1 \subseteq F_2$. Then $\approx_{F_1} \subseteq \approx_{F_2}$.

Proof. Since $F_1 \subseteq F_2$, for all $(a, b) \in \approx_{F_1}$, we have $a \sim b \in F_1$ and so $a \sim b \in F_2$, which yields that $(a, b) \in \approx_{F_2}$. Hence $\approx_{F_1} \subseteq \approx_{F_2}$. □

Theorem 3.4. Let F_1, F_2 be two filters of an EQ-algebra $\mathcal{E}$. Then $\approx_{F_1} \cap \approx_{F_2} = \approx_{F_1 \cap F_2}$.

Proof. For any $(a, b) \in E$, $(a, b) \in \approx_{F_1} \cap \approx_{F_2} \Leftrightarrow (a, b) \in \approx_{F_1}$ and $(a, b) \in \approx_{F_2} \Leftrightarrow a \sim b \in F_1$ and $a \sim b \in F_2 \Leftrightarrow a \sim b \in F_1 \cap F_2 \Leftrightarrow (a, b) \in \approx_{F_1 \cap F_2}$. □

3.2 Fuzzy Filters of EQ-algebras

Remark 3.5. At first, it is not difficult to prove that the item (ii) is a result of items (i), (iii) and (iv) in Definition 3.1. In fact, assume that items (i), (iii) and (iv) in Definition 3.1 hold and $a, b \in F$, it follows from Theorem 2.3 (ix) that $b \leq \mathbf{1} \to b$. By Theorem 2.2 (i), $b \to (\mathbf{1} \to b) = \mathbf{1}$. Furthermore $b \to (\mathbf{1} \to b) \in F$ from Definition 3.1 (i). According to Definition 3.1 (iii), $\mathbf{1} \to b \in F$ holds. By Definition 3.1 (iv), $\mathbf{1} \otimes a \to b \otimes a \in F$ holds. This implies that $a \to a \otimes b \in F$ holds. Applying Definition 3.1 (iii), $a \otimes b \in F$ holds, i.e. item (ii) in Definition 3.1 holds. Therefore the item (ii) in Definition 3.1 is not necessary and the conditions of Definition 3.1 can be reduced.

Definition 3.6. ([8]). Let $\mathcal{E}$ be an EQ-algebra. A fuzzy subset μ of E is said to be a fuzzy filter (or EQ-filter) of $\mathcal{E}$ if for all $a, b, c \in E$

(i) $\mu(\mathbf{1}) = 1$,
(ii) $\mu(a) \wedge \mu(a \to b) \leq \mu(b)$,
(iii) $\mu(a \to b) \leq \mu((a \otimes c) \to (b \otimes c))$.

One easy to see that the condition (i) in Definition 3.6 is excessively strict. We relevantly relax this restriction and propose a more general definition of a fuzzy filter of $\mathcal{E}$ as follows:

Definition 3.7. Let $\mathcal{E}$ be an EQ-algebra. A fuzzy subset μ of E is said to be a fuzzy filter of $\mathcal{E}$ if for all $a, b, c \in E$

(i) $\mu(a) \leq \mu(\mathbf{1})$,
(ii) $\mu(a) \wedge \mu(a \to b) \leq \mu(b)$,
(iii) $\mu(a \to b) \leq \mu((a \otimes c) \to (b \otimes c))$.

In Definition 3.7, if μ only satisfies items (i) and (ii), then μ is called a fuzzy prefilter of $\mathcal{E}$.

Example 3.8. A linearly ordered 6-element EQ-algebra $\mathcal{E}$ is the following [11]:

$$\begin{array}{c|cccccc} \otimes & \mathbf{0} & a & b & c & d & \mathbf{1} \\ \hline \mathbf{0} & \mathbf{0} & \mathbf{0} & \mathbf{0} & \mathbf{0} & \mathbf{0} & \mathbf{0} \\ a & \mathbf{0} & \mathbf{0} & \mathbf{0} & \mathbf{0} & \mathbf{0} & a \\ b & \mathbf{0} & \mathbf{0} & \mathbf{0} & \mathbf{0} & a & b \\ c & \mathbf{0} & \mathbf{0} & \mathbf{0} & a & a & c \\ d & \mathbf{0} & \mathbf{0} & a & a & a & d \\ \mathbf{1} & \mathbf{0} & a & b & c & d & \mathbf{1} \end{array} \qquad \begin{array}{c|cccccc} \sim & \mathbf{0} & a & b & c & d & \mathbf{1} \\ \hline \mathbf{0} & \mathbf{1} & c & b & a & \mathbf{0} & \mathbf{0} \\ a & c & \mathbf{1} & b & a & a & a \\ b & b & b & \mathbf{1} & b & b & b \\ c & a & a & b & \mathbf{1} & c & c \\ d & \mathbf{0} & a & b & c & \mathbf{1} & d \\ \mathbf{1} & \mathbf{0} & a & b & c & d & \mathbf{1} \end{array} \qquad \begin{array}{c|cccccc} \to & \mathbf{0} & a & b & c & d & \mathbf{1} \\ \hline \mathbf{0} & \mathbf{1} & \mathbf{1} & \mathbf{1} & \mathbf{1} & \mathbf{1} & \mathbf{1} \\ a & c & \mathbf{1} & \mathbf{1} & \mathbf{1} & \mathbf{1} & \mathbf{1} \\ b & b & b & \mathbf{1} & \mathbf{1} & \mathbf{1} & \mathbf{1} \\ c & a & a & b & \mathbf{1} & \mathbf{1} & \mathbf{1} \\ d & \mathbf{0} & a & b & c & \mathbf{1} & \mathbf{1} \\ \mathbf{1} & \mathbf{0} & a & b & c & d & \mathbf{1} \end{array}$$

By verification, we can know that a fuzzy subset

$$\mu = \frac{0.2}{\mathbf{0}} + \frac{0.2}{a} + \frac{0.2}{b} + \frac{0.2}{c} + \frac{0.6}{d} + \frac{0.8}{\mathbf{1}}$$

is a fuzzy filter of $\mathcal{E}$. Furthermore, all fuzzy subsets μ of E, which satisfy the conditions $\mu(\mathbf{0}) = \mu(a) = \mu(b) = \mu(c) \leq \mu(d) \leq \mu(\mathbf{1})$, are fuzzy filters of $\mathcal{E}$.

Definition 3.9. A fuzzy filter μ of $\mathcal{E}$ is said to be *normal* if there exists an $x \in E$ such that $\mu(x) = 1$.

Theorem 3.10. A fuzzy filter μ of $\mathcal{E}$ is *normal* if and only if $\mu(\mathbf{1}) = 1$.

Proof. It is trivial. □

For the sake of distinction, a fuzzy filter of $\mathcal{E}$ in Definition 3.6 is called a *normal* fuzzy filter of $\mathcal{E}$.

Theorem 3.11. Let μ be a fuzzy filter of $\mathcal{E}$. If there exists a sequence $\{a_n\}$ in E such that $\lim_{n\to\infty} \mu(a_n) = 1$, then $\mu(\mathbf{1}) = 1$.

Proof. Let μ is a fuzzy filter of $\mathcal{E}$. Then $\mu(a) \leq \mu(\mathbf{1})$ for all $a \in E$. It implies that $\mu(a_n) \leq \mu(\mathbf{1})$ for every positive integer n. Note that $1 = \lim_{n\to\infty} \mu(a_n) \leq \mu(\mathbf{1}) \leq 1$, hence $\mu(\mathbf{1}) = 1$. □

Applying Theorem 3.11, it is not difficult to obtain the following corollary.

Corollary 3.12. Let μ be a fuzzy filter of $\mathcal{E}$. If there exists a sequence $\{a_n\}$ in E such that $\lim_{n\to\infty} \mu(a_n) = 1$, then μ is *normal.*

Let μ be a fuzzy filter of $\mathcal{E}$. Then we can define a fuzzy subset μ^+ of E by

$$\mu^+(a) = \mu(a) + 1 - \mu(\mathbf{1}), \tag{5}$$

for all $a \in E$.

Theorem 3.13. Let μ be a fuzzy filter of $\mathcal{E}$. Then μ^+ is a *normal* fuzzy filter of $\mathcal{E}$ and $\mu \subseteq \mu^+$.

Proof. Obviously $\mu^+(\mathbf{1}) = \mu(\mathbf{1}) + 1 - \mu(\mathbf{1}) = 1$. Since μ is a fuzzy filter of $\mathcal{E}$, for all $a, b \in E$, we have $\mu^+(a) \wedge \mu^+(a \to b) = (\mu(a) + 1 - \mu(\mathbf{1})) \wedge (\mu(a \to b) + 1 - \mu(\mathbf{1})) = (\mu(a) \wedge \mu(a \to b)) + 1 - \mu(\mathbf{1}) \leq \mu(b) + 1 - \mu(\mathbf{1}) = \mu^+(b)$, i.e. $\mu^+(a) \wedge \mu^+(a \to b) \leq \mu^+(b)$. And also $\mu^+(a \to b) = \mu(a \to b) + 1 - \mu(\mathbf{1}) \leq \mu((a \otimes c) \to (b \otimes c)) + 1 - \mu(\mathbf{1}) = \mu^+((a \otimes c) \to (b \otimes c))$, i.e. $\mu^+(a \to b) \leq \mu^+((a \otimes c) \to (b \otimes c))$. By Definition 3.6, it follows that μ^+ is a *normal* fuzzy filter of $\mathcal{E}$. Clearly $\mu \subseteq \mu^+$. □

Theorem 3.14. A fuzzy filter μ of $\mathcal{E}$ is *normal* if and only if $\mu^+ = \mu$.

Proof. The sufficiency is obvious from *Eq.* (5). For the necessity, assume that μ a *normal* fuzzy filter of $\mathcal{E}$, then $\mu(\mathbf{1}) = 1$ by Theorem 3.10. Hence for all $a \in E$, one can obtain that $\mu^+(a) = \mu(a) + 1 - \mu(\mathbf{1}) = \mu(a)$. Thus $\mu^+ = \mu$. □

Theorem 3.15. Let μ be a fuzzy filter of $\mathcal{E}$. Then $(\mu^+)^+ = \mu^+$.

Proof. Since $\mu^+(\mathbf{1}) = 1$, for all $a \in E$, we have $(\mu^+)^+(a) = \mu^+(a) + 1 - \mu^+(\mathbf{1}) = \mu^+(a)$. Thus $(\mu^+)^+ = \mu^+$. □

Theorem 3.16. Let $\mathcal{E}$ be an EQ-algebra and μ be a fuzzy filter of $\mathcal{E}$. Then for all $a, b \in E$

(i) $a \leq b$ implies $\mu(a) \leq \mu(b)$;
(ii) $\mu(a) \wedge \mu(b) = \mu(a \wedge b)$;
(iii) $\mu(a) \wedge \mu(b) = \mu(a \otimes b)$;
(iv) $\mu(a \otimes b) \leq \mu(a \sim b) \leq \mu(a \to b)$;
(v) $\mu(a) \wedge \mu(a \otimes b) \leq \mu(a) \wedge \mu(a \sim b) \leq \mu(a) \wedge \mu(a \to b) \leq \mu(b)$.

Proof. (i) According to Theorem 2.2 (i), we can know that $a \leq b$ implies $a \to b = \mathbf{1}$. It holds from Definition 3.7 (i) that $\mu(a) \leq \mu(a \to b)$. One can obtain $\mu(a) = \mu(a) \wedge \mu(a \to b) \leq \mu(b)$ by Definition 3.7 (ii).

(ii) From Theorem 2.3 (i), we have $a \otimes b \leq a \wedge b$. Hence $\mu(a \otimes b) \leq \mu(a \wedge b)$ by item (i). It follows from item (ii) that $\mu(a) \wedge \mu(b) \leq \mu(a \wedge b)$. On the other hand, $a \wedge b \leq a$, $a \wedge b \leq b$, which imply that $\mu(a \wedge b) \leq \mu(a)$, $\mu(a \wedge b) \leq \mu(b)$. Therefore $\mu(a \wedge b) \leq \mu(a) \wedge \mu(b)$.

(iii) By Theorem 2.3 (ix), it holds that $b \leq \mathbf{1} \to b$. From item (i) and Definition 3.7 (iii), we have $\mu(b) \leq \mu(\mathbf{1} \to b) \leq \mu((\mathbf{1} \otimes a) \to (b \otimes a)) = \mu(a \to (b \otimes a))$. By Definition 3.7 (ii), $\mu(a) \wedge \mu(b) \leq \mu(a) \wedge \mu(a \to (b \otimes a)) \leq \mu(b \otimes a)$, i.e. $\mu(a) \wedge \mu(b) \leq \mu(a \otimes b)$. On the other hand, by Theorem 2.3 (i) and item (i), it follows immediately that $\mu(a \otimes b) \leq \mu(a), \mu(b)$, which implies that $\mu(a \otimes b) \leq \mu(a) \wedge \mu(b)$. Therefore $\mu(a) \wedge \mu(b) = \mu(a \otimes b)$.

(iv) Applying Definition 2.1 (E8), Theorem 2.3 (ii) and item (i), one can obtain that $\mu(a \otimes b) \leq \mu(a \sim b) \leq \mu(a \to b)$.

(v) It is trivial. □

Theorem 3.17. Let $\mathcal{E}$ be an EQ-algebra and μ be a fuzzy filter of $\mathcal{E}$. Then for all $a, b \in E$ $\mu(a \sim b) = \mu(a \leftrightarrow b) = \mu(a \overset{\circ}{\leftrightarrow} b)$.

Proof. By Theorem 3.16 (ii), (iii), (i) and Theorem 2.4 (ii), it holds that $\mu(a \leftrightarrow b) = \mu(a \to b \wedge b \to a) = \mu(a \to b) \wedge \mu(b \to a) = \mu(a \to b \otimes b \to a) = \mu(a \overset{\circ}{\leftrightarrow} b) \leq \mu(a \sim b) \leq \mu(a \leftrightarrow b)$. Hence $\mu(a \sim b) = \mu(a \leftrightarrow b) = \mu(a \overset{\circ}{\leftrightarrow} b)$. □

Theorem 3.18. Let μ be a fuzzy filter of an EQ-algebra $\mathcal{E}$. Then for all $a, b, c \in E$ the following properties hold:

(i) $\mu(a \sim b) \leq \mu((a \wedge c) \sim (b \wedge c))$;
(ii) $\mu(a \sim b) \leq \mu((a \otimes c) \sim (b \otimes c))$;
(iii) $\mu(a \sim b) \leq \mu((a \sim c) \sim (b \sim c))$;
(iv) $\mu(a \to b) \leq \mu((a \wedge c) \to (b \wedge c))$;
(v) $\mu(a \leftrightarrow b) \leq \mu((a \sim c) \to (b \sim c))$;
(vi) $\mu(a \to b) \leq \mu((b \to c) \to (a \to c))$ and $\mu(b \to a) \leq \mu((c \to b) \to (c \to a))$.

Proof. (i) By axioms (E3) and (E4), we get $a \sim b = \mathbf{1} \otimes (a \sim b) = ((a \wedge c) \sim (a \wedge c)) \otimes (a \sim b) \leq (a \wedge c) \sim (b \wedge c)$. So $\mu(a \sim b) \leq \mu((a \wedge c) \sim (b \wedge c))$ from Theorem 3.16 (i).

(ii) From Theorem 2.3 (ii), it follows that $a \sim b \leq a \to b$ and $a \sim b \leq b \to a$. By Theorem 3.16 (i), (iii), Definition 3.7 (iii) and Theorem 2.4 (ii), we have $\mu(a \sim b) \leq \mu(a \to b) \wedge \mu(b \to a) \leq \mu((a \otimes c) \to (b \otimes c)) \wedge \mu((b \otimes c) \to (a \otimes c)) = \mu((a \otimes c) \to (b \otimes c)) \otimes \mu((b \otimes c) \to (a \otimes c)) = \mu((a \otimes c) \overset{\circ}{\leftrightarrow} (b \otimes c)) \leq \mu((a \otimes c) \sim (b \otimes c))$, i.e. $\mu(a \sim b) \leq \mu((a \otimes c) \sim (b \otimes c))$.

(iii) By axioms (E3), (E5) and (E8), we have $a \sim b = (a \sim b) \otimes \mathbf{1} = (a \sim b) \otimes (c \sim c) \leq (a \sim c) \otimes (b \sim c) \leq (a \sim c) \sim (b \sim c)$. Hence $\mu(a \sim b) \leq \mu((a \sim c) \sim (b \sim c))$ from Theorem 3.16 (i).

(iv) By axiom (E7), we have $(a \wedge b) \sim a \leq (a \wedge b \wedge c) \sim (a \wedge c) = ((a \wedge c) \wedge (b \wedge c)) \sim (a \wedge c)$, which implies that $a \to b \leq (a \wedge c) \to (b \to c)$ by *Eq.* (1).

(v) According to Theorem 3.17, item (iii), Theorem 2.3 (ii) and Theorem 3.16 (i), it is not difficult to obtain that $\mu(a \leftrightarrow b) = \mu(a \sim b) \leq \mu((a \sim c) \sim (b \sim c)) \leq \mu((a \sim c) \to (b \sim c))$.

(vi) By Theorem 2.5 (v) and (vi), it follows immediately that two inequalities hold. □

Theorem 3.19. Let μ be a fuzzy subset of E. Then μ is a fuzzy filter of $\mathcal{E}$ if and only if for all $\lambda \in [0,1]$ such that the nonempty level set μ_λ is a filter of $\mathcal{E}$.

Proof. Let μ be a fuzzy filter of $\mathcal{E}$ and μ_λ be the nonempty level set. For all $\lambda \in [0,1]$ and $a \in \mu_\lambda$, it follows that $\lambda \leq \mu(a) \leq \mu(\mathbf{1})$, i.e. $\mathbf{1} \in \mu_\lambda$. If $a, a \to b \in \mu_\lambda$, then $\lambda \leq \mu(a)$ and $\lambda \leq \mu(a \to b)$. Thus $\lambda \leq \mu(a) \wedge \mu(a \to b) \leq \mu(b)$, i.e. $b \in \mu_\lambda$. If $a \to b \in \mu_\lambda$, then $\lambda \leq \mu(a \to b) \leq \mu(a \otimes c \to b \otimes c)$, that is $a \otimes c \to b \otimes c \in \mu_\lambda$. By Definition 3.1, for all $\lambda \in [0,1]$, μ_λ is a filter of $\mathcal{E}$.

Conversely, assume that for all $\lambda \in [0,1]$, the nonempty level set μ_λ is a filter of $\mathcal{E}$, then $\mathbf{1} \in \mu_1$, i.e. $\mu(\mathbf{1}) \geq 1$, which yields that $\mu(\mathbf{1}) = 1$. Hence $\mu(a) \leq \mu(\mathbf{1})$, for all $a \in E$. For all $a, b \in E$, there exists $\lambda \in [0,1]$ such that $a, a \to b \in \mu_\lambda$. Since $a, a \to b \in \mu_\lambda$ implies $b \in \mu_\lambda$, in other words, $\mu(a) \geq \lambda$, $\mu(a \to b) \geq \lambda$ implies $\mu(b) \geq \lambda$, it is easy to get that $\mu(a) \wedge \mu(a \to b) \leq \mu(b)$. For all $a, b, c \in E$, there exists $\lambda \in [0,1]$ such that $a \to b \in \mu_\lambda$. Since $a \to b \in \mu_\lambda$ implies $a \otimes c \to b \otimes c \in \mu_\lambda$, that is $\mu(a \to b) \geq \lambda$ implies $\mu(a \otimes c \to b \otimes c) \geq \lambda$, it means that $\mu(a \to b) \leq \mu(a \otimes c \to b \otimes c)$. By Definition 3.7, μ is a fuzzy filter of $\mathcal{E}$. □

Theorem 3.20. Let μ be a fuzzy filter of $\mathcal{E}$ and μ_{λ_1} and μ_{λ_2} (where $\lambda_1 < \lambda_2$) be two level filters of μ. Then $\mu_{\lambda_1} = \mu_{\lambda_2}$ if and only if there no exists $a \in E$ such that $\lambda_1 \leq \mu(a) < \lambda_2$.

Proof. Let $\mu_{\lambda_1} = \mu_{\lambda_2}$ $(\lambda_1 < \lambda_2)$. In contrary, suppose that there exists $a \in E$ such that $\lambda_1 \leq \mu(a) < \lambda_2$. Then μ_{λ_2} is a proper subset of μ_{λ_1}, which is a contradiction.

Conversely, assume that there no exists $a \in E$ such that $\lambda_1 \leq \mu(a) < \lambda_2$. If $\forall a \in \mu_{\lambda_1}$, i.e. $\mu(a) \geq \lambda_1$, then $\mu(a) \geq \lambda_2$ by hypotheses. Hence $a \in \mu_{\lambda_2}$ and $\mu_{\lambda_1} \subseteq \mu_{\lambda_2}$. On the other hand $\lambda_1 < \lambda_2$ implies $\mu_{\lambda_1} \supseteq \mu_{\lambda_2}$. Therefore $\mu_{\lambda_1} = \mu_{\lambda_2}$. □

Let μ be a fuzzy subset of E. Denote E_μ by the set

$$E_\mu := \{a \in E | \mu(a) = \mu(\mathbf{1})\}. \tag{6}$$

Theorem 3.21. If μ is a fuzzy filter of $\mathcal{E}$, then E_μ is a filter of $\mathcal{E}$.

Proof. It is easy to know that E_μ is a special level set of μ. By Theorem 3.19, E_μ is a filter of $\mathcal{E}$. □

Let F be a non-empty subset of E and $\alpha, \beta \in [0,1]$ such that $\alpha > \beta$. Define a fuzzy subset $\mu_F^{(\alpha,\beta)}$ in E by

$$\mu_F^{(\alpha,\beta)}(x) := \begin{cases} \alpha, & \text{if } x \in F, \\ \beta, & \text{otherwise.} \end{cases} \tag{7}$$

Obviously, $\beta \leq \mu_F^{(\alpha,\beta)}(x) \leq \alpha$, $\forall x \in E$. It is easy to see that $\mu_F^{(1,0)}$ is the characteristic function of F.

Theorem 3.22. Let F be a non-empty subset of E. Then the fuzzy subset $\mu_F^{(\alpha,\beta)}$ is a fuzzy filter of $\mathcal{E}$ if and only if F is a filter of $\mathcal{E}$. Moreover, in this case $E_{\mu_F^{(\alpha,\beta)}} = F$.

Proof. Let $\mu_F^{(\alpha,\beta)}$ be a fuzzy filter of $\mathcal{E}$. By Definition 3.7 (i), for all $a \in E$, $\mu_F^{(\alpha,\beta)}(a) \le \mu_F^{(\alpha,\beta)}(\mathbf{1})$. If $a \in F$, then $\alpha = \mu_F^{(\alpha,\beta)}(a) \le \mu_F^{(\alpha,\beta)}(\mathbf{1})$. Obviously, $\mu_F^{(\alpha,\beta)}(\mathbf{1}) \le \alpha$. So $\mu_F^{(\alpha,\beta)}(\mathbf{1}) = \alpha$, which implies that $\mathbf{1} \in F$, i.e. Definition 3.1 (i) holds. If $a, a \to b \in F$, then $\mu_F^{(\alpha,\beta)}(a) = \mu_F^{(\alpha,\beta)}(a \to b) = \alpha$. By Definition 3.7 (ii), $\alpha = \mu_F^{(\alpha,\beta)}(a) \wedge \mu_F^{(\alpha,\beta)}(a \to b) \le \mu_F^{(\alpha,\beta)}(b)$. Obviously, $\mu_F^{(\alpha,\beta)}(b) \le \alpha$, so $\mu_F^{(\alpha,\beta)}(b) = \alpha$, which yields that $b \in F$, i.e. Definition 3.1 (iii) holds. If $a \to b \in F$, then $\mu_F^{(\alpha,\beta)}(a \to b) = \alpha$. By Definition 3.7 (iii), $\alpha = \mu_F^{(\alpha,\beta)}(a \to b) \le \mu_F^{(\alpha,\beta)}(a \otimes c \to b \otimes c)$. On the other hand, it is obvious that $\mu_F^{(\alpha,\beta)}(a \otimes c \to b \otimes c) \le \alpha$. So $\mu_F^{(\alpha,\beta)}(a \otimes c \to b \otimes c) = \alpha$, i.e. $a \otimes c \to b \otimes c \in F$. It implies that Definition 3.1 (iv) holds. Therefore F is a filter of $\mathcal{E}$.

Conversely, assume that F is a filter of $\mathcal{E}$. Since $\mathbf{1} \in F$, it follows that for all $a \in E$, $\mu_F^{(\alpha,\beta)}(a) \le \alpha = \mu_F^{(\alpha,\beta)}(\mathbf{1})$, i.e. Definition 3.7 (i) holds.

For all $a, b \in E$, the following discussions are divided into two cases: (a) If $a, a \to b \in F$, then $b \in F$ by Definition 3.1 (iii). In other words, if $\mu_F^{(\alpha,\beta)}(a) = \mu_F^{(\alpha,\beta)}(a \to b) = \alpha$, then $\mu_F^{(\alpha,\beta)}(b) = \alpha$. Hence $\mu_F^{(\alpha,\beta)}(a) \wedge \mu_F^{(\alpha,\beta)}(a \to b) = \mu_F^{(\alpha,\beta)}(b)$. (b) If $a \notin F$ or $a \to b \notin F$, which means that $\mu_F^{(\alpha,\beta)}(a) = \beta$ or $\mu_F^{(\alpha,\beta)}(a \to b) = \beta$, then $\mu_F^{(\alpha,\beta)}(a) \wedge \mu_F^{(\alpha,\beta)}(a \to b) = \beta$. Since for any $b \in E$, $\beta \le \mu_F^{(\alpha,\beta)}(b) \le \alpha$, it is easy to see from the above discussions that $\mu_F^{(\alpha,\beta)}(a) \wedge \mu_F^{(\alpha,\beta)}(a \to b) \le \mu_F^{(\alpha,\beta)}(b)$ in two cases. Hence Definition 3.7 (ii) holds.

If $a \to b \in F$, then $a \otimes c \to b \otimes c \in F$ by Definition 3.1 (iv). It is that if $\mu_F^{(\alpha,\beta)}(a \to b) = \alpha$ then $\mu_F^{(\alpha,\beta)}(a \otimes c \to b \otimes c) = \alpha$. If $a \to b \notin F$, then $\mu_F^{(\alpha,\beta)}(a \to b) = \beta$. Since $\beta \le \mu_F^{(\alpha,\beta)}(a \otimes c \to b \otimes c) \le \alpha$, it follows that $\mu_F^{(\alpha,\beta)}(a \to b) \le \mu_F^{(\alpha,\beta)}(a \otimes c \to b \otimes c)$. In the above two cases, it holds that $\mu_F^{(\alpha,\beta)}(a \to b) \le \mu_F^{(\alpha,\beta)}(a \otimes c \to b \otimes c)$, that is, Definition 3.7 (iii) holds. Therefore $\mu_F^{(\alpha,\beta)}$ is a fuzzy filter of $\mathcal{E}$.

Now, we check that $E_{\mu_F^{(\alpha,\beta)}} = F$. From *Eq.* (7), it is easy to see that $F = \{a \in E | \mu_F^{(\alpha,\beta)}(a) = \alpha\}$. Since F is a filter of $\mathcal{E}$, we have $\mu_F^{(\alpha,\beta)}(\mathbf{1}) = \alpha$. By *Eq.* (6), we have $E_{\mu_F^{(\alpha,\beta)}} = \{a \in E | \mu_F^{(\alpha,\beta)}(a) = \mu_F^{(\alpha,\beta)}(\mathbf{1})\} = \{a \in E | \mu_F^{(\alpha,\beta)}(a) = \alpha\} = F$. □

Theorem 3.23. Let μ_i $(i \in I)$ be fuzzy filters of $\mathcal{E}$. Then $\bigcap_{i\in I} \mu_i$ is also a fuzzy filter of $\mathcal{E}$.

Proof. Let μ_i $(i \in I)$ be fuzzy filters of $\mathcal{E}$. Then for all $a \in E$, $\mu_i(a) \le \mu(\mathbf{1})$ $(i \in I)$ by Definition 3.7 (i). So $(\bigcap_{i\in I} \mu_i)(a) = \bigwedge_{i\in I} \mu_i(a) \le \mu(\mathbf{1})$. For all $a, b \in E$, $\mu_i(a) \wedge \mu_i(a \to b) \le \mu_i(b)$ $(i \in I)$ by Definition 3.7 (ii). $(\bigcap_{i\in I} \mu_i)(a) \wedge (\bigcap_{i\in I} \mu_i)(a \to b) = (\bigwedge_{i\in I} \mu_i(a)) \wedge (\bigwedge_{i\in I} \mu_i(a \to b)) = \bigwedge_{i\in I} (\mu_i(a) \wedge (a \to b)) \le \bigwedge_{i\in I} \mu_i(b) = (\bigcap_{i\in I} \mu_i)(b)$.

For all $a, b, c \in E$, $\mu_i(a \to b) \le \mu_i(a \otimes c \to b \otimes c)$ $(i \in I)$ by Definition 3.7 (iii). $(\bigcap_{i\in I} \mu_i)(a \to b) = \bigwedge_{i\in I} \mu_i(a \to b) \le \bigwedge_{i\in I} \mu_i(a \otimes c \to b \otimes c) = (\bigcap_{i\in I} \mu_i)(a \otimes c \to b \otimes c)$. So $\bigcap_{i\in I} \mu_i$ is a fuzzy filter of $\mathcal{E}$. □

Let μ be a fuzzy subset of E and $\alpha > 0$ be a real number. Then we can define a new fuzzy subset μ^α of E by

$$\mu^\alpha(x) = (\mu(x))^\alpha, \ x \in E. \tag{8}$$

Theorem 3.24. If μ is a fuzzy filter of $\mathcal{E}$, then μ^α is also a fuzzy filter of $\mathcal{E}$ and $E_{\mu^\alpha} = E_\mu$.

Proof. Since μ is a fuzzy filter of $\mathcal{E}$, for all $a, b, c \in E$, $\mu^\alpha(a) = (\mu(a))^\alpha \le (\mu(\mathbf{1}))^\alpha = \mu^\alpha(\mathbf{1})$, $\mu^\alpha(a) \wedge \mu^\alpha(a \to b) = (\mu(a))^\alpha \wedge (\mu(a \to b))^\alpha = (\mu(a) \wedge \mu(a \to b))^\alpha \le (\mu(b))^\alpha = \mu^\alpha(b)$, $\mu^\alpha(a \to b) \le (\mu(a \to b))^\alpha \le (\mu(a \otimes c \to b \otimes c))^\alpha = \mu^\alpha(a \otimes c \to b \otimes c)$. By Definition 3.7, μ^α is a fuzzy filter of $\mathcal{E}$.

By *Eq.* (6), we obtain $E_{\mu^\alpha} = \{a \in E | \mu^\alpha(a) = \mu^\alpha(\mathbf{1})\} = \{a \in E | (\mu(a))^\alpha = (\mu(\mathbf{1}))^\alpha\} = \{a \in E | \mu(a) = \mu(\mathbf{1})\} = E_\mu$. □

Let μ be a fuzzy subset of E and f be an endomorphism on E. Then we can define a new fuzzy subset μ_f of E by

$$\mu_f(x) = \mu(f(x)), \ x \in E. \tag{9}$$

Theorem 3.25. If μ is a fuzzy filter of $\mathcal{E}$, then μ_f is also a fuzzy filter of $\mathcal{E}$.

Proof. Let μ be a fuzzy filter of $\mathcal{E}$. Then for all $a \in E$, $\mu_f(a) = \mu(f(a)) \le \mu(\mathbf{1}) = \mu(f(\mathbf{1})) = \mu_f(\mathbf{1})$. For all $a, b \in E$, $\mu_f(a) \wedge \mu_f(a \to b) = \mu(f(a)) \wedge \mu(f(a \to b)) = \mu(f(a)) \wedge \mu(f(a) \to f(b)) \le \mu(f(b)) = \mu_f(b)$. For all $a, b, c \in E$, $\mu_f(a \to b) = \mu(f(a \to b)) = \mu(f(a) \to f(b)) \le \mu(f(a) \otimes f(c) \to f(b) \otimes f(c)) = \mu(f(a \otimes c) \to f(b \otimes c)) = \mu(f(a \otimes c \to b \otimes c)) = \mu_f(a \otimes c \to b \otimes c)$. By Definition 3.7, it holds that μ_f is a fuzzy filter of $\mathcal{E}$. □

Theorem 3.26. Let μ be a fuzzy filter of $\mathcal{E}$ and $\lambda \in [0, \mu(\mathbf{1}))$. Then $\mu \vee \lambda$ is also a fuzzy filter of $\mathcal{E}$, where $(\mu \vee \lambda)(a) = \mu(a) \vee \lambda$, $\forall a \in E$.

Proof. For any $a \in E$, according to hypothesis, we have $(\mu \vee \lambda)(a) = \mu(a) \vee \lambda \le \mu(\mathbf{1}) = (\mu \vee \lambda)(\mathbf{1})$. For all $a, b \in E$, $(\mu \vee \lambda)(a) \wedge (\mu \vee \lambda)(a \to b) = (\mu(a) \vee \lambda) \wedge (\mu(a \to b) \vee \lambda) = (\mu(a) \wedge \mu(a \to b)) \vee \lambda \le \mu(b) \vee \lambda = (\mu \vee \lambda)(b)$. For all $a, b, c \in E$, $(\mu \vee \lambda)(a \to b) = \mu(a \to b) \vee \lambda \le \mu(a \otimes c \to b \otimes c) \vee \lambda = (\mu \vee \lambda)(a \otimes c \to b \otimes c)$. By Definition 3.7, $\mu \vee \lambda$ is a fuzzy filter of $\mathcal{E}$. □

Theorem 3.27. Let F be a filter of $\mathcal{E}$ and $\lambda \in [0, 1]$. Then λF is a fuzzy filter of $\mathcal{E}$, where λF is the scaler product of λ and F, i.e.

$$(\lambda F)(a) = \lambda \wedge \chi_F(a) = \begin{cases} \lambda, \, a \in F, \\ 0, \, a \notin F. \end{cases}$$

Proof. Since $\mathbf{1} \in F$, we have $(\lambda F)(\mathbf{1}) = \lambda$. It yields that for all $a \in E$, $(\lambda F)(a) \leq (\lambda F)(\mathbf{1})$. For all $a, b \in E$, the following discussions are divided into two cases: (a) If $a, a \to b \in F$, then $b \in F$ by Definition 3.1 (iii). This means that $(\lambda F)(a) = (\lambda F)(a \to b) = \lambda$ implies $(\lambda F)(b) = \lambda$. Hence $(\lambda F)(a) \wedge (\lambda F)(a \to b) = \lambda = (\lambda F)(b)$. (b) If $a \notin F$ or $a \to b \notin F$, then $(\lambda F)(a) = 0$ or $(\lambda F)(a \to b) = 0$. It is easy to obtain that $(\lambda F)(a) \wedge (\lambda F)(a \to b) = 0 \leq (\lambda F)(b)$. In the above two cases, we can see that $(\lambda F)(a) \wedge (\lambda F)(a \to b) \leq (\lambda F)(b)$. Similarly, we can verify that for all $a, b, c \in E$, $(\lambda F)(a \to b) \leq (\lambda F)(a \otimes c \to b \otimes c)$. Therefore λF is a fuzzy filter of $\mathcal{E}$. □

Theorem 3.28. (*Representation theorem*) Let $\mathscr{N}(E) = \{F(\lambda) | \lambda \in [0,1]\}$ be a sets nest of filters of $\mathcal{E}$, i.e. $\mathscr{N}(E)$ satisfies that for all $\lambda \in [0,1]$ $F(\lambda)$ is a filter of $\mathcal{E}$ and $\forall\ \lambda_1 < \lambda_2 \ \Rightarrow\ F(\lambda_2) \subseteq F(\lambda_1)$. Then $\mu = \bigcup_{\lambda \in [0,1]} \lambda F(\lambda)$ is a fuzzy filter of $\mathcal{E}$ such that for all $\eta \in [0,1]$ $\mu_\eta = F(\eta)$.

Proof. Firstly, we prove that μ is a fuzzy filter of $\mathcal{E}$. Since for all $\lambda \in [0,1]$ $F(\lambda)$ is a filter of $\mathcal{E}$, it follows from Theorem 3.27 that $\lambda F(\lambda)$ is a fuzzy filter of $\mathcal{E}$. For any $a \in E$, $\mu(a) = \Big(\bigcup_{\lambda \in [0,1]} \lambda F(\lambda)\Big)(a) = \bigvee_{\lambda \in [0,1]} (\lambda F(\lambda))(a) \leq \bigvee_{\lambda \in [0,1]} (\lambda F(\lambda))(\mathbf{1}) = \Big(\bigcup_{\lambda \in [0,1]} \lambda F(\lambda)\Big)(\mathbf{1}) = \mu(\mathbf{1})$. For any $a, b \in E$, $\mu(a) \wedge \mu(a \to b) = \Big(\bigcup_{\lambda \in [0,1]} \lambda F(\lambda)\Big)(a) \wedge \Big(\bigcup_{\lambda \in [0,1]} \lambda F(\lambda)\Big)(a \to b) = \Big(\bigvee_{\lambda \in [0,1]} (\lambda F(\lambda))(a)\Big) \wedge \Big(\bigvee_{\lambda \in [0,1]} (\lambda F(\lambda))(a \to b)\Big) \leq \bigvee_{\lambda \in [0,1]} \Big((\lambda F(\lambda))(a) \wedge (\lambda F(\lambda))(a \to b)\Big) \leq \bigvee_{\lambda \in [0,1]} (\lambda F(\lambda))(b) = \Big(\bigcup_{\lambda \in [0,1]} \lambda F(\lambda)\Big)(b) = \mu(b)$. For any $a, b, c \in E$, $\mu(a \to b) = \Big(\bigcup_{\lambda \in [0,1]} \lambda F(\lambda)\Big)(a \to b) = \bigvee_{\lambda \in [0,1]} (\lambda F(\lambda))(a \to b) \leq \bigvee_{\lambda \in [0,1]} (\lambda F(\lambda))(a \otimes c \to b \otimes c) = \Big(\bigcup_{\lambda \in [0,1]} \lambda F(\lambda)\Big)(a \otimes c \to b \otimes c) = \mu(a \otimes c \to b \otimes c)$. Therefore μ is a fuzzy filter of $\mathcal{E}$.

Now we verify that all $\eta \in [0,1]$ $\mu_\eta = F(\eta)$. Assume that for any $\eta \in [0,1]$, $a \in \mu_\eta$, then $\mu(a) \geq \eta$, i.e. $\bigvee_{\lambda \in [0,1]} (\lambda F(\lambda))(a) \geq \eta$. Furthermore $\bigvee_{\lambda \in [0,1]} (\lambda \wedge \chi_{F(\lambda)}(a)) \geq \eta$, which implies that there exists $\lambda_0 \in [0,1]$ such that $\lambda_0 \wedge \chi_{F(\lambda_0)}(a) \geq \eta$, that is $\lambda_0 \geq \eta$ and $\chi_{F(\lambda_0)}(a) = 1$. This means that $a \in F(\lambda_0) \subseteq F(\eta)$. Therefore $\mu_\eta \subseteq F(\eta)$. On the other hand, suppose that for any $\eta \in [0,1]$, $a \in F(\eta)$, then $(\eta F(\eta))(a) = \eta$. Hence $\mu(a) = \Big(\bigcup_{\lambda \in [0,1]} \lambda F(\lambda)\Big)(a) = \bigvee_{\lambda \in [0,1]} (\lambda F(\lambda))(a) \geq \eta$, i.e. $a \in \mu_\eta$. Therefore $F(\eta) \subseteq \mu_\eta$. □

3.3 Fuzzy Prime Filters of EQ-algebras

A filter F of an EQ-algebra $\mathcal{E}$ is said to be proper if $F \neq E$. A proper filter F is prime if for all $a, b \in E$, $a \vee b \in F$ implies $a \in F$ or $b \in F$.

Definition 3.29. Let μ be a non-constant fuzzy filter of an EQ-algebra $\mathcal{E}$. Then μ is called a fuzzy prime filter of $\mathcal{E}$ if for all $a, b \in E$, $\mu(a \vee b) = \mu(a) \vee \mu(b)$.

Theorem 3.30. Let μ be a non-constant fuzzy filter of $\mathcal{E}$. Then μ is a fuzzy prime filter of $\mathcal{E}$ if and only if for all $\lambda \in [0,1]$ such that the nonempty level set μ_λ is a prime filter of $\mathcal{E}$.

Proof. Let μ be a non-constant fuzzy prime filter of $\mathcal{E}$. By Theorem 3.19, for all $\lambda \in [0,1]$, the nonempty level set μ_λ is a filter of $\mathcal{E}$. Now we prove that μ_λ is prime. For all $a, b \in E$, let $\mu(a \vee b) = \lambda$. Then $a \vee b \in \mu_\lambda$ and $\mu(a) \vee \mu(b) = \lambda$, which implies $\mu(a) \geq \lambda$ or $\mu(b) \geq \lambda$, that is $a \in \mu_\lambda$ or $b \in \mu_\lambda$. Therefore μ_λ is a prime filter of $\mathcal{E}$.

Conversely, suppose that for all $\lambda \in [0,1]$, the nonempty level set μ_λ is a prime filter of $\mathcal{E}$. By Theorem 3.19, μ is a fuzzy prime filter of $\mathcal{E}$. Now we verify that μ is prime. For all $a, b \in E$, let $\mu(a \vee b) = \lambda$. Then $a \vee b \in \mu_\lambda$, which implies $a \in \mu_\lambda$ or $b \in \mu_\lambda$, i.e. $\mu(a) \geq \lambda$ or $\mu(b) \geq \lambda$. Hence $\mu(a) \vee \mu(b) \geq \lambda = \mu(a \vee b)$. Obviously, $\mu(a \vee b) \geq \mu(a) \vee \mu(b)$. Therefore $\mu(a \vee b) = \mu(a) \vee \mu(b)$. This means that μ is a fuzzy prime filter of $\mathcal{E}$. □

Theorem 3.31. Let F be a non-empty subset of E. Then the fuzzy subset $\mu_F^{(\alpha,\beta)}$ is a fuzzy prime filter of $\mathcal{E}$ if and only if F is a prime filter of $\mathcal{E}$.

Proof. Let $\mu_F^{(\alpha,\beta)}$ be a fuzzy prime filter of $\mathcal{E}$. By Theorem 3.22, F is a filter of $\mathcal{E}$. Now we check that F is prime. For all $a, b \in E$, if $a \vee b \in F$, then $\mu_F^{(\alpha,\beta)}(a \vee b) = \alpha$ and so $\mu_F^{(\alpha,\beta)}(a) \vee \mu_F^{(\alpha,\beta)}(b) = \alpha$, which yields that $\mu_F^{(\alpha,\beta)}(a) = \alpha$ or $\mu_F^{(\alpha,\beta)}(b) = \alpha$, i.e. $a \in F$ or $b \in F$. Therefore F is a prime filter of $\mathcal{E}$.

Conversely, suppose that F is a prime filter of $\mathcal{E}$. By Theorem 3.22, $\mu_F^{(\alpha,\beta)}$ is a fuzzy filter of $\mathcal{E}$. Now we prove that $\mu_F^{(\alpha,\beta)}$ is prime. For all $a, b \in E$, if $a \vee b \in F$, then $a \in F$ or $b \in F$, in other words, if $\mu_F^{(\alpha,\beta)}(a \vee b) = \alpha$ then $\mu_F^{(\alpha,\beta)}(a) = \alpha$ or $\mu_F^{(\alpha,\beta)}(b) = \alpha$. This implies that $\mu_F^{(\alpha,\beta)}(a \vee b) = \alpha = \mu_F^{(\alpha,\beta)}(a) \vee \mu_F^{(\alpha,\beta)}(b)$. If $a \vee b \notin F$, then $\mu_F^{(\alpha,\beta)}(a \vee b) = \beta$. Since for all $a \in E$, $\beta \leq \mu_F^{(\alpha,\beta)}(a) \leq \alpha$, it follows immediately that $\mu_F^{(\alpha,\beta)}(a \vee b) \leq \mu_F^{(\alpha,\beta)}(a) \vee \mu_F^{(\alpha,\beta)}(b)$. Clearly, $\mu_F^{(\alpha,\beta)}(a \vee b) \geq \mu_F^{(\alpha,\beta)}(a) \vee \mu_F^{(\alpha,\beta)}(b)$. In the above two cases, it holds that $\mu_F^{(\alpha,\beta)}(a \vee b) = \mu_F^{(\alpha,\beta)}(a) \vee \mu_F^{(\alpha,\beta)}(b)$. Therefore $\mu_F^{(\alpha,\beta)}$ is a fuzzy prime filter of $\mathcal{E}$. □

Theorem 3.32. Let μ_i $(i \in I)$ be fuzzy prime filters of $\mathcal{E}$. Then $\bigcap_{i \in I} \mu_i$ is also a fuzzy prime filter of $\mathcal{E}$.

Proof. It suffices to prove from Theorem 3.23 that $\bigcap_{i \in I} \mu_i$ is prime. For all $a, b \in E$, $(\bigcap_{i \in I} \mu_i)(a \vee b) = \bigwedge_{i \in I} \mu_i(a \vee b) = \bigwedge_{i \in I} \mu_i(a) \vee \bigwedge_{i \in I} \mu_i(b) = (\bigcap_{i \in I} \mu_i)(a) \vee (\bigcap_{i \in I} \mu_i)(b)$. This implies that $\bigcap_{i \in I} \mu_i$ is a fuzzy prime filter of $\mathcal{E}$. □

Theorem 3.33. Let μ be a fuzzy prime filter of $\mathcal{E}$. Then μ^+ is a *normal* fuzzy prime filter of $\mathcal{E}$.

Proof. By Theorem 3.13, we only need to prove that μ^+ is prime. Since μ is prime, for all $a, b \in E$, it holds that $\mu^+(a \vee b) = \mu(a \vee b) + 1 - \mu(\mathbf{1}) = (\mu(a) \vee \mu(b)) + 1 - \mu(\mathbf{1}) = (\mu(a) + 1 - \mu(\mathbf{1})) \vee (\mu(b) + 1 - \mu(\mathbf{1})) = \mu^+(a) \vee \mu^+(b)$. This is implies that μ^+ is a *normal* fuzzy prime filter of $\mathcal{E}$. $\square$

Theorem 3.34. If μ is a fuzzy prime filter of $\mathcal{E}$, then μ^α is also a fuzzy prime filter of $\mathcal{E}$.

Proof. According to Theorem 3.24, it suffices to verify that μ^α is prime. Since μ is prime, for all $a, b \in E$, it follows that $\mu^\alpha(a \vee b) = (\mu(a \vee b))^\alpha = (\mu(a) \vee \mu(b))^\alpha = (\mu(a))^\alpha \vee (\mu(b))^\alpha = \mu^\alpha(a) \vee \mu^\alpha(b)$. Therefore μ^α is a fuzzy prime filter of $\mathcal{E}$. $\square$

Theorem 3.35. If μ is a fuzzy prime filter of $\mathcal{E}$, then μ_f is also a fuzzy prime filter of $\mathcal{E}$.

Proof. By Theorem 3.25, we only need to check that μ_f is prime. Since μ is prime, for all $a, b \in E$, one can obtain that $\mu_f(a \vee b) = \mu(f(a \vee b)) = \mu(f(a) \vee f(b)) = \mu(f(a)) \vee \mu(f(b)) = \mu_f(a) \vee \mu_f(b)$. This means that μ_f is a fuzzy prime filter of $\mathcal{E}$. $\square$

Theorem 3.36. Let μ be a fuzzy prime filter of $\mathcal{E}$ and $\lambda \in [0, \mu(\mathbf{1}))$. Then $\mu \vee \lambda$ is also a fuzzy prime filter of $\mathcal{E}$, where $(\mu \vee \lambda)(a) = \mu(a) \vee \lambda$, $\forall a \in E$.

Proof. According to Theorem 3.26, we only need to prove that $\mu \vee \lambda$ is prime. For all $a, b \in E$, $(\mu \vee \lambda)(a \vee b) = \mu(a \vee b) \vee \lambda = \mu(a) \vee \mu(b) \vee \lambda = (\mu(a) \vee \lambda) \vee (\mu(b) \vee \lambda) = (\mu \vee \lambda)(a) \vee (\mu \vee \lambda)(b)$. Therefore $\mu \vee \lambda$ is a fuzzy prime filter of $\mathcal{E}$. $\square$

4 Conclusion

In this paper, our main work is to redefine the concept of fuzzy filter of an EQ-algebra and to study in detail the properties of fuzzy filters and fuzzy congruences of EQ-algebras. Representation theorem on fuzzy filter is derived. We introduce the notion fuzzy prime filter of an EQ-algebra and discuss its basic characterizations. The further work is to define and characterize fuzzy implicative and fuzzy positive implicative prefilters of EQ-algebras.

Acknowledgements. Thanks to the support by National Nature Science Foundation of China (Grant Nos. 11401128, 11661028, 11661030), Nature Science Foundation of Guangxi, China (Grant Nos. 2016GXNSFAA380059, 2016GXNSFBA380077), Colleges Science Research Project of Guangxi, China (Grant Nos. KY2015LX118, KY2016YB196) and Scientific Research Start-up Foundation of Guilin University of Technology, China (Grant No. 002401003452).

References

1. Dyba, M., Novák, V.: On EQ-fuzzy logics with delta connective. In: EUSFLAT-LFA, pp. 156–162 (2011)
2. Dyba, M., Novák, V.: First-order EQ-logic. In: EUSFLAT-LFA, pp. 200–206 (2013)
3. Dyba, M., et al.: Non-commutative first-order EQ-logics. Fuzzy Sets Syst. **292**, 215–241 (2016)
4. Dyba, M., Novák, V.: EQ-logics with delta connective. Iran. J. Fuzzy Syst. **12**(2), 41–61 (2015)
5. El-Zekey, M.: Representable good EQ-algebras. Soft Comput. **14**, 1011–1023 (2010)
6. El-Zekey, M., et al.: On good EQ-algebras. Fuzzy Sets Syst. **178**, 1–23 (2011)
7. Liu, L., Zhang, X.: Implicative and positive implicative prefilters of EQ-algebras. J. Intell. Fuzzy Syst. **26**, 2087–2097 (2014)
8. Ma, Z.M., Hu, B.Q.: Fuzzy EQ-filters of EQ-algebras. In: World Scientific Proceedings Series on Computer Engineering and Information Science: Volume 5 Quantitative Logic and Soft Computing, Proceedings of the QL & SC 2012 Xi'an, China, 12–15 May 2012, pp. 528–535 (2012)
9. Mohtashamnia, N., Torkzadeh, L.: The lattice of prefilters of an EQ-algebra. Fuzzy Sets Syst. **311**, 86–98 (2017)
10. Novák, V.: EQ-algebras: primary concepts and properties. In: Proceedings of Czech-Japan Seminar, Ninth Meeting, Kitakyushu and Nagasaki, 18–22 August 2006, Graduate School of Information, Waseda University, pp. 219–223 (2006)
11. Novák, V., De Baets, B.: EQ-algebras. Fuzzy Sets Syst. **160**, 2956–2978 (2009)
12. Novák, V.: On fuzzy type theory. Fuzzy Sets Syst. **149**, 235–273 (2005)

The $(\in, \in \vee q_{(\lambda,\mu)})-$ fuzzy Subalgebras of $KU-$ algebras

Liu-hong Chen[1,2], Zu-hua Liao[1,2,4(✉)], Lun Li[1,2], Wei Song[3], Yong Li[1], Wei-long Liu[1], and Cui-cui Liao[1]

[1] School of Science, Jiangnan University, Wuxi 214122, People's Republic of China
liaozuhua57@163.com
[2] Honors School, Jiangnan University, Wuxi 214122, People's Republic of China
[3] School of Internet of Things Engineering, Jiangnan University, Wuxi 214122, People's Republic of China
[4] Institute of Intelligence System Network Computing, Jiangnan University, Wuxi 214122, China

Abstract. In this paper, we carry out a detailed investigation into the $(\in, \in \vee q_{(\lambda,\mu)})-$ fuzzy subalgebra of a $KU-$ algebra from the following aspects. Firstly, the concepts of the pointwise $(\in, \in \vee q_{(\lambda,\mu)})-$ fuzzy subalgebra and generalized fuzzy subalgebra of KU-algebras are introduced. Secondly, the equivalent descriptions of the $(\in, \in \vee q_{(\lambda,\mu)})-$ fuzzy subalgebra are given, including the level set, the $(\in, \in \vee q_{(\lambda,\mu)})-$ fuzzy subalgebra is better than $(\in, \in)-$ fuzzy subalgebra and $(\in, \in \vee q)-$ fuzzy subalgebra, which has rich hierarchy structure. Once more, we use methods of pointwise to discuss some basic properties of homomorphic image and homomorphic preimage of the $(\in, \in \vee q_{(\lambda,\mu)})-$ fuzzy subalgebra. Finally, we discuss the related properties about direct product and projection.

Keywords: $KU-$ algebra · $(\in, \in \vee q_{(\lambda,\mu)})-$ fuzzy Subalgebra
Generalized fuzzy subalgebra · Level set

1 Introduction

In 1965, Zadeh [1] has introduced a fuzzy set. The fuzzy set theory has been widely used in many branches of mathematics for more than 50 years. Rosenfeld [2] was inspired by the fuzzification of algebraic structures in 1971, and the definition of fuzzy subgroups was introduced, which created a new field of fuzzy algebra research.

Between 1992 and 1996, Bhakat et al. [3–6] gave the concept of $(\in, \in \vee q)-$ fuzzy subgroups by using the relationship of the belong to relation $(\in)$ and the quasi-coincident with relation (q) between a fuzzy point and a fuzzy sets, and made a series of research work. In 2006, Liao [7] extended the "q (quasi-coincident with)" relation to "$q_{(\lambda,\mu)}$ (generalized quasi-coincident with)" relation between fuzzy point and fuzzy set, and extended the $(\in, \in)-$ fuzzy algebra under the meaning of Rosenfeld and $(\in, \in \vee q)-$ fuzzy algebra, $(\overline{\in}, \overline{\in} \vee \overline{q})-$ fuzzy algebra

B.-Y. Cao and Y.-B. Zhong (Eds.): ICFIE 2017, AISC 872, pp. 55–67, 2019.
https://doi.org/10.1007/978-3-030-02777-3_6

to $(\in, \in \vee q_{(\lambda,\mu)})-$ fuzzy algebra under the meaning of Bharat, obtained a lot of meaningful results [7–17].

In 2009, Chanwit et al. [18] proposed KU-algebras based on logical algebra BCK/BCI and BCC, and made a series of studies on the ideal and congruence of KU-algebra. In 2011, Mostafa proposed KU-algebra. However, in the second inequality of the definition, y is equal to 0, we can see that the fuzzy set satisfying the condition can only be a constant function, and it is clear that the definition is unreasonable. In 2015, Muhammad in [20] extended the concept of fuzzy KU-subalgebra in [19], but it is also quite reasonable when the range of fuzzy sets is also a constant function in [0, 0.5]. The theorem 3 is also wrong in [20], the proof did not get the result of subalgebra.

On the basis of the above research work, we proposed the concept of $(\in, \in \vee q_{(\lambda,\mu)})-$ fuzzy subalgebra and generalized fuzzy subalgebra of $KU-$ algebras, and discussed their series of properties. The characterization of the level set $A_t \neq \emptyset$ of $(\in, \in \vee q_{(\lambda,\mu)})-$ fuzzy subalgebra A. The $(\in, \in \vee q_{(\lambda,\mu)})-$ fuzzy subalgebra has three hierarchical structures (see Theorem 3.8), while the $(\in, \in)-$ fuzzy algebra has only one hierarchical structure;$(\in, \in \vee q)-$ fuzzy algebra has only two hierarchical structures, so the $(\in, \in \vee q_{(\lambda,\mu)})-$ fuzzy subalgebra is better than $(\in, \in)-$ fuzzy algebra and $(\in, \in \vee q)-$ fuzzy algebra has a richer hierarchical structure. Also by the Example 3.1 to know $KU-$ algebra $(\in, \in \vee q)-$ fuzzy subalgebra is $(\in, \in)-$ fuzzy algebra, $(\in, \in \vee q)-$ fuzzy algebra of non-ordinary promotion.

2 Preliminary Notes

This section gives some knowledge of the $KU-$ algebras and $(\in, \in \vee q)-$ fuzzy subalgebra needed in the text.

Definition 2.1. [19] An algebra $G = (G, \cdot, 0)$ is called $KU-$ algebras, if $\forall x, y, z \in G$ satisfies the following condition:

(1) $(xy)((yz)(xz)) = 0$,
(2) $0x = x$,
(3) $x0 = 0$,
(4) $xy = 0 = yx \Rightarrow x = y$

Definition 2.2. [21] ($KU-$ subalgebras) Let $(G, \cdot, 0)$ be a $KU-$ algebras, and for $\emptyset \neq A \subseteq G$ satisfy: $\forall x, y \in A$, then $xy \in A$, then let A be the $KU-$ subalgebras of G.

Definition 2.3. [19] Let G_1, G_2 be a $KU-$ algebras, f is the mapping of G_1 to G_2, if satisfied $f(xy) = f(x)f(y), \forall x, y \in G_1$, then f if homomorphic mapping G_1 to G_2.

Definition 2.4. [19] Let G_1, G_2 be a $KU-$ algebras, then the calculation $\cdot$: $\forall (x_1, x_2), (y_1, y_2) \in G_1 \times G_2, (x_1, x_2)(y_1, y_2) = (x_1y_1, x_2y_2)$ is specified on $G_1 \times G_2$, Then $(G_1 \times G_2 \cdot, (0, 0))$ is also an $KU-$ algebra.

Definition 2.5. [7] If A is a fuzzy subset of G, and $A(y) = \begin{cases} \lambda(\neq 0), & y = x \\ 0, & y \neq x \end{cases}$" it is called A a fuzzy point, denoted as x_λ.

Definition 2.6. [7] Let A be a fuzzy subset of G. If $A(x) \geq t$, Then we said x_t is generalized to A, $x_t q_{(\lambda,\mu)} A$. If $x_t \in A$ or $x_t q_{(\lambda,\mu)} A$, denoted as $x_t q_{(\lambda,\mu)} A$.

Lemma 2.7. [22] If $\{A_i\}_{i \in I}$ is $(\in, \in \vee q_{(\lambda,\mu)})-$ fuzzy subalgebra family of G, and $\forall i, j \in I$, we have $A_i \subseteq A_j$ or $A_j \subseteq A_i$, so $\underset{i\in I}{\vee}(A_i(x) \wedge A_i(y) \wedge \mu) = (\underset{i\in I}{\cup} A_i)(x) \wedge (\underset{i\in I}{\cup} A_i)(y) \wedge \mu$.

Definition 2.8. [23] Let A be a fuzzy subset on a nonempty set G_1 and G_2 is a nonempty set, f is mapping from G_1 to G_2. When $\forall x, y \in G$ and $f(x) = f(y)$, we have $A(x) = A(y)$, Then that$A$ is$f-$ invariant.

Definition 2.9. [24] Let A, B be the fuzzy subsets of nonempty sets G_1, G_2 respectively, $A \times B : G_1 \times G_2 \to [0, 1]$,$(A \times B)4x, y) = A(x) \wedge B(y)$, then $A \times B$ is a fuzzy subset of $G_1 \times G_2$, then $A \times B$ is called a direct product over A and B.

Definition 2.10. [25] Let A, B be the fuzzy subsets of nonempty sets G_1, G_2 respectively. Define the fuzzy subsets of the projections of $A \times B$ on G_1 and G_2 respectively:

$A_1(x) = \underset{z\in G_2}{\vee} (A \times B)(x, z), \forall x \in G_1.$

$B_1(x) = \underset{z\in G_1}{\vee} (A \times B)(z, y), \forall y \in G_2.$

From now on, it is assumed that G is an $KU-$ algebras, $\lambda, \mu \in [0, 1]$ and $\lambda < \mu$.

3 The $(\in, \in \vee q_{(\lambda,\mu)})-$ fuzzy Subalgebras

Definition 3.1. Let A be a fuzzy subset of G, $\forall x, y \in G$, $\forall t_1, t_2 \in (\lambda, 1]$ and if $x_{t_1}, x_{t_2} \in A$ and $(xy)_{t_1 \wedge t_2} \in \vee q_{(\lambda,\mu)} A$, then A is the $(\in, \in \vee q_{(\lambda,\mu)})-$ fuzzy subalgebra of G.

Definition 3.2. Let A be a fuzzy subset of G, $A(xy) \vee \lambda \geq A(x) \wedge A(y) \wedge \mu, \forall x, y \in G$ then A be a generalized fuzzy subalgebra of G.

Theorems 3.1 and 3.2 give an equivalent characterization of $(\in, \in \vee q_{(\lambda,\mu)})-$ fuzzy subalgebras.

Theorem 3.1. Let A be a fuzzy subset of G, the following conditions are equivalent:

(i) A is the $(\in, \in \vee q_{(\lambda,\mu)})-$ fuzzy subalgebras of G.
(ii) A is a generalized fuzzy subalgebra of G.
(iii) $\forall t \in (\lambda, \mu]$, the nonempty sets $A_t = \{x | A(x) \geq t\}$ is the subalgebra of G.

Proof: $(i) \Rightarrow (ii)$

Assume that $\exists x, y \in G, A(xy) \vee \lambda < A(x) \wedge A(y) \wedge \mu$ is established, let $t = A(x) \wedge A(y) \wedge \mu$, then $\lambda < t \leq \mu$, $A(x) \geq t$, $A(y) \geq t$ and $A(xy) < \lambda$, so there are $x_t \in A$ and $y_t \in A$. Because A is the $(\in, \in \vee q_{(\lambda,\mu)})-$ fuzzy subalgebra of G, so $(xy)_t \in \vee q_{(\lambda,\mu)} A$,, so there is $(xy)_t \in A$ or $(xy)_t q_{(\lambda,\mu)} A$, then $A(xy) \geq t$ or $A(xy)+t > 2\mu$. But $A(xy) < t \leq \mu$ and $A(xy)+t < t+t \leq 2\mu$ are contradictory. So for all x, y, there is $A(xy) \vee \lambda \geq A(x) \wedge A(y) \wedge \mu$. Hence A is a generalized fuzzy subalgebra of G.

$(ii) \Rightarrow (i)$

For all $x, y \in G$ and $\forall t_1, t_2 \in (\lambda, 1]$ if $x_{t_1}, y_{t_2} \in A$, then $A(x) \geq t_1, A(y) \geq t_2$. Let $t = t_1 \wedge t_2$, because A is a generalized fuzzy subalgebra of G, so we have $A(xy) \vee \lambda \geq A(x) \wedge A(y) \wedge \mu \geq t \wedge \mu$. If $t \leq \mu$, because $\lambda < t$, so $A(xy) \geq t$. we have $(xy)_t \in A$. If $t > \mu$, then $A(xy) \geq \mu$. We have $A(xy)+t \geq \mu+t > 2\mu$, that means $(xy)_t q_{(\lambda,\mu)} A$, so $(xy)_t \in \vee q_{(\lambda,\mu)} A$. Hence A is the $(\in, \in \vee q_{(\lambda,\mu)})-$ fuzzy subalgebra of G.

$(ii) \Rightarrow (iii)$

For all $t \in (\lambda, \mu]$ and $A_t \neq \emptyset$, then $\forall x, y \in A_t$, we have $A(x) \geq t$ and $A(y) \geq t$. Because A is a generalized fuzzy subalgebra of G, so we obtain $A(xy) \vee \lambda \geq A(x) \wedge A(y) \wedge \mu \geq t \wedge \mu = t$. Because $\lambda < t$, so we have $A(xy) \geq t$,, $xy \in A_t$. Therefore $A_t = \{x | A(x) \geq t\}$ is the subalgebra of G.

$(iii) \Rightarrow (ii)$

Assume $\exists x, y \in G$ satisfying $A(xy) \vee \lambda < A(x) \wedge A(y) \wedge \mu$, let $t = A(x) \wedge A(y) \wedge \mu$, then $\lambda < t \leq \mu$, $A(x) \geq t$ and $A(y) \geq t$, so $x \in A_t, y \in A_t$. Because $A_t = \{x | A(x) \geq t\}$ is the subalgebra of G, so we have $xy \in A_t$, that means $A(xy) \geq t$. But $A(xy) < t$, hence A is a generalized fuzzy subalgebra of G.

Theorem 3.2. A necessary and sufficient condition is $\forall t \in [\lambda, \mu), A_{(t)} = \{x | A(x) > t\} \neq \emptyset$ is a subalgebra of G.

Proof: Sufficiency: For all $x, y \in A_{(t)}$, we have $A(x) > t$, $A(y) > t$,. Since A is the $(\in, \in \vee q_{(\lambda,\mu)})-$ fuzzy subalgebra of G, then $A(xy) \vee \lambda \geq A(x) \wedge A(y) \wedge \mu > t$. Because $\lambda \leq t$, so $A(xy) > t$, $A(xy) > t$. Hence $\forall t \in [\lambda, \mu), A_{(t)} = \{x | A(x) > t\} \neq \emptyset$ is a subalgebra of G.

Necessity: Assume there exists $x, y \in G$ satisfying $A(xy) \vee \lambda < A(x) \wedge A(y) \wedge \mu$, let $A(xy) \vee \lambda = t$, then $\lambda \leq t < \mu, A(x) > t, A(y) > t$ and $A(xy) \leq t$. So $x \in A_{(t)}$, $y \in A_{(t)}$,. Because $\forall t \in [\lambda, \mu), A_{(t)} = \{x | A(x) > t\} \neq \emptyset$ is a subalgebra of G, so $xy \in A_{(t)}$, that means $A(xy) > t$. But we have $A(xy) \leq t$, so A is a generalized fuzzy subalgebra of G. Learned by Theorem 3.1, A is the $(\in, \in \vee q_{(\lambda,\mu)})-$ fuzzy subalgebra of G.

Inference 3.3. Let A be a fuzzy subset of G, then the following conditions are equivalent:

(i) A is the $(\in, \in \vee q_{(\lambda,\mu)})-$ fuzzy subalgebras of G.
(ii) A is a generalized fuzzy subalgebra of G.
(iii) $\forall t \in (\lambda, \mu]$, the nonempty sets $A_t = \{x | A(x) \geq t\}$ is the subalgebra of G.

(iv) $\forall t \in [\lambda, \mu)$, the nonempty sets $A_{(t)} = \{x|A(x) > t\}$ is the subalgebra of G.

When $\lambda = 0, \mu = 1$, we get the conclusion in Rosenfeld sense.

Inference 3.4. Let A be a fuzzy subset of G, then the following conditions are equivalent:

(i) A is the $(\in, \in)-$ fuzzy subalgebra of G.
(ii) A is a generalized fuzzy subalgebra of G, so $\forall x, y \in G, A(xy) \geq A(x) \wedge A(y)$.
(iii) $\forall t \in (0, 1]$, the nonempty sets $A_t = \{x|A(x) \geq t\}$ is the subalgebra of G.
(iv) $\forall t \in [0, 1)$, the nonempty sets $A_{(t)} = \{x|A(x) > t\}$ is the subalgebra of G.

When $\lambda = 0, \mu = 0.5$, we get the corollary of $(\in, \in \vee q)-$ fuzzy subalgebra in Bhakat sense.

Inference 3.5. Let A be a fuzzy subset of G, then the following conditions are equivalent:

(i) A is the $(\in, \in \vee q)-$ fuzzy subalgebra of G.
(ii) A is a generalized fuzzy subalgebra of G, so $\forall x, y \in G$, $A(xy) \geq A(x) \wedge A(y) \wedge 0.5$.
(iii) $\forall t \in (0, 0.5]$, the nonempty sets $A_t = \{x|A(x) \geq t\}$ is the subalgebra of G.
(iv) $\forall t \in [0, 0.5)$, the nonempty sets $A_{(t)} = \{x|A(x) > t\}$ is the subalgebra of G.

Example 3.1. Let $G = \{0, 1, 2, 3\}$, The multiplication $\cdot$ on G is defined as follows.

$\cdot$	0	1	2	3
0	0	1	2	3
1	0	0	0	2
2	0	2	0	1
3	0	0	0	0

Then $G = \{0, 1, 2, 3\}$ is an $KU-$ algebra. Let $\lambda = \frac{1}{5}, \mu = \frac{2}{5}$ and $\frac{1}{5} < t_1 < t_2 = \frac{2}{5} < t_3 < \frac{1}{2} < t_4 < \frac{1}{5} < t_1 < t_2 = \frac{2}{5} < t_3 < \frac{1}{2} < t_4 < 1$, $A(0) = t_2, A(1) = t_1, A(2) = t_3, A(3) = t_4$.

(1) By definition it is verified that A is a $(\in, \in \vee q_{(\lambda,\mu)})-$ fuzzy subalgebra of G.
(2) Because $A(2 \cdot 3) = A(1) = t_1, A(2) = t_3, A(3) = t_4$, $A(2 \cdot 3) = t_1 < t_3 \wedge t_4 = A(2) \wedge A(3)$, so A is not $(\in, \in)-$ fuzzy subalgebra.
(3) Because $A(3 \cdot 3) = A(0) = t_2 < \frac{1}{2} = t_4 \wedge t_4 \wedge \frac{1}{2} = A(3) \wedge A(3) \wedge \frac{1}{2}$, so A is not $(\in, \in)-$ fuzzy subalgebra.

From (1) (2) (3), we can see that $(\in, \in \vee q_{(\lambda,\mu)})-$ fuzzy subalgebra is a new fuzzy algebra structure.

Theorem 3.6. Let A be a fuzzy subset of G, then for all $t \in (\mu, 1]$, the necessary and sufficient condition for the nonempty subset A_t to be the subalgebra of G is $\forall x, y \in G$, $A(x) \wedge A(y) \leq A(xy) \vee \mu$

Proof:

Sufficiency: For all $t \in (\mu, 1], A_t \neq \emptyset$, then $forall x, y \in A_t$, we have $A(x) \geq t, A(y) \geq t$. By the known conditions, $t \leq A(x) \wedge A(y) \leq A(xy) \vee \mu$. Because $t > \mu$, so $A(xy) \geq t$, $xy \in A_t$, hence A_t is a subalgebra of G.

Necessity: Assume there exists $x, y \in G$ satisfying $A(x) \wedge A(y) > A(xy) \vee \mu$. Let t satisfying $A(x) \wedge A(y) > t > A(xy) \vee \mu$, then $t \in (\mu, 1]$, $x, y \in A_t$. Since A_t is the subalgebra of G, then $xy \in A_t$, $A(xy) \geq t$. This is a contradiction with $A(xy) < t$. Hence $\forall x, y \in G$, $A(x) \wedge A(y) \leq A(xy) \vee \mu$.

Theorem 3.7. A is the $(\in, \in \vee q)-$ fuzzy subalgebra of G. Then $\forall t \in (0, \lambda]$, the necessary and sufficient condition for the nonempty subset A_t to be the subalgebra of G is $\forall x, y \in G, A(x) \wedge A(y) \wedge \lambda \leq A(xy)$

Proof:

Necessity: Assume there exists $x, y \in G$ satisfying $A(x) \wedge A(y) \wedge \lambda > A(xy)$. Let $A(x) \wedge A(y) \wedge \lambda = t$, then $A(x) \geq t, A(y) \geq t$ and $t \in (0, \lambda]$. By the known condition, A_t is the subalgebra of G, then we know $xy \in A_t$ from $x \in A_t$ and $y \in A_t$. This is a contradiction with $A(xy) < t$, hence $\forall x, y \in G, A(x) \wedge A(y) \wedge \lambda \leq A(xy)$.

Sufficiency: For all $t \in (0, \lambda], A_t \neq \emptyset$, then $forall x, y \in A_t$, we have $A(x) \geq t, A(y) \geq t$. By the known conditions, $A(xy) \geq A(x) \wedge A(y) \wedge \lambda$, then $A(xy) \geq t$, $xy \in A_t$, hence A_t is a subalgebra of G.

The theorem is given by Theorems 3.1, 3.6 and 3.7 Theorem shows that $(\in, \in \vee q_{(\lambda,\mu)})-$ fuzzy subalgebra has a richer hierarchical structure than $(\in, \in)-$ fuzzy algebra and $(\in, \in \vee q)-$ fuzzy algebra.

Theorem 3.8. A is the fuzzy subset of G, $A_t(t \in (0, 1])$ is the level set of A. The following conclusions are true.

(1) $\forall t \in (0, \lambda]$, the necessary and sufficient condition for the nonempty subset A_t to be the subalgebra of G is $A(x) \wedge A(y) \wedge \lambda \leq A(xy)$.
(2) $\forall t \in (\lambda, \mu]$, the necessary and sufficient condition for the nonempty subset A_t to be the subalgebra of G is $A(xy) \vee \lambda \geq A(x) \wedge A(y) \wedge \mu$.
(3) $\forall t \in (\mu, 1]$, the necessary and sufficient condition for the nonempty subset A_t to be the subalgebra of G is $A(x) \wedge A(y) \leq A(xy) \vee \mu$.

The properties of the intersection and union of $(\in, \in \vee q_{(\lambda,\mu)})-$ fuzzy subalgebras are given below.

Theorem 3.9. If $\{A_i\}_{i \in I}$ is $(\in, \in \vee q_{(\lambda,\mu)})-$ fuzzy subalgebra family of G, then $\bigcap_{i \in I} A_i$ is the $(\in, \in \vee q)-$ fuzzy subalgebra of G.

Proof:

For all $x, y \in G$, we have $(\underset{i\in I}{\cap} A_i)(xy) \vee \lambda = (\underset{i\in I}{\wedge} A_i(xy)) \vee \lambda = \underset{i\in I}{\wedge}(A_i(xy) \vee \lambda) \geq \underset{i\in I}{\wedge}(A_i\ (x) \underset{i\in I}{\wedge}(A_i(x) \wedge A_i(y) \wedge \mu) \geq (\underset{i\in I}{\wedge} A_i(x)) \wedge (\underset{i\in I}{\wedge} A_i(y)) \wedge \mu = (\underset{i\in I}{\cap} A_i)(x) \wedge (\underset{i\in I}{\cap} A_i)(y) \wedge \mu$. Hence $\underset{i\in I}{\cap} A_i$ is the $(\in, \in \vee q)-$ fuzzy subalgebra of G.

Theorem 3.10. If $\{A_i\}_{i\in I}$ is $(\in, \in \vee q_{(\lambda,\mu)})-$ fuzzy subalgebra family of G, and for all $i, j \in I$, we have $A_i \subseteq A_j$ or $A_j \subseteq A_i$, then $\underset{i\in I}{\cup} A_i$ is the $(\in, \in \vee q)-$ fuzzy subalgebra of G.

Proof:

By Lemma 2.7 we know that $\underset{i\in I}{\vee}(A_i(x) \wedge A_i(y) \wedge \mu) = (\underset{i\in I}{\cup} A_i)(x) \wedge (\underset{i\in I}{\cup} A_i)(y) \wedge \mu$. So $\forall x, y \in G$, we have $(\underset{i\in I}{\cup} A_i)(xy) \vee \lambda = (\underset{i\in I}{\vee} A_i(xy)) \vee \lambda = \underset{i\in I}{\vee}(A_i(xy) \vee \lambda) \geq \underset{i\in I}{\vee}(A_{(i)}(x) \wedge A_{(i)}\ (y) \wedge \mu) = (\underset{i\in I}{\cup} A_i)(x) \wedge (\underset{i\in I}{\cup} A_i)(y) \wedge \mu$. Hence $\underset{i\in I}{\cup} A_i$ is the $(\in, \in \vee q)-$ fuzzy subalgebra of G.

Theorem 3.11. Let A be a nonempty subset of G, B is the fuzzy subset of G defined as follows: $B_\alpha = \begin{cases} s,\ x \in A \\ t,\ x \notin A \end{cases}$

Among them $t < s, 0 \leq t < \mu, \lambda < s \leq 1$, Then the necessary and sufficient condition for B is the $(\in, \in \vee q_{(\lambda,\mu)})-$ fuzzy subalgebra of G is that A is the subalgebra of G.

Proof:

Since $0 \leq t \leq \lambda, \lambda < s < \mu$, we have $B_\alpha = \begin{cases} A,\ \lambda < \alpha \leq s \\ \emptyset,\ \ s < \alpha \leq \mu \end{cases}$.

Since $0 \leq t \leq \lambda, \mu \leq s \leq 1$, we have $\forall_\alpha \in (\lambda, \mu], B_\alpha = A$.

Since $\lambda < t < \mu, t < s < \mu$, we have $B_\alpha = \begin{cases} G,\ \lambda < \alpha \leq t \\ A,\ t < \alpha \leq s \\ \emptyset,\ \ s < \alpha \leq \mu \end{cases}$.

Since $\lambda < t < \mu, \mu \leq s \leq 1$, we have $B_\alpha = \begin{cases} G,\ \lambda < \alpha \leq t \\ A,\ t < \alpha \leq \mu \end{cases}$.

From the Theorem 3.1, the conclusion of the theorem is established.

Inference 3.6. A is a necessary and sufficient condition for the subalgebra of G is that the characteristic function χ_A of A is the $(\in, \in \vee q_{(\lambda,\mu)})-$ fuzzy subalgebra of G.

This inference illustrates the rationality of the definition of $(\in, \in \vee q_{(\lambda,\mu)})-$ fuzzy subalgebra of G.

4 Homomorphic image and preimage of the $(\in, \in \vee q_{(\lambda,\mu)})-$ fuzzy subalgebras

Theorem 4.1. Let f be the homomorphism of G to G, and B is the $(\in, \in \vee q_{(\lambda,\mu)})$ fuzzy subalgebra of G', then $f^{-1}(B)$ is the $(\in, \in \vee q_{(\lambda,\mu)})$ fuzzy subalgebra of G.

Proof:

For all $x, y \in G$, because B is the $(\in, \in \vee q_{(\lambda,\mu)})-$ fuzzy subalgebra of G', so $f^{-1}(B)(xy) \vee \lambda = B(f(xy)) \vee \lambda = B(f(x)f(y)) \vee \lambda \geq B(f(x)) \wedge B(f(y)) \wedge \mu = f^{-1}(B)(x) \wedge f^{-1}(B)(y) \wedge \mu$. Hence $f^{-1}(B)$ is the $(\in, \in \vee q_{(\lambda,\mu)})-$ fuzzy subalgebra of G.

Theorem 4.2. Let A be the $(\in, \in \vee q_{(\lambda,\mu)})-$ fuzzy subalgebra of G and f is the homomorphism mapping of G to G', then $f(A)$ is the $(\in, \in \vee q_{(\lambda,\mu)})-$ fuzzy subalgebra of G'.

Proof:

For all $x', y' \in G'$,if $f^{-1}(x') = \emptyset$ or $f^{-1}(y') = \emptyset$,so $f(A)(x) = 0$ or $f(A)(y) = 0$. We have $(xy) \vee \lambda \geq f(A)(x) \wedge f(A)(x) \wedge \mu$. If $f^{-1}(x') \neq \emptyset$ and $f^{-1}(y') \neq \emptyset$, then $\exists x, y \in G$, $f(x) = x', f(y) = y'$. Because f is the homomorphism mapping, so $f(xy) = f(x)f(y) = x'\ y', f^{-1}(x'y') \neq \emptyset$, then $f(A)(x'y') \vee \lambda = \vee\{A(z)|z \in G, f(z) = x'y'\} \vee \lambda \geq \vee\{A(xy)\ x', y' \in G', f(x) = x', f(y) = y'\} \vee \lambda = \vee\{A(xy) \vee \lambda|x', y' \in G', f(x) = x', f(y) = y'\} \geq \vee\{A(x) \wedge A(y) \wedge \mu|x', y' \in G', f(x) = x', f(y) = y'\} = \vee\{A(x) \wedge A(y)|x', y' \in G'\ f(x) = x', f(y) = y'\} \wedge \mu = (\bigvee_{f(x)=x'} A(x)) \wedge (\bigvee_{f(y)=y'} A(y)) \wedge \mu = f(A)(x') \wedge f(A)(y') \wedge \mu$. Hence $f(A)$ is the $(\in, \in \vee q_{(\lambda,\mu)})-$ fuzzy subalgebra of G'.

Theorem 4.3. Let A be a fuzzy subalgebra of G, f be a homomorphic mapping of G to G', and A is $f-$ invariant, Then the necessary and sufficient condition for $f(A)$ to be $(\in, \in \vee q_{(\lambda,\mu)})-$ fuzzy subalgebra of G' is A is the $(\in, \in \vee q_{(\lambda,\mu)})-$ fuzzy subalgebra of G.

Proof:

Necessity: For all $t_1, t_2 \in (\lambda, 1], \forall x_1, x_2 \in G$, let $f(x_1) = y_1$ and $f(x_2) = y_2$. Because f is a homomorphic mapping of G to G', so $f(x_1x_2) = f(x_1)f(x_2) = y_1y_2$. If $(x_1)_{t_1}, (x_2)_{t_2} \in A$, then $A(x_1) \geq t_1$ and $A(x_2) \geq t_2$. Because A is $f-$ invariant, so $f(A)(y_1) = \bigvee_{f(x)=y_1} A(x) = A(x_1) \geq t_1$ and $f(A)(y_2) = \bigvee_{f(x)=y_2} A(x) = A(x_2) \geq t_2$, then $(y_1)_{t_1} \in f(A)$ and $(y_2)_{t_2} \in f(A)$. Because $f(A)$ is $(\in, \in \vee q_{(\lambda,\mu)})-$ fuzzy subalgebra of G', then $(y_1y_2)_{t_1 \wedge t_2} \in f(A)$ or $(y_1y_2)_{t_1 \wedge t_2} q_{(\lambda,\mu)}\ f(A)$. And A is $f-$ invariant, we have $A(x_1x_2) = \bigvee_{f(x)=y_1y_2} A(x) \geq t_1 \wedge t_2$ or $A(x_1x_2) + t_1 \wedge t_2 = \bigvee_{f(x)=y_1y_2} A(x) \geq t_1 \wedge t_2 > 2\mu$, so $(x_1x_2)_{t_1 \wedge t_2} \in \vee q_{(\lambda,\mu)} A$, hence A is the $(\in, \in \vee q_{(\lambda,\mu)})-$ fuzzy subalgebra of G.

Sufficiency: From the Theorem 4.2, we know the theorem is established.

5 Direct product and projection of the $(\in, \in \vee q_{(\lambda,\mu)})-$ fuzzy subalgebras

This section discusses the properties of $(\in, \in \vee q_{(\lambda,\mu)})-$ fuzzy subalgebra direct product and projection.

Theorem 5.1. Let A and B be $(\in, \in \vee q_{(\lambda,\mu)})-$ fuzzy subalgebra of $KU-$ algebra G_1 and G_2 respectively, then $A \times B$ is the $(\in, \in \vee q_{(\lambda,\mu)})-$ fuzzy subalgebra of $G_1 \times G_2$.

Proof: Because A and B are $(\in, \in \vee q_{(\lambda,\mu)})-$ fuzzy subalgebra of $KU-$ algebra G_1 and G_2 respectively. From Theorem 3.1, A and B are the generalized fuzzy subalgebra of $KU-$ algebra G_1 and G_2 respectively. $\forall (x_1, y_1), (x_2, y_2) \in G_1 \times G_2$,

$$(A \times B)((x_1, y_1), (x_2, y_2)) \vee \lambda$$
$$= (A \times B)((x_1x_2, y_1y_2)) \vee \lambda$$
$$= (A(x_1x_2) \wedge B(y_1y_2)) \vee \lambda$$
$$= (A(x_1x_2) \vee \lambda) \wedge (B(y_1y_2) \vee \lambda) \geq (A(x_1) \wedge A(x_2) \wedge \mu) \wedge (B(y_1) \wedge B(y_2) \wedge \mu)$$
$$= (A(x_1) \wedge B(y_1)) \wedge (A(x_2) \wedge B(y_2)) \wedge \mu$$
$$= (A \times B)(x_1, y_1) \wedge (A \times B)(x_2, y_2) \wedge \mu.$$

So $A \times B$ is a generalized fuzzy subalgebra of $G_1 \times G_2$. And then by the Theorem 3.1 know $A \times B$ is the $(\in, \in \vee q_{(\lambda,\mu)})-$ fuzzy subalgebra of $G_1 \times G_2$.

Theorem 5.2. Let A and B be the fuzzy subsets of nonempty sets G_1 and G_2 respectively. If $A \times B$ is the fuzzy subalgebra of $G_1 \times G_2$, then A_1 is the $(\in, \in \vee q_{(\lambda,\mu)})-$ fuzzy subalgebra of G_1 and B_1 is the $(\in, \in \vee q_{(\lambda,\mu)})-$ fuzzy subalgebra of G_2.

Proof: Because $A \times B$ is the $(\in, \in \vee q_{(\lambda,\mu)})-$ fuzzy subalgebra of $G_1 \times G_2$. From the Theorem 3.1, we know $A \times B$ is a generalized fuzzy subalgebra of $G_1 \times G_2$. $\forall x, y \in G_1$, we have

$$A_1(xy) \vee \lambda = (\bigvee_{z \in G_2} (A \times B)(xy, z)) \vee \lambda \geq (\bigvee_{z_1, z_2 \in G_2} (A \times B)(xy, z_1z_2)) \vee \lambda \geq$$
$$(A \times B)((x, z_1)(y, z_2)) \vee \lambda \geq (A \times B)(x, z_1) \wedge (A \times B)(y, z_2) \wedge \mu$$

By the arbitrariness of z_1,

$$A_1(xy) \vee \lambda \geq \bigvee_{z_1 \in G_2} ((A \times B)(x, z_1) \wedge (A \times B)(y, z_2) \wedge \mu)$$
$$= (\bigvee_{z_1 \in G_2} (A \times B)(x, z_1)) \wedge (A \times B)(y, z_2) \wedge \mu$$
$$= A_1(x) \wedge (A \times B)(y, z_2) \wedge \mu$$

By the arbitrariness of z_2,

$$A_1(xy) \vee \lambda \geq \bigvee_{z_2 \in G_2} (A_1(x) \wedge (A \times B)(y, z_2) \wedge \mu)$$
$$= A_1(x) \wedge (\bigvee_{z_2 \in G_2} (A \times B)(y, z_1)) \wedge \mu$$
$$= A_1(x) \wedge A_1(y) \wedge \mu$$

Hence A_1 is a generalized fuzzy subalgebra of G, from the Theorem 3.1, we know A_1 is the $(\in, \in \vee q_{(\lambda,\mu)})-$ fuzzy subalgebra of G_1.

Similarly, B_1 is the $(\in, \in \vee q_{(\lambda,\mu)})-$ fuzzy subalgebra of G_2.

6 Fuzzy Subalgebra Degree

Definition 6.1. Let A be a fuzzy subset of G. Denote $m_{G(\lambda)}(A) = \bigwedge_{x,y \in G} \bigvee_{t \in (\lambda,1]} \{t | A(x) \wedge A(y) \wedge t \leq A(xy) \vee \lambda\}$

Let $m_{G(\lambda)}(A)$ be the fuzzy subalgebra of A.

Theorem 6.1. Let A be a fuzzy subset of G and $\mu > \lambda$, then the necessary and sufficient condition for $m_{G(\lambda)}(A) \geq \mu$ is:

$\forall x, y \in G, A(x) \wedge A(y) \wedge \mu \leq A(xy) \vee \lambda$

Necessity: If $m_{G(\lambda)}(A) \geq \mu > \lambda$, then $\bigwedge\limits_{x,y\in G} \bigvee\limits_{t\in(\lambda,1]} \{t|A(x) \wedge A(y) \wedge t \leq A(xy) \vee \lambda\} \geq \mu$, so $\forall x, y \in G$, we have $\bigvee\limits_{t\in(\lambda,1]} \{t|A(x) \wedge A(y) \wedge t \leq A(xy) \vee \lambda\} \geq \mu$. Let $\mu_1 = \bigvee\limits_{t\in(\lambda,1]} \{t|A(x) \wedge A(y) \wedge t \leq A(xy) \vee \lambda\}$, then $\forall \varepsilon > 0, \exists t_\varepsilon > \lambda$, we have $t_\varepsilon > \mu_1 - \varepsilon$, so $A(x) \wedge A(y) \wedge (\mu_1 - \varepsilon) \leq A(x) \wedge A(y) \wedge t_\varepsilon \leq A(xy) \vee \lambda$. By the arbitrariness of ε, $A(x) \wedge A(y) \wedge \mu_1 \leq A(xy) \vee \lambda$, hence $A(x) \wedge A(y) \wedge \mu \leq A(xy) \vee \lambda$.

Sufficiency: If $\forall x, y \in G, A(x) \wedge A(y) \wedge \mu \leq A(xy) \vee \lambda$, then $\forall x, y \in G$, we have $\bigvee\limits_{t\in(\lambda,1]} \{t|A(x) \wedge A(y) \wedge t \leq A(xy) \vee \lambda\} \geq \mu$, so $\bigwedge\limits_{x,y\in G} \bigvee\limits_{t\in(\lambda,1]} \{t|A(x) \wedge A(y) \wedge t \leq A(xy) \vee \lambda\} \geq \mu$, hence $m_{G(\lambda)}(A) \geq \mu$.

Theorem 6.2. Let A be a fuzzy subset of G, then $m_{G(\lambda)}(A) = \bigvee\limits_{t\in(\lambda,1]} \{t|A(x) \wedge A(y) \wedge t \leq A(xy) \vee \lambda, \forall x, y \in G\}$.

Proof: Let $m_{G(\lambda)}(A) = c$, then $\bigwedge\limits_{x,y\in G} \bigvee\limits_{t\in(\lambda,1]} \{t|A(x) \wedge A(y) \wedge t \leq A(xy) \vee \lambda\} = c$. So $\forall x, y \in G$, we have $\bigvee\limits_{t\in(\lambda,1]} \{t|A(x) \wedge A(y) \wedge t \leq A(xy) \vee \lambda\} \geq c$. And $b = \bigvee\limits_{t\in(\lambda,1]} \{t|A(x) \wedge A(y) \wedge t \leq A(xy) \vee \lambda\}$, then $b \geq c$ and $\forall \varepsilon > 0, \exists t_\varepsilon > \lambda$. since $t_\varepsilon > b - \varepsilon$ and $A(x) \wedge A(y) \wedge t_\varepsilon \leq A(xy) \vee \lambda$, so $A(x) \wedge A(y) \wedge (b - \varepsilon) \leq A(x) \wedge A(y) \wedge t_\varepsilon \leq A(xy) \vee \lambda$. By the arbitrariness of ε,$\forall x, y \in G$ $A(x) \wedge A(y) \wedge b \leq A(xy) \vee \lambda$ is obtained. From the Theorem 3.1, we know $c \geq b$, so $b = c$, then $m_{G(\lambda)}(A) = \bigvee\limits_{t\in(\lambda,1]} \{t|A(x) \wedge A(y) \wedge t \leq A(xy) \vee \lambda\}$.

Theorem 6.3. Let A be a fuzzy subset of G, then $m_{G(\lambda)}(A) = \bigvee\limits_{t\in(\lambda,1]} \{t|\forall b \in (\lambda, t]$, A_b be the subalgebra of G.

Proof: Let $m_{G(\lambda)} = c$,$B_1 = \{t \in (\lambda, 1]|\forall b(\lambda, t], A_b \neq \emptyset$ is the subalgebra of G $\}$. $B_2 = \{t|A(x) \wedge A(y) \wedge t \leq A(xy) \vee \lambda, \forall x, y \in G\}$. $\forall b_1 \in B_1$, then $\forall t \in (\lambda, b_1]$, A_t is the subalgebra of G. From the Theorem 3.1, $\forall x, y \in G, A(x) \wedge A(y) \wedge b_1 \leq A(xy) \vee \lambda$, so $b_1 \in B_2$, then $B_1 \subseteq B_2$. And $\forall b_2 \in B_2$, then $\forall x, y \in G$, we have $A(x) \wedge A(y) \wedge b_2 \leq A(xy) \vee \lambda$. And From the Theorem 3.1, we have $\forall t \in (\lambda, b_2], A_t \neq \emptyset$ is the subalgebra of G, so $b_2 \in B_1$, $B_2 \subseteq B_1$. Hence $B_1 = B_2$. From the Theorem 6.2, we know $m_{G(\lambda)}(A) = \bigvee\limits_{t\in(\lambda,1]} B_2 = \bigvee\limits_{t\in(\lambda,1]} B_1 \bigvee\limits_{t\in(\lambda,1]} \{t|\forall b \in (\lambda, t]$, A_b is the subalgebra of G $\}$.

Theorem 6.4. Let A be a fuzzy subset of G, then $m_{G(\lambda)}(A) = \bigvee\limits_{t\in(\lambda,1]} \{t|\forall b \in [\lambda, t), A_{(b)}$ is the subalgebra of G $\}$.

Proof: Let $m_{G(\lambda)} = c$, $B_1 = \{t \in (\lambda, 1]|\forall b(\lambda, t], A_b \neq \emptyset$ is the subalgebra of G $\}$. $B_2 = \{t|A(x)A(y) \wedge t \leq A(xy) \vee \lambda, \forall x, y \in G\}$. Let $b_1 \in B_1$, then$\forall t \in [\lambda, b_1), A_{(t)} \neq \emptyset$ is the subalgebra of G $\}$. From the Inference 3.3, we know $b_2 \in$

$B_1, B_2 \subseteq B_1$ for $\forall t \in [\lambda, b_2), A_{(t)} \neq \emptyset$ is the subalgebra of G. Hence $B_1 = B_2$, $m_{G(\lambda)}(A) = \bigvee_{t\in(\lambda,1]} B_2 = \bigvee_{t\in(\lambda,1]} B_1 = \bigvee_{t\in(\lambda,1]} \{t|\forall b \in [\lambda, t), A_{(b)}$ is the subalgebra of G.

Theorem 6.5. If $\{A_i\}_{i\in I}$ is $(\in, \in \vee q_{(\lambda,\mu)})-$ fuzzy subalgebra family of G, then $m_{G(\lambda)}(\bigcap_{i\in I} A_i) \geq \bigwedge_{i\in I} m_{G(\lambda)}(A_i)$.

Proof:

$$
\begin{aligned}
& m_{G(\lambda)}(\bigcap_{i\in I} A_i) = \bigwedge_{x,y\in G} \bigvee_{t\in(\lambda,1]} \{t|(\bigcap_{i\in I} A_i)(x) \wedge (\bigcap_{i\in I} A_i)(y) \wedge t \leq (\bigcap_{i\in I} A_i)(xy) \vee \lambda\} \\
&= \bigwedge_{x,y\in G} \bigvee_{t\in(\lambda,1]} \{t|(\bigwedge_{i\in I} A_i(x)) \wedge (\bigwedge_{i\in I} A_i(y)) \wedge t \leq (\bigwedge_{j\in I} A_j(xy)) \vee \lambda\} \\
&= \bigwedge_{x,y\in G} \bigvee_{t\in(\lambda,1]} \bigwedge_{j\in I} \{t|(\bigwedge_{i\in I} A_i(x)) \wedge (\bigwedge_{i\in I} A_i(y)) \wedge t \leq A_j(xy) \vee \lambda\} \\
&\geq \bigwedge_{x,y\in G} \bigvee_{t\in(\lambda,1]} \bigwedge_{j\in I} \{t| \bigwedge_{i\in I} (A_i(x) \wedge A_i(y)) \wedge t \leq A_j(xy) \vee \lambda\} \\
&\geq \bigwedge_{x,y\in G} \bigvee_{t\in(\lambda,1]} \bigwedge_{j\in I} \{t|A_j(x) \wedge A_j(y) \wedge t \leq A_j(xy) \vee \lambda\} \\
&= \bigwedge_{i\in I} \bigwedge_{x,y\in G} \bigvee_{t\in(\lambda,1]} \{t|A_i(x) \wedge A_i(y)) \wedge t \leq A_i(xy) \vee \lambda\} \\
&\geq \bigwedge_{i\in I} m_{G(\lambda)}(A_i)
\end{aligned}
$$

Theorem 6.6. Let B be a fuzzy subset of G', f be a homomorphic mapping of G to G', then $m_{G(\lambda)}(f^{-1}(B)) \geq m_{G(\lambda)}(B)$, among them $f^{-1}(B)(x) = B(f(x))$.

Proof: Let $f(x) = x', f(y) = y'$, because f be a homomorphic mapping of G to G', then $f(xy) = f(x)f(y) = x'y'$.

$$
\begin{aligned}
& m_{G(\lambda)}(f^{-1}(B)) = \bigwedge_{x,y\in G} \bigvee_{t\in(\lambda,1]} \{t|(f^{-1}(B)(x) \wedge f^{-1}(B)(y) \wedge t \leq f^{-1}(B)(xy) \vee \lambda\} \\
&= \bigwedge_{x,y\in G} \bigvee_{t\in(\lambda,1]} \{t|(B(f(x)) \wedge (B)(f(y)) \wedge t \leq B(f(xy)) \vee \lambda\} \\
&\geq \bigwedge_{x',y'\in G'} \bigvee_{t\in(\lambda,1]} \{t|B(x') \wedge (B)(y') \wedge t \leq B(x'y') \vee \lambda\} \\
&= m_{G(\lambda)}(B)
\end{aligned}
$$

Theorem 6.7. Let A_i be a fuzzy subset of $G_i, i = 1, 2, 3 \cdot \cdot\cdot, n$, then $m_{G(\lambda)}(\prod_{1\leq i\leq n} A_i) \geq \bigwedge_{1\leq i\leq n} m_{G_i(\lambda)}(A_i)$

Proof:

$\forall x, y \in \prod_{1\leq i\leq n} G_i = G, x = (x_1, x_2, \cdot\cdot\cdot, x_n), y = (y_1, y_2, \cdot\cdot\cdot, y_n)$

$$
\begin{aligned}
& m_{G(\lambda)}(\prod_{1\leq i\leq n} A_i) \\
&= \bigwedge_{x,y\in G} \bigvee_{t\in(\lambda,1]} \{t|(\bigwedge_{1\leq i\leq n} A_i(x_i)) \wedge (\bigwedge_{1\leq i\leq n} A_i(y_i)) \wedge t \leq (\bigwedge_{1\leq i\leq n} A_i(x_iy_i)) \vee \lambda\} \\
&= \bigwedge_{x,y\in G} \bigvee_{t\in(\lambda,1]} \bigwedge_{1\leq i\leq n} \{t|(\bigwedge_{1\leq i\leq n} A_i(x_i)) \wedge (\bigwedge_{1\leq i\leq n} A_i(y_i)) \wedge t \leq (A_i(x_iy_i)) \vee \lambda\} \\
&\geq \bigwedge_{x,y\in G} \bigvee_{t\in(\lambda,1]} \bigwedge_{1\leq i\leq n} \{t|(A_i(x_i) \wedge A_i(y_i) \wedge t \leq A_i(x_iy_i) \vee \lambda\} \\
&= \bigwedge_{1\leq i\leq n} \bigwedge_{x,y\in G} \bigvee_{t\in(\lambda,1]} \{t|(A_i(x_i) \wedge A_i(y_i) \wedge t \leq A_i(x_iy_i) \vee \lambda\} \\
&= \bigwedge_{1\leq i\leq n} \{ \bigwedge_{x,y\in G} \bigvee_{t\in(\lambda,1]} \{t|(A_i(x_i) \wedge A_i(y_i) \wedge t \leq A_i(x_iy_i) \vee \lambda\}\} \\
&= \bigwedge_{1\leq i\leq n} m_{G_i(\lambda)}(A_i)
\end{aligned}
$$

Theorem 6.8. Let $A_i (i \in I)$ be a fuzzy subset of G_i, then $m_{G(\lambda)}(\prod\limits_{i \in I} A_i) \geq \bigwedge\limits_{i \in I} m_{G_i(\lambda)}(A_i)$

Proof: $\forall x, y \in \prod\limits_{i \in I} G_i = G, x = \prod\limits_{i \in I} x_i, y = \prod\limits_{i \in I} y_i$, among them $x_i, y_i \in G_i$.

$m_{G(\lambda)}(\prod\limits_{1 \leq i \leq n} A_i)$

$= \bigwedge\limits_{x,y \in G} \bigvee\limits_{t \in (\lambda, 1]} \{t | (\bigwedge\limits_{i \in I} A_i(x_i)) \wedge (\bigwedge\limits_{i \in I} A_i(y_i)) \wedge t \leq (\bigwedge\limits_{i \in I} A_i(x_i y_i)) \vee \lambda\}$

By I for any index set,$a \to \bigwedge\limits_{i \in I} a_i = \bigwedge\limits_{i \in I} (a \to a_i), a_i \in [0, 1] (i \in I)$,then

$= \bigwedge\limits_{x,y \in G} \bigvee\limits_{t \in (\lambda, 1]} \bigwedge\limits_{i \in I} \{t | (\bigwedge\limits_{i \in I} A_i(x_i)) \wedge (\bigwedge\limits_{i \in I} A_i(y_i)) \wedge t \leq (A_i(x_i y_i)) \vee \lambda\}$

$\geq \bigwedge\limits_{x,y \in G} \bigvee\limits_{t \in (\lambda, 1]} \bigwedge\limits_{i \in I} \{t | (A_i(x_i) \wedge A_i(y_i) \wedge t \leq A_i(x_i y_i) \vee \lambda\}$

$= \bigwedge\limits_{i \in I} \bigwedge\limits_{x,y \in G} \bigvee\limits_{t \in (\lambda, 1]} \{t | (A_i(x_i) \wedge A_i(y_i) \wedge t \leq A_i(x_i y_i) \vee \lambda\}$

$= \bigwedge\limits_{i \in I} \{ \bigwedge\limits_{x_i, y_i \in G_i} \bigvee\limits_{t \in (\lambda, 1]} \{t | (A_i(x_i) \wedge A_i(y_i) \wedge t \leq A_i(x_i y_i) \vee \lambda\}\}$

$= \bigwedge\limits_{i \in I} m_{G_i(\lambda)}(A_i)$

7 Conclusion

In this paper, we give the algebraic structure of $KU-$ algebraic $(\in, \in \vee q_{(\lambda,\mu)})-$ fuzzy subalgebra and generalized fuzzy subalgebra, and discuss their series of basic properties. It is worth noting that $(\in, \in \vee q_{(\lambda,\mu)})-$ fuzzy subalgebra is better than $(\in, \in)-$ fuzzy subalgebra and $(\in, \in \vee q_{(\lambda,\mu)})-$ fuzzy subalgebras have a richer hierarchical structure, and these studies will have potential application value in information science. We will further study the $KU-$ algebraic $(\in, \in \vee q_{(\lambda,\mu)})-$ fuzzy k -subalgebra and $(\in, \in \vee q_{(\lambda,\mu)})-$ fuzzy convex Algebraic structure.

Acknowledgements. Thanks to the supported by the National Science Foundations of Chi-na (No:61673193,61170121,11401259), and the Natural Science Foundations of Jiangsu Prov-ince (No: BK2015117).

References

1. Zadeh, L.A.: Fuzzy set. Inf. Control. **8**(3), 338–353 (1965)
2. Rosenfeld, A.: Fuzzy groups. J. Math. Anal. Appl. **35**, 512–517 (1971)
3. Bhakat, S.K., Das, P.: On the definition of a fuzzy subgroup. Fuzzy Sets Syst. **51**, 235–241 (1992)
4. Bhakat, S.K., Das, P.: $(\in, \in \vee q)-$ fuzzy subgroup. Fuzzy Sets Syst. **80**, 359–368 (1996)
5. Bhakat, S.K.: $(\in, \in \vee q)-$ fuzzy normal quasi-normal and maximal subgroup. Fuzzy Sets Syst. **112**, 299–312 (2000)
6. Bhakat, S.K.: $(\in, \in \vee q)-$ fuzzy cyclic subgroup. J. Fuzzy Math. **8**, 597–606 (2000)
7. Liao, Z.H., Gu, H.: $(\in, \in \vee q_{(\lambda,\mu)})-$ fuzzy normal subgroup. Fuzzy Syst. Math. **20**(5), 47–53 (2006)

8. Rui, M., Liao, Z., Hu, M., Lu, J.: Generalized fuzzy subinclines and $(\in, \in \vee q_{(\lambda,\mu)})-$ fuzzy subinclines. Fuzzy Syst. Math. **25**(6), 60–68 (2011)
9. Zhang, Y., Liao, Z., Cao, S., Liu, C.: Homomorphic Properties of $(\in, \in \vee q_{(\lambda,\mu)})-$ fuzzy regular subsemigroups. Fuzzy Syst. Math. **26**(6), 21–25 (2012)
10. Liu, C., Liao, Z., Jang, X.: $(\in, \in \vee q_{(\lambda,\mu)})-$ fuzzy Boolean algebra. Fuzzy Syst. Math. **27**(4), 67–73 (2013)
11. Hu, M., Liao, Z., Yi, L., Rui, M.: $(\in, \in \vee q_{(\lambda,\mu)})-$ fuzzy Bi-ideals of semirings. Fuzzy Syst. Math. **25**(5), 69–75 (2011)
12. Fan, X., Liao, Z., Fan, Y., Chen, J., Zhang, Y., Zeng, J.: $(\in, \in \vee q_{(\lambda,\mu)})-$ fuzzy $\Gamma-$ completely Prime Ideals of $\Gamma-$ semigroups. Fuzzy Syst. Math. **28**(2), 52–61 (2014)
13. Zhang, Y., Liao, Z., Fan, Y., Zeng, J., Chen, J., Fan, X.: $(\in, \in \vee q_{(\lambda,\mu)})-$ fuzzy $\Gamma-$ ideals of $\Gamma-$ rings. Fuzzy Syst. Math. **28**(2), 62–68 (2014)
14. Jiang, X., Liao, Z., Liu, C., Shao, Y.: On generalized fuzzy ideals and $(\in, \in \vee q_{(\lambda,\mu)})-$ fuzzy $k-$ ideals in $\Gamma-$ hemi-rings. Fuzzy Syst. Math. **28**(6), 20–28 (2014)
15. Lu, T., Liao, Z., Liao, C., Yuan, W.: $(\overline{\in}, \overline{\in} \vee \overline{q}_{(\lambda,\mu)})$ fuzzy ideals of incline algebras. J. Front. Comput. Sci. Technol. **10**(08), 1191–1200 (2016)
16. Liao, Z., Hao, C., Chen, Y.: R-generalized fuzzy completely semiprime (prime) ideals of a ring. Inf. Int. Interidisciplinary J. **15**(6), 2539–2542 (2012)
17. Liao, Z., Zhou, J.: Rough convex cones and rough convex fuzzy cones. Soft. Comput. **16**, 2083–2087 (2012)
18. Chanwit, P., Utsanee, L.: On ideas and congruences $KU-$ algebras. Scientia Magna **5**(1), 54–57 (2009)
19. Mostafa, S.M.: Fuzzy ideals of algebras. Int. Math. Forum **6**(63), 3139–3149 (2011)
20. Muhammad, G., Muhammad, S., Sarfraz, A.: On $(\alpha, \beta)-$ fuzzy $KU-$ ideals $KU-$ algebras. Afr. Mat. **26**, 651–661 (2015)
21. Mostafa, S.M.: Interval fuzzy $KU-$ ideals in $KU-$ algebras. Int. Math. Forum **64**(6), 3151–3159 (2011)
22. Zhu, C., Liao, Z., Luo, X., Ye, L.: $(\overline{\in}, \overline{\in} \vee \overline{q}_{(\lambda,\mu)})$ fuzzy complemented semirings. Fuzzy Syst. Math. **27**(5), 48–54 (2013)
23. Mordeson, J.N.: Fuzzy Commutative Algebra. World Scientific Publishing, Singapore (1998)
24. Fu, X., Liao, Z.: $(\overline{\in}, \overline{\in} \vee \overline{q}_{(\lambda,\mu)})$ fuzzyprime filter of lattice implication algebra. J. Front. Comput. Sci. Technol. **9**(2), 227–233 (2015)
25. Fu, X., Liao, Z.: $(\in^{\delta}, \in \delta \vee {q^{\delta}}_{(\lambda,\mu)})-$ fuzzy ideals of N (2,2,0) algebras. J. Front. Comput. Sci. Technol. **7**(11), 323–332 (2016)

On Derivations of FI-Algebras

Jian-xiang Rong[1], Zu-hua Liao[1(✉)], Yue Xi[1], Wei Song[2], Lun Li[1], and Yong Li[3]

[1] Honor School, School of Science, Jiangnan University, Wuxi 214122, China
liaozuhua57@163.com

[2] Honor School, School of IoT Engineering, Jiangnan University, Wuxi 214122, China

[3] School of Science, School of IoT Engineering, Jiangnan University, Wuxi 214122, China

Abstract. In this paper, firstly, the concept of a new type of derivations on FI-algebras is introduced. The existence of it is verified by an example and a program. Then, the concepts of different kinds of derivations on FI-algebras are given. The properties of derivations on FI-algebras and the relationship between derivations and ideal are investigated. The equivalent conditions of identity derivation and the equivalent conditions of isotone derivation are proved. Finally, the concept of $a-$principal derivations on DFI-algebras is given. The existence of $a-$principal derivations is verified by an example and a program.

Keywords: Derivations · FI-algebra · Mapping · Ideal

1 Introduction

The notion of derivation came from the analytic theory. In the classical calculus, the derivation is defined as the linear part of the change. In the modern definition, the derivation is the mapping from the change of the independent variable to the linear part of the change. This mapping is also called tangent mapping. Because of the great success of derivation analysis theory, it is also important to introduce the derivation structure into other types of algebraic structures. In 1957, Posner E introduced the concept of derivation on the prime rings and used it to study the structure of prime rings. [15] Many scholars have studied the derivation and structure of the rings and the near-rings. [19] In 1975, scholars Szász introduced the concept of derivations on lattice. [17] Entering the new century, the concept of derivation has been introduced into many other algebraic structures, and Xin Xiaolong, a scholar in China, has studied the derivations in $BCI-$algebra, lattice and $\lambda-$lattice. [1,2,9] then different types of derivation on $MV-$algebra and $BL-$algebra are introduced by Xin, and he has done a lot of research on them. [3,13,14] In 2008, Çeven introduced the concept of f-derivations on lattice. [20] In 2010, Alshehri introduced derivations in $MV-$algebra,[16] and then Ghorbani, Motamed and other scholars introduced two new types of derivations in $MV-$algebra and studied their properties. [18] In 2012, Li Qingguo introduced the concept of derivations on Quantales theory, [12] and discussed the properties of derivations.

B.-Y. Cao and Y.-B. Zhong (Eds.): ICFIE 2017, AISC 872, pp. 68–80, 2019.
https://doi.org/10.1007/978-3-030-02777-3_7

In 1990, Chinese scholar Wu Ming introduced the concept of fuzzy implication algebra,[4] which used the implication connectives of logical value on $[0, 1]$ as an algebraic abstraction. Subsequently, a large number of scholars have discussed its properties, Li Zhiwei and other scholars have discussed the properties of FI-algebra, and given the concept of upper and lower bound on it. [5] In 2003, Chen Shilian gave the concept of ideal and quotient algebra of FI-algebra. [6] In 2008, Wu Hongbo, Dai Jianyun and other scholars gives the concept of DFI-algebras and investigate its properties, [7] then gives the concept of MT ideals of RFI-algebra. [8] In 2009, Liu Chunhui introduced the ideals lattices of RFI algebras. [10] Nowadays, great achievements have been made on the research of FI-algebras.

The structure of this paper is as follows: in section second, we will give the basic knowledge related to FI-algebraic system. In the third section, we give the definition of derivation on FI-algebra, and discuss the properties of the derivation on different types of FI-algebras. Then, we will give the equivalent conditions of the identity derivation on FI-algebra and the equivalent conditions of the isotone derivation on DFI-algebra, and the simple relation between the differential and the ideal. At last, The definition of $a-$principal derivation on FI-algebra is given.

2 Preliminaries

In this section, we will give a brief introduction of FI-algebra.

Definition 2.1 [4]. Let X be a universe set, and $0 \in X, \rightarrow$ be a binary operation on X, A (2,0)-type algebra $(X, \rightarrow, 0)$ is called a fuzzy implication algebra, shortly, FI-algebra, if the following five conditions hold for all $x, y, z \in X$:

(I_1) $x \rightarrow (y \rightarrow z) = y \rightarrow (x \rightarrow z)$
(I_2) $(x \rightarrow y) \rightarrow [(y \rightarrow z) \rightarrow (x \rightarrow z)] = 1$
(I_3) $(x \rightarrow x) = 1$
(I_4) if $x \rightarrow y = y \rightarrow x = 1$, then $x = y$
(I_5) $0 \rightarrow x = 1$

Where $1 = 0 \rightarrow 0$.

Definition 2.2 [4]. We defined a binary relation $\leq$ as follows:
$x \leq y$ if and only if $x \rightarrow y = 1$, for all $x, y \in X$.

Theorem 2.1 [4]. Let X be an FI-algebra. The relation "$\leq$" is a partial ordering on X, where 1 is the maximal element and 0 is the minimum element.

Theorem 2.2 [4,5]. Let X be an FI-algebra. The relation "$\leq$" is the partial ordering on X defined by Definition 1.2. Then the following conditions hold for all $x, y \in X$:

(I_6) $x \leq 1$, also $x \rightarrow 1 = 1$;
(I_7) $1 \rightarrow x = x$;
(I_8) if $x \leq y$, then $z \rightarrow x \leq z \rightarrow y$ and $y \rightarrow z \leq x \rightarrow z$;

(I_9) if $x \leq y$ and $y \leq z$, then $x \leq z$;
(I_{10}) $x \leq y \to x$;
(I_{11}) $x \leq y \to z$ if and only if $y \leq x \to z$;
(I_{12}) $(x \to y) \to y$ and $(y \to x) \to x$ are upper bounds of x and y.

Definition 2.3 [4]. Let X be an FI-algebra. We defined an operator C as follows:for all $x \in X$,

$$C(x) = x \to 0$$

And C is called the pseudo-complement operator on X.

Theorem 2.3 [4]. Let X be an FI-algebra and C be the pseudo-complement operator on X. Then the following conditions hold for all $x, y \in X$:

(C_1) $C(1) = 0$,$C(0) = 1$;
(C_2) $x \leq CC(x)$;
(C_3) if $x \leq y$, then $C(y) \leq C(x)$;
(C_4) $CCC(x) = C(x)$.

Definition 2.4 [4]. Let X be an FI-algebra and C be the pseudo-complement operator on X. Then X is regular, or, an RFI-algebra, if $CC(x) = x$ for all $x \in X$.

Theorem 2.4 [4]. Let X be an RFI-algebra, then $C(x) \to y = C(y) \to x$ for all $x, y \in X$

Definition 2.5 [4] Let X be an RFI-algebra. We defined two operators $T, \perp$ as follows: for all $x, y \in X$,

(*i*) $xTy = C(x \to C(y))$,
(*ii*) $x \perp y = C(x) \to y$.

Theorem 2.5 [4]. Let X be an RFI-algebra. Two operators $T, \perp$ are defined by Definition 2.5. Then the following conditions hold for all $x, y, z \in X$:

(T_1) $xTy = yTx, x \perp y = y \perp x$;
(T_2) $(xTy)Tz = xT(yTz), (x \perp y) \perp z = x \perp (y \perp z)$;
(T_3) if $x \leq y$, then $xTz \leq yTz, x \perp z \leq y \perp z$;
(T_4) $xT1 = x, x \perp 0 = x$;
(T_5) $x \perp y = C(C(x)TC(y))$;
(T_6) $x \to (y \to z) = (xTy) \to z$;
(T_7) $x \perp C(x) = 1$,$xTC(x) = 0$.

Definition 2.6 [7]. Let X be an RFI-algebra. X is said to be an DFI-algebra if one of the following conditions hold for all $x, y, z \in X$:

(*i*) $xT(y \perp z) = (xTy) \perp (xTz)$
(*ii*) $x \perp (yTz) = (x \perp y)T(x \perp z)$

Theorem 2.6 [7]. Let X be an DFI-algebra. Then $(X, T, \perp, 0, 1)$ is a Boolean algebra.

Theorem 2.7 [7]. Let X be an DFI-algebra. Then $x \perp y$ is the supremum of x and y, xTy is the infimum of x and y. That is $x \perp y = x \vee y, xTy = x \wedge y$.

Definition 2.7 [11]. Let X be an RFI-algebra. We defined an operator $\circ$ as follows: for all $x, y \in X$,

$$x \circ y = (x \to y) \to y$$

And X is said to be an CFI-algebra if the condition $x \circ y = y \circ x$ holds for all $x, y \in X$

Theorem 2.8 [11]. Let X be an RFI-algebra. Then:

(i) For all $x, y \in X$,$x \perp y$ is the upper bound of x and y,xTy is the lower bound of x and y;
(ii) $x \perp y$ is the supremum of x and y if and only if xTy is the infimum of x and y.

Theorem 2.9 [11]. Let X be an RFI-algebra. Then $(X, \leq)$ is a lattice and the following conditions hold for all $x, y, z \in X$:

(i) $x \vee y = (x \to y) \to y$,
(ii) $x \wedge y = C((C(x) \to C(y)) \to C(y))$.

Definition 2.8 [6]. Let X be an FI-algebra. Then A is an ideal of X if A is an nonempty subset of X and the following conditions hold for all $x, y \in X$:

(i) $1 \in A$, where $1 = (0 \to 0)$;
(ii) $x \in A, x \to y \in A \Rightarrow y \in A$.

Theorem 2.10 [6]. Let X be an FI-algebra and A be an ideal of X. If $x \in A$,$y \in X$ and $x \leq y$, then $y \in A$.

3 On Derivations of FI-Algebras

The following definition introduces the notion of derivations of FI-algebras.

Definition 3.1. Let X be an FI-algebra and d:$X \to X$ be a function. We call d a derivation on X, if the following condition hold for all $x, y, z \in X$:

$$d(x \to y) = (d(x) \to y) \perp (x \to d(y)).$$

Example 3.1. Let $X = \{0, a, 1\}$ be the FI-algebra of the following table (Table 1):

we define a function d on X by:

$$d(x) = \begin{cases} a, x = 0 \\ 1, x = a, 1 \end{cases} \quad (1)$$

Table 1. FI-algebra with order 3

$\rightarrow$	0	a	1
0	1	1	1
a	a	1	1
1	0	a	1

Then we can see that d is a derivation of X with the help of C language programming (In appendix 1).

As we give the definition of derivations of FI-algebras, then we research the properties of derivations on FI-algebras.

Theorem 3.1. Let X be an FI-algebra and d be a derivation of X. Then the following conditions hold for all $x, y \in X$:

(*i*) $d(1) = 1$;
(*ii*) $d(x) = x \bot d(x) = x \bot (d(C(x)) \rightarrow d(0))$;
(*iii*) $d(C(x)) = x \rightarrow d(C(x)) = x \rightarrow (d(x) \rightarrow d(0))$;
(*iv*) $x \leq d(x) \leq d(y \rightarrow x)$;
(*v*) $d(x) \rightarrow y \leq x \rightarrow d(y)$;
(*vi*) $d(x) \rightarrow d(y) \leq d(x \rightarrow y)$.

Proof.

(*i*). $d(1) = d(0 \rightarrow x) = (d(0) \rightarrow x) \bot (0 \rightarrow d(x)) = C(d(0) \rightarrow x) \rightarrow 1 = 1$.
(*ii*). $d(x) = d(1 \rightarrow x) = (d(1) \rightarrow x) \bot (1 \rightarrow d(x)) = C(1 \rightarrow x) \rightarrow (1 \rightarrow d(x)) = x \bot d(x)$. Also $d(x) = d(CC(x)) = d(C(x) \rightarrow 0) = (d(C(x)) \rightarrow 0) \bot (C(x) \rightarrow d(0)) = CC(d(C(x))) \rightarrow (C(x) \rightarrow d(0)) = d(C(x)) \rightarrow (C(x) \rightarrow d(0)) = C(x) \rightarrow (d(C(x)) \rightarrow d(0)) = x \bot (d(C(x)) \rightarrow d(0))$.
(*iii*). In (*ii*) let $x = C(x)$ then $d(C(x)) = C(x) \bot d(C(x)) = x \rightarrow d(C(x)) d(C(x)) = C(x) \bot d(C(C(x)) \rightarrow d(0)) = x \rightarrow (d(x) \rightarrow d(0))$.
(*iv*). By (*ii*): $x \rightarrow d(x) = x \rightarrow (x \bot d(x)) = x \rightarrow (d(x) \bot x) = x \rightarrow (C(d(x)) \rightarrow x) = C(d(x)) \rightarrow (x \rightarrow x) = C(d(x)) \rightarrow 1 = 1$ Then $x \leq d(x)$; Also $d(y \rightarrow x) = (d(y) \rightarrow x) \bot (y \rightarrow d(x)) \geq y \rightarrow d(x) \geq d(x)$ so $d(x) \leq d(y \rightarrow x)$.
(*v*). By (*iv*) and (I_8): $d(x) \rightarrow y \leq x \rightarrow y \leq x \rightarrow d(y)$.
(*vi*). $d(x \rightarrow y) = (d(x) \rightarrow y) \bot (x \rightarrow d(y)) \geq x \rightarrow d(y) \geq d(x) \rightarrow d(y)$.

Theorem 3.2. Let X be an FI-algebra and let A be a ideal of X, if $x \in A$, then $d(x) \in A$.

We can prove it by Theorem 3.1(*iv*) and Theorem 2.10.

Definition 3.2. Let X be an FI-algebra and d be a derivation of X. The following definitions are given:

(*i*) d is said to be an identical derivation of X, if $d(x) = x$ for all $x \in X$;
(*ii*) d is said to be the pseudo-complement derivation of X, if $d(x) = C(x)$ for all $x \in X$;

(iii) d is said to be a regular derivation of X, if $d(0) = 0$;
(iv) d is said to be a surjective derivation of X, if d is a surjective mapping;
(v) d is said to be an injective derivation of X, if d is a injective mapping;
(vi) d is said to be an isotone derivation of X, if $x \leq y \Rightarrow d(x) \leq d(y)$ for all $x, y \in X$;

Theorem 3.3. Let X be an FI-algebra and d be a regular derivation of X. Then the following conditions hold for all $x, y \in X$:

(i) $C(d(x)) \leq d(C(x))$;
(ii) $d(x \bot y) \leq d(x) \bot d(y)$;
(iii) $C(d(C(x))Td(C(y)) \leq d(xTy)$.

Proof.

(i). By applying Theorem 2.1(ii) : $d(C(x)) = x \to (d(x) \to d(0))$ and $d(0) = 0$. So $d(C(x)) = x \to (d(x) \to 0) = x \to C(d(x)) \geq C(d(x))$.
(ii). $d(x \bot y) = d(C(x) \to y) = (d(C(x)) \to y) \bot (C(x) \to d(y)) \leq (C(d(x)) \to y) \bot (C(x) \to d(y)) = (d(x) \bot y) \bot (x \bot d(y)) = d(x) \bot d(y)$.
(iii). $d(xTy) = d(C(C(x)TC(y))) \geq C(d(C(x) \bot C(y)))$. By applying $d(C(x) \bot C(y)) \leq d(C(x)) \bot d(C(y))$; We get $C(d(C(x) \bot C(y))) \geq C(d(C(x)) \bot d(C(y)))$. So $C(d(C(x))Td(C(y)) \leq d(xTy)$.

Theorem 3.4. Let X be an FI-algebra and d be a regular derivation of X. That is $d(0) = 0$. Then the following conditions are equivalent for all $x, y \in X$:

(i) d is an identical derivation;
(ii) $x \to d(y) = d(x) \to y$;
(iii) d is an injective derivation.

Proof.

$(i) \Rightarrow (ii)$,$(i) \Rightarrow (iii)$ are obviously.
$(ii) \Rightarrow (i)$. If $x \to d(y) = d(x) \to y$ for all $x, y \in X$: $d(x) = 1 \to d(x) = d(1) \to x = 1 \to x = x$, if d is an identical derivation.
$(iii) \Rightarrow (i)$. If d is an injective derivation, Then for all $x \in X$: $d(d(x) \to x) = (d(d(x)) \to x) \bot (d(x) \to d(x)) = C(d(d(x)) \to x) \to 1 = 1 = d(1)$. So $d(x) \to x = 1$ then $d(x) \leq x$. By applying Theorem 2.1(iv) we have:$\forall x \in X$,$d(x) = x$, So d is an identical derivation.

Theorem 3.5. Let X be an DFI-algebra and d be a derivation of X. So $x \bot y$ is the supremum of x and y, Then the following conditions hold for all $x, y \in X$:

(i) $d(x \to y) = (d(x) \to y) \vee (x \to d(y))$;
(ii) $d(x \to y) = x \to d(y)$;
(iii) $d(x \bot y) = x \bot d(y) = y \bot d(x)$;
(iv) if $d(0) = 0$, then $d(x) = x$.

Proof.

(i) is Obviously.
(ii). As $d(x \to y) = (d(x) \to y) \vee (x \to d(y)) \leq (x \to d(y)) \vee (x \to d(y)) = x \to d(y)$ and $d(x \to y) = (d(x) \to y) \vee (x \to d(y)) \geq x \to d(y)$, Then we have $x \to d(y) \leq d(x \to y) \leq x \to d(y)$. So $d(x \to y) = x \to d(y)$.
(iii). By applying (ii): $d(x \bot y) = d(C(x) \to y) = C(x) \to d(y) = x \bot d(y)$, Then we have: $d(x \bot y) = d(y \bot x) = y \bot d(x)$.
(iv). By applying (ii): $d(x) = d(CC(x)) = d(C(x) \to 0) = C(x) \to d(0) = C(x) \to 0 = CC(x) = x$.

Theorem 3.6. Let X be an DFI-algebra and d be a derivation of X. Then the following conditions are equivalent for all $x, y \in X$:

(i) d is an isotone derivation;
(ii) $d(0) \leq d(x)$;
(iii) $d(x) = d(0) \bot x$;
(iv) $d(x \vee y) = d(x) \vee y = x \vee d(y)$;
(v) $d(x \wedge y) = d(x) \wedge d(y)$;
(vi) $d(x \to y) = d(C(x)) \bot y = d(y) \bot C(x)$.

Proof.

$(i) \Rightarrow (ii)$. Note that $0 \leq x$, So it is straightforward.
$(ii) \Rightarrow (iii)$. Note that $x \leq d(x)$. So $d(0) \bot x \leq d(0) d(0) \vee d(x) = d(x)$and $d(x) = d(C(x)) \to (x \bot d(0)) \leq C(x) \to (x \bot d(0)) = (x \bot x) \bot d(0) = x \bot d(0)$. We have $d(0) \bot x \leq d(x) \leq d(0) \bot x$ then $d(x) = d(0) \bot x$.
$(iii) \Rightarrow (iv)$. $d(x \vee y) = d(x \bot y) = d(0) \bot (x \bot y) = d(0) \bot x \bot y = d(x) \bot y = x \bot d(y)$.
$(iii) \Rightarrow (v)$. $d(x \wedge y) = d(xTy) = d(0) \bot (xTy) = (d(0) \bot x) T (d(0) \bot y) = d(x) T d(y) = d(x) \wedge d(y)$.
$(iv) \Rightarrow (i)$. Assume $x \leq y$, then $d(y) = d(x \vee y) = d(x) \vee y \geq d(x)$. So d is an isotone derivation.
$(v) \Rightarrow (i)$. Assume $x \leq y$, then $d(x) = d(x \wedge y) = d(x) \wedge d(y) \leq d(y)$, So d is an isotone derivation.
$(iv) \Leftrightarrow (vi)$. In (iv) let $x = C(x)$then we get (vi), also in (vi) let $x = C(x)$then we get (iv)

Theorem 3.7. Let X be an CFI-algebra and d be a derivation of X. Then we have $d(x) = (d(x) \to x) \to x$.

Proof. Since $(x \to d(x)) \to d(x) = 1 \to d(x) = d(x)$ and X is an CFI-algebra. So we have $(x \to d(x)) \to d(x) = (d(x) \to x) \to x$. Then $d(x) = (d(x) \to x) \to x$.

Theorem 3.8. Let X be an DFI-algebra, $a \in X$, and define a mapping $d_a : X \to X$, for all $x \in X$, $d_a(x) = a \bot x$, then d_a is a derivation of X.

Proof.
For all $x, y \in X$: $(d_a(x) \rightarrow y) \perp (x \rightarrow d_a(y)) = (C(d_a(x)) \perp y) \perp (C(x) \perp d_a(y)) = (C(d_a(x)) \perp C(x)) \perp (d_a(y) \perp y) = (C(d_a(x)) \vee C(x)) \perp (d_a(y) \vee y)$; By (I_{10}), we know that $x \leq C(a) \rightarrow x$, then $x \leq d_a(x)$, so $C(x) \geq C(d_a(x))$. Then we have $(d_a(x) \rightarrow y) \perp (x \rightarrow d_a(y)) = C(x) \perp d_a(y) = C(x) \perp (a \perp y) = a \perp (C(x) \perp y) = a \perp (x \rightarrow y) = d_a(x \rightarrow y)$.

By Definition 3.1:d_a is a derivation of X.

Definition 3.3. Let X be an DFI-algebra, $a \in X$, and define a mapping $d_a : X \rightarrow X$, for all $x \in X$, $d_a(x) = a \perp x$, then d_a is said to be a $a-$principal derivation of X

Example 3.2. With the help of C language programming, we can find out every DFI-algebra with order 4 (In Appendix 2). We select one of the DFI-algebras (Table 2):

Table 2. DFI-algebra with order 4

$\rightarrow$	0	a	b	1
0	1	1	1	1
a	b	1	b	1
b	a	a	1	1
1	0	a	b	1

Then we define a mapping d_a on X by $d_a(x) = a \perp x$, so we have:

$$d_a(x) = \begin{cases} a, x = 0, a \\ 1, x = b, 1 \end{cases} \tag{2}$$

With the help of C language programming, we can see that d_a is a derivation of X (In Appendix 3).

4 Conclusion

In this paper, combining the notion of FI-algebra and derivation, we introduced the notion of derivation on FI-algebras. And we research the properties of derivations on FI-algebras.

In future, we can keep on introducing more notions such as f- derivation and generalized derivation on FI-algebras. Based on these new notions, we will discuss more properties on FI-algebras.

Acknowledgements. Thanks to the support by Natural Science Foundation of China (Grant Nos. 61170121; 61673193; 11401259), Natural Science Foundation of Jiangsu Province (Grant No. BK20151117) and the Undergraduate Innovation and Entrepreneurship Training Program of China (Grant No. 201610295005).

A Appendix

Appendix 1. The C language program of Example 3.1:

```
#include<iostream>
#define N 3
using namespace std;

void pri(int m[N][N])
{
    int i,j,im,jm;
    for(i=0;i<N;i++)
    {
        im=0;
        if(i>=1&&i<N-1)
            im=i+1;
        if(i==N-1)
            im=1;
        for(j=0;j<N;j++)
        {
            jm=0;
            if(j>=1&&j<N-1)
                jm=j+1;
            if(j==N-1)
                jm=1;
            cout<<m[im][jm]<<" ";
        }
        cout<<endl;
    }
}

int main()
{

    int d1[N]={2,1,1},d2[N][N],a[3][3]={{1,1,1},{0,1,2},{2,1,1}},b[3][3],c[N],i,j,im,
jm;
    for(i=0;i<N;i++)
        c[i]=a[i][0];
    for(i=0;i<N;i++)
        for(j=0;j<N;j++)
            d2[i][j]=a[c[a[d1[i]][j]]][a[i][d1[j]]];
    pri(d2);
    cout<<endl;
    for(i=0;i<N;i++)
        for(j=0;j<N;j++)
            b[i][j]=d1[a[i][j]];
    pri(b);
    return 0;
}
```

Appendix 2. Table 3 is one of the DFI-algebras with order 4:

And the C language program to find it:

```
#include <stdio.h>
#include <math.h>
#define N 4
void pri(int m[N][N],int k)
{
    int i,j,im,jm;
    printf("
    for(i=0;i<N;i++)
    {
        if(i!=0) printf(" ");
        im=0;
        if(i>=1&&i<N-1)
            im=i+1;
        if(i==N-1)
            im=1;
        for(j=0;j<N;j++)
        {
            jm=0;
            if(j>=1&&j<N-1)
                jm=j+1;
            if(j==N-1)
                jm=1;
            printf("
        }
        printf("");
    }
}

int lo(int x,int y,int a[N][N])
{
    return a[a[x][a[y][0]]][0];
}

int ol(int x,int y,int a[N][N])
{
    return a[a[x][0]][y];
}

int DFI(int a[N][N])
{
    int x,y,z,k1=0,k2=1,k3=0,c[N];
    for(x=0;x<N;x++)
        c[x]=a[x][0];
    for(x=0;x<N;x++)
    {
```

```
            if(a[x][x]==1&&a[0][x]==1) k1++;
            else
                break;
            for(y=0;y<N;y++)
            {
                if(a[x][y]==a[y][x])
                {
                    if(x!=y) k2=0;
                }
                for(z=0;z<N;z++)
                {
                    if(a[a[x][y]][a[a[y][z]][a[x][z]]]==1&&a[x][a[y][z]]==a[y][a[x][z]]&&
ol(lo(y,z,a),x,a)==lo(ol(y,x,a),ol(z,x,a),a))
                        k3++;
                    else
                        break;
                }
            }
        }
        if(k1==N&&k2==1&&k3==N*N*N)
            return 1;
        else
            return 0;
}

int main()
{
        int a[N][N],d[N],i,j,k=1;
        for(i=0;i<N;i++)
        {
            a[i][1]=1;
            a[1][i]=i;
            a[i][i]=1;
            a[0][i]=1;
        }
        for(a[2][0]=0;a[2][0]<N;a[2][0]++)
        for(a[3][0]=0;a[3][0]<N;a[3][0]++)
        for(a[2][3]=0;a[2][3]<N;a[2][3]++)
        for(a[3][2]=0;a[3][2]<N;a[3][2]++)
            {
                if(DFI(a)==1)
                {
                    pri(a,k);
                    k++;
                    printf("");
                }
            }
        return 0;
}
```

Table 3. DFI-algebra with order 4

$\rightarrow$	0	a	b	1
0	1	1	1	1
a	b	1	b	1
b	a	a	1	1
1	0	a	b	1

Appendix 3. The C language program of Example 3.2:

```
#include<iostream>
#define N 4
using namespace std;
void pri(int m[N][N]) {
    int i,j,im,jm;
    for(i=0;i<N;i++)
    {
        im=0;
        if(i>=1&&i<N-1)
            im=i+1;
        if(i==N-1)
            im=1;
        for(j=0;j<N;j++)
        {
            jm=0;
            if(j>=1&&j<N-1)
                jm=j+1;
            if(j==N-1)
                jm=1;
            cout<<m[im][jm]<<" ";
        }
        cout<<endl;
    }
}

int main()
{
    int d1[4]={2,1,2,1},d2[4][4],a[4][4]={{1,1,1,1},{0,1,2,3},{3,1,1,3},{2,1,2,1}},b[4]
,c[4],i,j,im,jm;
    for(i=0;i<N;i++)
        c[i]=a[i][0];
    for(i=0;i<N;i++)
        d1[i]=a[c[2]][i];
    for(i=0;i<N;i++)
        for(j=0;j<N;j++)
            d2[i][j]=a[c[a[d1[i]][j]]][a[i][d1[j]]];
```

```
        pri(d2);
        cout<<endl;
        for(i=0;i<N;i++)
            for(j=0;j<4;j++)
                b[i][j]=d1[a[i][j]];
        pri(b);
        return 0;
}
```

References

1. Jun, Y.B., Xin, X.L.: On derivations of BCI-algebras. Inf. Sci. **159**, 167–176 (2004)
2. Xin, X.L., Li, T.Y., Lu, J.H.: On derivations of lattices. Inf. Sci. **178**(2), 307–316 (2008)
3. Wang, J.T., Xin, X.L., He, P.F.: On($\rightarrow$,$\oplus$)-derivations of MV-algebras. J. Shaanxi Norm. Univ. (Nat. Sci. Ed.) **43**(4), 16–21 (2015)
4. Wu, W.M.: Fuzzy Implication algebras. Fuzzy Syst. Math. **4**(1), 56–64 (1990)
5. Li, Z.W., Li, G.H.: Some properties of fuzzy implication algebras. Fuzzy Syst. Math. **14**, 19–21 (2000)
6. Chen, S.L.: Ideal and quotient algebra of FI-algebra. J. Kunming Univ. Sci. Technol. (Sci. Technol.) **28**(5), 155–158 (2003)
7. Dai, J.Y., Wu, H.B.: Distributive fuzzy implication algebra. Fuzzy Syst. Math. **22**(1), 26–32 (2008)
8. Wu, H.B., Wang, N.: MT ideals of regular FI algebras with their applications. Acta Electron. Sin. **41**(7), 1389–1394 (2013)
9. Li, Y.J., Xin, X.L.: Derivation of λ-lattice. Comput. Eng. Appl. **46**(6), 46–47 (2010)
10. Liu, C.H.: Ideals lattices of regular fuzzy implication algebras. J. Inn. Mong. Norm. Univ. (Nat. Sci. Ed.) **38**(1), 5–9 (2009)
11. Li, Z.W., Sun, L.M., Zheng, C.Y.: Regular fuzzy implication algebras. Fuzzy Syst. Math. **16**(2), 23–27 (2002)
12. Xiao, Q.M., Li, Q.G.: Derivation of quantales. J. Hunan Univ. (Nat. Sci.) **39**(8), 87–89 (2012)
13. Xin, X.L., Zou, X.Y.: On generalized derivations of MV-algebras. Fuzzy Syst. Math. **29**(2), 18–23 (2015)
14. Xin, X.L., Feng, M., Yang, Y.W.: On $\odot$-derivations of BL-algebras. J. Math. **36**(3), 552–558 (2016)
15. Porner, E.: Derivations in prime rings. Proceeding Am. Math. Soc. **8**, 1093–1100 (1957)
16. Alshehri N.O.: Derivations of MV-algebras. Int. J. Math. Math. Sci. (2010). https://doi.org/10.1155/2010/312027
17. Szász, G.: Derivations of lattices. Acta Sci. Math. (Szeged) **37**, 149–154 (1957)
18. Ghorbani, S.H., Torkzadeh, L., Motamed, S.: $(\odot, \oplus)$-derivations and $(\ominus, \odot)$-derivations on MV-algebras. Iran. J. Math. Sci. Inform. **8**, 75–90 (2013)
19. Bell, H.E., Gordon, M.: On derivations in near-rings. In: betsch, G. (eds.) Near-Rings and Near-Fields, North-Holland, Amsterdam, pp. 31–35 (1987)
20. Çeven, Y., Öztürk, M.A.: On f-derivations of lattices. Bull. Korean Math. Soc. **45**(4), 701–707 (2008)

A New Type Soft Prime Ideal of KU-algebras

Yue Xi[1,2], Zu-hua Liao[1,2,3(✉)], Xiao-hao Wang[1], Xin-meng Chen[1], Wei Song[2,4], Shu-zhong Wu[1], and Yong Li[1]

[1] School of Science, Jiangnan University, Wuxi 214000, China
liaozuhua57@163.com
[2] Honors School, Jiangnan University, Wuxi 214000, China
[3] Institute of Intelligence System and Network Computing, Jiangnan University, Wuxi 214000, China
[4] School of Internet of Things Engineering, Jiangnan University, Wuxi 214000, China

Abstract. Through combining the soft set with KU-algebras, this paper introduces the concept of a new type soft prime ideal of KU-algebras and investigates its properties. Firstly, we give the equivalent descriptions of the new type prime soft ideal of KU-algebras. Then, we show the differences between the new type soft prime ideal of KU-algebras and the common soft prime ideal of KU-algebras by giving examples. After then, studies about the equivalent description of the new type soft prime ideal and the new type soft ideal prove that the algebraic structure of dual soft set is different from the algebraic structure of α-level set. Besides, we define the new concept of the projection of AND operation of soft set, and the obtain that the projection of AND operation of two soft set is also the new type soft prime ideal, if the AND operation is a new type soft prime ideal of KU-algebras. Finally, we explore the properties of the new type soft prime ideal of KU-algebras about the image and inverse image.

Keywords: KU-algebra · Soft set · New type soft prime ideal
Dual soft set · α-level set

1 Introduction

Uncertainty is a main characteristic of information. In 1999, Molodtsov a Russian scholar, proposed the concept of soft set [1], which provides a new mathematical framework for uncertainty problem in term of the parameterization. His research has widely drawn attention of domestic and foreign scholars, as this theory fills the gap of dealing with uncertain problems with the theory of probability, fuzzy set [2], and rough set [3]. In 2003, Maji et al. [4] introduced several basic operations of soft set. In 2007, Aktas et al. [5] announced the definition of soft group and investigated the properties of algebra. Then, in 2009, Ali et al. [6] proposed some new calculations for soft sets.

B.-Y. Cao and Y.-B. Zhong (Eds.): ICFIE 2017, AISC 872, pp. 81–90, 2019.
https://doi.org/10.1007/978-3-030-02777-3_8

KU-algebra [7] is the new logical algebra system introduced by Chanwit et al. in 2009. It is the expansion of some algebraic systems, such as BCK/BCI and BCC. In the same year, they [8] also provided and investigated the homomorphism of KU-algebras. In 2011, Mostafa et al. provided the concepts as well as investigated the relevant properties of fuzzy ideals [9], intuitionistic fuzzy ideals [10] and interval-valued fuzzy ideals [11] of KU-algebras. In 2013 and 2014, Muhammad et al. proposed (α, β)- fuzzy ideals [12] and bipolar fuzzy subalgebras/ideals [13] of KU-algebras.

In 2008, Xue-hai Yuan [14] supervised his student Yong-chuan Wen introducing the set of parameters to the algebraic structure of group and resulting in some relevant characteristics. Such results are more profound compared to the results of soft algebra in general context. Zhu-hua Liao's team also did a series of studies based on this methodology [15–21]. However, there is few researches of combining KU-algebras and soft set, we can only see that Muhammad et al. [22] proposed and studied the common soft subalgebra. Through this way, In [23], we introduced the parameter set to the structure of KU-algebra and provided a new type soft ideal of KU-algebra as well as discussed its basic properties. In this paper, we use this method to give the concept of a new type soft prime ideal of KU-algebras and investigate its series of basic properties. Example 4.1 displays the differences between the new type soft prime ideal and common soft prime ideal. Besides, the differences between dual soft set and α-level set of soft set are also showed through Theorems 4.2 and 4.3. Thus, it implies that the algebras of L-fuzzy set [24] and new type soft set are not the same. Generally, there is no element in lattice algebraic structure, instead, the element of lattice $P(U)$ has algebraic structure. Thus, the results of new type soft set algebras is more significant than the results of L-fuzzy algebras.

The rest of the paper is organized as follows. Section 2 introduces the fundamental knowledge of KU-algebras and soft set. Section 3 provides the concepts of prime ideal of KU-algebras and proves its basic properties. Section 4 introduces the parameters to the structure of KU-algebras and provides the concepts as well as basic properties of the new type soft prime ideal of KU-algebras. Apart from this, Sect. 3 also proves that α-level set is the sufficient conditions of the new type soft prime ideal of KU-algebras and give the equivalent characterization of the new type soft prime ideal. Meanwhile, we also provide the new concept of the projection of AND operation of soft set, and get the conclusion that the projection of AND operation of two soft set is also the new type soft prime ideal, if the AND operation is a new type soft prime ideal of KU-algebra. Section 5 discusses the characteristics between image and inverse image of the new type soft prime ideal of KU-algebras.

2 Preliminaries

In this section, some elementary aspects about KU-algebras, ideal of KU-algebras and soft set are introduced.

Definition 2.1 [7]. (KU-algebras) A $(2,0)$ KU-algebra $X = (X, \cdot, 0)$ is called a KU-algebra if for $\forall\ x, y, z \in X$, it satisfies the following axioms:

(1) $(xy)((yz)(xz)) = 0$;
(2) $0x = x$;
(3) $x0 = 0$;
(4) $xy = 0$ and $yx = 0$ implies $x = y$.

Definition 2.2 [7]. Let $(X, \cdot, 0)$ be a KU-algebra. On the KU-algebra X, we define a binary relation $\leq$:$x \leq y$ only if $yx = 0$, and $x, y \in X$, $\leq$ is called the relation leaded by "$\cdot$".

Lemma 2.1 Let $(X, \cdot, 0)$ be a KU-algebra, $\leq$ is the relation leaded by "$\cdot$". Then, it is easy to prove that $(X, \cdot, 0)$ is a partially ordered set and 0 is its smallest element.

Definition 2.3 [7]. Let $(X, \cdot, 0)$ be a KU-algebra. A non-empty subset A of KU-algebras is said to be an implicative ideal of A, if for $\forall x, y \in A$, we have $xy \in A$. A is an implicative ideal of X, obviously $0 \in A$.

Definition 2.4 (Cartesian product). Let A, B be two non-empty subsets, $A \times B = \{(x, y)|x \in A, y \in B\}$ is called the Cartesian product of A, B.

Definition 2.5 [7]. Let X_1 and X_2 be KU-algebras, a homomorphism is a map of $f : X_1 \rightarrow X_2$, satisfying $f(xy) = f(x)f(y)$ for $\forall x, y \in X_1$. An injective homomorphism is called a monomorphism, and a surjective homomorphism is called an eqimorphism.

Under Definition 2.5, it is easy to know $f(0) = 0$.

Definition 2.6 [7]. Let $(X, \cdot, 0)$ be a KU-algebra, $A \subseteq X$ is called as ideal if it satisfies the following axioms:

(1) $0 \in A$;
(2) $\forall x, y, z \in X$, $x(yz) \in A$ and $y \in A$ imply $xz \in A$.

Definition 2.7 [1]. (Soft set) Order U be a initial complete set and E be a set of parameters, $P(U)$ is a power set of U, $A \subset E.(F, A)$ is called a soft set of U or a soft set of A, satisfying a map $F : A \rightarrow P(U)$.

Definition 2.8 [4]. Let (F, A) and (G, B) be two soft sets in U. (F, A) is called a soft subset of (G, B), recorded as $(F, A)\tilde{\subseteq}(G, B)$, if it satisfies the following axioms:

(1) $A \subset B$;
(2) $\forall x \in A, F(x) = G(x)$.

Definition 2.9 [6]. (Restricted intersection of soft sets) Let (F, A) and (G, B) be two soft sets in U. (H, C) is called as a restricted intersection of soft set (F, A) and soft set (G, B), and recorded as $(F, A)\cap_R(G, B) = (H, C)$, if (H, C) satisfies the following axioms:

(1) $C = A \cap B$;
(2) $H(x) = F(x) \cap G(x)$, for $\forall x \in C$.

Definition 2.10 [6]. (Extended intersection of soft sets) Let (F, E_1) and (G,E_2) be two soft sets in U. (H, E) is called as a extended intersection of soft set (F, E_1) and soft set (G, E_2), recorded as $(H, E) = (F, E_1)\cap_E(G, E_2)$, if (H, E) satisfies the following axioms:

(1) $E = E_1 \cup E_2$;

(2) For $\forall g \in E$, $H(g) = \begin{cases} F(g), g \in E_1 - E_2 \\ G(g), g \in E_2 - E_1 \\ F(g) \cap G(g), g \in E_1 \cap E_2 \end{cases}$.

Definition 2.11 [4]. Let (F, A) and (G, B) be two soft sets in U. Order $(F, A) \wedge (G, B) = (H, C)$,$C = A \times B$, in which $H(x, y) = F(x) \cap G(y)$, then we can call $(H, A \times B)$ is a soft set and operation of (F, A) and (G, B), recorded as $(F, A)\wedge(G, B)$.

Definition 2.12 [14] (The duality of soft sets). Let (H, E) be a soft set in X.(A_H, X) is called a duality of (H, E), if for $\forall x \in X$,(A_H, X) satisfies $A_H : X \to P(E)$,$x \mapsto A_H(x) = \{g\in E|x \in H(g)\}$.

If $A : X \to P(E)$ is a soft set, implying $H_A : E \to g \mapsto H(g) = \{x|g \in A(x)\}$ is called a duality of A.

Definition 2.13 [16] (The image and original image of soft set). Let X_1 and X_2 be KU-algebras, U be a initial complete set, $P(U)$ is a power set of U, a map $f : X_1 \to X_2$,$H_1 : X_1 \to P(U)$ and $H_2 : X_2 \to P(U)$ are two soft sets. For $\forall x_1 \in X_1$,$x_2 \in X_2$, we define:

$$f(H_1)(x_2) = \begin{cases} \bigcup\limits_{f(x)=x_2} H_1(x) & f^{-1}(x_2) \neq \emptyset \\ \emptyset & f^{-1}(x_2) = \emptyset \end{cases} \quad \text{and } f^{-1}(H_2)(x_1) = H_2(f(x_1)),$$

implying $f(H_1)$ is a soft set of X_1 and $f^{-1}(H_2)$ is a soft set of X_2, then $f(H_1)$ is called as a image of H_1 and $f^{-1}(H_2)$ is called as a original image of H_2.

Theorem 2.1 [9]. Let X_1 and X_2 be KU-algebras, then we define "$\cdot$" on $X_1 \times X_2$, for $\forall(x_1, x_2)(y_1, y_2) \in X_1 \times X_2$,$(x_1, x_2)(y_1, y_2) = (x_1y_1, x_2y_2)$, then $(X_1 \times X_2, \cdot, (0, 0))X_2, \cdot, (0, 0))$ also is a KU-algebra.

Theorem 2.2. Let I be a non-empty subset and $\{A_i\}_{i\in I}$ is a subalgebra of KU-algebra X, and thus $\cap_{i\in I}A_i$ is also subalgebra of X.

Definition 2.14 [18]. Let X_1 and X_2 be sets, map $f : X_1 \to X_2$ is a map from X_1 to X_2.H is a soft set in X_1, $H(x) = H(y)$ if $H(x) = H(y)$ for $\forall x, y \in X_1$, then H is called as $f-$invariant.

Definition 2.15 [23]. Let $(X, \cdot, 0)$ be a KU-algebra, U be a initial complete set and $H : X \to P(U)$ is a soft set. H is called as a new soft ideal of X, if for $\forall x, y, z \in X$ it satisfies the following axioms:

(1) $H(0) \supseteq H(x)$;
(2) $H(xz) \supseteq H(x(yz)) \cap H(y)$.

3 Prime Ideal of KU-algebras

In this section, some concepts about prime ideal and some of its basic properties are introduced.

Definition 3.1. Let $X, \cdot, 0)$ be a KU-algebra. $y \in A$ or $xz \in A$ if $x(yz) \in A$, for $\forall x, y, z \in X$, then A is called as prime ideal of X.

The following example shows the existence of the prime ideal of a KU-algebra and it is different from the ideal of a KU-algebra.

Example 3.1. Let $U = X = \{0, 1, 2, 3, 4\}$, we define "$\cdot$" on X as following:

$\cdot$	0	1	2	3	4
0	0	1	2	3	4
1	0	0	2	2	0
2	0	1	0	3	4
3	0	4	0	0	0
4	0	0	2	0	0

By the definition of the prime ideal, we can know $\{1\}$, $\{3\}$, $\{1, 3\}$, $\{2, 3\}$, $\{1, 2, 3\}$, $\{1, 2, 4\}$, $\{1, 3, 4\}$, $\{2, 3, 4\}$, $\{1, 2, 3, 4\}$ are the prime ideals of KU-algebras but not ideals of KU-algebras X.

Theorem 3.1. Let I be a non-empty subset and $\{A_i\}_{i \in I}$ is a cluster of algebra of KU-algebras X, then, $\cup_{i \in I} A_i$ is a prime ideal of X.

Proof. If $(yz) \in \cup_{i \in I} A_i$, then $x(yz) \in A_i$ for $\exists i \in I$. Since A_i is a prime ideal of X, then $y \in A_i$ or $xz \in A_i$. This implies $y \in \cup_{i \in I} A_i$ or $xz \in \cup_{i \in I} A_i$.

Therefore, the prime ideal of KU-algebras and ideal of KU-algebras are two different types of algebraic structures.

4 A New Soft Prime Ideal of KU-algebras

This section gives the definition of the new soft prime ideal of KU-algebras, then illustrates the existence of new soft prime ideal of KU-algebras, and gives its equivalent characterizations and related properties.

Definition 4.1. Let E be a set of parameters, X be a KU-algebra and $H : X \to P(U)$ is a soft set, then H is called as a soft prime ideal of E.

Definition 4.2. Let $(X, \cdot, 0)$ be a KU-algebra, U be a initial complete set and $H : X \to P(U)$ is a soft set. We can call H is a new soft prime ideal of KU-algebras, for $\forall x, y, z \in X$ if it satisfies $H(x(yz)) \subseteq H(xz) \bigcup H(y)$.

In the case of no confusion, hereinafter referred to as the soft prime ideal.

The following example shows the existence of the new soft prime ideal and its difference from the common soft prime ideal.

Example 4.1. Exactly as Example 3.1, $U = X = \{0,1,2,3,4\}$ is a KU-algebra, $\{1\}$, $\{3\}$, $\{1,3\}$, $\{2,3\}$, $\{1,2,3\}$, $\{1,2,4\}$, $\{1,3,4\}$, $\{2,3,4\}$, $\{1,2,3,4\}$, $\{0,1,2,3,4\}$ are prime ideals of KU-algebra X.

Let $H : X \to P(U)$, $H(0) = \{0\}$, $H(1) = \{0,2,3,4\}$, $H(2) = \{0,1,2\}$, $H(3) = \{0,1,3,4\}$, $H(4) = \{0,1,2,4\}$, then from definition, we know H is a new soft prime ideal of X. And $H(0) = \{0\}$, $H(1) = \{0,2,3,4\}$, $H(2) = \{0,1,2\}$, $H(3) = \{0,1,3,4\}$, $H(4) = \{0,1,2,4\}$ are not prime ideal of X, therefore, H is not a prime ideal in the usual sense. This implies the new soft prime ideal of KU-algebras is a novel algebraic structure that differs from common soft prime ideal.

Theorem 4.1. *Let X be a KU algebra, $H : X \to P(U)$ be a soft set, $\forall \alpha \in P(U)$ and $H_\alpha \neq \emptyset$ are the ideas of X is a sufficient and necessary condition for H is a new soft prime ideal of X.*

Proof. (1) Necessity: (i) for $\forall \alpha \in P(U)$, if $H_\alpha \neq \emptyset$, then $\exists x \in H_\alpha$ for $\alpha \subseteq H(x)$, H is a new soft ideal, therefore $H(0) \supseteq H(x) \supseteq \alpha$, this implies $0 \in H_\alpha$.

(ii) For $\forall x, y, z \in X$, if $x(yz) \in H_\alpha$ and $y \in H_\alpha$, then $\alpha \subseteq H(x(yz))$ and $\alpha \subseteq H(y)$, therefore $\alpha \subseteq H(x(yz)) \bigcap H(y) \subseteq H(xz)$, this implies $xz \in H_\alpha$.

From (i) and (ii), we know H_α is a ideal of X.

(2) Sufficiency: (i) for $\forall x \in X$, let $\alpha = H(x)$, then $x \in H_\alpha$, therefore, $H_\alpha \neq \emptyset$. By knowH_α is a ideal of X, therefore $0 \in H_\alpha$, this implies $H(0) \supseteq \alpha = H(x)$.

(ii) For $\forall x, y, z \in X$, assume $\alpha = H(x(yz)) \bigcap H(y)$, therefore $x(yz) \in H_\alpha$ and $y \in H_\alpha$, besides H_α is a ideal of X, therefore $xz \in H_\alpha$, so $H(xz) \supseteq \alpha = H(x(yz)) \bigcap H(y)$.

From (i) and (ii) we know H_α is a new soft ideal of X.

Definition 4.3. Let be a KU algebra, U be a initial complete set, $H : X \to P(U)$ be a soft set, $H_\alpha = \{x | \alpha \subseteq H(x), \alpha \subseteq P(U)\}$ is called as $\alpha-$level set of H.

Theorem 4.2. *Let X be a KU algebra, $H : X \to P(U)$ be a soft set, if for $\forall \alpha \in P(U)$,$H_\alpha \neq \emptyset$ is a prime ideal of X, then H is a new soft prime ideal of X.*

Proof. For $\forall x, y, z \in X$,if$H(x(yz)) = \emptyset$, then $H(x(yz)) \subseteq H(xz) \bigcup H(y)$; if $H(x(yz)) \neq \emptyset$, let $H(x(yz)) = \alpha$, therefore $x(yz) \in H_\alpha$. Since H is a new soft prime ideal of X, then $xz \in H_\alpha$ or $y \in H_\alpha$, therefore $H(xz) \supseteq \alpha$ or $H(y) \supseteq \alpha$, then $H(xz) \bigcup H(y) \supseteq \alpha = H(x(yz))$. This implies that H is a new soft prime ideal of X.

The inverse proposition of this theorem is not tenable.

Example 4.2. U, X and H are the same as Example 4.1, H is a new soft prime ideal of KU-algebra, assume $\alpha = \{0,1,2\}$, therefore $H_\alpha = \{x | \alpha \subseteq H(x)\} = \{2,4\}$ is not a prime ideal of KU-algebra X.

Therefore by Theorems 4.1 and 4.2, we know the properties of new soft prime ideal and new soft ideal are very different.

Theorem 4.3. Let $(X, \cdot, 0)$ be a KU-algebra, $H : X \to P(U)$ be a soft set, then:

(1) For $\forall u \in U, A_H(u) \neq \emptyset$ is a prime idea of X is a sufficient and necessary condition for H is a new soft prime ideal of X.

(2) If $A : U \to P(X)$ is a soft set, then for $\forall u \in U$, $A(u) \neq \emptyset$ is a prime idea of X is a sufficient and necessary condition for H_A is a new soft prime ideal of X.

Proof. (1) Sufficiency: For $\forall x, y, z \in X$, if $H(xz) \bigcup H(y) = \emptyset$, suppose $H(x(yz)) \neq \emptyset$, then $\exists u \in H(x(yz))$, this implies $x(yz) \in A_H(u)$, since $A_H(u)$ is a prime ideal of X, $xz \in A_H(u)$ or $y \in A_H(u)$, then $u \in H(xz)$ or $u \in H(y)$, this implies $\exists u \in H(xz) \bigcup H(y)$, then $H(xz) \bigcup H(y) \neq \emptyset$contradicts the hypothesis, therefore $H(x(yz)) = \emptyset$. So $H(x(yz)) = H(xz) \bigcup H(y)$. Obviously, $H(x(yz)) \subseteq H(xz) \bigcup H(y)$. If $H(xz) \bigcup H(y) \neq \emptyset$, when $H(x(yz)) = \emptyset$, $H(x(yz)) \subseteq H(xz) \bigcup H(y)$; when $H(x(yz)) \neq \emptyset$, for $\forall u \in H(x(yz))$,$x(yz) \in A_H(u)$, since $A_H(u)$ is a prime ideal of X, then $xz \in A_H(u)$or $y \in A_H(u)$, this implies$u \in H(xz)$or$u \in H(y)$, therefore $u \in H(xz) \bigcup H(y)$, $H(x(yz)) \subseteq H(xz) \bigcup H(y)$. Therefore H is a new soft prime ideal of X.

Necessity: For $\forall u \in U$, if $A_H(u) \neq \emptyset$, then $\forall x, y, z \in X$, as $x(yz) \in A_H(u)$, this implies $u \in H(x(yz))$. Since H is a new soft prime ideal of X, $H(x(yz)) \subseteq H(xz) \bigcup H(y)$, therefore, $u \in H(xz) \bigcup H(y)$, this implies $u \in H(xz)$ or $u \in H(y)$. Thus, $xz \in A_H(u)$ or $y \in A_H(u)$. Therefore $A_H(u)$ is a prime ideal of X.

(2) Sufficiency: For $\forall u \in U$, when $A(u) \neq \emptyset$, $\forall x, y, z \in X$, if $x(yz) \in A(u)$, then $u \in H_A(x(yz))$, since H_A is a new soft prime ideal of X, then $H_A(x(yz)) \subseteq H_A(xz) \bigcup H_A(y)$, this implies $u \in H_A(xz) \bigcup H_A(y)$, then $u \in H_A(xz)$ or $u \in H_A(y)$, therefore $xz \in A(u)$ or $y \in A(u)$. This implies $A(u)$ is a prime ideal of X.

Necessity: For $\forall x, y, z \in X$, if $H_A(xz) \bigcup H_A(y) = \emptyset$, suppose $H_A(x(yz)) \neq \emptyset$, then $\exists u \in H_A(x(yz))$, this implies $x(yz) \in A(u)$, since $A(u)$ is a prime ideal of X, therefore $xz \in A(u)$ or $y \in A(u)$, and thus, $u \in H_A(xz)$ or $u \in H_A(y)$, this implies $\exists u \in H_A(xz) \bigcup H_A(y)$, then $H_A(xz) \bigcup H_A(y) \neq \emptyset$ contradict the hypothesis, therefore, $H(x(yz)) = \emptyset$. Therefore, $H_A(x(yz)) = H_A(xz) \bigcup H_A(y)$. If $H_A(xz) \bigcup H_A(y) \neq \emptyset$, when $H_A(x(yz)) = \emptyset$, then $H_A(x(yz)) \subseteq H_A(xz) \bigcup H_A(y)$; when $H_A(x(yz)) \neq \emptyset$, then $\forall u \in H_A(x(yz))$, we have $x(yz) \in A(u)$, since $A(u)$ is a prime ideal of X, therefore $xz \in A(u)$ or $y \in A(u)$, this implies $u \in H_A(xz)$ or $u \in H_A(y)$, then $u \in H_A(xz) \bigcup H_A(y)$, therefore $H_A(x(yz)) \subseteq H(xz) \bigcup H(y)$. This implies that H_A is a new soft prime ideal of X.

From Theorems 4.2 and 4.3, we know the dual soft set of soft set and the level set of soft set are two different algebraic structures.

Theorem 4.4. Let X_1, X_2 as two KU-subalgebras of KU-algebra X. H_1, H_2 are the new soft ideals ofX_1, X_2 respectively. And $(H, X_1 \bigcap X_2) = (H_1, X_1) \bigcap_R (H_2, X_2)$, then H is the new soft prime ideals of $X_1 \bigcap X_2$.

Proof. From Theorem 2.2, we know that $X_1 \bigcap X_2$ is also a subalgebra of X. For $\forall x \in X_1 \cap X_2$, since H_1, H_2 are the soft prime ideals of X_1, X_2, respectively, $H(x(yz)) = H_1(x(yz)) \cap H_2(x(yz)) \subseteq [H_1(xz) \cup H_1(y)] \cap [H_2(xz) \cup H_1(y)]$.

By definition, we know that H is the new soft prime ideal of KU-algebra $X_1 \bigcap X_2$.

Definition 4.4. Let X_1, X_2 as KU-algebras, $(H, X)= (F, X_1) \wedge (G, X_2)$, in which $X = X_1 \times X_2$. Define $H_{X_1} : X_1 \to P(U)$, $H_{X_1}(x) = \bigcup\limits_{y \in X_2} H(x, y), \forall x \in X_1$, define H_{X_1} as the projection of (H, X)onX_1. In the same vein, we can define H_{X_2} as the projection of (H, X).

Theorem 4.5. If (F, X_1) and (G, X_2) are the soft sets of X_1 and X_2, respectively, and $(H, X) = (F, X_1) \wedge (G, X_2)$ is the new soft prime ideal of KU-algebra $X = X_1 \times X_2$, then (H_{X_1}, X_1) and (H_{X_2}, X_2) are the new soft prime ideals of X_1 and X_2 respectively.

Proof. $H_{X_1}(xz) \cup H_{X_1}(y) = (\bigcup\limits_{\omega \in X_2} H((xz, \omega))) \cup (\bigcup\limits_{\omega_2 \in X_2} H((y, \omega_2))) \supseteq H((xz, \omega)) \cup H((y, 0)) \overset{\forall \omega \in X_2}{=} H((xz, 0\omega)) \cup H((y, 0)) \supseteq H((x, 0)((y, 0)(z, \omega))) = H((x, 0)((yz), 0\omega)) = H((x(yz), 0(0\omega))) = H((x(yz), \omega))\ \forall \omega \in X_2$.

Thus, $H_{X_1}(xz) \bigcup H_{X_1}(y) \supseteq \bigcup\limits_{\omega \in X_2} H((x(yz), \omega)) = H_{X_1}(x(yz))$.

Therefore, (H_{X_1}, X_1) is the new soft prime ideal of X_1.

Similarly, (H_{X_2}, X_2) is the new soft prime ideal of X_2.

5 The Image and Inverse Image of the New Type Soft Prime Ideal

Theorem 5.1. Let X_1, X_2 as two KU-algebras and U is a initial complete set. $f : X_1 \to X_2$ is the homomorphic map of the KU-algebras, $H_1 : X_1 \to P(U)$, $H_2 : X_2 \to P(U)$ are two soft sets and H_1 is $f-$invariant. Thus:

(1) f is a full homomorphic map. If H_1 is the new soft prime ideal of X_1, then $f(H_1)$ is the new soft prime ideal of X_2.

(2) If H_2 is the new soft ideal of X_2, then $f^{-1}(H_2)$ is the new soft ideal of X_1.

Proof. (1) $\forall x_2, y_2, z_2 \in X_2$ because f is full map. Hence, $\exists x_1, y_1, z_1 \in X_1$, which makes $f(x_1) = x_2, f(y_1) = y_2$, $f(z_1) = z_2.f(x_1z_1) = f(x_1)f(z_1) = x_2z_2$,$x_2(y_2z_2) = f(x_1)(f(y_1)f(z_1)) = f(x_1)(f(y_1z_1)) = f(x_1(y_1z_1))$. Since H is $f-$invariant, so $f(H_1)(x_2(y_2z_2)) = \bigcup\limits_{f(x)=x_2(y_2z_2)} H_1(x) = H_1(x_1(y_1z_1))$, $f(H_1)(y_2) = \bigcup\limits_{f(y)=y_2} H_1\ (y) = H_1(y_1)$,$f(H_1)(x_2z_2) = \bigcup\limits_{f(x)=x_2z_2} H_1(x) = H_1(x_1z_1)$. And since H_1 is the new soft prime ideal of X_1, $H(x_1(y_1z_1)) \subseteq H_1(x_1z_1) \bigcup H_1(y_1)$.

Therefore, $f(H_1)(x_2(y_2z_2)) \subseteq f(H_1)(x_2z_2) \bigcup f(H_1)(y_2)$.

Hence, $f(H_1)$ is the new soft prime ideal of X_2.

(2) $\forall x_1, y_1, z_1 \in X_1$, since f is the homomorphic map of KU-algebras and H_2 is the new soft prime ideal of X_2, so $f^{-1}(H_2)(x_1(y_1z_1)) = H_2(f(x_1(y_1z_1))) = H_2(f(x_1)(f(y_1)f(z_1)) \subseteq H_2(f(x_1z_1)) \bigcup H_2(f(y_2)) = f^{-1}(H_2)(x_1z_1) \bigcup f^{-1}(H_2)(y_1)$.

Therefore, $f^{-1}(H_2)$ is the new soft prime ideal of X_1.

Theorem 5.2. Let X_1, X_2 as two KU-algebras and U is a initial complete set. $f : X_1 \to X_2$ is a full homomorphic map, $H_1 : X_1 \to P(U)$ is soft set, and H_1 is $f-$invariant. Then, the necessary condition of H_1 being the new soft prime ideal of X_1 is that $f(H_1)$ is the new soft prime ideal of X_2.

Proof. Necessity can be seen from Theorem 4.3(1).

Sufficiency: $\forall x_1, y_1, z_1 \in X_1$ makes $f(x_1) = x_2, f(y_1) = y_2$, $f(z_1) = z_2$, so $f(x_1(y_1z_1)) = x_2(y_2z_2)$ and $f(x_1z_1) = x_2z_2$. Since $f(H_1)$ is the new soft prime ideal of X_2 and $H_1$is $f-$invariant, $H_1(x_1(y_1z_1)) = \bigcup\limits_{f(x)=x_2(y_2z_2)} H_1(x) = f(H_1)(x_2(y_2z_2)) \subseteq f(H_1)(x_2z_2) \bigcup f(H_1)(y_2) = (\bigcup\limits_{f(x)=x_2z_2} H_1(x)) \bigcup (\bigcup\limits_{f(x)=y_2} H_1(x)) = H_1(x_1z_1) \bigcup H_1(y_1)$.

Therefore, H_1 is the new soft prime ideal of X_1.

Theorem 5.3. Let X_1, X_2 as two KU-algebras, $f : X_1 \to X_2$ is a homomorphic map, $H_2 : X_2 \to P(U)$ is soft set, then the necessary and sufficient condition for H_2 being the new soft prime ideal of X_2 is that $f^{-1}(H_2)$ is the new soft prime ideal of X_1.

Proof. Necessity can be seen from Theorem 4.3(2).

Sufficiency: $\forall x_2, y_2, z_2 \in X_2$, since f is full homomorphic map, so $\exists x_1, y_1, z_1 \in X_1$, which makes $x_2 = f(x_1)$, $y_2 = f(y_1)$, $z_2 = f(z_1)$, $x_2z_2 = f(x_1)f(z_1) = f(x_1z_1)$, $x_2(y_2z_2) = f(x_1)(f(y_1)f(z_1)) = f(x_1(y_1z_1))$. Also, since $f^{-1}(H_2)$ is the new soft prime ideal of X_1, $H_2(x_2(y_2z_2)) = H_2(f(x_1(y_1z_1))) = f^{-1}(H_2)(x_1(y_1z_1)) \subseteq (f^{-1}(H_2)(x_1z_1)) \bigcup (f^{-1}(H_2)(y_1)) = H_2(f(x_1z_1)) \bigcup H_2(f(y_1)) = H_2(x_2z_2) \bigcup H_2(y_2)$.

Therefore, H_2 is the new soft prime ideal of X_2.

6 Conclusion

The theory of soft set continues to attract research attention from scholars as it can solve the insufficiency of parameter tool theory when dealing with uncertain problems. And as a type of logical algebra, KU-algebras link closely with artificial intelligence. In this paper, the parameter set is introduced to the structure of KU-algebras and the concept and basic properties of the new type soft prime ideal of KU-algebras are defined or proved. Furthermore, giving the definition of other soft algebraic structures and investigating its related properties, such as the soft convex ideals of KU-algebras, are important future research directions.

Acknowledgements. Thanks to the support by National Natural Science Foundation of China (No. 61170121, No. 11401259 and No. 61673193), Natural Science Foundation of Jiangsu Province (No. BK20151117) and Undergraduate Innovation and Entrepreneurship Training Program of China (No. 201610295005).

References

1. Molodtsow, D.: Soft set theory-first results. Comput. Math. Appl. **37**(4–5), 19–31 (1999)
2. Zadeh, L.A.: Fuzzy sets. Inf. Control. **8**, 338–353 (1965)
3. Pawlak, Z.: Rough sets. Int. J. Inf. Comput. Sci. **11**, 341–356 (1982)
4. Maji, P.K., Biswas, R., Roy, A.R.: Soft set theory. Comput. Math. Appl. **45**, 555–562 (2003)
5. Aktas, H., Çagman, N.: Soft sets and soft groups. Inf. Sci. **177**(13), 2726–2735 (2007)
6. Ali, M.I., Feng, F., Liu, X., et al.: On some new operations in soft set theory. Comput. Math. Appl. **57**(9), 1547–1553 (2009)
7. Chanwit, P., Utsanee, L.: On ideas and congruences in KU-algebras. Scientia Magna **5**(1), 54–57 (2009)
8. Chanwit, P., Utsanee, L.: On isomorphisms of KU-algebras. Scientia Magna **5**(3), 25–31 (2009)
9. Mostafa, S.M.: Fuzzy ideals of KU-algebras. Int. Math. Forum **6**(63), 3139–3149 (2011)
10. Mostafa, S.M., Abdel, M.A., Elgendy, O.R.: Intuitionistic fuzzy KU-ideals in KU-algebras. Int. J. Math. Sci. Appl. **1**(3), 1379–1384 (2011)
11. Mostafa, S.M.: Interval-valued fuzzy KU-ideals in KU-algebras. Int. Math. Forum **6**(64), 3151–3159 (2011)
12. Muhammad, G., Muhammad, S., Sarfraz, A.: On (α, β)-fuzzy KU-ideals of KU-algebras. Afrika Matematika **26**, 651–661 (2015)
13. Muhiuddin, G.: Bipolar fuzzy KU-subalgebras/ideals of KU-algebras. Ann. Fuzzy Math. Inform. **8**(3), 409–418 (2014)
14. Wen, Y.C.: The Study on Soft Set. Liaoning Normal University, Dalian (2008)
15. Liao, Z.H., Rui, M.L.: Soft inclines. Comput. Eng. Appl. **48**(2), 30–32 (2012)
16. Zheng, G.P., Liao, Z.H., Wang, N.N.: Soft lattice implication subalgebras. Appl. Math. Inf. Sci. **7**(3), 1181–1186 (2013)
17. Yin, X., Liao, Z.H.: Study on soft groups. J. Comput. **8**(4), 960–967 (2013)
18. Zhang, L.X., Liao, Z.H., Wang, Q.: New type of soft fields. CAAI Trans. Intell. Syst. **10**(6), 858–864 (2015)
19. Wu, X.Z., Liao, Z.H., Liu, W.L.: Anti-soft subsemigroups. Fuzzy Syst. Math. **29**(5), 45–50 (2015)
20. Liao, C.C., Liao, Z.H., Zhang, L.X.: New type of soft weak BCI-algebras. Comput. Sci. **43**(1), 22–24 (2016)
21. Tong, J., Liao, Z.H., Lu, T., et al.: Anti-soft subinclines. J. Jilin Univ. (Sci. Edn.) **54**(2), 215–221 (2016)
22. Muhammad, G., Muhammad, S.: On soft KU-algebras. J. Algebr. Number Theory Adv. Appl. **11**(1), 1–20 (2014)
23. Xi, Y., Liao, Z.H., Chen, X.M.: A new type soft ideal of KU algebra. Fuzzy Syst. Math. **31**(2), 13–21 (2017)
24. Wu, Z.J., Qin, K.Y.: L-fuzzy rough set based on complete residuated lattice. J. Southwest Jiaotong Univ. (Engl. Ed.) **1**, 95–98 (2008)

On $(\in, \in \vee q_{(\lambda,\mu)})$-fuzzy ideals of KU-algebras

Lun Li[1,2], Zu-hua Liao[1,2,3(✉)], Zhen-yu Liao[5], Liu-hong Chen[1,2,3], Wei Song[4], and Yong Li[1]

[1] School of Science, Jiangnan University, Wuxi 214122, People's Republic of China
liaozuhua57@163.com
[2] Honors School, Jiangnan University, Wuxi 214122, People's Republic of China
[3] Institute of Intelligence System and Network Computing, Jiangnan University, Wuxi 214122, People's Republic of China
[4] School of Internet of Things Engineering, Jiangnan University, Wuxi 214122, People's Republic of China
[5] Computer Science Department, Boston University, Boston, USA

Abstract. First, new concepts of pointwise $(\in, \in \vee q_{(\lambda,\mu)})$-fuzzy ideals and generalized fuzzy ideals of KU-algebras are defined. By using inequalities, level sets and characteristic functions, some equivalent characterizations of $(\in, \in \vee q_{(\lambda,\mu)})$-fuzzy ideals of KU-algebras are studied, a richer hierarchical structure of this fuzzy ideal is presented, and some properties are discussed using the partial order of KU-algebras. Second, it is proven that the intersections, unions (under certain conditions), homomorphic image and homomorphic preimage of $(\in, \in \vee q_{(\lambda,\mu)})$-fuzzy ideals of KU-algebras are also $(\in, \in \vee q_{(\lambda,\mu)})$-fuzzy ideals. Then, the direct product and projection of the $(\in, \in \vee q_{(\lambda,\mu)})$-fuzzy ideals of KU-algebras are also investigated. Finally, a new concept of the descending (ascending) chain conditions of the ideals of KU-algebras is introduced and is studied using the properties of $(\in, \in \vee q_{(\lambda,\mu)})$-fuzzy ideals.

Keywords: KU-algebra · $(\in, \in \vee q_{(\lambda,\mu)})$-fuzzy ideals · Homomorphic mapping · Projection · Descending (ascending) chain condition

1 Introduction

In 1965, in order to solve the problem that the membership of the element is not clear to the set, Zadeh [1] introduced the concept of fuzzy sets, which is widely applied to describe fuzzy phenomena and investigate fuzzy problems. By introducing the concept of fuzzy sets into algebra, Rosenfeld [2] inspired the fuzzification of algebraic structures and defined the notion of fuzzy subgroups in 1971. Since then, fuzzy algebra came into being. Kuroki [3–5] performed research on fuzzy ideals, fuzzy bi-ideals, fuzzy semiprime ideals and fuzzy generalized bi-ideals in semigroups. Pu and Liu [6,7] introduced the notions of the "belong to" relation ($\in$) and the "quasi-coincident with" relation (q) between a fuzzy point

B.-Y. Cao and Y.-B. Zhong (Eds.): ICFIE 2017, AISC 872, pp. 91–110, 2019.
https://doi.org/10.1007/978-3-030-02777-3_9

x_λ and a fuzzy set A in 1980. Then, using these two notions, Bhakat and Das [8,9] presented the concept of an $(\in, \in \vee q)$-fuzzy subgroup in 1992 and 1996. Liu [10,11] introduced the notions of $(\in, \in \vee q)$-fuzzy prime filters of BL-algebras and $(\in, \in \vee q)$-fuzzy soft ideals in BCK/BCI-algebras, and investigated several properties of them.

In 2003, Yuan et al. [12] presented the concept of generalized fuzzy subgroups. In 2006, Liao et al. [13] extended the common "quasi-coincident with" to "generalized quasi-coincident with" and presented $(\in, \in \vee q_{(\lambda,\mu)})$-fuzzy algebra, which is the unity and generalization of $(\in, \in)$-fuzzy algebra, $(\in, \in \vee q)$-fuzzy algebra and $(\bar{\in}, \bar{\in} \vee \bar{q})$-fuzzy algebra. When $\lambda = 0$ and $\mu = 1$, we have $(\in, \in)$-fuzzy algebra introduced by Rosenfeld. When $\lambda = 0$ and $\mu = 0.5$, we get the $(\in, \in \vee q)$-fuzzy algebra introduced by Bhakat and Das. When $\lambda = 0.5$ and $\mu = 1$, we have the $(\bar{\in}, \bar{\in} \vee \bar{q})$-fuzzy algebra. Later, his research team [14–24] performed a series of studies in this direction.

In 2009, based on BCK/BCI-algebras, Prabpayak et al. [25,26] introduced the concept of KU-algebras and the ideals of KU-algebras. KU-algebras have potential application value in computer coding and other fields. In [27], Rezaei et al. proved that KU-algebras are equivalent to commutative self-distributive BE-algebras, and a self-distributive KU-algebra is equivalent to the Hilbert algebra, and the relations among BE-algebras, dual BCK-algebras, Hilbert algebras, implication algebras, CI-algebras and KU-algebras were also given. Moreover, the axioms of FI-algebras [28] are similar to those of KU-algebras, and the axioms of lattice implication algebras [29] are also similar to those of KU-algebras, so they have some common characters.

Mostafa et al. [30] presented the notion of the fuzzy ideals of KU-algebras in 2011. Gulistan et al. [31,32] studied the properties of the $(\in, \in \vee q_k)$-fuzzy ideals and (α, β)-fuzzy ideals of KU-algebras in 2014 and 2015. Muhiuddin investigated the bipolar fuzzy ideals of KU-algebras [33]. Senapati et al. [34] investigated the fundamental properties of the Atanassov's intuitionistic fuzzy bi-normed ideals of a KU-algebra in 2016.

In 1921, Noether [35] introduced the ascending chain conditions of a ring, and Artin [36] presented the concept of a descending chain to distinguish rings in 1927. Yi presented the concept of the $(\in, \in \vee q_{(\lambda,\mu)})$-fuzzy k-ideals of semigroups and used it to characterize the chain conditions of the semigroups [16].

Based on the research above, we present the concepts of the $(\in, \in \vee q_{(\lambda,\mu)})$-fuzzy ideals, generalized fuzzy ideals and chain conditions of KU-algebras and investigate their properties. From our investigations, we obtain the results that the hierarchical structure of $(\in, \in \vee q_{(\lambda,\mu)})$-fuzzy ideals is richer than those of other types of fuzzy ideals. For instance, there are three hierarchical structures in the characterization of level set $A_t \neq \varnothing$ of $(\in, \in \vee q_{(\lambda,\mu)})$-fuzzy ideal of KU-algebras: $(\in, \in \vee q_{(0,\lambda)})$-fuzzy ideal, $(\in, \in \vee q_{(\lambda,\mu)})$-fuzzy ideal and $(\in, \in \vee q_{(\mu,1)})$-fuzzy ideal. From Example 1, we can see that the $(\in, \in \vee q_{(\lambda,\mu)})$-fuzzy ideal is a nontrivial generalization of the $(\in, \in)$-fuzzy ideal and the $(\in, \in \vee q)$ -fuzzy ideal.

In this paper, we recall some basic definitions and results of KU-algebras and fuzzy sets in Sect. 2. In Sect. 3, the new definitions of the $(\in, \in \vee q_{(\lambda,\mu)})$-fuzzy ideals of KU-algebras and generalized fuzzy ideals of KU-algebras are given, their equivalent characterizations are described, and the properties of unions, intersections and partial orders are investigated. In Sect. 4, we present the result that the homomorphic image and homomorphic preimage (under certain conditions) of the $(\in, \in \vee q_{(\lambda,\mu)})$-fuzzy ideals of KU-algebras are also $(\in, \in \vee q_{(\lambda,\mu)})$-fuzzy ideals. In Sect. 5, we obtain some properties: a direct product of an $(\in, \in \vee q_{(\lambda,\mu)})$-fuzzy ideal is an $(\in, \in \vee q_{(\lambda,\mu)})$-fuzzy ideal, and a projection of an $(\in, \in \vee q_{(\lambda,\mu)})$-fuzzy ideal is also an $(\in, \in \vee q_{(\lambda,\mu)})$-fuzzy ideal. Finally, in Sect. 6, we characterize the chain conditions of the ideals of KU-algebras by using the properties of $(\in, \in \vee q_{(\lambda,\mu)})$-fuzzy ideals.

2 Preliminaries

In this section, we introduce some concepts and results related to KU-algebras and fuzzy sets.

First, we introduce some concepts and results related to KU-algebras and general sets.

Definition 1. [25,26] An algebra $G = (G, *, 0)$ is called a KU-algebra if it satisfies the following axioms:

(1) $(x * y) * ((y * z) * (x * z)) = 0$;
(2) $0 * x = x$;
(3) $x * 0 = 0$;
(4) $x * y = 0$ and $y * x = 0$ implies $x = y$.

By (1), we get $(0 * 0) * ((0 * x) * (0 * x)) = 0$. It follows that $x * x = 0$ for all $x \in G$, and if we put $y = 0$ in (1), then we obtain $z * (x * z) = 0$ for all $x, z \in G$.

On the KU-algebra $(G, *, 0)$, we define a binary relation $\leq$ on G by putting $x \leq y \Leftrightarrow y * x = 0$. Then, it is easy to prove that $(G; \leq)$ is a partially ordered set and 0 is its smallest element.

In the further discussion of this article, we denote a KU-algebra by G and G'.

Definition 2. [25] A non-empty subset A of a KU-algebra is said to be an ideal of G if

(1) $0 \in A$;
(2) $\forall x, y, z \in G$, $x * (y * z) \in A$ and $y \in A$, implying $x * z \in A$.

Theorem 1. [30] Let G and G' be KU-algebras, then we define $''*''$ on $G \times G'$, $\forall (x_1, x_2), (y_1, y_2) \in G \times G'$, $(x_1, x_2) * (y_1, y_2) = (x_1 * y_1, x_2 * y_2)$, then $(G \times G', *, (0, 0'))$ is a KU-algebra.

Definition 3. [26] Let $(G, *, 0)$ and $(G', *', 0')$ be KU-algebras, a homomorphism is a map $f : G \to G'$, satisfying $f(x * y) = f(x) *' f(y), \forall x, y \in G$. An injective homomorphism is called a monomorphism, and a surjective homomorphism is called an epimorphism.

Definition 4. [38] Let $(S, \leq)$ be a partially ordered set, where $\leq$ is the partial order of the set. For every non-empty subset of S, if there exists a least element under the order, then $\leq$ is called a well-ordered relation, and $(S, \leq)$ is called a well-ordered set.

Second, we introduce some concepts and results related to fuzzy sets.

Definition 5. [13] A fuzzy subset A of G of the form

$$A(y) = \begin{cases} \lambda(\neq 0), & y = x \\ 0, & y \neq x \end{cases}$$

is said to be a fuzzy point with support x, and value λ is denoted by x_λ.

Definition 6. [13] Let $t, \lambda, \mu \in [0,1]$ and $\lambda < \mu$. Let A be a fuzzy subset of G. x_t is said to belong to A if $A(x) \geq t$ is denoted by $x_t \in A$. A fuzzy point x_t is said to belong to A in the generalization if $t > \lambda$ and $A(x) + t > 2\mu$, denoted by $x_t q_{(\lambda,\mu)} A$. If $x_t \in A$ or $x_t q_{(\lambda,\mu)} A$, then it is denoted by $x_t \in \vee q_{(\lambda,\mu)} A$.

When $\mu = 1$, $A(x) + t > 2\mu$ does not hold. Therefore $x_t q_{(\lambda,\mu)} A$ does not hold. So $x_t \in \vee q_{(\lambda,\mu)} A$ degenerates to $x_t \in A$.

Theorem 2. [18] Let $\{A_i\}_{i \in I}$ be a family of fuzzy subsets of G such that for all $i, j \in I$, $A_i \subseteq A_j$ or $A_j \subseteq A_i$, $\forall x, y \in G, \bigvee\limits_{i \in I} (A_i(x) \wedge A_i(y) \wedge \mu) = (\bigcup\limits_{i \in I} A_i)(x) \wedge (\bigcup\limits_{i \in I} A_i)(y) \wedge \mu$.

Definition 7. [37] Let A and B be two fuzzy sets of non-empty sets G and G', respectively. The direct product $A \times B$ of A and B is a fuzzy subset that is defined by $A \times B : G \times G' \to [0,1]$, $(A \times B)(x,y) = A(x) \wedge B(y), \forall (x,y) \in G \times G'$.

Definition 8. [37] Let G and G' be two non-empty sets. Let $A \times B$ be a fuzzy subset of $G \times G'$ such that the projection of $A \times B$ on G (resp.G')is the fuzzy subset A_G (resp.$B_{G'}$) of G (resp.G'), which is defined by

$$A_G(x) = \bigvee_{z \in G'} (A \times B)(x,z), \forall x \in G,$$

$$(\text{resp.} B_{G'}(y) = \bigvee_{z \in G} (A \times B)(z,y), \forall y \in G').$$

Definition 9. [39] Let f be a homomorphism of a set G into a set G'. Then, a fuzzy subset A of G is called f−invariant if for all $x, y \in G$, $A(x) = A(y)$.

Definition 10. [40] Let $f : G \to G'$ be a mapping. If A is a fuzzy subset of G, then fuzzy subset B of G', defined by

$$B(y) = \begin{cases} \bigvee\limits_{f(x)=y} A(x), & f^{-1}(y) = \{x \in G, f(x) = y\} \neq \varnothing \\ 0, & f^{-1}(y) = \varnothing \end{cases}$$

is said to be the image of A under f.

Similarly, if B is a fuzzy subset of G', then the fuzzy subset defined by $A(x) = f^{-1}(B)(x) = B(f(x))$ for all $x \in G$ is said to be the preimage of B under f.

3 $(\in, \in \vee q_{(\lambda,\mu)})$-fuzzy ideals

In this section, we present the new concept of an $(\in, \in \vee q_{(\lambda,\mu)})$-fuzzy ideal of KU-algebras and investigate some equivalent characterizations and basic properties of it.

Definition 11. A fuzzy subset A of KU-algebra G is said to be an $(\in, \in \vee q_{(\lambda,\mu)})$-fuzzy ideal of G if the following conditions are satisfied:

(I) $\forall t \in (\lambda, 1]$, $\forall x \in G$, if $x_t \in A$, then $0_t \in \vee q_{(\lambda,\mu)} A$;
(II) $\forall t_1, t_2 \in (\lambda, 1]$, $\forall x, y, z \in G$, if $(x * (y * z))_{t_1} \in A$, $y_{t_2} \in A$, then $(x * z)_{t_1 \wedge t_2} \in \vee q_{(\lambda,\mu)} A$.

Definition 12. A fuzzy subset A of KU-algebra G is said to be a generalized fuzzy ideal of G if $\forall x, y, z \in G$, the following conditions are satisfied:

(III) $A(0) \vee \lambda \geq A(x) \wedge \mu$;
(VI) $A(x * z) \vee \lambda \geq A(x * (y * z)) \wedge A(y) \wedge \mu$.

In the following theorems, we discuss the relations between an $(\in, \in \vee q_{(\lambda,\mu)})$-fuzzy ideal, a generalized fuzzy ideal and a level set.

Theorem 3. For a fuzzy subset A of G, the following statements are equivalent.

(1) A is an $(\in, \in \vee q_{(\lambda,\mu)})$-fuzzy ideal of G;
(2) A is a generalized fuzzy ideal of G;
(3) For all $t \in (\lambda, \mu]$, $A_t = \{x | A(x) \geq t\} \neq \varnothing$ is an ideal of G.

Proof. (1)⇒(2)

(I)⇒(III) Assume that there exists $x \in G$ such that $A(0) \vee \lambda < A(x) \wedge \mu$. Choose $t = A(x) \wedge \mu$ and $\lambda < t \leq \mu$, $A(x) \geq t$. Therefore, $x_t \in A$. Then, based on Definition 11, we have $0_t \in \vee q_{(\lambda,\mu)} A$. But $A(0) < t \leq \mu$ and $A(0) + t < t + t \leq 2\mu$, which is contradictory.

(II)⇒(IV) Assume that there exist $x, y, z \in G$ such that $A(x * z) \vee \lambda < A(x * (y * z)) \wedge A(y) \wedge \mu$. Choose $t = A(x * (y * z)) \wedge A(y) \wedge \mu$. Then, we can have $t \in (\lambda, \mu]$, $(x * (y * z))_t \in A$, $y_t \in A$ and $(x * z)_t \notin A$. Based on (II), $(x * z)_t \in \vee q_{(\lambda,\mu)} A$, and we can have $(x * z)_t q_{(\lambda,\mu)} A$ and $A(x * z) + t > 2\mu$. $A(x * z) > 2\mu - t \geq \mu \geq t$, so $(x * z)_t \in A$. But $(x * z)_t \notin A$, a contradiction. Therefore, (IV) holds.

(2)⇒(1)

(III)⇒(I) $\forall x \in G$, $\forall t \in (\lambda, 1]$, if $x_t \in A$, then $A(x) \geq t$. Since A is a generalized fuzzy ideal of G, $A(0) \vee \lambda \geq A(x) \wedge \mu \geq t \wedge \mu$. If $t \leq \mu$, since $\lambda < t$, then $A(0) \geq t$, so $0_t \in A$; if $t > \mu$, then $A(0) \vee \lambda \geq \mu$. Since $\lambda < \mu$, we have $A(0) \geq \mu$. Therefore, $A(0) + t \geq \mu + t > 2\mu$, so $0_t q_{(\lambda,\mu)} A$. Hence, $0_t \in \vee q_{(\lambda,\mu)} A$.

(IV)⇒(II) For all $t_1, t_2 \in (\lambda, 1]$, $\forall x, y, z \in G$, if $(x * (y * z))_{t_1} \in A$, $y_{t_2} \in A$, then $A(x * (y * z)) \geq t_1$ and $A(y) \geq t_2$. Since A is a generalized fuzzy ideal, we have $A(x * z) \vee \lambda \geq A(x * (y * z)) \wedge A(y) \wedge \mu$. Therefore, $A(x * z) \vee \lambda \geq (t_1 \wedge t_2) \wedge \mu$. If $(t_1 \wedge t_2) \leq \mu$, by $\lambda < t_1$ and $\lambda < t_2$, then $A(x * z) \geq t_1 \wedge t_2$, so $(x * z)_{t_1 \wedge t_2} \in A$;

if $(t_1 \wedge t_2) > \mu$, then $A(x*z) \geq \mu$. So, $A(x*z) + (t_1 \wedge t_2) \geq \mu + (t_1 \wedge t_2) > 2\mu$. Hence, $(x*z)_{t_1 \wedge t_2} q_{(\lambda,\mu)} A$. Therefore, $(x*z)_{t_1 \wedge t_2} \in \vee q_{(\lambda,\mu)} A$.

(2)⇒(3)

$\forall t \in (\lambda, \mu]$, if $A_t \neq \varnothing$, then $\exists x \in A_t$, $A(x) \geq t$. Based on (III), $A(0) \vee \lambda \geq A(x) \wedge \mu \geq t \wedge \mu = t$. Since $t > \lambda$, we have $A(0) \geq t$, so $0 \in A_t$. $\forall x*(y*z), y \in A_t$, we have $A(x*(y*z)) \geq t$ and $A(y) \geq t$. By (IV), $A(x*z) \vee \lambda \geq A(x*(y*z)) \wedge A(y) \wedge \mu \geq t \wedge \mu = t$. Since $t > \lambda$, we have $A(x*z) \geq t$, so $x*z \in A_t$. Therefore, A_t is an ideal of G.

(3)⇒(2)

Assume that there exists $x \in G$ such that $A(0) \vee \lambda < A(x) \wedge \mu$. Choose $t = A(x) \wedge \mu$, then $\lambda < t \leq \mu$, $A(x) \geq t$ and $A(0) < t$. Therefore, $x \in A_t$, $A_t \neq \varnothing$. Since A_t is an ideal of G, hence $0 \in A_t$, so $A(0) \geq t$. But $A(0) < t$, which is a contradiction. Therefore, $\forall x \in G$, we have$A(0) \vee \lambda \geq A(x) \wedge \mu$.

Assume that there exist $x, y, z \in G$ such that $A(x*z) \vee \lambda < A(x*(y*z)) \wedge A(y) \wedge \mu$, then choose $t = A(x*(y*z)) \wedge A(y) \wedge \mu$. So, $t \in (\lambda, \mu]$, $A(x*(y*z)) \wedge A(y) \geq t$ and $A(x*z) < t$. Therefore, $x*(y*z)$, $y \in A_t$. Since A_t is an ideal of G, we have $x*z \in A_t$, so $A(x*z) \geq t$. But $A(x*z) < t$, a contradiction. Therefore, $\forall x, y, z \in G$, we have $A(x*z) \vee \lambda \geq A(x*(y*z)) \wedge A(y) \wedge \mu$.

Hence, A is a generalized fuzzy ideal of G.

When $\lambda = 0, \mu = 1$, we can obtain the results of $(\in, \in)$-fuzzy ideals of KU-algebras.

Corollary 1. For a fuzzy subset A of G, the following statements are equivalent.

(1) A is an $(\in, \in)$-fuzzy ideal of G;
(2) $A(0) \geq A(x)$ and $A(x*z) \geq A(x*(y*z)) \wedge A(y)$, $\forall x, y, z \in G$;
(3) For all $t \in (0, 1]$, $A_t = \{x | A(x) \geq t\} \neq \varnothing$ is an ideal of G.

When $\lambda = 0$, $\mu = 0.5$, we can obtain the results of $(\in, \in \vee q)$-fuzzy ideals of KU-algebras.

Corollary 2. For a fuzzy subset A of G, the following statements are equivalent.

(1) A is an $(\in, \in \vee q)$-fuzzy ideal of G;
(2) $A(0) \geq A(x) \wedge 0.5$ and $A(x*z) \geq A(x*(y*z)) \wedge A(y) \wedge 0.5$, $\forall x, y, z \in G$;
(3) For all $t \in (0, 0.5]$, $A_t = \{x | A(x) \geq t\} \neq \varnothing$ is an ideal of G.

Example 1. Let $G = \{0, 1, 2, 3, 4, 5\}$, and the operation "$*$" is given by the Table 1.

Then, $(G, *, 0)$ is a KU-algebra.

(1) $A_1 : G \to [0, 1]$, $A_1(5) = t_0$, $A_1(0) = t_1 = 0.4$, $A_1(1) = t_2$, $A_1(2) = t_3$, $A_1(3) = t_4$, $A_1(4) = t_5$, $\lambda = 0.1$, $\mu = 0.4$, then $\lambda < t_0 < t_1 = \mu < t_2 < 0.5 < t_3 < t_4 < t_5 < 1$. Then, it is easy to show that A_1 is an $(\in, \in \vee q_{(\lambda,\mu)})$-fuzzy ideal of G. Since $A_1(0) = t_1 < A_1(2) \wedge 0.5 = 0.5$, A_1 is not an $(\in, \in \vee q)$-fuzzy ideal of G. By $A_1(0) = t_1 < A_1(3) = t_4$, then A_1 is not an $(\in, \in)$-fuzzy ideal of G.

Table 1. Operation "$*$"

$*$	0	1	2	3	4	5
0	0	1	2	3	4	5
1	0	0	2	2	4	5
2	0	0	0	1	4	5
3	0	0	0	0	4	5
4	0	0	0	1	0	5
5	0	0	0	0	0	0

(2) $A_2 : G \to [0,1]$, $A_2(5) = t_1$, $A_2(0) = t_2 = 0.5$, $A_2(1) = t_3$, $A_2(2) = t_4$, $A_2(3) = t_5$, $A_2(4) = t_6$. Then, $0 < \lambda < t_1 < t_2 = 0.5 < t_3 < \mu < t_4 < t_5 < t_6 < 1$. Based on Corollary 2, we have that A_2 is an $(\in, \in \vee q)$- fuzzy ideal of G. Since $A_2(0) \vee \lambda = t_2 < t_3 = A_2(1) \wedge \mu$, A_2 is not an $(\in, \in \vee q_{(\lambda,\mu)})$- fuzzy ideal of G.

Since $A_2(0) = t_2 < t_3 = A_2(1)$, A_2 is not an $(\in, \in)$-fuzzy ideal of G.

(3) $A_3 : G \to [0,1]$, $A_3(5) = t_1$, $A_3(2) = t_2, A_3(3) = t_3$, $A_3(4) = t_4$, $A_3(1) = t_5$, $A_3(0) = t_6$, $\lambda = 0.1$, $\mu = 0.5$. Then, $0 < \lambda < t_1 < \mu = 0.5 < t_2 = t_3 < t_4 = t_5 < t_6 = 1$. Hence, $\forall x, y, z \in G$, A_3 is an $(\in, \in)$- fuzzy ideal of G.

Therefore, an $(\in, \in \vee q_{(\lambda,\mu)})$-fuzzy ideal is a new type of fuzzy algebraic structure that is different from the $(\in, \in)$-fuzzy ideal or $(\in, \in \vee q)$-fuzzy ideal.

Theorem 4. *A is an $(\in, \in \vee q_{(\lambda,\mu)})$-fuzzy ideal of G if and only if when $A_{(t)} = \{x | A(x) > t\} \neq \varnothing$, $A_{(t)}$ is an ideal of G.*

Proof. $\Rightarrow \forall t \in [\lambda, \mu)$, if $A_{(t)}$ is non-empty, then there exists $x \in A_{(t)}$, so $A(x) > t$. Since A is an $(\in, \in \vee q_{(\lambda,\mu)})$-fuzzy ideal of G, we have $A(0) \vee \lambda \geq A(x) \wedge \mu > t$. By $t \geq \lambda$, then $A(0) > t$, so $0 \in A_{(t)}$. Then, $\forall x, y, z \in G$, if $x * (y * z) \in A_{(t)}$, $y \in A_{(t)}$, since A is an $(\in, \in \vee q_{(\lambda,\mu)})$-fuzzy ideal, we have $A(x * z) \vee \lambda \geq A(x * (y * z)) \wedge A(y) \wedge \mu > t$. By $t \geq \lambda$, then $A(x * z) > t$, so $x * z \in A_{(t)}$. Therefore, $A_{(t)}$ is an ideal of G.

$\Leftarrow$ Assume that there exists $x \in G$ such that $A(0) \vee \lambda < A(x) \wedge \mu$. Choose $t = A(0) \vee \lambda$, then $t \in [\lambda, \mu)$ and $A(x) > t, x \in A_{(t)}$. Since $A_{(t)}$ is an ideal of G, then $0 \in A_{(t)}$, so $A(0) > t$. But $A(0) \leq t$, which is a contradiction. Therefore, $\forall x \in G$, $A(0) \vee \lambda \geq A(x) \wedge \mu$.

Assume that there exist $x, y, z \in G$ such that $A(x * z) \vee \lambda < A(x * (y * z)) \wedge A(y) \wedge \mu$. Choose $t = A(x * z) \vee \lambda$, then $A(y) > t$, $A(x * (y * z)) > t$ and $\lambda \leq t < \mu$, so y, $x * (y * z) \in A_{(t)}$ and $t \in [\lambda, \mu)$. Since $A_{(t)}$ is an ideal of G, we have $x * z \in A_{(t)}$, so $A(x * z) > t$. But $A(x * z) \leq t$, which is a contradiction. Therefore, $\forall x, y, z \in G, A(x * z) \vee \lambda \geq A(x * (y * z)) \wedge A(y) \wedge \mu$.

Hence, A is an $(\in, \in \vee q_{(\lambda,\mu)})$-fuzzy ideal of G.

In the following Theorem 5–Theorem 7, we discuss the hierarchical structure of $(\in, \in \vee q_{(\lambda,\mu)})$-fuzzy ideals.

Theorem 5. For a fuzzy subset A of G, $\forall t \in (\mu, 1]$, $\varnothing \neq A_t$ is an ideal of G if and only if the following conditions hold.

(1) $A(x) \leq A(0) \vee \mu$;
(2) $A(x*(y*z)) \wedge A(y) \leq A(x*z) \vee \mu$.

Proof. $\Rightarrow$ If (1) does not hold, assume that there exists $x \in G$ such that $A(x) > A(0) \vee \mu$. Choose $t = A(x)$, then $x \in A_t$, $0 \notin A_t$, $t \in (\mu, 1]$. Since A_t is an ideal of G, $0 \in A_t$, which is a contradiction. Therefore, $\forall x \in G$, $A(x) \leq A(0) \vee \mu$.

If (2) does not hold, then there exist $x, y, z \in G$ such that $A(x*(y*z)) \wedge A(y) > A(x*z) \vee \mu$. Choose $t = A(x*(y*z)) \wedge A(y)$, then $t \in (\mu, 1]$, $A(x*(y*z)) \geq t$, $A(y) \geq t$ and $A(x*z) < t$. Hence, $x*(y*z) \in A_t$ and $y \in A_t$. Since A_t is an ideal of G, we have $x*z \in A_t$, so $A(x*z) \geq t$, but $A(x*z) < t$, a contradiction. Therefore, $\forall x, y, z \in G$, $A(x*(y*z)) \wedge A(y) \leq A(x*z) \vee \mu$.

$\Leftarrow$ $\forall t \in (\mu, 1]$ and $\varnothing \neq A_t$, there exists $x \in A_t$. Based on (1), $A(x) \leq A(0) \vee \mu$, so $A(0) \vee \mu \geq t$. By $t > \mu$, then $A(0) \geq t$, $0 \in A_t$. $\forall x, y, z \in G$, if $x*(y*z) \in A_t$, $y \in A_t$, then $A(x*(y*z)) \geq t$, $A(y) \geq t$. Based on (2), $A(x*z) \vee \mu \geq A(x*(y*z)) \wedge A(y) \geq t$. By $t > \mu$, then $A(x*z) \geq t$, so $x*z \in A_t$. Therefore, A_t is an ideal of G.

Corollary 3. For a fuzzy subset A of G, $\forall t \in (\mu, 1]$, $\varnothing \neq A_t$ is an ideal of G if and only if A is an $(\in, \in \vee q_{(\mu,1)})$-fuzzy ideal.

Theorem 6. For a fuzzy subset A of G, $\forall t \in (0, \lambda]$, $\varnothing \neq A_t$ is an ideal of G if and only if the following conditions hold.

(1) $A(x) \wedge \lambda \leq A(0)$;
(2) $A(x*(y*z)) \wedge A(y) \wedge \lambda \leq A(x*z)$.

Proof. $\Rightarrow$ If (1) does not hold, assume that there exists $x \in G$ such that $A(x) \wedge \lambda > A(0)$. Choose $t = A(x) \wedge \lambda$, then $x \in A_t$, $t \in (0, \lambda]$. Since A_t is an ideal of G, we have $0 \in A_t$, so $A(0) \geq t$, but $A(0) < t$, a contradiction. Therefore, $\forall x \in G$, $A(x) \wedge \lambda \leq A(0)$.

If (2) does not hold, then there exist $x, y, z \in G$ such that $A(x*(y*z)) \wedge A(y) \wedge \lambda > A(x*z)$. Choose $t = A(x*(y*z)) \wedge A(y) \wedge \lambda$, then $t \in (0, \lambda]$, $A(x*(y*z)) \geq t$ and $A(y) \geq t$. Hence, $x*(y*z), y \in A_t$. Since A_t is an ideal of G, then $x*z \in A_t$, so $A(x*z) \geq t$, but $A(x*z) < t$, a contradiction. Therefore, (2) holds.

$\Leftarrow$$\forall t \in (0, \lambda]$, if $\varnothing \neq A_t$, then $\exists x \in A_t$. Therefore, $A(x) \geq t$. Since (1) holds, we have $A(0) \geq A(x) \wedge \lambda \geq t \wedge t = t$, so $0 \in A_t$. Additionally, $\forall x, y, z \in G$, if $x*(y*z)$, $y \in A_t$, then based on (2), $A(x*z) \geq A(x*(y*z)) \wedge A(y) \wedge \lambda \geq t$, so $x*z \in A_t$. Therefore, A_t is an ideal of G.

Corollary 4. Let A be a fuzzy subset of G, $\forall t \in (0, \lambda]$, $\varnothing \neq A_t$ is an ideal of G if and only if A is an $(\in, \in \vee q_{(0,\lambda)})$-fuzzy ideal.

Based on Theorems 3, 5 and 6, we can have Theorem 7 below, which shows that the hierarchical structure of $(\in, \in \vee q_{(\lambda,\mu)})$-fuzzy ideals is richer than those of $(\in, \in)$-fuzzy ideals or $(\in, \in \vee q)$-fuzzy ideals.

Theorem 7. Let A be a fuzzy subset of G, so $A_t(t \in (0,1])$ is a level set of A, and then the following conclusions hold.

(1) $\forall t \in (0, \lambda], A_t \neq \varnothing$ is an ideal of G if and only if A is an $(\in, \in \vee q_{(0,\lambda)})$-fuzzy ideal.
(2) $\forall t \in (\lambda, \mu], A_t \neq \varnothing$ is an ideal of G if and only if A is an $(\in, \in \vee q_{(\lambda,\mu)})$-fuzzy ideal.
(3) $\forall t \in (\mu, 1], A_t \neq \varnothing$ is an ideal of G if and only if A is an $(\in, \in \vee q_{(\mu,1)})$-fuzzy ideal.

Theorem 8. If A is an $(\in, \in \vee q_{(\lambda,\mu)})$-fuzzy ideal of G, then $A(x*(x*y)) \vee \lambda \geq A(y) \wedge \mu$.

Proof. $A(x*(x*y))\vee\lambda = (A(x*(x*y))\vee\lambda)\vee\lambda \geq (A(x*(y*(x*y))))\wedge A(y)\wedge\mu)\vee\lambda = (A(x*0)\wedge A(y)\wedge\mu)\vee\lambda = (A(0)\wedge A(y)\wedge\mu)\vee\lambda = (A(0)\vee\lambda)\wedge(A(y)\vee\lambda)\wedge(\mu\vee\lambda) \geq (A(y)\wedge\mu)\wedge A(y)\wedge\mu = A(y)\wedge\mu$.

In the two theorems below, we discuss the properties of the union and intersection of the family of $(\in, \in \vee q_{(\lambda,\mu)})$-fuzzy ideals.

Theorem 9. If $\{A_i\}_{i\in I}$ is a family of $(\in, \in \vee q_{(\lambda,\mu)})-$ fuzzy ideals of G, then $\bigcap_{i\in I} A_i$ is an $(\in, \in \vee q_{(\lambda,\mu)})$-fuzzy ideal of G.

Proof. $\forall x \in G$, we have $(\bigcap_{i\in I} A_i)(0) \vee \lambda = (\bigwedge_{i\in I} A_i)(0) \vee \lambda = \bigwedge_{i\in I} (A_i(0) \vee \lambda) \geq \bigwedge_{i\in I} (A_i(x) \wedge \mu) = (\bigwedge_{i\in I} A_i(x)) \wedge \mu = (\bigcap_{i\in I} A_i)(x) \wedge \mu$. Hence, (III) holds. Since $\forall x, y, z \in G, (\bigcap_{i\in I} A_i)(x * z) \vee \lambda = (\bigwedge_{i\in I} A_i(x * z)) \vee \lambda = \bigwedge_{i\in I} (A_i(x * z) \vee \lambda) \geq \bigwedge_{i\in I} (A_i(x*(y*z))\wedge A_i(y)\wedge\mu) \geq (\bigwedge_{i\in I} A_i(x*(y*z)))\wedge(\bigwedge_{i\in I} A_i(y))\wedge\mu = (\bigcap_{i\in I} A_i)(x*(y*z)) \wedge (\bigcap_{i\in I} A_i)(y) \wedge \mu$. Therefore, (IV) holds.

Based on Definition 12, we can know that $\bigcap_{i\in I} A_i$ is an $(\in, \in \vee q_{(\lambda,\mu)})$-fuzzy ideal of G.

Theorem 10. If $\{A_i\}_{i\in I}$ is a family of $(\in, \in \vee q_{(\lambda,\mu)})$-fuzzy ideals of G, and for all $i, j \in I$, $A_i \subseteq A_j$ or $A_j \subseteq A_i$, then $\bigcup_{i\in I} A_i$ is an $(\in, \in \vee q_{(\lambda,\mu)})$-fuzzy ideal of G.

Proof. $\forall x \in G$, then we have $(\bigcup_{i\in I} A_i)(0) \vee \lambda = \bigvee_{i\in I} (A_i(0)) \vee \lambda = \bigvee_{i\in I} (A_i(0) \vee \lambda) \geq \bigvee_{i\in I} (A_i(x) \wedge \mu) = (\bigcup_{i\in I} A_i)(x) \wedge \mu$. Hence, (III) holds.

$\forall x, y, z \in G$, based on Theorem 2, we have $(\bigcup_{i\in I} A_i)(x * z) \vee \lambda = (\bigvee_{i\in I} A_i(x * z) \vee \lambda) = \bigvee_{i\in I} (A_i(x * z) \vee \lambda) \geq \bigvee_{i\in I} (A_i(x * (y * z)) \wedge A_i(y) \wedge \mu) = (\bigcup_{i\in I} A_i)(x * (y * z)) \wedge (\bigcup_{i\in I} A_i)(y) \wedge \mu$. Therefore, (IV) holds.

Hence, $\bigcup_{i\in I} A_i$ is an $(\in, \in \vee q_{(\lambda,\mu)})$-fuzzy ideal of G.

Theorem 11. Let A be a non-empty subset of G. Let B be a fuzzy subset of G defined by

$$B(x)=\begin{cases} s, x\in A \\ t,\ otherwise \end{cases},$$

where $t<s$, $0\le t<\mu$, $\lambda<s\le 1$. Then, B is an $(\in,\in\vee q_{(\lambda,\mu)})$-fuzzy ideal of G if and only if A is an ideal of G.

Proof. $\forall\alpha\in(\lambda,\mu]$, when $0\le t\le\lambda$, $\lambda<s<\mu$,

$$B_\alpha=\begin{cases} A, \lambda<\alpha\le s \\ \varnothing, s<\alpha\le\mu \end{cases};$$

when $0\le t\le\lambda$, $\mu\le s\le 1$,

$$B_\alpha=A;$$

when $\lambda<t<\mu$, $\lambda<s<\mu$,

$$B_\alpha=\begin{cases} G, \lambda<\alpha\le t; \\ A, t<\alpha\le s; \\ \varnothing, s<\alpha\le\mu; \end{cases}$$

when $\lambda<t<\mu$,$\mu\le s<1$,

$$B_\alpha=\begin{cases} G, \lambda<\alpha\le t; \\ A, t<\alpha\le s; \end{cases}$$

Based on Theorem 3 (3), B is an $(\in,\in\vee q_{(\lambda,\mu)})$-fuzzy ideal of G.

Corollary 5. A is an ideal of G if and only if the characteristic function χ_A of A is an $(\in,\in\vee q_{(\lambda,\mu)})$-fuzzy ideal of G.

Corollary 5 shows that the definition of the $(\in,\in\vee q_{(\lambda,\mu)})$-fuzzy ideal is reasonable.

In the following, we discuss some properties of $(\in,\in\vee q_{(\lambda,\mu)})$-fuzzy ideals by using the partial ordering relation of G.

Theorem 12. Let A be an $(\in,\in\vee q_{(\lambda,\mu)})$-fuzzy ideal of G, and if the inequality $y*z\le x$ holds in G, then $A(z)\vee\lambda\ge A(x)\wedge A(y)\wedge\mu$.

Proof. If $y*z\le x$, then $x*(y*z)=0$. Since $A(z)\vee\lambda=A\ (0*z)\vee\lambda\ge A(0*(y*z))\wedge A(y)\wedge\mu=A(y*z)\wedge A(y)\ \wedge\mu$.

Therefore $A(z)\vee\lambda=(A(z)\vee\lambda)\vee\lambda\ge(A(y*z)\wedge A\ (y)\wedge\mu)\vee\lambda=(A(y*z)\vee\lambda)\wedge(A(y)\vee\lambda)\wedge(\lambda\vee\mu)\ge(A(y*(x*z))\wedge A(x)\wedge\mu)\wedge A(y)\wedge\mu=A(x*(y*z))\wedge A(x)\wedge A(y)\wedge\mu=A(0)\wedge A(x)\wedge A(y)\wedge\mu$

Hence, $A(z)\vee\lambda=(A(z)\vee\lambda)\vee\lambda\ge(A(0)\wedge A\ (x)\wedge A(y)\wedge\mu)\vee\lambda=(A(0)\vee\lambda)\wedge(A(x)\vee\lambda)\wedge(A(y)\vee\lambda)\wedge(\lambda\vee\mu)\ge A(x)\wedge\mu\wedge A(x)\wedge A(y)\ \wedge\mu=A(x)\wedge A(y)\wedge\mu$.

Theorem 13. Let A be an $(\in,\in\vee q_{(\lambda,\mu)})$-fuzzy ideal of G, and if $x\le y$, then $A(x)\vee\lambda\ge A(y)\wedge\mu$.

Proof. If $x\le y$, then $y*x=0$, therefore $A(x)\vee\lambda=(A(0*x)\vee\lambda)\vee\lambda\ge(A(0*(y*x))\wedge A(y)\wedge\mu)\vee\lambda=(A(0*0)\wedge A(y)\wedge\mu)\vee\lambda=(A(0)\vee\lambda)\wedge((A(y)\wedge\mu)\vee\lambda)\ge(A(y)\wedge\mu)\wedge((A(y)\wedge\mu)\vee\lambda)\ge A(y)\wedge\mu$.

4 Image and Preimage of $(\in, \in \vee q_{(\lambda,\mu)})$-fuzzy ideals

In this section, we investigate the properties of the homomorphism image and homomorphism preimage of an $(\in, \in \vee q_{(\lambda,\mu)})$-fuzzy ideal.

Theorem 14. Let $f : G \to G'$ be a homomorphism. Let B be an $(\in, \in \vee q_{(\lambda,\mu)})$-fuzzy ideal of G', then $f^{-1}(B)$ is an $(\in, \in \vee q_{(\lambda,\mu)})$-fuzzy ideal of G.

Proof. We first show that $f^{-1}(B)$ satisfies (I). $\forall t \in (\lambda, 1]$, $\forall x \in G$, if $x_t \in f^{-1}(B)$, then $f^{-1}(B)(x) = B(f(x)) \geq t$, so $(f(x))_t \in B$. Since B is an $(\in, \in \vee q_{(\lambda,\mu)})$-fuzzy ideal of G', we have $(0')_t \in \vee q_{(\lambda,\mu)} B$. Since f is a homomorphism from G to G', then $f(0) = f(0 * 0) = f(0) *' f(0) = 0'$. If $(0')_t \in B$, so $B(0') \geq t$, then $f^{-1}(B)(0) = B(f(0)) = B(0') \geq t$, hence $0_t \in f^{-1}(B)$. If $(0')_t q_{(\lambda,\mu)} B$, then $f^{-1}(B)(0) + t = B(f(0)) + t = B(0') + t > 2\mu$. So $0_t q_{(\lambda,\mu)} f^{-1}(B)$. Therefore, $0_t \in \vee q_{(\lambda,\mu)} f^{-1}(B)$.

Then, we show that $f^{-1}(B)$ satisfies (II)

$\forall t_1, t_2 \in (\lambda, 1]$, $\forall x, y, z \in G$, if $y_{t_1} \in f^{-1}(B)$, $(x * (y * z))_{t_2} \in f^{-1}(B)$, then $f^{-1}(B)(y) = B(f(y)) \geq t_1$, $f^{-1}(B)(x * (y * z)) = B(f(x * (y * z))) = B(f(x) *' (f(y) *' f(z))) \geq t_2$. Hence, $(f(y))_{t_1} \in B$, $(f(x) *' (f(y) *' f(z)))_{t_2} \in B$. Since B is an $(\in, \in \vee q_{(\lambda,\mu)})$-fuzzy ideal of G', then $(f(x) *' f(z))_{t_1 \wedge t_2} \in \vee q_{(\lambda,\mu)} B$. If $(f(x) *' f(z))_{t_1 \wedge t_2} \in B$, then $f^{-1}(B)(x * z) = B(f(x * z)) = B(f(x) *' f(z)) \geq t_1 \wedge t_2$, hence $(x * z)_{t_1 \wedge t_2} \in f^{-1}(B)$; if $(f(x) *' f(z))_{t_1 \wedge t_2} q_{(\lambda,\mu)} B$, then $f^{-1}(B)(x * z) + t_1 \wedge t_2 = B(f(x * z)) + t_1 \wedge t_2 = B(f(x) *' f(z)) + t_1 \wedge t_2 > 2\mu$. So, $(x * z)_{t_1 \wedge t_2} q_{(\lambda,\mu)} f^{-1}(B)$. Therefore, $(x * z)_{t_1 \wedge t_2} \in \vee q_{(\lambda,\mu)} f^{-1}(B)$.

Hence, $f^{-1}(B)$ is an $(\in, \in \vee q_{(\lambda,\mu)})$-fuzzy ideal of G.

Theorem 15. Let A be a fuzzy subset of G. Let $f : G \to G'$ be a homomorphism. A is $f-$invariant. If $f(A)$ is an $(\in, \in \vee q_{(\lambda,\mu)})$-fuzzy ideal of G', then A is an $(\in, \in \vee q_{(\lambda,\mu)})$-fuzzy ideal of G.

Proof. (1) $\forall t \in (\lambda, 1]$, $\forall x \in G$, if $x_t \in A$, then $A(x) \geq t$. Put $y = f(x)$. Then $f(A)(y) = \bigvee_{f(w)=y} A(w) = A(x) \geq t$, so $y_t \in f(A)$. Since $f(A)$ is an $(\in, \in \vee q_{(\lambda,\mu)})$-fuzzy ideal of G', we have $0'_t \in \vee q_{(\lambda,\mu)} f(A)$. If $0'_t \in f(A)$, so $f(A)(0') \geq t$. Therefore, $A(0) = \bigvee_{f(w)=0'} A(w) = f(A)(0') \geq t$. Hence, $0_t \in A$. If $0'_t q_{(\lambda,\mu)} f(A)$, then we have $A(0) + t = \bigvee_{f(w)=0'} A(w) + t = f(A)(0') + t > 2\mu$. Hence, $0_t \in \vee q_{(\lambda,\mu)} A$

(2) $\forall x, y, z \in G$, $\forall t_1, t_2 \in (\lambda, 1]$, if $y_{t_1} \in A$, $(x * (y * z))_{t_2} \in A$, then $A(y) \geq t_1$, $A(x * (y * z)) \geq t_2$. Put $f(x) = x'$, $f(y) = y'$, $f(z) = z'$. Since f is a homomorphism, then $f(x * (y * z)) = x' *' (y' *' z')$. Since A is $f-$invariant, we have $t_1 \leq A(y) = \bigvee_{f(w)=y'} A(w) = f(A)(y')$, $t_2 \leq A(x * (y * z)) = \bigvee_{f(w)=x' *' (y' *' z')} A(w) = f(A)(x' *' (y' *' z'))$. Hence,

$y'_{t_1} \in f(A)$ and $(x' *' (y' *' z'))_{t_2} \in f(A)$. Since $f(A)$ is an $(\in, \in \vee q_{(\lambda,\mu)})$-fuzzy ideal of G', then $(x' *' z')_{t_1 \wedge t_2} \in \vee q_{(\lambda,\mu)} f(A)$, so $(x' *' z')_{t_1 \wedge t_2} \in f(A)$ or $(x' *' z')_{t_1 \wedge t_2} q_{(\lambda,\mu)} f(A)$. Therefore, $A(x * z) = \bigvee_{f(w)=x' *' z'} A(w) = f(A)(x' *' z') \geq t_1 \wedge t_2$ or $A(x * z) + t_1 \wedge t_2 = \bigvee_{f(w)=x' *' z'} A(w) + t_1 \wedge t_2 = f(A)(x' *' z') + t_1 \wedge t_2 > 2\mu$. Therefore, $(x * z)_{t_1 \wedge t_2} \in \vee q_{(\lambda,\mu)} A$.
Hence, A is an $(\in, \in \vee q_{(\lambda,\mu)})$-fuzzy ideal of G.

Theorem 16. Let $f : G \to G'$ be an epimorphism. A is $f-$invariant. Then, A is an $(\in, \in \vee q_{(\lambda,\mu)})$-fuzzy ideal of G if and only if $f(A)$ is an $(\in, \in \vee q_{(\lambda,\mu)})$-fuzzy ideal of G'.

Proof. $\Rightarrow$ (1) $\forall y \in G'$, $\forall t \in (\lambda, 1]$, since f is an epimorphism, then $\exists x_0 \in G$ such that $f(x_0) = y$. If $y_t \in f(A)$, then $t \leq f(A)(y) = \bigvee_{f(x)=y} A(x) = A(x_0)$. Hence, $(x_0)_t \in A$. Since A is an $(\in, \in \vee q_{(\lambda,\mu)})$-fuzzy ideal of G, then $0_t \in \vee q_{(\lambda,\mu)} A$. If $0_t \in A$, then we have $f(A)(0') = \bigvee_{f(x)=0'} A(x) = A(0) \geq t$, thus $(0')_t \in f(A)$. If $0_t q_{(\lambda,\mu)} A$, then we have $f(A)(0') + t = \bigvee_{f(x)=0'} A(x) + t = A(0) + t > 2\mu$. Therefore, $(0')_t \in \vee q_{(\lambda,\mu)} f(A)$.

(2) $\forall x', y', z' \in G$, $\forall t_1, t_2 \in (\lambda, 1]$, since f is an epimorphism, then there $\exists x, y, z \in G$, such that $f(x) = x'$, $f(y) = y'$, $f(z) = z'$ and $f(x * (y * z)) = x' *' (y' *' z')$. If $y'_{t_1} \in f(A)$, $(x' *' (y' *' z'))_{t_2} \in f(A)$, then $A(y) = \bigvee_{f(w)=y'} A(w) = f(A)(y') \geq t_1$, $A(x * (y * z)) = \bigvee_{f(w)=x' *' (y' *' z')} A(w) = f(A)(x' *' (y' *' z')) \geq t_2$, therefore, $y_{t_1} \in A$ and $(x * (y * z))_{t_2} \in A$. By A is an $(\in, \in \vee q_{(\lambda,\mu)})$-fuzzy ideal of G, thus $(x * z)_{t_1 \wedge t_2} \in \vee q_{(\lambda,\mu)} A$. Therefore, $f(A)(x' *' z') = \bigvee_{f(w)=x' *' z'} A(w) = A(x * z) \geq t_1 \wedge t_2$, or $f(A)(x' *' z') + t_1 \wedge t_2 = \bigvee_{f(w)=x' *' z'} A(w) + t_1 \wedge t_2 = A(x * z) + t_1 \wedge t_2 > 2\mu$. So, we have $(x' *' z')_{t_1 \wedge t_2} \in \vee q_{(\lambda,\mu)} f(A)$. Therefore, $f(A)$ is an $(\in, \in \vee q_{(\lambda,\mu)})$-fuzzy ideal of G'.

$\Leftarrow$Based on Theorem 15, we can obtain that A is an $(\in, \in \vee q_{(\lambda,\mu)})$-fuzzy ideal of G.

5 Direct Product and Projection of $(\in, \in \vee q_{(\lambda,\mu)})$-fuzzy ideals

In this section, we investigate the properties of the direct product and projection of $(\in, \in \vee q_{(\lambda,\mu)})$-fuzzy ideals.

Theorem 17. Let A and B be the $(\in, \in \vee q_{(\lambda,\mu)})$-fuzzy ideals of G and G', respectively, then $A \times B$ is an $(\in, \in \vee q_{(\lambda,\mu)})$-fuzzy ideal of $G \times G'$.

Proof. Since A and B are $(\in, \in \vee q_{(\lambda,\mu)})$-fuzzy ideals of G and G', respectively, A and B are generalized fuzzy ideals of G and G', respectively. Then, $\forall(x, y) \in G \times G'$, where $x \in G$ and $y \in G'$, we have $(A \times B)(0, 0) \vee \lambda = (A(0) \wedge B(0)) \vee \lambda = (A(0) \vee \lambda) \wedge (B(0) \vee \lambda) \geq (A(x) \wedge \mu) \wedge (B(y) \wedge \mu) = A(x) \wedge B(y) \wedge \mu = (A \times B)(x, y) \wedge \mu$.

By $\forall(x, y), (x', y'), (z_1, z_2) \in G \times G'$, where x, x', $z_1 \in G$ and y, y', $z_2 \in G'$, then $(A \times B)((x, y) * (z_1, z_2)) \vee \lambda$

$= (A \times B)(x * z_1, y * z_2) \vee \lambda$
$= (A(x * z_1) \wedge B(y * z_2)) \vee \lambda$
$= (A(x * z_1) \vee \lambda) \wedge (B(y * z_2) \vee \lambda)$
$\geq (A(x * (x' * z_1)) \wedge A(x') \wedge \mu) \wedge (B(y * (y' * z_2)) \wedge B(y') \wedge \mu)$
$= (A(x * (x' * z_1)) \wedge B(y * (y' * z_2))) \wedge (A(x') \wedge B(y')) \wedge \mu$
$= (A \times B)(x * (x' * z_1), y * (y' * z_2)) \wedge (A \times B)(x', y') \wedge \mu$
$= (A \times B)((x, y) * ((x', y') * (z_1, z_2))) \wedge (A \times B)(x', y') \wedge \mu.$

Therefore, $A \times B$ is an $(\in, \in \vee q_{(\lambda,\mu)})$-fuzzy ideal of $G \times G'$.

Theorem 18. Let $A \times B$ be an $(\in, \in \vee q_{(\lambda,\mu)})$-fuzzy ideal of $G \times G'$, then A_G is an $(\in, \in \vee q_{(\lambda,\mu)})$-fuzzy ideal of G and $B_{G'}$ is an $(\in, \in \vee q_{(\lambda,\mu)})$-fuzzy ideal of G'.

Proof. Since $A \times B$ is an $(\in, \in \vee q_{(\lambda,\mu)})$-fuzzy ideal of $G \times G'$, based on Theorem 3, we can determine that $A \times B$ is a generalized fuzzy ideal of $G \times G'$.

(1) $\forall x \in G$, $A_G(0) \vee \lambda = (\bigvee_{z \in G'} (A \times B)(0, z)) \vee \lambda = (\bigvee_{z \in G'} (A(0) \wedge B(z))) \vee \lambda = \bigvee_{z \in G'} ((A(0) \vee \lambda) \wedge (B(z) \vee \lambda)) \geq \bigvee_{z \in G'} ((A(x) \wedge \mu) \wedge B(z)) = \bigvee_{z \in G'} (A(x) \wedge B(z)) \wedge \mu = \bigvee_{z \in G'} ((A \times B)(x, z)) \wedge \mu = A_G(x) \wedge \mu$.

(2) $\forall x, y, z \in G$, $A_G(x * z) \vee \lambda = (\bigvee_{w \in G'} (A \times B)(x * z, w)) \vee \lambda \vee \lambda = \bigvee_{w \in G'} ((A \times B)(x * z, w) \vee \lambda) \vee \lambda \geq ((A \times B)(x * z, 0') \vee \lambda) \vee \lambda = ((A \times B)((x, 0') * (z, 0')) \vee \lambda) \vee \lambda \geq [(A \times B)((x, 0') * ((y, w_1) * (z, 0'))) \wedge (A \times B)(y, w_1) \wedge \mu] \vee \lambda = ((A \times B)(x * (y * z), 0') \wedge (A \times B)(y, w_1) \wedge \mu) \vee \lambda = ((A(x * (y * z)) \wedge B(0')) \wedge ((A \times B)(y, w_1) \wedge \mu)) \vee \lambda = (A(x * (y * z)) \vee \lambda) \wedge (B(0') \vee \lambda) \wedge ((A \times B)(y, w_1) \vee \lambda) \wedge (\mu \vee \lambda) \geq A(x * (y * z)) \wedge (B(w_2) \wedge \mu) \wedge ((A \times B)(y, w_1) \wedge \mu) = (A \times B)(x * (y * z), w_2) \wedge (A \times B)(y, w_1) \wedge \mu$.

By the arbitrariness of w_2, we can have that

$A_G(x * z) \vee \lambda \geq \bigvee_{w_2 \in G'} ((A \times B)(x * (y * z), w_2) \wedge (A \times B)(y, w_1) \wedge \mu) = (\bigvee_{w_2 \in G'} (A \times B)(x * (y * z), w_2)) \wedge (A \times B)(y, w_1) \wedge \mu = A_G(x * (y * z)) \wedge (A \times B)(y, w_1) \wedge \mu$.

By the arbitrariness of w_1, we can have that

$A_G(x * z) \vee \lambda \geq \bigvee_{w_1 \in G'} (A_G(x * (y * z)) \wedge (A \times B)(y, w_1) \wedge \mu) = A_G(x * (y * z)) \wedge (\bigvee_{w_1 \in G'} (A \times B)(y, w_1)) \wedge \mu = A_G(x * (y * z)) \wedge A_G(y) \wedge \mu$.

From the above, A_G is an $(\in, \in \vee q_{(\lambda,\mu)})$-fuzzy ideal of G.

Similarly, to prove that $B_{G'}$ is an $(\in, \in \vee q_{(\lambda,\mu)})$-fuzzy ideal of G'.

6 Chain Conditions

In this section, we introduce the new conception of the chain condition of the ideals of KU-algebras, and depict the chain conditions by using the properties of $(\in, \in \vee q_{(\lambda,\mu)})$-fuzzy ideals.

Definition 13. Let $A_i (i = 1, 2, \ldots)$ be ideals of KU-algebra G. If there exist finite terms in every descending sequence of ideals of G

$$A_1 \supset A_2 \supset \cdots A_n \supset \cdots$$

that is to say, for every descending sequence of ideals containing infinite terms, $A_1 \supseteq A_2 \supseteq \cdots A_n \supseteq \cdots$, there necessarily exists a positive integer m such that after A_m, all the ideals are equal, so $A_m = A_{m+1} = \cdots$. Then, we state that G satisfies the descending chain conditions with respect to ideals (or G is Artinian with respect to ideals).

Definition 14. Let set $(S, \leq)$ be a partially ordered set. For every non-empty subset S, if there exists a maximal element under the partial order of the set, then $\leq$ is called a dual well-ordered relation and $(S, \leq)$ is called a dual well-ordered set.

Theorem 19. Let G be a KU-algebra, $\{A_i\}_{i \in I}$ is a family of ideals of G. $\forall i, j \in I$, we have $A_i \subseteq A_j$ or $A_j \subseteq A_i$, then $A = \bigcup_{i \in I} A_i$ is an ideal of G.

Proof. Since $A_i (i = 1, 2, \ldots)$ are ideals of G, then $0 \in A_i$, so $0 \in \bigcup_{i \in I} A_i = A$. $\forall x, y, z \in G$, if $x * (y * z) \in A$ and $y \in A$, then there exist i_1, i_2 such that $x * (y * z) \in A_{i_1}$, $y \in A_{i_2}$. Since $A_{i_1} \subseteq A_{i_2}$ or $A_{i_2} \subseteq A_{i_1}$, we can assume that $A_{i_1} \subseteq A_{i_2}$. Therefore, $x * (y * z) \in A_{i_1} \subseteq A_{i_2}$. Because A_{i_2} is an ideal, $x * z \in A_{i_2} \subseteq A$. Hence, $A = \bigcup_{i \in I} A_i$ is an ideal of G.

Theorem 20. Let G be a KU-algebra, $\{A_i\}_{i \in I}$ is a family of ideals of G. Then, $A = \bigcap_{i \in I} A_i$ is an ideal of G.

Proof. $\forall i \in I$, since A_i is an ideal of G, then $0 \in A_i$, so $0 \in \bigcap_{i \in I} A_i = A$. $\forall x, y, z \in G$, if $x * (y * z) \in A$ and $y \in A$, then $\forall i \in I$, we have $x * (y * z) \in A_i$ and $y \in A_i$. Since A_i is an ideal of G, $x * z \in A_i$. Thus, $x * z \in A$. Therefore, $A = \bigcap_{i \in I} A_i$ is an ideal of G.

Theorem 21. Let G satisfy the descending chain condition with respect to ideals. Let A be an $(\in, \in \vee q_{(\lambda,\mu)})$-fuzzy ideal of G. Then, there does not exist an infinite ascending sequence in $\mathrm{Im}(A) \bigcap [\lambda, \mu]$.

Proof. Let $\{t_i | i = 0, 1, 2, \ldots\}$ be an infinite ascending sequence in $\mathrm{Im}(A) \bigcap [\lambda, \mu]$, then $t_0 < t_1 < \cdots < \mu$. Therefore, $t_1 < t_2 < \cdots < \mu$ is a strictly infinite ascending chain, where $t_i \in (\lambda, \mu], i = 1, 2, \ldots$, and $\exists x_i \in G$ such that $A(x_i) = t_i$. By Theorem 3, we can determine that level sets A_{t_i} are ideals of G. Since

$A(x_{i-1}) = t_{i-1} < t_i$, so $x_{i-1} \notin A_{t_i}$. Thus $A_{t_i} \subset A_{t_{i-1}}$. Therefore, $A_{t_1} \supset A_{t_2} \supset \cdots$ is an infinite descending chain of ideals of G. But the ideals of G satisfy the descending chain condition, which is a contradiction.

Theorem 22. Let G be a KU-algebra, and then G is Noetherian with respect to an ideal if and only if for every $(\in, \in \vee q_{(\lambda,\mu)})$-fuzzy ideal A of G, $\mathrm{Im}(A) \bigcap [\lambda, \mu]$ is a well-ordered set.

Proof. $\Rightarrow$Let A be an $(\in, \in \vee q_{(\lambda,\mu)})$-fuzzy ideal of G. Based on Theorem 3, we can determine that A is a generalized fuzzy ideal of G. Assuming that $\mathrm{Im}(A) \bigcap [\lambda, \mu]$ is not a well-ordered set, then there exists a strictly descending sequence $\{t_i\}$ in $\mathrm{Im}(A) \bigcap [\lambda, \mu]$, and $t_i \neq \lambda$(otherwise, there does not exist an infinite descending chain), $i = 1, 2, \ldots$, and there exists $x_i \in G$ such that $A(x_i) = t_i$. Therefore, $A_{t_1} \subseteq A_{t_2} \subseteq A_{t_3} \subseteq \cdots$. By $t_i \in (\lambda, \mu]$, based on Theorem 3, we can have that level sets A_{t_i} are ideals. If $t_{i_1} > t_{i_2}$, then $A(x_{i_2}) = t_{i_2} < t_{i_1}$, so $x_{i_2} \notin A_{t_{i_1}}$. Hence, $A_{t_{i_1}} \subset A_{t_{i_2}}$.

Therefore, $A_{t_1} \subset A_{t_2} \subset A_{t_3} \subset \cdots$ is a strictly ascending chain of G with respect to ideals. But G is Noetherian, which is a contradiction.

$\Leftarrow$For every generalized fuzzy ideal A of G, $\mathrm{Im}(A) \bigcap [\lambda, \mu]$ is a well-ordered set. If G is not Noetherian, then there exists a strictly ascending chain $A_1 \subset A_2 \subset A_3 \subset \cdots$, where A_i are ideals of G. Based on Theorem 19, we can have that $\bigcup\limits_{i=1}^{\infty} A_i$ is an ideal of G.

Define a fuzzy subset by

$$B(x) = \begin{cases} \lambda, & x \notin \bigcup\limits_{i=1}^{\infty} A_i \\ \lambda + \dfrac{1}{i_x}(\mu - \lambda), & i_x = \min\{i \in N | x \in A_i\} \end{cases}$$

Then, $\mathrm{Im}(B) \subseteq [\lambda, \mu]$, and in $\mathrm{Im}(B) \bigcap [\lambda, \mu] = \mathrm{Im}(B)$ there is no least element in the subset $\{\lambda + \frac{1}{i_x}(\mu - \lambda),\ i_x = \min\{i \in N | x \in A_i\}\}$. Therefore, $\mathrm{Im}(B) \bigcap [\lambda, \mu]$ is not a well-ordered set.

In the following, we prove that B is a generalized fuzzy ideal of G.

Since A_i is an ideal, $0 \in A_i (i = 1, 2, 3...)$, $i_0 = 1$. Therefore, $\forall x \in G$, $B(0) \vee \lambda = (\lambda + (\mu - \lambda)) \vee \lambda = \mu \geq B(x) \wedge \mu$.

$\forall x, y, z \in G$, the following are discussed in four cases:

(1) $x * (y * z) \notin \bigcup\limits_{i=1}^{\infty} A_i$, $y \notin \bigcup\limits_{i=1}^{\infty} A_i$, then $B(x * (y * z)) = B(y) = \lambda$. So, $B(x * z) \vee \lambda \geq \lambda = B(x * (y * z)) \wedge B(y) \wedge \mu$.

(2) $x * (y * z) \notin \bigcup\limits_{i=1}^{\infty} A_i$, $y \in \bigcup\limits_{i=1}^{\infty} A_i$, then $B(x * (y * z)) = \lambda$. So, $B(x * z) \vee \lambda \geq B(x * (y * z)) \wedge B(y) \wedge \mu$.

(3) $x * (y * z) \in \bigcup\limits_{i=1}^{\infty} A_i$, $y \notin \bigcup\limits_{i=1}^{\infty} A_i$, then $B(y) = \lambda$. So $B(x * z) \vee \lambda \geq B(x * (y * z)) \wedge B(y) \wedge \mu$.

(4) $x * (y * z) \in \bigcup\limits_{i=1}^{\infty} A_i$, $y \in \bigcup\limits_{i=1}^{\infty} A_i$, then $x * (y * z) \in A_{i_{x*(y*z)}}$, $y \in A_{i_y}$.

① If $i_{x*(y*z)} = i_y$, assume $m = i_{x*(y*z)}$. Since $x*(y*z) \in A_m$, $y \in A_m$, and because A_m is an ideal of G, hence $x*z \in A_m$. By the definition of B, we have $i_{x*z} \le m$, so $B(x*z) \vee \lambda = (\lambda + \frac{1}{i_{x*z}}(\mu-\lambda)) \vee \lambda \ge (\lambda + \frac{1}{m}(\mu-\lambda)) \vee \lambda \ge B(x*(y*z)) \wedge B(y) \wedge \mu$.

② If $i_{x*(y*z)} > i_y$, since $A_{i_{x*(y*z)}} \supset A_{i_y}$, we have $y \in A_{i_{x*(y*z)}}$. Therefore, $x*z \in A_{i_{x*(y*z)}}$. By the definition of B, we have $i_{x*z} \le i_{x*(y*z)}$. Therefore, $B(x*z) \vee \lambda = (\lambda + \frac{1}{i_{x*z}}(\mu-\lambda)) \vee \lambda \ge (\lambda + \frac{1}{i_{x*(y*z)}}(\mu-\lambda)) \vee \lambda \ge B(x*(y*z)) \wedge B(y) \wedge \mu$.

③ If $i_{x*(y*z)} < i_y$, similarly, to prove $B(x*z) \vee \lambda \ge B(x*(y*z)) \wedge B(y) \wedge \mu$.

From the above, we can determine that B is an $(\in, \in \vee q_{(\lambda,\mu)})$-fuzzy ideal of G, a contradiction. Hence, G is Noetherian with respect to an ideal.

Corollary 6. For every $(\in, \in \vee q_{(\lambda,\mu)})$-fuzzy ideal A of G, if its range is a finite set, then G is Noetherian with respect to an ideal.

Proof. Since $\mathrm{Im}(B) \bigcap [\lambda, \mu] = \mathrm{Im}(B)$ is a finite set, then it is a well-ordered set. By Theorem 22, we have that G is Noetherian with respect to an ideal.

Theorem 23. For every $(\in, \in \vee q_{(\lambda,\mu)})$-fuzzy ideal A of G, if its range is a finite set, then G is Artinian with respect to an ideal.

Proof. Assuming that G is not Artinian with respect to an ideal, then there exists a strictly infinite descending chain of ideals of G : $A_1 \supset A_2 \supset \cdots$. If $A_1 \ne G$, choose $A_0 = G$, and then we have a strictly descending chain: $G = A_0 \supset A_1 \supset A_2 \supset \cdots$. Therefore, we can always have a strictly descending chain: $G = A_0 \supset A_1 \supset A_2 \supset \cdots$.

Define a fuzzy subset B of G:

$$B(x) = \begin{cases} \dfrac{n}{n+1}, x \in A_n - A_{n+1}, n = 0, 1, 2... \\ 1 \quad\quad , x \in \bigcap\limits_{n=0}^{\infty} A_n \end{cases}$$

By Theorem 20, we have that $\bigcap\limits_{n=0}^{\infty} A_n$ is an ideal of G, so $0 \in \bigcap\limits_{n=0}^{\infty} A_n$. Hence, $\forall x \in G$, $B(0) \vee \lambda = 1 \ge B(x) \wedge \mu$.

$\forall x, y, z \in G$, the following are discussed in four cases:

① $x*(y*z)$, $y \in \bigcap\limits_{n=0}^{\infty} A_n$, so $x*z \in \bigcap\limits_{n=0}^{\infty} A_n$. Hence $B(x*z) = 1$, therefore $B(x*z) \vee \lambda \ge B(x*(y*z)) \wedge B(y) \wedge \mu$.

② $x*(y*z) \in \bigcap\limits_{n=0}^{\infty} A_n$, $y \notin \bigcap\limits_{n=0}^{\infty} A_n$, then there exists k such that $y \in A_k - A_{k+1}$, so $B(y) = \frac{k}{k+1}$. Since $x*(y*z) \in \bigcap\limits_{n=0}^{\infty} A_n \subset A_k$, $y \in A_k - A_{k+1} \subset A_k$ and A_k is an ideal of G, so $x*z \in A_k$. If $x*z \in \bigcap\limits_{n=0}^{\infty} A_n$, then $B(x*z) = 1$. Hence, $B(x*z) \vee \lambda \ge B(x*(y*z)) \wedge B(y) \wedge \mu$; if $x*z \notin \bigcap\limits_{n=0}^{\infty} A_n$, by $A_k \supset A_{k+1} \supset A_{k+2} \supset \cdots$, so there exists $k^{'}$ such that $A_k \supset \cdots \supset A_{k^{'}} \supset A_{k^{'}+1} \supset \cdots$, $x*z \in A_{k^{'}} - A_{k^{'}+1}$ and $k^{'} \geq k$. So $B(x*z) = \frac{k^{'}}{k^{'}+1} \geq \frac{k}{k+1}$. Therefore, $B(x*z) \vee \lambda \geq \frac{k}{k+1} \geq B(x*(y*z)) \wedge B(y) \wedge \mu$.

③ $x * (y * z) \notin \bigcap_{n=0}^{\infty} A_n$, $y \in \bigcap_{n=0}^{\infty} A_n$, it is similar to prove $B(x * z) \vee \lambda \geq B(x * (y * z)) \wedge B(y) \wedge \mu$.

④ $x * (y * z) \notin \bigcap_{n=0}^{\infty} A_n, y \notin \bigcap_{n=0}^{\infty} A_n$, there exist $k, k^{'}$ such that $x * (y * z) \in A_k - A_{k+1}$, $y \in A_{k^{'}} - A_{k^{'}+1}$. Assume $k \geq k^{'}$. By $A_k - A_{k+1} \subset A_k \subseteq A_{k^{'}}$, thus $x * (y * z) \in A_{k^{'}}$. Since $A_{k^{'}}$ is an ideal, so $x * z \in A_{k^{'}}$. If $x * z \in \bigcap_{n=0}^{\infty} A_n$, then $B(x * z) = 1$. Hence, $B(x * z) \vee \lambda \geq B(x * (y * z)) \wedge B(y) \wedge \mu$; if $x * z \notin \bigcap_{n=0}^{\infty} A_n$, by $A_{k^{'}} \supset A_{k^{'}+1} \supset A_{k^{'}+2} \supset \cdots$, there exists $k^{''}$ such that $A_{k^{'}} \supset \cdots \supset A_{k^{''}} \supset A_{k^{''}+1} \supset \cdots$, $x*z \in A_{k^{''}} - A_{k^{''}+1}$ and $k^{''} \geq k^{'}$. Since $B(x*(y*z)) \wedge B(y) \wedge \mu = \frac{k}{k+1} \wedge \frac{k^{'}}{k^{'}+1} \wedge \mu \leq \frac{k^{'}}{k^{'}+1} \leq \frac{k^{''}}{k^{''}+1}$, hence $B(x * z) \vee \lambda \geq B(x * (y * z)) \wedge B(y) \wedge \mu$. Therefore, B is a generalized fuzzy ideal of G. By Theorem 3, we can obtain that B is an $(\in, \in \vee q_{(\lambda,\mu)})$-fuzzy ideal of G. But the range of B is an infinite set, which is a contradiction.

Corollary 7. If the range of each $(\in, \in \vee q_{(\lambda,\mu)})$-fuzzy ideal of G is a finite set, then G is Artinian and Noetherian with respect to ideals.

Proof. By Theorem 23, we have that G is Artinian. By Corollary 6, we have that G is Noetherian.

Theorem 24. Let G be a KU-algebra, and then G is Artinian with respect to an ideal if and only if for every $(\in, \in \vee q_{(\lambda,\mu)})$-fuzzy ideal A of G, $\mathrm{Im}(A) \bigcap [\lambda, \mu]$ is a dual well-ordered set.

Proof. $\Rightarrow$ Let A be an $(\in, \in \vee q_{(\lambda,\mu)})$-fuzzy ideal of G. By Theorem 3, we have that A is a generalized fuzzy ideal of G. Assuming that $\mathrm{Im}(A) \bigcap [\lambda, \mu]$ is not a dual well-ordered set, then there exists a strictly ascending sequence in $\mathrm{Im}(A) \bigcap [\lambda, \mu]$, denoted by $\{t_i\}(i = 0, 1, 2, \ldots)$. So, $t_i > t_0 \geq \lambda (i = 1, 2, \ldots)$ and $\{t_i\}(i = 1, 2, \ldots)$ is also a strictly ascending sequence, $t_i \in (\lambda, \mu](i = 1, 2, \ldots)$. Therefore, $A_{t_1} \supseteq A_{t_2} \supseteq A_{t_3} \supseteq \cdots$.

By Theorem 3 and $t_i \in (\lambda, \mu]$, we have that level sets A_{t_i} $(i = 1, 2, \ldots)$ are ideals. Since $t_i \in \mathrm{Im}(A)$, then there exists $x_i \in G$ such that $A(x_i) = t_i$. If $i < j$, then $A(x_i) = t_i < t_j$. Thus $x_i \notin A_{t_j}$. Hence, $A_{t_i} \supset A_{t_j}$.

Therefore, $A_{t_1} \supset A_{t_2} \supset A_{t_3} \supset \cdots$ is a strictly descending chain with respect to the ideals of G. But G is Artinian, which is a contradiction.

$\Leftarrow$For every generalized fuzzy ideal A of G, if $\mathrm{Im}(A) \bigcap [\lambda, \mu]$ is a dual well-ordered set but G is not Artinian, then there exists a strictly descending chain $A_1 \supset A_2 \supset A_3 \supset \cdots$, where A_i are ideals of G. By Theorem 19, $\bigcup_{i=1}^{\infty} A_i$ is an ideal of G.

Define a fuzzy subset:

$$B(x)=\begin{cases}\mu, & x=0\\ \lambda, & x\notin\bigcup_{i=1}^{\infty}A_i\\ \mu-\frac{1}{i_x}(\mu-\lambda), & \\ i_x=\max\{i\in N|x\in A_i\ and\ x\neq 0\} & \end{cases}$$

Then, $\mathrm{Im}(B)\bigcap[\lambda,\mu]=\mathrm{Im}(B)$. In $\{\mu-\frac{1}{i_x}(\mu-\lambda),\ i_x=\max\{i\in N|x\in A_i\ and\ x\neq 0\}\}$, there does not exist a maximal element. Therefore, $\mathrm{Im}(B)\bigcap[\lambda,\mu]$ is not a dual well-ordered set.

In the following, we prove that B is a generalized fuzzy ideal of G. Since A_i is an ideal, then $0\in A_i(i=1,2,3...)$. Therefore, $\forall x\in G$, $B(0)\vee\lambda=\mu\geq B(x)\wedge\mu$. $\forall x,y,z\in G$, the following are discussed in two cases:

(1) $x*(y*z)\notin\bigcup_{i=1}^{\infty}A_i$ or $y\notin\bigcup_{i=1}^{\infty}A_i$, then $B(x*(y*z))=\lambda$ or $B(y)=\lambda$.

Therefore, $B(x*z)\vee\lambda\geq\lambda\geq B(x*(y*z))\wedge B(y)\wedge\mu$.

(2) $x*(y*z)\in\bigcup_{i=1}^{\infty}A_i$ and $y\in\bigcup_{i=1}^{\infty}A_i$, then $x*(y*z)\in A_{i_{x*(y*z)}}$, $y\in A_{i_y}$.

① If $i_{x*(y*z)}=i_y$, assume $m=i_{x*(y*z)}$. Since $x*(y*z)\in A_m$, $y\in A_m$, and A_m is an ideal of G, then $x*z\in A_m$. By the definition of B, we have $i_{x*z}\geq m$. Therefore $B(x*z)\vee\lambda=(\mu-\frac{1}{i_{x*z}}(\mu-\lambda))\vee\lambda\geq(\mu-\frac{1}{m}(\mu-\lambda))\vee\lambda=B(y)\vee\lambda\geq B(x*(y*z))\wedge B(y)\wedge\mu$.

② If $i_{x*(y*z)}>i_y$, then $A_{i_{x*(y*z)}}\subset A_{i_y}$. Therefore $x*(y*z)\in A_{i_y}$ and $y\in A_{i_y}$. Since A_{i_y} is an ideal, we have $x*z\in A_{i_y}$. By the definition of B, we have $i_{x*z}\geq i_y$. Therefore $B(x*z)\vee\lambda=(\mu-\frac{1}{i_{x*z}}(\mu-\lambda))\vee\lambda\geq(\mu-\frac{1}{i_y}(\mu-\lambda))\vee\lambda\geq B(x*(y*z))\wedge B(y)\wedge\mu$.

③ If $i_{x*(y*z)}<i_y$, it is similar to prove $B(x*z)\vee\lambda\geq B(x*(y*z))\wedge B(y)\wedge\mu$.

From the above, B is an $(\in,\in\vee q_{(\lambda,\mu)})$-fuzzy ideal of G. However, there does not exist a maximal element in $\{\mu-\frac{1}{i_x}(\mu-\lambda),\ i_x=\max\{i\in N|x\in A_i\ and\ x\neq 0\}\}\subseteq\mathrm{Im}(B)\bigcap[\lambda,\mu]$, so $\mathrm{Im}(B)\bigcap[\lambda,\mu]$ is not a dual well-ordered set, which is contradictory.

Therefore, G is Artinian with respect to ideals.

7 Conclusion

In this paper, we propose the concept of $(\in,\in\vee q_{(\lambda,\mu)})$-fuzzy ideals of KU-algebras and consider their equivalent characterizations and other fundamental properties by using the pointwise method. Moreover, the rich hierarchical structure of $(\in,\in\vee q_{(\lambda,\mu)})$-fuzzy ideals of KU-algebras is also noted. The generalized fuzzy algebra structure was introduced in [12], which is an original study. But it is quite different from the usual algebra research methods. The pointwise method is more similar to the classical mathematical methods, which is a feature of this paper. By using the properties of the $(\in, \in \vee q_{(\lambda,\mu)})$-fuzzy ideals of KU-algebras, the chain conditions of the ideals of KU-algebras are characterized, where there needs to be weaker conditions than using the fuzzy ideals of KU-algebras to characterize the chain conditions (it needs that $\mathrm{Im}(A)$ is a well-ordered set when the chain conditions are characterized by fuzzy ideals, but in this paper, it only needs $\mathrm{Im}(A) \bigcap [\lambda, \mu]$ is a well-ordered set, the condition is weaker). This is a new idea to investigate classical algebra by using the methods of fuzzy mathematics. These investigations promote research on the theories of KU-algebras and enrich the study of fuzzy algebra. Further research is needed to introduce and characterize other types of $(\in, \in \vee q_{(\lambda,\mu)})$-fuzzy ideals of KU-algebras.

Acknowledgements. This work is supported by the National Natural Science Foundation of China (No. 611702121, No. 11401259 and No. 61673193), the Natural Science Foundation of Jiangsu Province (No. BK2015117) and the Undergraduate Innovation and Entrepreneurship Training Program of China (No. 201610295005).

References

1. Zadeh, L.A.: Fuzzy sets. Inf. Control. **8**(3), 338–353 (1965)
2. Rosenfeld, A.: Fuzzy groups. J. Math. Anal. Appl. **35**(3), 512–517 (1971)
3. Kuroki, N.: On fuzzy ideals and fuzzy bi-ideals in semigroups. Fuzzy Sets Syst. **5**(2), 203–215 (1981)
4. Kuroki, N.: Fuzzy semiprime ideals in semigroups. Fuzzy Sets Syst. **8**(1), 71–79 (1982)
5. Kuroki, N.: Fuzzy generalized bi-ideals in semigroups. Inf. Sci. **66**(3), 235–243 (1992)
6. Pu, P.P., Liu, Y.M.: Fuzzy topology (I): neighbourhood structure of a fuzzy point and Moore-Smith cover. J. Math. Anal. Appl. **76**(2), 571–599 (1980)
7. Pu, P.P., Liu, Y.M.: Fuzzy topology (II): product and quotient spaces. J. Math. Anal. Appl. **77**(1), 20–37 (1980)
8. Bhakat, S.K., Das, P.: On the definition of a fuzzy subgroup. Fuzzy Sets Syst. **51**(2), 235–241 (1992)
9. Bhakat, S.K., Das, P.: $(\in, \in \vee q)$-fuzzy subgroup. Fuzzy Sets Syst. **80**(3), 359–368 (1996)
10. Liu, C.H.: $(\in, \in \vee q)$-fuzzy prime filters in BL-algebras. J. Zhejiang Univ. Sci. Ed. **41**(5), 489–493 (2014)
11. Liu, C.H.: $(\in, \in \vee q)$-fuzzy soft ideals in BCK/BCI-algebras. J. Zhejiang Univ. Sci. Ed. **43**(3), 256–260 (2015)
12. Yuan, X., Zhang, C., Ren, Y.: Generalized fuzzy groups and many-valued implications. Fuzzy Sets Syst. **138**(1), 205–211 (2003)
13. Liao, Z.H., Gu, H.: $(\in, \in \vee q_{(\lambda,\mu)})$-fuzzy normal subgroup. Fuzzy Syst. Math. **20**(5), 47–53 (2006)
14. Gu, H., Liao, Z.: $(\in, \in \vee q_{(\lambda,\mu)})$-covex fuzzy set. Fuzzy Syst. Math. **21**(1), 92–96 (2007)
15. Chen, M., Liao, Z.: $(\bar{\in}, \bar{\in} \vee \bar{q}_{(\lambda,\mu)})$-fuzzy subnear rings and ideals. Math. Pract. Theory **39**(4), 225–230 (2009)

16. Yi, L.H.: Several kinds of $(\in, \in \vee q_{(\lambda,\mu)})$-fuzzy ideals of semigroups and the characterization of Drazin semigroups. Jiangnan University, Wuxi (2010)
17. Liu, C.Z., Liao, Z.H., Jiang, X.: $(\in, \in \vee q_{(\lambda,\mu)})$-fuzzy Boolean algebra. Fuzzy Syst. Math. **27**(4), 67–73 (2013)
18. Zhu, C., Liao, Z.H., Luo, X., Ye, L.: $(\in, \in \vee q_{(\lambda,\mu)})$- fuzzy complemented semirings. Fuzzy Syst. Math. **27**(5), 48–54 (2013)
19. Zhang, J.Z., Fu, X.B., Liao, Z.H.: The $(\in, \in \vee q_{(\lambda,\mu)})$- fuzzy implicative ideals of a N(2,2,0) algebra. Math. Pract. Theory **44**(4), 278–285 (2014)
20. Fan, X.W., Liao, Z.H., Fan, Y.Y.: $(\in, \in \vee q_{(\lambda,\mu)})$-fuzzy Γ- completely prime ideals of Γ- semigroups. Fuzzy Syst. Math. **28**(2), 52–61 (2014)
21. Zhang, J.Z., Fu, X.B., Liao, Z.H.: $(\in, \in \vee q_{(\lambda,\mu)})$-fuzzy associative ideal of N(2,2,0) algebra. Comput. Eng. Appl. **50**(12), 54–58 (2014)
22. Fu, X.B., Liao, Z.H., Zheng, G.: $(\in, \in \vee q_{(\lambda,\mu)})$-fuzzy LI-ideals of lattice implication algebra. Fuzzy Syst. Math. **28**(5), 41–50 (2014)
23. Lu, T., Liao, Z.H., Liao, C.C., Yuan, W.G.: $(\bar{\in}, \bar{\in} \vee \bar{q}_{(\lambda,\mu)})$- fuzzy ideals of incline algebra. J. Front. Comput. Sci. Technol. **10**(8), 1191–1200 (2016)
24. Fu, X.B., Liao, Z.H.: $(\in, \in \vee q_{(\lambda,\mu)})$- fuzzy prime filter of lattice implication algebra. J. Front. Comput. Sci. Technol. **9**(2), 227–233 (2015)
25. Prabpayak, C., Leerawat, U.: On ideas and congruences of KU-algebras. Scientia Magna **5**(1), 54–57 (2009)
26. Prabpayak, C., Leerawat, U.: On isomorphisms of KU-algebras. Scientia Magna **5**(3), 25–31 (2009)
27. Rezaei, A., Saeid, A.B.: KU-algebras are equivalent to commutative self-distributive BE-algebras. Bollettino di Matematica Pura ed Applicata **7**, 1–8 (2014)
28. Wu, W.M.: Fuzzy implication algebras. Fuzzy Syst. Math. **4**(1), 56–64 (1990)
29. Xu, Y., Ruan, D., Qin, K.: Lattice-Valued Logic. Springer, Germany (2003)
30. Mostafa, S.M.: Fuzzy ideals of KU-algebras. Int. Math. Forum **6**(63), 3139–3149 (2011)
31. Gulistan, M., Shahzad, M., Yaqoob, N.: On $(\in, \in \vee q_k)$-fuzzy KU-ideals of KU-algebras. Acta Univ. Apulensis **39**, 75–83 (2014)
32. Gulistan, M., Shahzad, M., Ahmed, S.: On (α, β) -fuzzy KU-ideals of KU-algebras. Afr. Mat. **26**(3), 651–661 (2015)
33. Muhiuddin, G.: Bipolar fuzzy KU-subalgebras/ideals of KU-algebras. Ann. Fuzzy Math. Inform. **8**(3), 409–418 (2014)
34. Senapati, T., Shum, K.P.: Atanassov's intuitionistic fuzzy bi-normed KU-ideals of a KU-algebra. J. Intell. Fuzzy Syst. **30**(2), 1169–1180 (2016)
35. Noether, E.: Ideal theorie in ringbereiche. Math. Ann **83**, 24–66 (1921)
36. Artin, E.: Zur Theorie der hyperkomplexen Zahlen. Abh. Math. Sem. Univ. Hamburg **5**, 251–260 (1927)
37. Zhang, J.Z., Fu, X.B., Liao, Z.H.: $(\in, \in \vee q_{(\lambda,\mu)})$- fuzzy ideal of N(2,2,0) algebra. J. Front. Comput. Sci. Technol. **7**(11), 1048–1056 (2013)
38. Meng, D.J., Wang, L., Yuan, L.: Abstract algebra(III)-Commutative Algebra. Science Press (2016)
39. Mordeson, J.N., Malik, D.S.: Fuzzy Commutative Algebra. World Scientific Publishing, Singapore (1998)
40. Hu, B.S.: Foundations of Fuzzy Theory, 2nd edn. Wuhan University Press, Wuhan (2010)

The Equivalence Between Stratified L-neighborhood Groups and Stratified L-neighborhood Topological Groups

Yan-rui Lv, Qiu Jin, Fang-fang Zhao, and Ling-qiang Li(✉)

College of Mathematical Science, Liaocheng University, Liaocheng 252059, China
lilingqiang0614@126.com

Abstract. In this paper, considering L being an arbitrary complete residuated lattice, we present an equivalent characterization on stratified L-neighborhood groups defined by Ahsanullah et al., and then prove that a stratified L-neighborhood group is a stratified L-neighborhood topological group. Thus stratified L-neighborhood groups and stratified L-neighborhood topological groups are equivalent notions. It should be pointed out that if not together the group structure, a stratified L-neighborhood system is not a stratified L-neighborhood topological system in general.

Keywords: Residuated lattice · Fuzzy set
Fuzzy neighborhood group · Fuzzy neighborhood topological group

1 Introduction

In 1979, Foster [8] first introduced the notion of fuzzy topological group by taking Lowen's fuzzy topology [17]. Later, with the development of fuzzy topology, a number of fuzzy topological groups were defined and discussed [7,9,18,21]. Meanwhile, replacing fuzzy topologies by more general lattice-valued topologies [11,16,20,23], many kinds of lattice-valued topological groups were proposed [1–4,22,24,25]. According to the above literatures, it is easily observed that Pu-Liu's quasi-coincident neighborhood systems [16] and its generalizations [6], Wang's remote-neighborhood system [20], Shi's neighborhood system [19] and Höhel-Šostak's neighborhood system [11] are important tools to study lattice-valued topological groups. In particular, taking L to be a complete Heyting algebra or a divisible complete residuated lattice, Ahsanullah et al. [1,2] defined an enriched lattice-valued topological groups-called stratified L-neighborhood topological groups by considering Höhel-Šostak's stratified L-neighborhood topological systems [11]. By dropping the topological condition in stratified L-neighborhood topological systems, Ahsanullah et al. [2] also defined a notion of stratified L-neighborhood groups. It is well known that a crisp neighborhood system is not a neighborhood topological system (i.e., neighborhood system of topology). But

B.-Y. Cao and Y.-B. Zhong (Eds.): ICFIE 2017, AISC 872, pp. 111–118, 2019.
https://doi.org/10.1007/978-3-030-02777-3_10

there is no difference between a crisp neighborhood group and a crisp neighborhood topological group (since the topological condition can be concluded from the continuity of group operation), and they are equivalent notions. Thus a natural question arises–is a stratified L-neighborhood group and a stratified L-neighborhood topological group equivalent? In [2], Ahsanullah et al. did not answer this question. In this paper, we shall affirm this question. Specifically, we will give an equivalent characterization on stratified L-neighborhood groups, and then by using this characterization we prove that a stratified L-neighborhood group and a stratified L-neighborhood topological group are equivalent notions.

In this paper, if not otherwise specified, $L = (L, *, \to, \wedge, \vee, \bot, \top)$ [10,11] is always a complete residuated lattice. The basic properties of complete residuated lattice are collected as below.

(I1) $a * b = b * a$, $a \to b = \top \Leftrightarrow a \leq b$, $\top \to a = a$; (I2) $a * b \leq c \Leftrightarrow a \leq b \to c$;
(I3) $a * (a \to b) \leq b$; (I4) $a \to (b \to c) = (a * b) \to c = b \to (a \to c)$;
(I5) $a * (\bigvee_{t \in T} b_t) = \bigvee_{t \in T} (a * b_t)$, (I5) $\Rightarrow$ (I5') $b \leq c \Rightarrow a * b \leq a * c$;
(I6) $(\bigvee_{t \in T} a_t) \to b = \bigwedge_{t \in T} (a_t \to b)$, (I6)$\Rightarrow$ (I6') $a \leq b \Rightarrow a \to c \geq b \to c$;
(I7) $a \to (\bigwedge_{t \in T} b_t) = \bigwedge_{t \in T} (a \to b_t)$, (I7)$\Rightarrow$ (I7') $b \leq c \Rightarrow a \to b \leq a \to c$.

We call a function $\mu : X \to L$ as an L-fuzzy set in X. We use L^X to denote the set of all L-fuzzy sets in X. The operators $\vee, \wedge, *, \to$ on L can be translated onto L^X in a pointed wise. That is, for any $\mu, \nu, \mu_t (t \in T) \in L^X$, $\mu \leq \nu \Leftrightarrow \mu(x) \leq \nu(x), \forall x \in X$, thus

$$(\bigvee_{t \in T} \mu_t)(x) = \bigvee_{t \in T} \mu_t(x), (\bigwedge_{t \in T} \mu_t)(x) = \bigwedge_{t \in T} \mu_t(x),$$

$$(\mu * \nu)(x) = \mu(x) * \nu(x), (\mu \to \nu)(x) = \mu(x) \to \nu(x).$$

We make no difference between a constant function and its value since no confusion will arise. Also, we make no difference between a subset of X and its characteristic function. In particular, for any $x \in X$, we use x to denote x for simplicity.

Let $f : X \to Y$ be a function. We define Zadeh type function $f^{\to} : L^X \to L^Y$ and $f^{\leftarrow} : L^X \to L^Y$ [11] as follows:

$$\forall \mu \in L^X, y \in Y, f^{\to}(\mu)(y) = \bigvee_{f(x)=y} \mu(x); \forall \nu \in L^Y, x \in X, f^{\leftarrow}(\nu)(x) = \nu(f(x)).$$

The function $S : L^X \times L^X \to L$ is defined as: $\forall \mu, \nu \in L^X, S(\mu, \nu) = \bigwedge_{x \in X} (\mu(x) \to \nu(x))$, where $S(\mu, \nu)$ is interpreted as the degree of μ being a subset of ν [5]. It is easily seen that $S(\mu, \nu) = \top \Leftrightarrow \mu \leq \nu$ and $\mu * S(\mu, \nu) \leq \nu$ [12–15].

Let $(X, \cdot)$ be a group, and $m : X \times X \to X, (x, y) \mapsto xy$ and $r : X \to X, x \mapsto x^{-1}$ are the multiplication and inverse operations, respectively. For any $x \in X$, let

$$\mathcal{L}_x : X \to X, y \mapsto xy; \mathcal{R}_x : X \to X, y \mapsto yx,$$

be the left translation and right translation, respectively. For any $\mu, \nu, \omega \in L^X$. Define μ^{-1}, $\mu \odot \nu \in L^X$, $\mu \times \nu \in L^{X \times X}$ as

$$\forall z \in X, \mu^{-1}(z) = \mu(z^{-1}), (\mu \odot \nu)(z) = \bigvee_{xy=z} (\mu(x) * \nu(y)) = \bigvee_{x \in X} (\mu(x) * \nu(x^{-1}z)),$$

$$\forall x = (x_1, x_2) \in X \times X, (\mu \times \nu)(x) = \mu(x_1) * \nu(x_2).$$

Lemma 1. [2,3] *Let $(X, \cdot)$ be a group, e be the identity element of the group, and $\mu, \nu, \omega \in L^X$. Then*
(1) $\forall x \in X, x \odot \mu = \mathcal{L}_x^{\rightarrow}(\mu) = \mathcal{L}_{x^{-1}}^{\leftarrow}(\mu)$, $\mu \odot x = \mathcal{R}_x^{\rightarrow}(\mu) = \mathcal{R}_{x^{-1}}^{\leftarrow}(\mu)$, $e \odot \mu = \mu = \mu \odot e$,
(2) $\mu \odot \nu = m^{\rightarrow}(\mu \times \nu)$, $r^{\rightarrow}(\mu) = r^{\leftarrow}(\mu) = \mu^{-1}$,
(3) $(\mu \odot \nu)^{-1} = \nu^{-1} \odot \mu^{-1}$,
(4) $(\mu \odot \nu) \odot \omega = \mu \odot (\nu \odot \omega)$,
(5) $\forall a \in L, ((a * \mu) \odot \nu) = (\mu \odot (a * \nu)) = a * (\mu \odot \nu)$,
(6) $\forall x, y \in X, (xy) \odot \mu = x \odot y \odot \mu$, $(x \odot \mu)^{-1} = \mu^{-1} \odot x^{-1}$,
(7) $\forall x \in X, S(\mu, \nu) \leq S(x \odot \mu, x \odot \nu)$.

2 The Main Results

In [1,2], Ahsanullah et al., introduced the notion of stratified L-neighborhood (topological) group. In the following, we shall present an equivalent characterization on the stratified L-neighborhood groups, and then prove that a stratified L-neighborhood group and a stratified L-neighborhood topological group are equivalent notions.

Definition 1. [11] *The pair $(X, \mathscr{R})$ is called a stratified L-neighborhood space, if the function $\mathscr{R} : X \times L^X \rightarrow L$ satisfies the following conditions* (R1)-(Rs).
(R1) $\mathscr{R}(x, \top) = \top$, (R2) $\mu \leq \nu \Rightarrow \mathscr{R}(x, \mu) \leq \mathscr{R}(x, \nu)$, (R3) $\mathscr{R}(x, \mu) * \mathscr{R}(x, \nu) \leq \mathscr{R}(x, \mu * \nu)$,
(R4) $\mathscr{R}(x, \mu) \leq \mu(x)$, (Rs) $\mathscr{R}(x, b * \mu) \geq b * \mathscr{R}(x, \mu), \forall b \in L$.

And $(X, \mathscr{R})$ is called topological (stratified L-neighborhood topological space) if $\mathscr{R}$ satisfies condition (R5).
(R5) $\mathscr{R}(x, \mu) \leq \bigvee\{\mathscr{R}(x, \nu) | \nu \in L^X, \nu \leq \mathscr{R}(-, \mu)\}$.
It is easily seen that in the present of (R2), the condition (R5) is equivalent to *(R5′)*: $\mathscr{R}(x, \mu) \leq \mathscr{R}(x, \mathscr{R}(-, \mu))$.
Let $(X, \mathscr{R})$ be a stratified L-neighborhood space. Then the pair $(X \times X, \mathscr{R}^2)$ defined by for any $\mu \in L^{X \times X}$, $x = (x_1, x_2) \in X \times X$,

$$\mathscr{R}^2(x, \mu) = \vee\{\mathscr{R}(x_1, \mu_1) * \mathscr{R}(x_2, \mu_2) \mid \mu_1, \mu_2 \in L^X, \mu_1 \times \mu_2 \leq \mu\},$$

is a stratified L-neighborhood space. And if $(X, \mathscr{R})$ is topological then so is $(X \times X, \mathscr{R}^2)$[2].
Let $f : (X, \mathscr{R}^X) \rightarrow (Y, \mathscr{R}^Y)$ be a function of the stratified L-neighborhood (topological) space. Then f is said continuous if for any $x \in X$, $\nu \in L^Y$ we have $\mathscr{R}^Y(f(x), \nu) \leq \mathscr{R}^X(x, f^{\leftarrow}(\nu))$.

Definition 2. [1,2] *Let $(X, \mathscr{R})$ be a stratified L-neighborhood(topological) space. Then the triple $(X, \cdot, \mathscr{R})$ is called a stratified L-neighborhood (topological) group if*

(LFRG1) *The function $m : (X \times X, \mathscr{R}^2) \to (X, \mathscr{R}), (x, y) \mapsto xy$ and*
(LFRG2) *the function $r : (X, \mathscr{R}) \to (X, \mathscr{R}), x \mapsto x^{-1}$ are continuous.*

Lemma 2. [1,2] *Let $(X, \cdot, \mathscr{R})$ be a stratified L-neighborhood group and $x, y \in X$. Then*

(1) *Both the left translation $\mathcal{L}_x : X \to X, z \mapsto xz$ and the right translation $\mathscr{R}_x : X \to X, z \mapsto zx$ are homeomorphism.*
(2) *For any $\mu \in L^X, \mathscr{R}(x, \mu) = \mathscr{R}(e, x^{-1} \odot \mu) = \mathscr{R}(e, \mu \odot x^{-1})$.*

The condition (2) in the above lemma shows that a stratified L-neighborhood group is determined by the neighborhood of the identity element e of the group. The next theorem gives an equivalent characterization on stratified L-neighborhood group.

Theorem 1. *Let $(X, \mathscr{R})$ be an L-neighborhood space. Then the triple $(X, \cdot, \mathscr{R})$ forms a stratified L-neighborhood group if and only if for any $\mu \in L^X, a \in L$,*

(1) $\mathscr{R}(e, \mu) \leq \vee\{\mathscr{R}(e, \mu_1) * \mathscr{R}(e, \mu_2) \mid \mu_1, \mu_2 \in L^X, \mu_1 \odot \mu_2 \leq \mu\} = \vee\{\mathscr{R}(e, \mu_1) * \mathscr{R}(e, \mu_2) * S(\mu_1 \odot \mu_2, \mu) \mid \mu_1, \mu_2 \in L^X\}$,
(2) $\forall x \in X, \mathscr{R}(e, \mu) \leq \vee\{\mathscr{R}(e, \nu) \mid \nu \in L^X, x \odot \nu \odot x^{-1} \leq \mu\} = \vee\{\mathscr{R}(e, \nu) * S(x \odot \nu \odot x^{-1}, \mu) \mid \nu \in L^X\}$,
(3) $\mathscr{R}(e, \mu) \leq \mathscr{R}(e, \mu^{-1})$,
(4) $\forall x \in X, \mathscr{R}(x, \mu) = \mathscr{R}(e, x^{-1} \odot \mu) = \mathscr{R}(e, \mu \odot x^{-1})$.

Proof. Necessity. Let $(X, \cdot, \mathscr{R})$ be a stratified L-neighborhood group.
(1) By the continuity of m we get that

$$\begin{aligned}
\mathscr{R}(e, \mu) &= \mathscr{R}(m(e, e), \mu) \leq \mathscr{R}^2((e, e), m^{\leftarrow}(\mu)) \\
&= \vee\{\mathscr{R}(e, \mu_1) * \mathscr{R}(e, \mu_2) \mid \mu_1, \mu_2 \in L^X, \mu_1 \times \mu_2 \leq m^{\leftarrow}(\mu)\} \\
&\leq \vee\{\mathscr{R}(e, \mu_1) * \mathscr{R}(e, \mu_2) \mid \mu_1, \mu_2 \in L^X, m^{\rightarrow}(\mu_1 \times \mu_2) \leq m^{\rightarrow} \circ m^{\leftarrow}(\mu)\} \\
&\leq \vee\{\mathscr{R}(e, \mu_1) * \mathscr{R}(e, \mu_2) \mid \mu_1, \mu_2 \in L^X, m^{\rightarrow}(\mu_1 \times \mu_2) \leq \mu\}, \text{ by Lemma 1(2)} \\
&= \vee\{\mathscr{R}(e, \mu_1) * \mathscr{R}(e, \mu_2) \mid \mu_1, \mu_2 \in L^X, \mu_1 \times \mu_2 \leq \mu\} \\
&= \vee\{\mathscr{R}(e, \mu_1) * \mathscr{R}(e, \mu_2) * S(\mu_1 \odot \mu_2, \mu) \mid \mu_1, \mu_2 \in L^X, S(\mu_1 \odot \mu_2, \mu) = 1\} \\
&\leq \vee\{\mathscr{R}(e, \mu_1) * \mathscr{R}(e, \mu_2) * S(\mu_1 \odot \mu_2, \mu) \mid \mu_1, \mu_2 \in L^X\} \text{ by (Rs)} \\
&\leq \vee\{\mathscr{R}(e, \mu_1) * \mathscr{R}(e, S(\mu_1 \odot \mu_2, \mu) * \mu_2 \mid \mu_1, \mu_2 \in L^X\} \\
&\leq \vee\{\mathscr{R}(e, \mu_1) * \mathscr{R}(e, \nu) \mid \mu_1, \mu_2 \in L^X, \nu = S(\mu_1 \odot \mu_2, \mu) * \mu_2\} \\
&\leq \vee\{\mathscr{R}(e, \mu_1) * \mathscr{R}(e, \nu) \mid \mu_1, \nu \in L^X, \mu_1 * \nu \leq \mu\},
\end{aligned}$$

where the last inequality holds by Lemma 1(5) and

$$\mu_1 \odot \nu = \mu_1 \odot (S(\mu_1 \odot \mu_2, \mu) * \mu_2) = S(\mu_1 \odot \mu_2, \mu) * (\mu_1 \odot \mu_2) \leq \mu.$$

(2) By the continuity of $\mathcal{L}_x \circ \mathscr{R}_{x^{-1}}$,

$$\begin{aligned}
\mathscr{R}(e,\mu) &= \mathscr{R}(\mathcal{L}_x \circ \mathscr{R}_{x^{-1}}(e),\mu) \leq \mathscr{R}(e,(\mathcal{L}_x \circ \mathscr{R}_{x^{-1}})^{\leftarrow}(\mu)), \text{ by Lemma 1(1)} \\
&= \mathscr{R}(e,(x^{-1}\odot\mu\odot x)) \overset{(\mathrm{R2})}{=} \vee\{\mathscr{R}(e,\nu) \mid \nu \in L^X, \nu \leq x^{-1}\odot\mu\odot x\}, \text{ by Lemma 1(4)} \\
&= \vee\{\mathscr{R}(e,\nu) \mid \nu \in L^X, x\odot\nu\odot x^{-1} \leq \mu\} \\
&= \vee\{\mathscr{R}(e,\nu) * S(x\odot\nu\odot x^{-1},\mu) \mid \nu \in L^X, S(x\odot\nu\odot x^{-1},\mu) = 1\} \\
&\leq \vee\{\mathscr{R}(e,\nu) * S(x\odot\nu\odot x^{-1},\mu) \mid \nu \in L^X\}, \text{by (Rs)} \\
&\leq \vee\{\mathscr{R}(e,S(x\odot\nu\odot x^{-1},\mu) * \nu) \mid \nu \in L^X\} \\
&= \vee\{\mathscr{R}(e,\omega) \mid \nu \in L^X, \omega = S(x\odot\nu\odot x^{-1},\mu) * \nu\} \\
&\leq \vee\{\mathscr{R}(e,\omega) \mid \omega \in L^X, x\odot\omega\odot x^{-1} \leq \mu\},
\end{aligned}$$

where the last inequality holds by Lemma 1(5) and

$$x\odot\omega\odot x^{-1} = x\odot(S(x\odot\nu\odot x^{-1},\mu)*\nu)\odot x^{-1} = S(x\odot\nu\odot x^{-1},\mu)*(x\odot\nu\odot x^{-1}) \leq \mu.$$

(3) By the continuity of r, we have that

$$\mathscr{R}(e,\mu) = \mathscr{R}(r(e),\mu) \leq \mathscr{R}(e,r^{\leftarrow}(\mu)) = \mathscr{R}(e,\mu^{-1}).$$

(4) It has been proved in Lemma 2(2).

Sufficiency. Let $(X,\mathscr{R})$ satisfies the condition (1)-(4).

(LENG1) For any $x=(x_1,x_2) \in X \times X, \mu \in L^X, a \in L$,

$$\begin{aligned}
&\mathscr{R}(m(x),\mu) = \mathscr{R}(x_1x_2,\mu) \overset{(4)}{=} \mathscr{R}(e,(x_1x_2)^{-1}\odot\mu) \\
\overset{(1)}{\leq}\ & \vee\{\mathscr{R}(e,\mu_1) * \mathscr{R}(e,\mu_2) \mid \mu_1,\mu_2 \in L^X, \mu_1\odot\mu_2 \leq (x_1x_2)^{-1}\odot\mu\} \\
\overset{(2)}{\leq}\ & \bigvee_{\mu_1,\mu_2\in L^X,\mu_1\odot\mu_2\leq(x_1x_2)^{-1}\odot\mu} \Big(\bigvee_{\nu_1\in L^X, x_2^{-1}\odot\nu_1\odot x_2\leq\mu_1} \mathscr{R}(e,\nu_1) * \bigvee_{\nu_2\in L^X, x_2^{-1}\odot\nu_2\odot x_2\leq\mu_2} \mathscr{R}(e,\nu_2)\Big) \\
=\ & \bigvee_{\mu_1,\mu_2\in L^X,\mu_1\odot\mu_2\leq(x_1x_2)^{-1}\odot\mu} \ \bigvee_{\nu_1,\nu_2\in L^X, x_2^{-1}\odot\nu_1\odot x_2\leq\mu_1, x_2^{-1}\odot\nu_2\odot x_2\leq\mu_2} (\mathscr{R}(e,\nu_1) * \mathscr{R}(e,\nu_2)) \\
\leq\ & \bigvee_{\nu_1,\nu_2\in L^X,(x_1\odot\nu_1)\odot(\nu_2\odot x_2)\leq\mu} (\mathscr{R}(e,\nu_1) * \mathscr{R}(e,\nu_2)) \\
\overset{(4)}{=}\ & \bigvee_{\nu_1,\nu_2\in L^X,(x_1\odot\nu_1)\odot(\nu_2\odot x_2)\leq\mu} (\mathscr{R}(x_1,x_1\odot\nu_1) * \mathscr{R}(x_2,x_2\odot\nu_2)) \\
\leq\ & \bigvee_{\omega_1,\omega_2\in L^X,\omega_1\odot\omega_2\leq\mu} (\mathscr{R}(x_1,\omega_1) * \mathscr{R}(x_2,\omega_2)) \text{ by Lemma 1(2)} \\
=\ & \bigvee_{\omega_1,\omega_2\in L^X,m^{\rightarrow}(\omega_1\times\omega_2)\leq\mu} (\mathscr{R}(x_1,\omega_1) * \mathscr{R}(x_2,\omega_2)) \\
=\ & \bigvee_{\omega_1,\omega_2\in L^X,\omega_1\times\omega_2\leq m^{\leftarrow}(\mu)} (\mathscr{R}(x_1,\omega_1) * \mathscr{R}(x_2,\omega_2)) = \mathscr{R}^2(x,m^{\leftarrow}(\mu)),
\end{aligned}$$

the third inequality in the above proof holds by

$$x_2^{-1}\odot\nu_2\odot x_2 \leq \mu_2 \Rightarrow \nu_2\odot x_2 \leq x_2\odot\mu_2, \quad x_2^{-1}\odot\nu_1\odot x_2 \leq \mu_1 \Rightarrow \nu_1\odot x_2 \leq x_2\odot\mu_1,$$

$$\mu_1\odot\mu_2 \leq (x_1x_2)^{-1}\odot\mu \Rightarrow (x_1x_2)\odot\mu_1\odot\mu_2 \leq \mu,$$

and then by Lemma 1 (4), (6)

$$\begin{aligned}
(x_1\odot\nu_1)\odot(\nu_2\odot x_2) &\leq (x_1\odot\nu_1)\odot(x_2\odot\mu_2) = x_1\odot(\nu_1\odot x_2)\odot\mu_2 \leq x_1\odot(x_2\odot\mu_1)\odot\mu_2 \\
&= (x_1\odot x_2)\odot(\mu_1\odot\mu_2) = (x_1x_2)\odot(\mu_1\odot\mu_2) \leq \mu.
\end{aligned}$$

(LFNG2) For any $x \in X, \mu \in L^X, a \in L$, by Lemma 1 (2), (6)

$$\mathscr{R}(x^{-1}, \mu) \overset{(4)}{=} \mathscr{R}(e, x \odot \mu) \overset{(3)}{\leq} \mathscr{R}(e, (x \odot \mu)^{-1}) = \mathscr{R}(e, \mu^{-1} \odot x^{-1}) \overset{(4)}{=} \mathscr{R}(x, \mu^{-1}).$$

In general, a stratified L-neighborhood space is not a stratified L-neighborhood topological space. The following theorem shows that a stratified L-neighborhood group is stratified L-neighborhood topological group.

Theorem 2. *Let $(X, \cdot, \mathscr{R})$ be a stratified L-neighborhood group. Then $(X, \cdot, \mathscr{R})$ is a stratified L-neighborhood topological group.*

Proof. We need only check $(R5')$. For any $x \in X, \mu \in L^X, a \in L$, it follows by Theorem 1 (4) and (1) that

$$\mathscr{R}(x, \mu) = \mathscr{R}(e, x^{-1} \odot \mu) \leq \vee\{\mathscr{R}(e, \mu_1) * \mathscr{R}(e, \mu_2) * S(\mu_1 \odot \mu_2, x^{-1} \odot \mu) \mid \mu_1, \mu_2 \in L^X\}.$$

For any $\mu_1, \mu_2 \in L^X$,

$$\begin{aligned}
S(\mu_1 \odot \mu_2, x^{-1} \odot \mu) &= \bigwedge_{y \in X}((\mu_1 \odot \mu_2)(y) \to (x^{-1} \odot \mu)(y)) \\
&= \bigwedge_{y \in X}((\bigvee_{z \in X} \mu_1(z) * \mu_2(z^{-1}y)) \to (x^{-1} \odot \mu)(y)) \\
&= \bigwedge_{y \in X}\bigwedge_{z \in X}((\mu_1(z) * \mu_2(z^{-1}y)) \to (x^{-1} \odot \mu)(y)) \\
&= \bigwedge_{y \in X}\bigwedge_{z \in X}(\mu_1(z) \to (\mu_2(z^{-1}y) \to (x^{-1} \odot \mu)(y))) \\
&= \bigwedge_{y \in X}\bigwedge_{z \in X}(\mu_1(z) \to ((z \odot \mu_2)(y) \to (x^{-1} \odot \mu)(y))) \\
&= \bigwedge_{z \in X}(\mu_1(z) \to \bigwedge_{y \in X}((z \odot \mu_2)(y) \to (x^{-1} \odot \mu)(y))) \\
&= \bigwedge_{z \in X}(\mu_1(z) \to S(z \odot \mu_2, x^{-1} \odot \mu)).
\end{aligned}$$

Then it follows that

$$\begin{aligned}
\mathscr{R}(e, \mu_2) * S(\mu_1 \odot \mu_2, x^{-1} \odot \mu) &= \mathscr{R}(e, \mu_2) * \bigwedge_{z \in X}(\mu_1(z) \to S(z \odot \mu_2, x^{-1} \odot \mu)) \\
&\leq \bigwedge_{z \in X}(\mathscr{R}(e, \mu_2) * (\mu_1(z) \to S(z \odot \mu_2, x^{-1} \odot \mu))) \\
&\leq \bigwedge_{z \in X}(\mu_1(z) \to (\mathscr{R}(e, \mu_2) * S(z \odot \mu_2, x^{-1} \odot \mu))), \text{ by Lemma 1(7)} \\
&\leq \bigwedge_{z \in X}(\mu_1(z) \to (\mathscr{R}(e, \mu_2) * S(\mu_2, z^{-1} \odot x^{-1} \odot \mu))) \\
&\overset{(Rs)}{\leq} \bigwedge_{z \in X}(\mu_1(z) \to \mathscr{R}(e, [\mu_2 * S(\mu_2, z^{-1} \odot x^{-1} \odot \mu)])), \text{ by Lemma 1(6)} \\
&\leq \bigwedge_{z \in X}(\mu_1(z) \to \mathscr{R}(e, [(xz)^{-1} \odot \mu])) = \bigwedge_{z \in X}(\mu_1(z) \to \mathscr{R}(xz, \mu)) \\
&= \bigwedge_{z \in X}((x \odot \mu_1)(xz) \to \mathscr{R}(xz, \mu)) \leq \bigwedge_{y \in X}((x \odot \mu_1)(y) \to \mathscr{R}(y, \mu)) \\
&= S(x \odot \mu_1, \mathscr{R}(-, \mu)),
\end{aligned}$$

and so

$$\begin{aligned}\mathscr{R}(x,\mu) &\leq \vee\{\mathscr{R}(e,\mu_1) * \mathscr{R}(e,\mu_2) * S(\mu_1 \odot \mu_2, x^{-1} \odot \mu) \mid \mu_1,\mu_2 \in L^X\} \\ &\leq \vee\{\mathscr{R}(e,\mu_1) * S(x \odot \mu_1, \mathscr{R}(-,\mu)) \mid \mu_1 \in L^X\}, \text{ by Lemma 1(7)} \\ &\leq \vee\{\mathscr{R}(e,\mu_1) * S(\mu_1, x^{-1} \odot \mathscr{R}(-,\mu)) \mid \mu_1 \in L^X\} \\ &\overset{\text{(Rs)}}{\leq} \vee\{\mathscr{R}(e,[\mu_1 * S(\mu_1, x^{-1} \odot \mathscr{R}(-,\mu))]) \mid \mu_1 \in L^X\} \\ &\leq \{\mathscr{R}(e,[x^{-1} \odot \mathscr{R}(-,\mu)]) = \mathscr{R}(x,\mathscr{R}(-,\mu)).\end{aligned}$$

3 Conclusions

It is well known that, neighborhood group and neighborhood topological group are an equivalent notions. Stratified L-neighborhood group and stratified L-neighborhood topological group [1,2] are their lattice-valued generalization. In this paper, we give an equivalent characterization on stratified L-neighborhood group, and prove that a stratified L-neighborhood group and a stratified L-neighborhood topological group are equivalent notions.

Acknowledgements. Thanks to the support by National Natural Science Foundation of China (11501278, 11471152) and Shandong Provincial Natural Science Foundation, China (ZR2013AQ011, ZR2014AQ011) and the Ke Yan Foundation of Liaocheng University (318011505).

References

1. Al-Mufarrij, J., Ahsanullah, T.M.G.: On the category of fixed basis frame valued topological groups. Fuzzy Sets Syst. **159**, 2529–2551 (2008)
2. Ahsanullah, T.M.G., Gauld, D., Al-Mufarrij, J., Al-Thukair, F.: Enriched lattice-valued topological groups. New Math. Nat. Comput. **1**, 27–53 (2014)
3. Ahsanullah T. M. G, Jäger G.: Stratified LMN-convergence tower groups and their stratified LMN-uniform convergence tower structures. Fuzzy Sets Syst. https://doi.org/10.1016/j.fss.2017.01.011
4. Bayoumi, F.: Global L-neighborhood groups. Fuzzy Sets Syst. **159**, 605–619 (2008)
5. Bělohlávek, R.: Fuzzy Relational Systems: Foundations and Principles. Kluwer Academic Publishers, New York (2002)
6. Fang, J.M.: Categories isomorphic to L-FTOP. Fuzzy Sets Syst. **157**, 820–831 (2006)
7. Fang, J.X.: On fuzzy topological groups. Chin. Sci. Bull. **29**, 727–730 (1984)
8. Foster, D.H.: Fuzzy topological groups. J. Math. Anal. Appl. **67**, 549–564 (1979)
9. Ganster, M., Georgiou, D.N., Jafari, S.: On fuzzy topological groups and fuzzy continuous functions. Hacettepe J. Math. Stat. **34**, 35–43 (2005)
10. Hájek, P.: Metamathematics of Fuzzy Logic. Kluwer Academic Publishers, Dordrecht (1998)
11. Höhle, U., Rodabaugh, S.E.: Mathematics of Fuzzy Sets: Logic, Topology and Measure Theory. The Handbooks of Fuzzy Sets Series. Kluwer Academic Publishers, Boston (1999)

12. Jin, Q., Li, L.Q.: One-axiom characterizations on lattice-valued closure (interior) operators. J. Intell. Fuzzy Syst. **31**, 1679–1688 (2016)
13. Li, L.Q., Jin, Q.: On adjunctions between Lim, SL-Top, and SL-Lim. Fuzzy Sets Syst. **182**, 66–78 (2011)
14. Li, L.Q., Li, Q.G.: On enriched L-topologies: base and subbase. J. Intell. Fuzzy Syst. **28**, 2423–2432 (2015)
15. Li, L.Q., Jin, Q., Hu, K., Zhao, F.F.: The axiomatic characterizations on L-fuzzy covering-based approximation operators. Int. J. Gen. Syst. **46**(4), 332–353 (2017)
16. Liu, Y., Luo, M.: Fuzzy Topology. World Scientific Publishing, Singapore (1997)
17. Lowen, R.: Fuzzy topological spaces and fuzzy compactness. J. Math. Anal. Appl. **56**, 621–633 (1976)
18. Ma, J.L., Yu, C.H.: Fuzzy topological groups. Fuzzy Sets Syst. **12**, 289–299 (1984)
19. Shi, F.G.: L-fuzzy interiors and L-fuzzy closures. Fuzzy Sets Syst. **160**, 1218–1232 (2009)
20. Wang, G.J.: Theory of L-Fuzzy Topological Spaces. Shanxi Normal University Press, Xi'an (1988)
21. Yan, C.H., Guo, S.Z.: I-fuzzy topological groups. Fuzzy Sets Syst. **161**, 2166–2180 (2010)
22. Yu, C.H., Ma, J.L.: L-fuzzy topological groups. Fuzzy Sets Syst. **44**, 83–91 (1991)
23. Zhang, D.X.: An enriched category approach to many valued topology. Fuzzy Sets Syst. **158**, 349–366 (2007)
24. Zhao, H., Li, S.G., Chen, G.X.: (L, M)-fuzzy topological groups. J. Intell. Fuzzy Syst. **26**, 1517–1526 (2014)
25. Zhang, H.P., Fang, J.X.: $I(L)$-topological groups andits level L-topological groups. Fuzzy Sets Syst. **158**, 1504–1510 (2007)

On $(\in, \in\vee q_k)$-fuzzy Filters in R_0-algebras

Chun-hui Liu(✉)

Department of Mathematics and Statistics, Chifeng University, Chifeng 024000, China
chunhuiliu1982@163.com

Abstract. In the present paper, the $(\in, \in\vee q_k)$-fuzzy filter theory in R_0-algebras is further studied. Some new properties of $(\in, \in\vee q_k)$-fuzzy filters are given. Representation theorem of $(\in, \in\vee q_k)$-fuzzy filter which is generated by a fuzzy set is established. It is proved that the set consisting of all $(\in, \in\vee q_k)$-fuzzy filters on a given R_0-algebra, under the partial order $\sqsubseteq$, forms a complete distributive lattice.

Keywords: Fuzzy logic · R_0-algebra · $(\in, \in\vee q_k)$-fuzzy filter
Complete distributive lattice

1 Introduction

To make the computers simulate beings in dealing with certainty and uncertainty in information is one important task of artificial intelligence. Logic appears in a "sacred" (resp., a "profane") form which is dominant in proof theory (resp., model theory). The role of logic in mathematics and computer science is twofold–as a tool for applications in both areas, and a technique for laying the foundations. Nonclassical logic [1] including many-valued logic and fuzzy logic takes the advantage of classical logic to handle information with various facets of uncertainty [2], such as fuzziness and randomness. At present, nonclassical logic has become a formal and useful tool for computer science to deal with fuzzy information and uncertain information. R_0-algebra is an important class of non-classical fuzzy logical algebras which was introduced by Wang in [3] by providing an algebra proof of the completeness theorem of the formal deductive system $\mathcal{L}^*$. From then, R_0-algebras has been extensively investigated by many researchers. Among them, Jun and Liu studied the theory of filters in R_0-algebras in [4]. The concept of fuzzy sets is introduced firstly by Zadeh in [5]. Liu and Li in [6] proposed the concept of fuzzy filters of R_0-algebras and discussed some their properties by using fuzzy sets theory. As an extension of the concept of fuzzy filter, Ma et al. dealt with the notion of $(\in, \in\vee q)$-fuzzy filters and investigated their properties in R_0-algebras [7]. In [8], Jun et al. introduced a more general notion of $(\in, \in\vee q)$-fuzzy filters, named $(\in, \in\vee q_k)$-fuzzy filters, and investigated their related properties. At the same time, some results about $(\in, \in\vee q)$-fuzzy

B.-Y. Cao and Y.-B. Zhong (Eds.): ICFIE 2017, AISC 872, pp. 119–129, 2019.
https://doi.org/10.1007/978-3-030-02777-3_11

filters and $(\in, \in\vee q_k)$-fuzzy filters in other logic algebras also have been obtained by many scholars in [9–15].

In this paper, we will further research the properties of $(\in, \in\vee q_k)$-fuzzy filters in R_0-algebras. The rest of this article is organized as follows. In Sect. 2, we review related basic knowledge of R_0-algebras and fuzzy sets. In Sect. 3, we give several new properties of $(\in, \in\vee q_k)$-fuzzy filters. In Sect. 4, we firstly introduce the concept of $(\in, \in\vee q_k)$-fuzzy filter which is generated by an fuzzy set and establish its representation theorem. And then, we investigate the lattice structural feature of the set containing all of $(\in, \in\vee q_k)$-fuzzy filters in a given R_0-algebra. Finally, we conclude this paper in Sect. 5.

2 Preliminaries

Definition 1 (cf. [3]). Let M be an algebra of type $(\neg, \vee, \rightarrow)$, where $\neg$ is a unary operation, $\vee$ and $\rightarrow$ are binary operations. $(M, \neg, \vee, \rightarrow, 1)$ is called an R_0-algebra if there is a partial order $\leqslant$ such that $(M, \leqslant, 1)$ is a bounded distributive lattice with the greatest element 1, $\vee$ is the supremum operation with respect to $\leqslant$, $\neg$ is an order-reversing involution, and the following conditions hold for every $a, b, c \in M$:

(M1) $\neg a \rightarrow \neg b = b \rightarrow a$;
(M2) $1 \rightarrow a = a, a \rightarrow a = 1$;
(M3) $b \rightarrow c \leqslant (a \rightarrow b) \rightarrow (a \rightarrow c)$;
(M4) $a \rightarrow (b \rightarrow c) = b \rightarrow (a \rightarrow c)$;
(M5) $a \rightarrow (b \vee c) = (a \rightarrow b) \vee (a \rightarrow c), a \rightarrow (b \wedge c) = (a \rightarrow b) \wedge (a \rightarrow c)$;
(M6) $(a \rightarrow b) \vee ((a \rightarrow b) \rightarrow (\neg a \vee b)) = 1$.

Lemma 1 (cf. [3]). Let M be an R_0-algebra, $a, b, c \in M$. Then the following properties hold.

(P1) $a \leqslant b$ if and only if $a \rightarrow b = 1$;
(P2) $a \leqslant b \rightarrow c$ if and only if $b \leqslant a \rightarrow c$;
(P3) $(a \vee b) \rightarrow c = (a \rightarrow c) \wedge (b \rightarrow c), (a \wedge b) \rightarrow c = (a \rightarrow c) \vee (b \rightarrow c)$;
(P4) If $b \leqslant c$, then $a \rightarrow b \leqslant a \rightarrow c$, and if $a \leqslant b$, then $b \rightarrow c \leqslant a \rightarrow c$;
(P5) $a \rightarrow b \geqslant \neg a \vee b$ and $a \wedge \neg a \leqslant b \vee \neg b$;
(P6) $(a \rightarrow b) \vee (b \rightarrow a) = 1$ and $a \vee b = ((a \rightarrow b) \rightarrow b) \wedge ((b \rightarrow a) \rightarrow a)$;
(P7) $a \rightarrow (b \rightarrow a) = 1$ and $a \rightarrow (\neg a \rightarrow b) = 1$;
(P8) $a \rightarrow b \leqslant a \vee c \rightarrow b \vee c$ and $a \rightarrow b \leqslant a \wedge c \rightarrow b \wedge c$;
(P9) $a \rightarrow b \leqslant (a \rightarrow c) \vee (c \rightarrow b)$.

Lemma 2 (cf. [3]). Let M be an R_0-algebra. Define a new operator $\otimes$ on M such that $a \otimes b = \neg(a \rightarrow \neg b)$, for every $a, b, c \in M$. Then the following properties hold.

(P10) $(M, \otimes, 1)$ is a commutative monoid with the multiplicative unit element 1;
(P11) If $a \leqslant b$, then $a \otimes c \leqslant b \otimes c$;

(P12) $0 \otimes a = 0$ and $a \otimes \neg a = 0$;
(P13) $a \otimes b \leqslant a \wedge b$ and $a \otimes (a \to b) \leqslant b$ and $a \leqslant b \to (a \otimes b)$;
(P14) $a \otimes b \to c = a \to (b \to c)$ and $a \otimes (b \vee c) = (a \otimes b) \vee (a \otimes c)$.

In the unit interval $[0, 1]$ equipped with the natural order, $\vee = \max$ and $\wedge = \min$. Let $X \neq \emptyset$, A mapping $f : X \to [0, 1]$ is called a fuzzy set on X (see [5]). Let f and g be two fuzzy sets on X. We define $f \cap g$ and $f \cup g$ as follows:

(i) $(f \cap g)(x) = f(x) \wedge g(x)$, for all $x \in X$;
(ii) $(f \cup g)(x) = f(x) \vee g(x)$, for all $x \in X$.

Let $X \neq \emptyset$, $x \in X$ and $t \in (0, 1]$. A fuzzy set f on X with the form

$$f(y) = \begin{cases} t, & y = x, \\ 0, & y \neq x, \end{cases}$$

for all $y \in X$, is said to be a fuzzy point with support x and value t and is denoted by x_t. For a fuzzy point x_t and a fuzzy set f on X, x_t is said to belong to (resp. be quasi-coincident with) the fuzzy set f, written as $x_t \in f$(resp. $x_t q f$), we mean that $f(x) \geqslant t$(resp. $f(x) + t > 1$). And to say that $x_t \in\vee q f$(resp. $x_t \in\wedge q f$), we mean that $x_t \in f$ or $x_t q f$(resp. $x_t \in f$ and $x_t q f$).

3 Some New Properties of $(\in, \in\vee q_k)$-fuzzy Filters

According to [8], we let k denote an arbitrary element of $[0, 1)$ if no other statements. And for any fuzzy set f on non-empty set X, we let that:

(1) $x_t q_k f$ denotes $f(x) + t + k > 1$;
(2) $x_t \in\vee q_k f$ denotes $x_t \in f$ or $x_t q_k f$;
(3) $x_t \overline{\theta} f$ denotes $x_t \theta f$ does not hold, where $\theta \in \{\in, q_k, \in\vee q_k\}$.

Definition 2 (cf. [8]). Let M be an R_0-algebra. A fuzzy set f on M is said to be an $(\in, \in\vee q_k)$-fuzzy filter of M, if it for all $t, r \in (0, 1]$ and $a, b \in M$, it satisfies the following conditions:

(FF1) $a_t \in f$ and $a \leqslant b$ imply $b_t \in\vee q_k f$;
(FF2) $a_t \in f$ and $b_r \in f$ imply $(a \otimes b)_{\min\{t,r\}} \in\vee q_k f$.

The set of all $(\in, \in\vee q_k)$-fuzzy filters of M is denoted by **FFil**(M).

Theorem 1 (cf. [8]). Let M be an R_0-algebra. A fuzzy set f on M is an $(\in, \in\vee q_k)$-fuzzy filter of M if and only if it satisfies the following conditions:

(FF3) $a \leqslant b$ implies $f(b) \geqslant f(a) \wedge \dfrac{1-k}{2}$ for all $a, b \in M$;
(FF4) $f(a \otimes b) \geqslant f(a) \wedge f(b) \wedge \dfrac{1-k}{2}$ for all $a, b \in M$.

Theorem 2 (cf. [8]). Let M be an R_0-algebra. A fuzzy set f on M is an $(\in, \in\vee q_k)$-fuzzy filter of M if and only if it satisfies the following conditions:

(FF3) $f(1) \geqslant f(a) \wedge \dfrac{1-k}{2}$ for all $a \in M$;

(FF4) $f(b) \geqslant f(a) \wedge f(a \to b) \wedge \dfrac{1-k}{2}$ for all $a, b \in M$.

Definition 3. Let M be an R_0-algebra and f a fuzzy set on M. A fuzzy set f^λ on M is defined as follows:

$$f^\lambda(a) = \begin{cases} f(a), & a \neq 1, \\ f(1) \vee \lambda, & a = 1, \end{cases} \tag{1}$$

for all $a \in M$, where $\lambda \in [0, 1]$.

Theorem 3. Let M be an R_0-algebra and $f \in \mathbf{FFil}(M)$. Then $f^\lambda \in \mathbf{FFil}(M)$ for all $\lambda \in [0, 1]$.

Proof. Firstly, for all $a, b \in M$, let $a \leqslant b$, we consider the following two cases:

(i) Assume that $b = 1$. If $a = 1$, we have that $f^\lambda(b) = f^\lambda(1) = f(1) \vee \lambda = f^\lambda(a) \geqslant f^\lambda(a) \wedge \dfrac{1-k}{2}$. If $a \neq 1$, by using $f \in \mathbf{FFil}(M)$ and (FF5), we have that $f^\lambda(b) = f^\lambda(1) = f(1) \vee \lambda \geqslant f(1) \geqslant f(a) \wedge \dfrac{1-k}{2} = f^\lambda(a) \wedge \dfrac{1-k}{2}$.

(ii) Assume that $b \neq 1$, then $a \neq 1$. It follows that $f^\lambda(b) = f(b) \geqslant f(a) \wedge \dfrac{1-k}{2} = f^\lambda(a) \wedge \dfrac{1-k}{2}$ from $f \in \mathbf{FFil}(M)$ and (FF3).

Summarize these two cases, we conclude that $a \leqslant b$ implies $f^\lambda(b) \geqslant f^\lambda(a) \wedge \dfrac{1-k}{2}$, for all $a, b \in M$. i.e., f^λ satisfies (FF3).

Secondly, for all $a, b \in M$, we consider the following two cases:

(i) Assume that $a \otimes b = 1$. If $a = b = 1$, it is obvious that $f^\lambda(a \otimes b) = f(1) \vee \lambda = f^\lambda(a) \wedge f^\lambda(b) \geqslant f^\lambda(a) \wedge f^\lambda(b) \wedge \dfrac{1-k}{2}$.

If $a = 1, b \neq 1$ or $a \neq 1, b = 1$, then $a \otimes b \neq 1$, it is a contradiction.

If $a \neq 1$ and $b \neq 1$, it follows that $f^\lambda(a) \wedge f^\lambda(b) \wedge \dfrac{1-k}{2} = f(a) \wedge f(b) \wedge \dfrac{1-k}{2} \leqslant f(a \otimes b) = f(1) \leqslant f(1) \vee \lambda = f^\lambda(a \otimes b)$. from $f \in \mathbf{FFil}(M)$, (FF4) and (1).

(ii) Assume that $a \otimes b \neq 1$. If $a = b = 1$, it is obvious a contradiction.

If $a = 1, b \neq 1$ or $a \neq 1, b = 1$, let's assume that $a = 1, b \neq 1$, then $a \otimes b = \neg(1 \to \neg b) = b$, and so $f^\lambda(a) \wedge f^\lambda(b) \wedge \dfrac{1-k}{2} \leqslant f^\lambda(b) = f(b) = f(a \otimes b) = f^\lambda(a \otimes b)$.

If $a \neq 1$ and $b \neq 1$, it follows that $f^\lambda(a \otimes b) = f(a \otimes b) \geqslant f(a) \wedge f(b) \wedge \dfrac{1-k}{2} = f^\lambda(a) \wedge f^\lambda(b) \wedge \dfrac{1-k}{2}$ from $f \in \mathbf{FFil}(M)$ and (FF4).

Summarize these two cases, we conclude that $f^\lambda(a \otimes b) \geqslant f^\lambda(a) \wedge f^\lambda(b) \wedge \dfrac{1-k}{2}$, for all $a, b \in M$. i.e., f^λ satisfies (FF4).

Thus it follows that $f^\lambda \in \mathbf{FFil}(M)$ from Theorem 1.

Definition 4. Let M be an R_0-algebra and f, g two fuzzy sets on M. Fuzzy sets f^g and g^f on M are defined as follows: for all $a \in M$,

$$f^g(a) = \begin{cases} f(a), & a \neq 1, \\ f(1) \vee g(1), & a = 1, \end{cases} \quad \text{and} \quad g^f(a) = \begin{cases} g(a), & a \neq 1, \\ g(1) \vee f(1), & a = 1. \end{cases} \tag{2}$$

Corollary 1. Let M be an R_0-algebra and $f, g \in \mathbf{FFil}(M)$. Then $f^g, g^f \in \mathbf{FFil}(M)$.

Proof. It is straightforward form Theorem 3 and Definition 4.

Definition 5. Let M be an R_0-algebra and f, g two fuzzy sets on M. A fuzzy set $f \uplus g$ on M is defined as follows: for all $a, x, y \in M$,

$$(f \uplus g)(a) = \bigvee_{x \otimes y \leqslant a} \left[f(x) \wedge g(y) \wedge \frac{1-k}{2} \right]. \tag{3}$$

Theorem 4. Let M be an R_0-algebra and $f, g \in \mathbf{FFil}(M)$. Then $f^g \uplus g^f \in \mathbf{FFil}(M)$.

Proof. Firstly, for all $a, b \in M$, let $a \leqslant b$, then $\{x \otimes y | x \otimes y \leqslant a\} \subseteq \{x \otimes y | x \otimes y \leqslant b\}$, and so

$$\begin{aligned} \left(f^g \uplus g^f\right)(b) &= \bigvee_{x \otimes y \leqslant b} \left[f^g(x) \wedge g^f(y) \wedge \frac{1-k}{2} \right] \geqslant \bigvee_{x \otimes y \leqslant a} \left[f^g(x) \wedge g^f(y) \wedge \frac{1-k}{2} \right] \\ &= \frac{1-k}{2} \wedge \bigvee_{x \otimes y \leqslant a} \left[f^g(x) \wedge g^f(y) \wedge \frac{1-k}{2} \right] = \left(f^g \uplus g^f\right)(a) \wedge \frac{1-k}{2}, \end{aligned}$$

Hence $f^g \uplus g^f$ satisfies (FF3). Secondly, for all $a, b \in M$, we have that

$$\begin{aligned} \left(f^g \uplus g^f\right)(a \otimes b) &= \bigvee_{x \otimes y \leqslant a \otimes b} \left[f^g(x) \wedge g^f(y) \wedge \frac{1-k}{2} \right] \\ &\geqslant \bigvee_{x_1 \otimes x_2 \leqslant a \text{ and } y_1 \otimes y_2 \leqslant b} \left[f^g(x_1 \otimes y_1) \wedge g^f(x_2 \otimes y_2) \wedge \frac{1-k}{2} \right] \\ &\geqslant \bigvee_{x_1 \otimes x_2 \leqslant a \text{ and } y_1 \otimes y_2 \leqslant b} \left[f^g(x_1) \wedge f^g(y_1) \wedge g^f(x_2) \wedge g^f(y_2) \wedge \frac{1-k}{2} \right] \\ &= \bigvee_{x_1 \otimes x_2 \leqslant a} \left[f^g(x_1) \wedge g^f(x_2) \wedge \frac{1-k}{2} \right] \wedge \bigvee_{y_1 \otimes y_2 \leqslant b} \left[f^g(y_1) \wedge g^f(y_2) \wedge \frac{1-k}{2} \right] \\ &= \frac{1-k}{2} \wedge \bigvee_{x_1 \otimes x_2 \leqslant a} \left[f^g(x_1) \wedge g^f(x_2) \wedge \frac{1-k}{2} \right] \wedge \\ &\qquad \frac{1-k}{2} \wedge \bigvee_{y_1 \otimes y_2 \leqslant b} \left[f^g(y_1) \wedge g^f(y_2) \wedge \frac{1-k}{2} \right] \\ &= \left(f^g \uplus g^f\right)(a) \wedge \left(f^g \uplus g^f\right)(b) \wedge \frac{1-k}{2}, \end{aligned}$$

and so $f^g \uplus g^f$ also satisfies (FF4). Hence $f^g \uplus g^f \in \mathbf{FFil}(M)$ by Theorem 1.

4 The Lattice of $(\in, \in\vee q_k)$-fuzzy Filters in an R_0-algebra

In this section, we investigate the lattice structural feature of the set $\mathbf{FFil}(M)$.

Definition 6. Let M be an R_0-algebra and f, g two fuzzy sets on M. The binary relation $\sqsubseteq$ is defined as follows:

$$f \sqsubseteq g \Longleftrightarrow f(a) \wedge \frac{1-k}{2} \leqslant g(a), \text{for all } a \in M, \tag{4}$$

$$f = g \Longleftrightarrow f \sqsubseteq g \text{ and } g \sqsubseteq f. \tag{5}$$

Remark 1. Let M be an R_0-algebra and f, g two fuzzy sets on M. Then according to Definition 6, we have that

$$f \sqsubseteq g \text{ implies } f(a) \wedge \frac{1-k}{2} \leqslant g(a) \wedge \frac{1-k}{2}, \text{for all } a \in M. \tag{6}$$

At the same time, it is easy to verify that the following assertions are hold:

(i) By the definition of $\sqsubseteq$, $f \sqsubseteq f$;
(ii) By the definition of $\sqsubseteq$, $f \sqsubseteq g$ and $g \sqsubseteq f$ imply $f = g$;
(iii) $f \sqsubseteq g$ and $g \sqsubseteq h$ imply $f \sqsubseteq h$.

Hence we can see that $\sqsubseteq$ is a partial order on the set of all fuzzy sets on M.

Definition 7. Let M be an R_0-algebra and f a fuzzy set on M. The intersection of all $(\in, \in\vee q_k)$-fuzzy filters of M containing f is called the generated $(\in, \in\vee q_k)$-fuzzy filter by f, denoted $\langle f \rangle$.

Theorem 5. Let M be an R_0-algebra and f a fuzzy set on M. A fuzzy set g on M is defined as follows:

$$g(a) = \bigvee \left\{ f(x_1) \wedge \cdots f(x_n) \wedge \frac{1-k}{2} | x_1, \cdots, x_n \in M \text{ and } x_1 \otimes \cdots \otimes x_n \leqslant a \right\} \tag{7}$$

for all $a \in M$. Then $g = \langle f \rangle$.

Proof. Firstly, we prove that $g \in \mathbf{FFil}(M)$. For all $a, b \in M$, let $a \leqslant b$, then $g(a) \wedge \frac{1-k}{2} = \frac{1-k}{2} \wedge \bigvee \left\{ f(x_1) \wedge \cdots f(x_n) \wedge \frac{1-k}{2} | x_1, \cdots, x_n \in M \text{ and } x_1 \otimes \cdots \otimes x_n \leqslant a \right\} = \bigvee \left\{ f(x_1) \wedge \cdots f(x_n) \wedge \frac{1-k}{2} | x_1, x_2, \cdots, x_n \in M \text{ and } x_1 \otimes x_2 \otimes \cdots \otimes x_n \leqslant a \right\} \leqslant \bigvee \left\{ f(x_1) \wedge \cdots f(x_n) \wedge \frac{1-k}{2} | x_1, \cdots, x_n \in M \text{ and } x_1 \otimes \cdots \otimes x_n \leqslant b \right\} = g(b)$. Thus g satisfies (FF3). Assume that $\exists x_1, x_2, \cdots, x_n \in M$ and $y_1, y_2, \cdots, y_m \in M$ such that $x_1 \otimes x_2 \otimes \cdots \otimes x_n \leqslant a$ and $y_1 \otimes y_2 \otimes \cdots \otimes y_m \leqslant b$, we have that $x_1 \otimes x_2 \otimes \cdots \otimes x_n \otimes y_1 \otimes y_2 \otimes \cdots \otimes y_m \leqslant a \otimes b$ by (P11). Thus, we can obtain that $g(a) \wedge g(b) \wedge \frac{1-k}{2} = \frac{1-k}{2} \wedge \bigvee \left\{ f(x_1) \wedge \cdots f(x_n) \wedge \frac{1-k}{2} | x_1, \cdots, x_n \in \right.$

M and $x_1\otimes\cdots\otimes x_n \leqslant a\Big\}\wedge\bigvee\Big\{f(y_1)\wedge\cdots f(y_m)\wedge\frac{1-k}{2}|y_1,\cdots,y_m \in M$ and $y_1\otimes\cdots\otimes y_m \leqslant b\Big\} = \bigvee\Big\{f(x_1)\wedge\cdots f(x_n)\wedge\frac{1-k}{2}|x_1,x_2,\cdots,x_n \in M$ and $x_1\otimes x_2\otimes\cdots\otimes x_n \leqslant a\Big\}\wedge\bigvee\Big\{f(y_1)\wedge\cdots f(y_m)\wedge\frac{1-k}{2}|y_1,\cdots,y_m \in M$ and $y_1\otimes\cdots\otimes y_m \leqslant b\Big\} = \bigvee\Big\{f(x_1)\wedge\cdots\wedge f(x_n)\wedge f(y_1)\wedge\cdots\wedge f(y_m)\wedge\frac{1-k}{2}|x_1,\cdots,x_n,y_1,\cdots,y_m \in M$ such that $x_1\otimes x_2\otimes\cdots\otimes x_n \leqslant a$ and $y_1\otimes y_2\otimes\cdots\otimes y_m \leqslant b\Big\} \leqslant \bigvee\Big\{f(x_1)\wedge\cdots\wedge f(x_n)\wedge f(y_1)\wedge\cdots\wedge f(y_m)\wedge\frac{1-k}{2}|x_1,\cdots,x_n,y_1,\cdots,y_m \in M$ such that $x_1\otimes x_2\otimes\cdots\otimes x_n\otimes y_1\otimes y_2\otimes\cdots\otimes y_m \leqslant a\otimes b\Big\} = g(a\otimes b)$. Hence f also satisfies (FF4). It follows from Theorem 1, that $g \in \mathbf{FFil}(M)$.

Secondly, For any $a \in M$, it follows from $a \leqslant a$ and the definition of g that $f(a)\wedge\frac{1-k}{2} \leqslant g(a)$. This means that $f \sqsubseteq g$.

Finally, assume that $h \in \mathbf{FFil}(M)$ with $f \sqsubseteq h$. Then for any $a \in M$, we have $g(a)\wedge\frac{1-k}{2} = \frac{1-k}{2}\wedge\bigvee\Big\{f(x_1)\wedge\cdots f(x_n)\wedge\frac{1-k}{2}|x_1,\cdots,x_n \in M$ and $x_1\otimes\cdots\otimes x_n \leqslant a\Big\} \leqslant \frac{1-k}{2}\wedge\bigvee\Big\{h(x_1)\wedge\cdots h(x_n)|x_1,x_2,\cdots,x_n \in M$ and $x_1\otimes x_2\otimes\cdots\otimes x_n \leqslant a\Big\} = \bigvee\Big\{h(x_1)\wedge\cdots h(x_n)\wedge\frac{1-k}{2}|x_1,x_2,\cdots,x_n \in M$ and $x_1\otimes x_2\otimes\cdots\otimes x_n \leqslant a\Big\} = \frac{1-k}{2}\wedge\bigvee\Big\{h(x_1)\wedge\cdots h(x_n)\wedge\frac{1-k}{2}|x_1,\cdots,x_n \in M$ and $x_1\otimes\cdots\otimes x_n \leqslant a\Big\} \leqslant \frac{1-k}{2}\wedge\bigvee\Big\{h(x_1\otimes\cdots\otimes x_n)|x_1,x_2,\cdots,x_n \in M$ and $x_1\otimes x_2\otimes\cdots\otimes x_n \leqslant a\Big\} = \bigvee\Big\{h(x_1\otimes\cdots\otimes x_n)\wedge\frac{1-k}{2}|x_1,x_2,\cdots,x_n \in M$ and $x_1\otimes x_2\otimes\cdots\otimes x_n \leqslant a\Big\} \leqslant \bigvee\Big\{h(a)\Big\} = h(a)$. Hence $g \sqsubseteq h$ holds. To sum up, we have that $g = \langle f\rangle$.

Theorem 6. Let M be an R_0-algebra. Then $(\mathbf{FFil}(M), \sqsubseteq, 0_M, 1_M)$ is a complete lattice.

Proof. For any $\{f_\alpha\}_{\alpha\in\Lambda} \subseteq \mathbf{FFil}(M)$, where Λ is an indexed set. It is easy to verify that $\bigcap\limits_{\alpha\in\Lambda} f_\alpha$ is infimum of $\{f_\alpha\}_{\alpha\in\Lambda}$, where $\left(\bigcap\limits_{\alpha\in\Lambda} f_\alpha\right)(a) = \bigwedge\limits_{\alpha\in\Lambda} f_\alpha(a)$ for all $a \in M$. i. e., $\bigwedge\limits_{\alpha\in\Lambda} f_\alpha = \bigcap\limits_{\alpha\in\Lambda} f_\alpha$. Define $\bigcup\limits_{\alpha\in\Lambda} f_\alpha$ such that $\left(\bigcup\limits_{\alpha\in\Lambda} f_\alpha\right)(a) = \bigvee\limits_{\alpha\in\Lambda} f_\alpha(a)$ for all $a \in M$. Then $\left\langle\bigcup\limits_{\alpha\in\Lambda} f_\alpha\right\rangle$ is supermun of $\{f_\alpha\}_{\alpha\in\Lambda}$, where $\left\langle\bigcup\limits_{\alpha\in\Lambda} f_\alpha\right\rangle$ is the $(\in, \in\vee q_k)$-fuzzy filter generated by $\bigcup_{\alpha\in\Lambda} f_\alpha$ of M. i. e., $\bigvee\limits_{\alpha\in\Lambda} f_\alpha = \left\langle\bigcup\limits_{\alpha\in\Lambda} f_\alpha\right\rangle$. Therefor $(\mathbf{FFil}(M), \sqsubseteq, 0_M, 1_M)$ is a complete lattice.

Remark 2. Let M be an R_0-algebra. For all $f, g \in \mathbf{FFil}(M)$, by Theorem 6 we know that $f \wedge g = f \cap g$ and $f \vee g = \langle f \cup g\rangle$.

Theorem 7. Let M be an R_0-algebra. Then for all $f, g \in \mathbf{FFil}(M)$, $f \vee g = \langle f \cup g \rangle = f^g \uplus g^f$ in the complete lattice $(\mathbf{FFil}(M), \sqsubseteq, 0_M, 1_M)$.

Proof. For all $f, g \in \mathbf{FFil}(M)$, it is obvious that $f \sqsubseteq f^g \uplus g^f$ and $g \sqsubseteq f^g \uplus g^f$, that is, $f(a) \wedge \frac{1-k}{2} \leqslant \left(f^g \uplus g^f\right)(a)$ and $g(a) \wedge \frac{1-k}{2} \leqslant \left(f^g \uplus g^f\right)(a)$ for all $a \in M$. Thus $(f \cup g)(a) \wedge \frac{1-k}{2} = [f(a) \vee g(a)] \wedge \frac{1-k}{2} = \left[f(a) \wedge \frac{1-k}{2}\right] \vee \left[g(a) \wedge \frac{1-k}{2}\right] \leqslant \left(f^g \uplus g^f\right)(a)$, that is, $f \cup g \sqsubseteq f^g \uplus g^f$, and thus $\langle f \cup g \rangle \sqsubseteq f^g \uplus g^f \in \mathbf{FFil}(M)$ by Theorem 4. Let $h \in \mathbf{FFil}(M)$ such that $f \cup g \sqsubseteq h$. For all $a \in M$, we consider the following two cases:

(i) If $a=1$, then $\left(f^g \uplus g^f\right)(1) \wedge \frac{1-k}{2} = \frac{1-k}{2} \wedge \bigvee_{x \otimes y \leqslant 1} \left[f^g(x) \wedge g^f(y) \wedge \frac{1-k}{2}\right] = f^g(1) \wedge g^f(1) \wedge \frac{1-k}{2} = [f(1) \vee g(1)] \wedge \frac{1-k}{2} = (f \cup g)(1) \wedge \frac{1-k}{2} \leqslant h(1)$,

(ii) If $a < 1$, then we have $\left(f^g \uplus g^f\right)(a) \wedge \frac{1-k}{2} = \frac{1-k}{2} \wedge \bigvee_{x \otimes y \leqslant a} \left[f^g(x) \wedge g^f(y) \wedge \frac{1-k}{2}\right] = \bigvee_{x \otimes y \leqslant a, x \neq 1, y \neq 1} \left[f^g(x) \wedge g^f(y) \wedge \frac{1-k}{2}\right] \vee \bigvee_{x \leqslant a} \left\{f(x) \wedge \left[f(1) \vee g(1)\right] \wedge \frac{1-k}{2}\right\} \vee \bigvee_{y \leqslant a} \left\{\left[f(1) \vee g(1)\right] \wedge g(y) \wedge \frac{1-k}{2}\right\} = \bigvee_{x \otimes y \leqslant a, x \neq 1, y \neq 1} \left[f(x) \wedge g(y) \wedge \frac{1-k}{2}\right] \vee \bigvee_{x \leqslant a} \left[f(x) \wedge \frac{1-k}{2}\right] \vee \bigvee_{y \leqslant a} \left[g(y) \wedge \frac{1-k}{2}\right] \leqslant \bigvee_{x \otimes y \leqslant a, x \neq 1, y \neq 1} \left[h(x) \wedge h(y) \wedge \frac{1-k}{2}\right] \vee \bigvee_{x \leqslant a} \left[h(x) \wedge \frac{1-k}{2}\right] \vee \bigvee_{y \leqslant a} \left[h(y) \wedge \frac{1-k}{2}\right] = \bigvee_{x \otimes y \leqslant a} \left[h(x) \wedge h(y) \wedge \frac{1-k}{2}\right] = \frac{1-k}{2} \wedge \bigvee_{x \otimes y \leqslant a} \left[h(x) \wedge h(y) \wedge \frac{1-k}{2}\right] \leqslant \bigvee_{x \otimes y \leqslant a} \left[h(x \otimes y) \wedge \frac{1-k}{2}\right] \leqslant h(a)$, thus $f^g \uplus g^f \sqsubseteq h$ for above two cases.

By Definition 7 and Theorem 5 we have that $f \vee g = \langle f \cup g \rangle = f^g \uplus g^f$.

Theorem 8. Let M be an R_0-algebra. Then $(\mathbf{FFil}(M), \sqsubseteq, 0_M, 1_M)$ is a distributive lattice, where, $f \wedge g = f \cap g$ and $f \vee g = \langle f \cup g \rangle$, for all $f, g \in \mathbf{FFil}(M)$.

Proof. To finish the proof, it suffices to show that $h \wedge (f \vee g) = (h \wedge f) \vee (h \wedge g)$, for all $f, g, h \in \mathbf{FFil}(M)$. Since the inequality $(h \wedge f) \vee (h \wedge g) \sqsubseteq h \wedge (f \vee g)$ holds automatically in a lattice, we need only to show the inequality $h \wedge (f \vee g) \sqsubseteq (h \wedge f) \vee (h \wedge g)$, i.e., $\left(h \cap \left(f^g \uplus g^f\right)\right)(a) \wedge \frac{1-k}{2} \leqslant \left((h \cap f)^{h \cap g} \uplus (h \cap g)^{h \cap f}\right)(a)$, for all $a \in M$. For these, we consider the following two cases:

(i) If $a = 1$, we have

$$\left(h \cap \left(f^g \uplus g^f\right)\right)(1) \wedge \frac{1-k}{2} = h(1) \wedge \left(f^g \uplus g^f\right)(1) \wedge \frac{1-k}{2}$$
$$= h(1) \wedge \bigvee_{x \otimes y \leqslant 1} \left[f^g(x) \wedge g^f(y) \wedge \frac{1-k}{2}\right] = h(1) \wedge \left[f^g(1) \wedge g^f(1) \wedge \frac{1-k}{2}\right]$$
$$= h(1) \wedge [f(1) \vee g(1)] \wedge \frac{1-k}{2} = \{[h(1) \wedge f(1)] \vee [h(1) \wedge g(1)]\} \wedge \frac{1-k}{2}$$
$$= [(h \cap f)(1) \wedge (h \cap g)(1)] \wedge \frac{1-k}{2}$$
$$= \left((h \cap f)^{(h \cap g)}\right)(1) \wedge \left((h \cap g)^{(h \cap f)}\right)(1) \wedge \frac{1-k}{2}$$
$$= \bigvee_{x \otimes y \leqslant 1} \left[\left((h \cap f)^{(h \cap g)}\right)(x) \wedge \left((h \cap g)^{(h \cap f)}\right)(y) \wedge \frac{1-k}{2}\right]$$
$$= \left((h \cap f)^{h \cap g} \uplus (h \cap g)^{h \cap f}\right)(1).$$

(ii) If $a < 1$, we have

$$\left(h \cap \left(f^g \uplus g^f\right)\right)(a) \wedge \frac{1-k}{2} = h(a) \wedge \left(f^g \uplus g^f\right)(a) \wedge \frac{1-k}{2}$$
$$= h(a) \wedge \bigvee_{x \otimes y \leqslant a} \left[f^g(x) \wedge g^f(y) \wedge \frac{1-k}{2}\right] = \bigvee_{x \otimes y \leqslant a} \left[h(a) \wedge f^g(x) \wedge g^f(y) \wedge \frac{1-k}{2}\right]$$
$$= \bigvee_{x \otimes y \leqslant a, x \neq 1, y \neq 1} \left[h(a) \wedge f^g(x) \wedge g^f(y) \wedge \frac{1-k}{2}\right] \vee$$
$$\bigvee_{y \leqslant a} \left[h(a) \wedge f^g(1) \wedge g(y) \wedge \frac{1-k}{2}\right] \vee \bigvee_{x \leqslant a} \left[h(a) \wedge f(x) \wedge g^f(1) \wedge \frac{1-k}{2}\right]$$
$$= \bigvee_{x \otimes y \leqslant a, x \neq 1, y \neq 1} \left\{[h(a) \wedge f(x)] \wedge \left[h(a) \wedge g(y) \wedge \frac{1-k}{2}\right]\right\} \vee$$
$$\bigvee_{y \leqslant a} \left\{\left[h(a) \wedge \frac{1-k}{2} \wedge f^g(1)\right] \wedge [h(a) \wedge g(y)] \wedge \frac{1-k}{2}\right\} \vee$$
$$\bigvee_{x \leqslant a} \left\{[h(a) \wedge f(x)] \wedge \left[h(a) \wedge \frac{1-k}{2} \wedge g^f(1)\right] \wedge \frac{1-k}{2}\right\}$$

$$\leqslant \bigvee_{x \otimes y \leqslant a, x \neq 1, y \neq 1} \left\{[h(a \vee x) \wedge f(a \vee x)] \wedge [h(a \vee y) \wedge g(a \vee y)] \wedge \frac{1-k}{2}\right\} \vee$$
$$\bigvee_{y \leqslant a} \left\{[h(1) \wedge (f(1) \vee g(1))] \wedge [h(a \vee y) \wedge g(a \vee y)] \wedge \frac{1-k}{2}\right\} \vee$$
$$\bigvee_{x \leqslant a} \left\{[h(a \vee x) \wedge f(a \vee x)] \wedge [h(1) \wedge (g(1) \vee f(1))] \wedge \frac{1-k}{2}\right\}$$

$$= \bigvee_{x\otimes y\leqslant a, x\neq 1, y\neq 1} \left[(h\cap f)(a\vee x)\wedge(h\cap g)(a\vee y)\wedge\frac{1-k}{2}\right]\vee$$

$$\bigvee_{y\leqslant a}\left\{[(h\cap f)(1)\vee(h\cap g)(1)]\wedge(h\cap g)(a\vee y)\wedge\frac{1-k}{2}\right\}\vee$$

$$\bigvee_{x\leqslant a}\left\{(h\cap f)(a\vee x)\wedge[(h\cap g)(1)\vee(h\cap f)(1)]\wedge\frac{1-k}{2}\right\}$$

$$= \bigvee_{x\otimes y\leqslant a, x\neq 1, y\neq 1} \left[(h\cap f)^{h\cap g}(a\vee x)\wedge(h\cap g)^{h\cap f}(a\vee y)\wedge\frac{1-k}{2}\right]\vee$$

$$\bigvee_{y\leqslant a}\left[(h\cap f)^{h\cap g}(1)\wedge(h\cap g)^{h\cap f}(a\vee y)\wedge\frac{1-k}{2}\right]\vee$$

$$\bigvee_{x\leqslant a}\left[(h\cap f)^{h\cap g}(a\vee x)\wedge(h\cap g)^{h\cap f}(1)\wedge\frac{1-k}{2}\right]$$

$$= \bigvee_{x\otimes y\leqslant a}\left[(h\cap f)^{h\cap g}(a\vee x)\wedge(h\cap g)^{h\cap f}(a\vee y)\wedge\frac{1-k}{2}\right].$$

Let $a\vee x = u$ and $a\vee y = v$, since $x\otimes y\leqslant a$, using Lemma 2 we get that $u\otimes v = (a\vee x)\otimes(a\vee y) = ((a\vee x)\otimes a)\vee((a\vee x)\otimes y) = (a\otimes a)\vee(a\otimes x)\vee(a\otimes y)\vee(x\otimes y)\leqslant a\vee a\vee a\vee(x\otimes y) = a\vee(x\otimes y)\leqslant a\vee a = a$. Hence we can conclude that

$$\begin{aligned}
&\left(h\cap\left(f^g\uplus g^f\right)\right)(a)\wedge\frac{1-k}{2}\\
&\leqslant \bigvee_{x\otimes y\leqslant a}\left[(h\cap f)^{h\cap g}(a\vee x)\wedge(h\cap g)^{h\cap f}(a\vee y)\wedge\frac{1-k}{2}\right]\\
&\leqslant \bigvee_{u\otimes v\leqslant a}\left[(h\cap f)^{h\cap g}(u)\wedge(h\cap g)^{h\cap f}(v)\wedge\frac{1-k}{2}\right]\\
&= \left((h\cap f)^{h\cap g}\uplus(h\cap g)^{h\cap f}\right)(a).
\end{aligned}$$

To sum up, we have that $\left(h\cap\left(f^g\uplus g^f\right)\right)(a)\wedge\frac{1-k}{2}\leqslant\left((h\cap f)^{h\cap g}\uplus(h\cap g)^{h\cap f}\right)$ (a), for all $a\in M$. The proof is completed.

5 Conclusion

As well known, filters is an important concept for studying the structural features of R_0-algebras. In this paper, the $(\in,\in\vee q_k)$-fuzzy filter theory in R_0-algebras is further studied. Some new properties of $(\in,\in\vee q_k)$-fuzzy filters are given. Representation theorem of $(\in,\in\vee q_k)$-fuzzy filter which is generated by a fuzzy set is established. It is proved that the set consisting of all $(\in,\in\vee q_k)$-fuzzy filters in an R_0-algebra, under the partial order $\sqsubseteq$, forms a complete distributive lattice. Results obtained in this paper not only enrich the content of $(\in,\in\vee q_k)$-fuzzy filter theory in R_0-algebras, but also show interactions of algebraic technique and fuzzifying method in the studying logic problems. We hope that more links of fuzzy sets and logics emerge by the stipulating of this work.

Acknowledgements. Thanks to the support by Higher School Research Foundation of Inner Mongolia, China (No. NJSY14283, NJZY18206).

Recommender: Shu-hai Li, a professor of Chifeng University.

References

1. Wang, G.J.: Non-classical Mathematical Logic and Approximate Reasoning. Science Press, Beijing (2003)
2. Zadeh, L.A.: Toward a generalized theory of uncertainty (GTU)-an outline. Inf. Sci. **172**(1–2), 1–40 (2005)
3. Wang, G.J.: Introduction to Mathematical Logic and Resolution Principle. Science Press, Beijing (2003)
4. Jun, Y.B., Liu, L.Z.: Filters of R_0-algebras. Int. J. Math. Math. Sci. **35**(2), 1–9 (2006)
5. Zadeh, L.A.: Fuzzy Sets. Inf. Control **8**, 338–353 (1965)
6. Liu, L.Z., Li, K.T.: Fuzzy implicative and Boolean filters in R_0-algebras. Inf. Sci. **171**(1–3), 61–71 (2005)
7. Ma, X.L., Zhan, J.M., Jun, Y.B.: On $(\in, \in\vee q)$-fuzzy filters of R_0-algebras. Math. Log. Q. **55**(5), 493–508 (2009)
8. Jun, Y.B., Song, S.Z., Zhan, J.M.: Generalizations of $(\in, \in\vee q)$-fuzzy filters in R_0-algebras. Int. J. Math. Math. Sci. **2010**, 1–19 (2010). Article no. 918656
9. Peng, J.Y.: $(\in, \in\vee q)$-fuzzy filters of lattice implication algebras. Commun. Comput. Inf. Sci. **227**, 397–404 (2011)
10. Wu, Y., Xin, X.L.: $(\in, \in\vee q)$-fuzzy filters of residuated lattices. Comput. Eng. Appl. **47**(31), 38–39 (2011)
11. Namdar, A., Saeid, A.B., Jabbari, G.: On $(\in, \in\vee q)$-fuzzy filters of CI-algebras. Ann. Fuzzy Math. Inform. **7**(5), 851–858 (2014)
12. Yang, Y.W., Xin, X.L., He, P.F.: $(\in_\gamma, \in_\gamma \vee q_\delta)$-intuitionistic fuzzy (soft) filter of BL-algebras. Chin. Q. J. Math. **29**(1), 65–75 (2014)
13. Liu, C.H.: $(\in, \in\vee q)$-fuzzy prime filters in BL-algebras. J. ZheJiang Univ. (Sci. Ed.) **41**(5), 489–493 (2014)
14. Fu, X.B., Liao, Z.H.: $(\in, \in\vee q_{(\lambda,\mu)})$-fuzzy prime filters of lattice implication algebra. J. Front. Comput. Sci. Technol. **9**(2), 227–233 (2015)
15. Liu, C.H.: $(\in, \in\vee q)$-fuzzy filters theory in BL-algebras. Fuzzy Syst. Math. **29**(1), 50–58 (2015)

Classification and Recognition

A Classification Method by Using Fuzzy Neural Network and Ensemble Learning

Shining Ding[1], Wenyan Song[2], De-gang Wang[1](✉), and Hong-xing Li[1]

[1] School of Control Science and Engineering, Dalian University of Technology, Dalian 116024, People's Republic of China
wangdg@dlut.edu.cn

[2] School of Economics, Dongbei University of Finance and Economics, Dalian 116025, People's Republic of China

Abstract. In this paper, a new multi-class classification technology combining fuzzy neural network and ensemble learning technology is proposed. Several fuzzy neural networks, including fuzzy TS neural network and fuzzy wavelet neural network, are considered as sub-classifiers. Restricted Boltzmann machines and gradient descent algorithm are used to train the parameters of fuzzy neural network. Then, adaptive boosting (AdaBoost) algorithm based on hypothesis margin and particle swarm optimization is provided to determine the weights of sub-classifiers. Numerical simulation results illustrate the validity of proposed method.

Keywords: Fuzzy neural network · Ensemble learning · Classification

1 Introduction

Classification is an intelligible method of distributing individuals with common or similar features into classes or categories. It can help to make a large number of complicated objects structured and systematic. Besides, it can help to discover and grasp general law of development for things. Hence, how to establish classification algorithm to achieve higher accuracy has gained much more attentions by scholars.

Some scholars utilize fuzzy neural network to solve classification problem. In [1], a fuzzy binary neural network based on binary matrix memories and fuzzy logic is proposed for interpretable classifications. In [2], an enhanced discriminability recurrent fuzzy network is proposed for temporal classification problems. In [3], a methodology through a rough-fuzzy artificial neural network is proposed for biological image classification. In [4], the modified fuzzy min-max neural network classification model is proposed for data with mixed attributes.

On the other hand, some scholars attempt to use ensemble learning technology to handle data classification. In [5], a new classifier combination technique which uses principal component analysis (PCA) to do feature axis rotation for each base classifier was developed. In [6], a method combining Restricted Boltzmann Machines (RBM) is presented to generate classifiers. In [7], deep support vector machine was applied to produce the weight for each feature. In [8], discriminant deep belief network is

B.-Y. Cao and Y.-B. Zhong (Eds.): ICFIE 2017, AISC 872, pp. 133–142, 2019.
https://doi.org/10.1007/978-3-030-02777-3_12

proposed for image classification, in which ensemble learning and deep belief network are combined to discriminate features.

From existing results, we find that the selection of model influences corresponding classification accuracy definitively. And a single model often could not be suitable for various datasets with different features. Hence, how to combine several sub-classifiers to improve classification accuracy is an interesting problem.

In this paper, a novel classification method combining fuzzy neural networks and ensemble learning method is proposed. The rest of this paper is organized as follows. Three sub-classifiers models and RBM model are briefly introduced in Sect. 2. Section 3 introduces the whole proposed algorithm. Section 4 conducts some experiments to verify the validity of proposed methods. Conclusions are offered in Sect. 5.

2 Preliminary Knowledge

In this section, we will introduce some preliminaries about fuzzy neural network and RBM.

In the following we will summarize the mathematical representation of fuzzy neural networks at first. In this paper, TS fuzzy neural networks (TS-FNN), fuzzy neural networks based on second-order Taylor expansion (Taylor-FNN), and wavelet fuzzy neural networks (WFNN, [9]) are chosen as sub-classifiers.

The construction and process of these networks are similar. They include the following layers. The first layer receives the input variables. The second layer divides received variables with Gaussian fuzzy function. The third layer calculates the membership degree of each fuzzy rule. The fourth layer normalizes the membership degrees. The fifth layer calculates the corresponding output for each rule. The sixth layer calculates the total output of the network.

The fuzzy rule of each sub-classifier is shown as:

R_l: IF x_1 is F_{1l}, …, x_q is F_{ql}, then $\hat{y}$ is ω_l

where $x_i(i = 1, \cdots, q)$ is the i-th input variable, F_{il} is the fuzzy set corresponding to x_i, and L is the number of rules, $l = 1, \ldots, L$.

The outputs of fuzzy rule in three sub-classifiers are as follows:

Fuzzy TS neural networks

$$\omega_l = p_0^l + p_1^l x_1 + p_2^l x_2 + \cdots + p_q^l x_q$$

Fuzzy TS neural networks based on second-order Taylor expansion

$$\omega_l = p_0^l + p_1^l x_1 + \cdots + p_q^l x_q + p_{11}^l x_1^2 + \cdots + p_{qq}^l x_q^2 + p_{12}^l x_1 x_2 + \cdots + p_{(q-1)q}^l x_{q-1} x_q$$

Wavelet Fuzzy neural networks

$$\omega_l = \left(\sum_{i=1}^{q} \varphi_{il}\right) w_l = \left(\sum_{i=1}^{q} -\vartheta_{il} \exp(-\vartheta_{il}^2/2)\right) w_l$$

where $-\vartheta_{il} = (x_i - b_{il})/a_{il}$, b_{il} and a_{il} are the translation and dilation of the wavelet, and w_l is the parameter of the node.

The Gaussian fuzzy function is defined as follows:

$$\mu_{il} = \exp\left(-(x_i - m_{il})^2 \Big/ 2\sigma_{il}^2\right)$$

where m_{il} and σ_{il} are the center and the width of the Gaussian function.

Then, we will introduce the basic idea of RBM. RBM is a parametric generation model that represents a probability distribution consisting of a visible layer and a hidden layer. The basic structure of it is shown in Fig. 1.

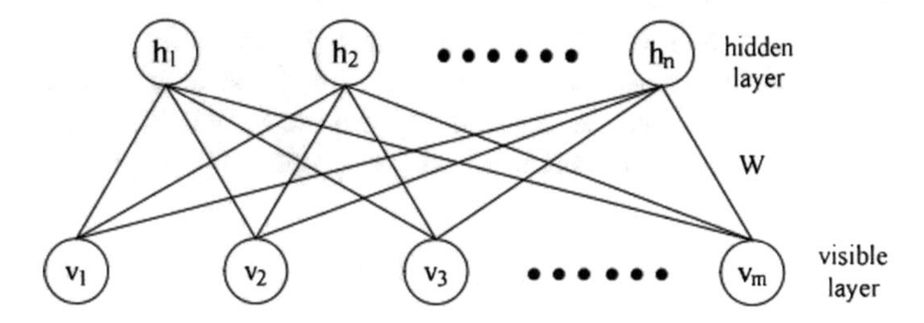

Fig. 1. The structure of RBM model

The energy function of RBM is defined as:

$$E(v,h) = -\sum_{e=1}^{m} a_e v_e - \sum_{g=1}^{n} b_g h_g - \sum_{e=1}^{m}\sum_{g=1}^{n} v_e w_{eg} h_g \tag{1}$$

where w is weight matrix, a_e is visible unit biases, b_g is hidden unit biases, v_e is visible variable, h_g is hidden variable.

Based on expression (1), the joint distribution of v and h is written by

$$P(v,h) = \exp(-E(v,h))/Z$$

where $Z = \sum_{v,h} \exp(-E(v,h))$. v is the set of v_e, h is the set of h_g.

There is no connection between nodes in the same layer in RBM, so the activation conditions of each neuron are independent. The conditional probability distributions of the hidden layer and visible layer are as follows:

$$P(h_g = 1|v) = sigmoid\left(b_g + \sum_{e=1}^{m} v_e w_{eg}\right)$$

$$P(v_e = 1|h) = sigmoid\left(a_e + \sum_{g=1}^{n} h_g w_{eg}\right)$$

where $sigmoid(u) = 1/(1 + e^{-u})$.

3 Classification Algorithm Based on Fuzzy Neural Networks and Ensemble Learning

In this section, we will introduce the proposed classification algorithm.

The framework of the whole algorithm is shown in Fig. 2. Three sub-classifiers are TS-FNN, Taylor-FNN and WFNN. With each sub-classifies, the input data can be classified into some subsets respectively. Further, AdaBoost with hypothesis margin (AdaBoost.HM, [10]) is chosen as ensemble algorithm, in which the results of sub-classifiers are given weights and summed with corresponding weights to output the final hypothesis.

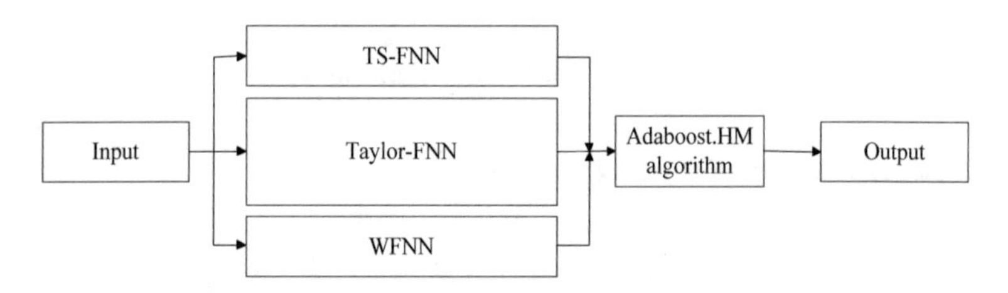

Fig. 2. Structure of the classification algorithm

Next, we will introduce how to determine parameters in sub-classifiers.

For wavelet fuzzy neural networks, parameters $m_{il}, \sigma_{il}, b_{il}, a_{il}$ and w_l are updated by the gradient descent algorithm with adaptive learning rate. The detailed formulas are shown as follows.

$$m_{il}(k+1) = m_{il}(k) + \eta(k)(y(k) - \hat{y}(k))(o_l(k) - \hat{y}(k))v_l \frac{(x_i - m_{il})}{\sigma_{il}^2}$$

$$\sigma_{il}(k+1) = \sigma_{il}(k) + \eta(k)(y(k) - \hat{y}(k))(o_l(k) - \hat{y}(k))v_l \frac{(x_i - m_{il})^2}{\sigma_{il}^3}$$

$$b_{il}(k+1) = b_{il}(k) + \eta(k)(y(k) - \hat{y}(k))v_l w_l \varphi_{il} \frac{(\vartheta_{il}^2 - 1)}{\vartheta_{il} a_{il}}$$

$$a_{il}(k+1) = a_{il}(k) + \eta(k)(y(k) - \hat{y}(k))v_l w_l \varphi_{il} \frac{(\vartheta_{il}^2 - 1)}{a_{il}}$$

$$w_l(k+1) = w_l(k) + \eta(k)(y(k) - \hat{y}(k))v_l\left(\sum_{i=1}^{n} -\vartheta_{il}\exp(-\vartheta_{il}^2/2)\right)$$

where $\eta(k)$ is adaptive learning rate in the k-th iteration, i.e.,

$$\eta(k) = ((100 - ACC(k))/VAL)^{\tau}\cdot\eta(k-1)$$

where $ACC(k)$ is the classification accuracy of the training sample in the k-th iteration, VAL is an constant, and $\tau \in [0, 1]$. It can be seen that the value of $\eta(k)$ is related to the value of $ACC(k)$. In the course of iterative optimization, when classification accuracy is high, the learning rate could be decreased; on the contrary, when classification accuracy is low, the learning rate should be increased. In addition, $\hat{y}(k)$ is the output of WFNN, $y(k)$ is the label of actual output, and $v_l = \mu_l \Big/ \sum_{l=1}^{L} \mu_l$.

For TS-FNN and Taylor-FNN, parameters p_i^l and p_{ij}^l are updated by the following equations:

$$p_i^l(k+1) = p_i^l(k) - \eta(t)(y(k) - \hat{y}(k))v_l * x_i$$

$$p_{ij}^l(k+1) = p_{ij}^l(k) - \eta(t)(y(k) - \hat{y}(k))v_l * x_i x_j, j = 1, \cdots, q$$

The number of fuzzy rules of each sub-classifier is determined as below. First, an estimated classification accuracy rate is preseted. Then the number of fuzzy rules is increased from 5 to 30 one by one, until the results of corresponding network exceeds the estimated value. In this case, stop the iteration loop and appropriate value as the number of fuzzy rules can be determined.

Then we will introduce the parameters determination of the integration model. The process of the proposed algorithm is shown as below.

Suppose the following conditions are known: training sample set is $(x_1, y_1), \cdots, (x_S, y_S), x_s \in \chi, y \in Y, s = 1, \cdots, S$ and the initial weight for each sample is $D_1(s) = 1/S$, where S is the number of sample.

The subsequent implementation step 1 to step 4 are performed for each sub-classifier, and a modified AdaBoost.HM method is designed to get the final classification results.

Step 1. Calculate the classification accuracy in sub-classifier by formula (2):

$$u_t(x_s) = \begin{cases} 1, & y_s = f_t^{y_s} \\ -1, & y_s \neq f_t^{y_s} \end{cases} \tag{2}$$

where $f_t^{y_s}$ is the output of t-th sub-classifier, y_s is the label data. When $y_s = f_t^{y_s}$, the classification in this sub-classifier is correct, else the classification is wrong.

Step 2. Determine the coefficients of sub-classifiers by (3):

$$\alpha_t = \lambda \ln((1 + r_t)/(1 - r_t)), r_t = \sum_s D_t(s) u_t(x_s), \lambda \in (0, 1] \quad (3)$$

where the value of this coefficient λ affects the update of the weights. The optimal value of coefficient is optimized by the particle swarm algorithm (PSO).
Step 3. Update the weights by (4):

$$D_{t+1}(s) = (D_t(s) \exp(-\alpha_t u_t(x_s)))/Z_{t+1} \quad (4)$$

where the denominator Z_{t+1} is the normalized function of molecular.
Step 4. Finally, formula (5) is used to output the final hypothesis:

$$H(x) = \arg\max \sum_{t=1}^{T} \alpha_t h_t(x) \quad (5)$$

where T is the number of sub-classifiers, and h_t is the direct output of the sub-classifier, which is composed of number closing to 0 or 1, and its length is the number of categories.

4 Experimental Studies

In this section, some datasets are provided to demonstrate the effectiveness of proposed model.

Example 1. Classification of UCI machine learning repository.

In this experiment, seven datasets are selected from the UCI machine learning repository (http://archive.ics.uci.edu/ml/), and the main features of these datasets used in our experiments are summarized in Table 1.

Table 1. Summary of characteristics for the used datasets

Datasets	Size	Number of features	Classes
German	1000	24	2
Heart	270	13	2
Sonar	208	60	2
Pima	768	8	2
WDBC	569	30	2
WPBC	194	32	2
Haberman	306	3	2

The general process of this experiment is as follows. 10 fold cross validation is used for these datasets. Therefore, we randomly divide dataset into 10 subsets with approximately equal sizes. Then 9 of 10 subsets are used as training sets, and the rest part is testing set. Next, each feather of training sets is normalized to have lower limit 0 and upper limit 1, and the same normalization process is performed on the testing set.

Using training sets to obtain results of three sub-classifiers, then the coefficients of sub-classifiers are calculated. And the results of sub-classifiers are summed with weights in training sets and testing sets. Finally classification results are counted. Alternating 10 subsets until all of these subsets are used to be tested once. The final error rate is the average error rate of subsets, and the standard deviation of error rate is calculated as follows:

$$\sigma = \sqrt{\frac{1}{M}\left(\sum_{i=1}^{M}(x_i - \mu)^2\right)}$$

where M is the number of fold, and μ is the average of error rate.

Now, take Haberman dataset as an example, the parameter setting is briefly introduced. For TS-FNN, the initial value of parameter p_i^l is 0.25. For Taylor-FNN, the initial value of parameter p_{ij}^l is 0.2. The number of fuzzy rules is 5, $VAL = 30$, and $\tau = 0.1$. The initial learning rate and momentum factors are set to 0.05 and 0.01. The number of input nodes and output nodes is set to 3 and 2, and the number of iterations is set to 100. In this example, RBM is used to initialize parameters m_{il}, σ_{il}, b_{il} and a_{il}.

The results of three sub-networks and integrated classifier are shown in Table 2. In this table, the rows in the table represent different data sets, and the columns represent different algorithms. The statistical results are the mean and standard deviation of the classification error rate, which expressed as a percentage. The lower the mean value of error rate is, the better the classification is. The lower the standard deviation of error rate is, the lower the degree of dispersion is. The best classification results in the table are marked with bold font.

Table 2. Means and standard deviations of error rate with random initialization (%)

Datasets	TS-FNN	Taylor-FNN	WFNN	AdaBoost. HM	Proposed method
German	24.70 ± 3.16	24.60 ± 3.32	24.30 ± 3.97	23.30 ± 3.58	**23.10 ± 3.56**
Heart	16.30 ± 7.45	15.93 ± 6.84	16.67 ± 9.26	15.56 ± 8.25	**15.19 ± 8.36**
Sonar	19.26 ± 9.16	20.26 ± 10.17	23.95 ± 6.91	18.31 ± 8.35	**17.83 ± 8.43**
Pima	23.83 ± 3.36	23.83 ± 2.64	25.52 ± 2.90	23.05 ± 2.83	**22.79 ± 3.28**
WDBC	3.51 ± 2.72	3.69 ± 2.98	3.86 ± 3.49	3.34 ± 2.41	**3.34 ± 2.41**
WPBC	22.16 ± 8.38	21.66 ± 9.20	21.76 ± 11.96	20.66 ± 9.00	**20.13 ± 8.80**
Haberman	25.49 ± 6.46	25.49 ± 5.37	25.84 ± 5.96	23.86 ± 5.42	**23.52 ± 5.55**

It can be seen from Table 2 that the proposed method can achieve higher classification performance than three sub-classifiers.

Besides, the results of proposed method compared to references [11] and [12] are shown in Table 3 and Table 4. The best classification results in the table are marked with bold font.

Table 3. The results compared to other literatures (%)

Datasets	German	Heart	Sonar
CART + 1bag [6]	23.48 ± 3.37	18.56 ± 7.64	19.87 ± 9.32
CART + 2bag [6]	23.50 ± 3.05	17.41 ± 7.03	18.41 ± 9.46
CART + 3bag [6]	23.72 ± 3.30	17.67 ± 7.44	18.09 ± 9.33
OS-ELM-RBF [11]	24.54 ± 3.90	18.20 ± 6.09	26.80 ± 3.87
SRBF [11]	25.20 ± 3.82	18.15 ± 8.97	21.62 ± 9.03
Q-Gaussian [11]	24.75 ± 2.98	15.93 ± 7.20	23.96 ± 13.56
Proposed method	**23.10 ± 3.56**	**15.19 ± 8.36**	**17.83 ± 8.43**

Table 4. The results compared to other methods (%)

Datasets	Pima	WDBC	WPBC	Haberman
CART + 1bag [6]	23.99 ± 3.87	4.76 ± 2.74	25.80 ± 6.04	30.45 ± 6.69
CART + 2bag [6]	24.92 ± 4.12	4.01 ± 2.51	24.54 ± 5.60	28.92 ± 6.17
CART + 3bag [6]	25.25 ± 4.05	4.08 ± 2.65	24.68 ± 5.28	29.76 ± 5.78
C4.5 [12]	26.1 ± 4.8	6.4 ± 3.2	23.4 ± 2.2	28.3 ± 4.8
M-SVM [12]	25.7 ± 4.8	5.9 ± 2.3	21.2 ± 7.6	24.2 ± 8.0
AbDG [12]	23.5 ± 4.6	5.0 ± 2.9	20.8 ± 5.2	24.7 ± 5.8
Proposed method	**22.79 ± 3.28**	**3.34 ± 2.41**	**20.13 ± 8.80**	**23.52 ± 5.55**

Example 2. Classification of hemodialysis datasets.

We will utilize the proposed method to handle the blood dialysis dataset. The datasets consist of chronic kidney disease 1 (CDK1), chronic kidney disease 2 (CDK2), chronic kidney disease 3 (CDK3), and chronic kidney disease 4 (CDK4). Each of them contains 100 samples, which are collected according to some patients with kidney disease. Each record is composed of nine features for one patient, including gender, age, diabetes, heart failure, hemoglobin, albumin, urea nitrogen, serum creatinine and phosphorus. These samples are mixed together to be classified.

The parameter setting of randomly initialization is briefly introduced below. For TS-FNN, the initial value of parameter p_i^l is 0.25. For Taylor-FNN, the initial value of parameter p_{ij}^l is 0.02. And in these two models, the parameters m_{il}, σ_{il} of Gaussian function are randomly initialized between 0.5 and 1.5, the number of fuzzy rules is set to 10, $VAL = 30$, and $\tau = 0.1$. For WFNN, the parameters m_{il}, σ_{il}, b_{il}, a_{il}, w_l are random initialized between 0.3 and 1.3. The number of fuzzy rules is 15, $VAL = 30$, and $\tau = 0.1$. The initial learning rate and momentum factors of these three subnets are set to 0.05 and 0.01, the number of input nodes and output nodes is set to 9 and 4, and the number of iterations is set to 100. The results are shown in Table 5, where 10-fold cross validation is used. As can be seen from this table, the proposed method has better performance.

Table 5. Experimental results in measured medical dataset (%)

Dataset	TS-FNN	Taylor-FNN	WFNN	AdaBoost. HM	Proposed method
CDK1-CDK2-CDK3-CDK4	83.50 ± 4.06	83.00 ± 4.30	82.5 ± 4.33	84.75 ± 3.94	**85.75 ± 5.01**

5 Conclusion

In this paper, an improved AdaBoost algorithm is applied as ensemble strategy to weighting and summing the results of fuzzy TS neural networks, fuzzy TS neural networks based on second order Taylor formula and wavelet fuzzy neural networks. The validity of proposed method is verified by some empirical datasets from UCI machine learning repository and a hemodialysis database.

Acknowledgements. This work is supported by the National Natural Science Foundation of China (No. 61773088, No. 71571035 and No. 61374118) and the Research Special Fund for Public Welfare Industry of Health (Grant No. 201502023).

References

1. Meyer, R., O'Keefe, S.: A fuzzy binary neural network for interpretable classifications. Neurocomputing **121**, 401–415 (2013)
2. Wu, G.-D., Zhu, Z.-W.: An enhanced discriminability recurrent fuzzy neural network for temporal classification problems. Fuzzy Sets and Syst. **237**, 47–62 (2014)
3. Affonso, C., Sassi, R.J., Barreiros, R.M.: Biological image classification using rough-fuzzy artificial neural network. Expert Syst. Appl. **42**, 9482–9488 (2015)
4. Shinde, S., Kulkarni, U.: Extracting classification rules from modified fuzzy min–max neural network for data with mixed attributes. Appl. Soft Comput. **40,** 364–378 (2016)
5. Rodriguez, J.J., Kuncheva, L.I., Alonso, C.J.: Rotation forest: a new classifier ensemble method. IEEE Trans. Pattern Anal. Mach. Intell. **28**(10), 1619–1630 (2006)
6. Zhang, C.-X., Zhang, J.-S., Ji, N.-N., Guo, G.: Learning ensemble classifiers via restricted Boltzmann machines. Pattern Recognit. Lett. **36**, 161–170 (2014)
7. Qi, Z., Wang, B., Tian, Y., Zhang, P.: When ensemble learning meets deep learning: a new deep support vector machine for classification. Knowl.-Based Syst. **107**, 54–60 (2016)
8. Zhao, Z., Jiao, L., Zhao, J., Gu, J., Zhao, J.: Discriminant deep belief network for high-resolution SAR image classification. Pattern Recognit. **61**, 686–701 (2017)
9. Lu, C.-H.: Wavelet fuzzy neural networks for identification and predictive control of dynamic systems. IEEE Trans. Industr. Electron. **58**(7), 3046–3058 (2011)
10. Jin, X., Hou, X., Liu, C.-L.: Multi-class AdaBoost with hypothesis margin. International Conference on Pattern Recognition (2010)

11. Fernandez-Navarro, F., Hervas-Martinez, C., Gutierrez, P.A., Pena-Barragan, J.M.: Francisca Lopez-Granados. Parameter estimation of Q-Gaussian radial basis functions neural networks with a hybrid algorithm for binary classification. Neurocomputing **75**, 123–134 (2012)
12. Junior, J.R.B., do Carmo Nicoletti, M., Zhao, L.: Attribute-based decision graphs: a framework for multiclass data classification. Neural Netw. **85**, 69–84 (2017)

Power Load Pattern Classification Based on Threshold and Cloud Improved Fuzzy Clustering

Yun-dong Gu(✉), Hong-chao Cheng, and Shuang Zhang

School of Mathematics and Physics, North China Electric Power University, Beijing 102206, China
guyund@126.com

Abstract. A pattern classification method for power load analysis is proposed based on threshold and cloud improved fuzzy clustering algorithm (for short, TACIFCA). Firstly, the classic FCM clustering algorithm were improved by introducing a threshold to recognize the in-homogeneous datum and atypical homogeneous datum of each cluster and reduce their affects on the forming of cluster center. Then, cloud description of each cluster is given and the weights of each homogeneous data in same cluster were determined by the correlation coefficient which indicates the typical degree of the sample data for the cluster. The experimental result shows that the new method has better performance than traditional fuzzy c-means clustering.

Keywords: Power load pattern · Cloud model · Fuzzy c-means clustering Membership degree threshold · Typical degree

1 Introduction

The power load analysis is important for the operation and management of the modern power system [1–3]. In the past decades, many methods, such as support vector clustering [4, 5], self-organizing map [6], clustering algorithm [7–9], mixture model clustering and Markov models [10–12], etc., had been proposed for load pattern classification. Fuzzy c-means clustering had been widely used in load pattern classification [13–18]. However, it still has many limitations, such as too sensitive for the initial conditions and easy to fall into local optimum etc. In this paper, a pattern classification method for power load analysis is proposed based on threshold and cloud improved Fuzzy clustering algorithm. The structure is displayed as follows: Sect. 2 gives the cloud improved fuzzy clustering method for power load pattern classification, Sect. 3 verify the efficiency of it by example analysis and Sect. 4 concludes the paper.

B.-Y. Cao and Y.-B. Zhong (Eds.): ICFIE 2017, AISC 872, pp. 143–148, 2019.
https://doi.org/10.1007/978-3-030-02777-3_13

2 Threshold and Cloud Improved Fuzzy Clustering Algorithm for Power Load Pattern Classification

First, we introduce the main contributions of the threshold and cloud improved fuzzy clustering algorithm.

2.1 The Main Ideas

The main contribution of the threshold and cloud improved fuzzy clustering algorithm for power load pattern classification mainly includes two parts: (1) introduce a membership degree threshold for cluster center calculation to reduce the affect of in-homogeneous and atypical homogeneous data, (2) set up a cloud model for each cluster and weight the data in same cluster by its typical degree.

2.2 Membership Degree Threshold to Reduce the Affect of Atypical Data for Cluster Center

Suppose the data set is $X = \{x_i = (x_{i1}, x_{i2} \cdots x_{im}),\ i = 1, 2, \cdots, N\}$, and $\{c_j\}_{j=1}^{c}$ are the cluster centers, where c is the number of cluster, $d_{ij} = \|x_i - c_j\|$ is the Euclidean distance between x_i and the center of the j-th cluster, u_{ij} denotes the membership degree of x_i to the j-th cluster, which satisfy that $1 \leq u_{ij} \leq 1$ for $i = 1, 2, \cdots, N;\ j = 1, 2, \cdots, c$, and $\sum_{j=1}^{c} u_{ij} = 1$. Traditional fuzzy c-means clustering is carried out through iterative update the membership degree u_{ij} and the cluster centers c_j by using the Formula (1) and (2) till the cluster center has no big changes.

$$u_{ij} = \left(\sum_{k=1}^{c} \left(d_{ij}/d_{ik} \right)^{2/(q-1)} \right)^{-1}, \tag{1}$$

$$c_j = \sum_{i=1}^{n} u_{ij}^{q} x_i / \sum_{i=1}^{n} u_{ij}^{q} \tag{2}$$

where q is a parameter controlling the ambiguity of the fuzzy partition matrix. In order to reduce the affect of in-homogeneous and atypical homogeneous data on the forming of cluster center, we introduce a membership threshold $\alpha > 0$, i.e. let

$$u'_{ij} = \begin{cases} u_{ij}, & u_{ij} \geq \alpha; \\ 0, & u_{ij} < \alpha. \end{cases} \tag{3}$$

2.3 Data Weighting Method Based on Cloud Modeling of Clusters

In this section, we consider the user's possible power loads as a cloud and set up a cloud model for each cluster. First, calculate the most possible power load value, the variance, the Entropy En and the Hyper entropy He of the jth cluster, get its cloud model [17] (Ec_j, En_j, He_j), $j = 1, 2, \cdots, c$.

$$Ec_j = \sum_{i=1}^{n} u_{ij}'^{q} x_i \Big/ \sum_{i=1}^{n} u_{ij}'^{q}, \tag{4}$$

$$s_j^2 = \sum_{i=1}^{N} \left(u_{ij}' \left\| x_{ij} - Ec_j \right\| \right) \Big/ \sum_{i=1}^{N} u_{ij}' \tag{5}$$

$$En_j = \sqrt{\frac{\pi}{2}} \sum_{i=1}^{N} \left(u_{ij}' \left\| x_{ij} - Ec_j \right\| \right) \Big/ \sum_{i=1}^{N} u_{ij}' \tag{6}$$

$$He_j = \sqrt{s_j^2 - En_j^2} \tag{7}$$

Then, the weight of data x_i can be determined by it's typical degree in the jth cluster, i.e.

$$v_{ij} = \exp\left(- \frac{\left\| x_{ij} - Ec_j \right\|^2}{2En_j^2} \right) \tag{8}$$

2.4 Threshold and Cloud Improved Fuzzy Clustering Algorithm for Power Load Pattern Classification

The threshold and cloud improved fuzzy clustering algorithms (for short, TACI-FCA) for power load pattern classification can be summarized as follows:

Step 1: Input the number of clusters c, set $q = 2$ and the threshold $\alpha > 0, \beta > 0$, random initialize the cluster center $c_j^0 (j = 1, 2, \cdots, c)$.
Step 2: Calculate u_{ij}', $(i = 1, 2, \cdots n; j = 1, 2, \cdots c)$ by using Formula (1) and (3).
Step 3: Produce the cloud model (Ec_j, En_j, He_j), $j = 1, 2, \cdots, c$.
Step 4: Calculate the typical degree v_{ij}, and let $v_{ij}' = 0$, if $v_{ij} \le \beta$; $v_{ij}' = v_{ij}$ if $v_{ij} > \beta$.
Step 5: Update the cluster center $c_j^1 = \sum_{i=1}^{n} v_{ij}' u_{ij}'^{q} x_i / \sum_{i=1}^{n} u_{ij}'^{q}$, $j = 1, 2, \cdots, c$.
Step 6: If $\max_{1 \le j \le c} \left(\left\| c_j^0 - c_j^1 \right\| / \left\| c_j^1 \right\| \right) \le \varepsilon$ then go to Step 7; else let $c_j^0 = c_j^1 (j = 1, 2, \cdots, c)$ and go to Step 2.
Step 7: Output the clustering results: cluster centers and the elements of each cluster, i.e., c_j^0, $C_j = \left\{ x_i \middle| u_{ij} \ge \alpha,\ i = 1, 2, \cdots, N \right\}$, $j = 1, 2, \cdots, c$.

3 Simulation Analysis

To evaluate the proposed method, some electricity customers of Jinmen city connected to the medium voltage distribution system have been considered. The representative power load pattern of each customer is obtained by averaging the power load of customers sampling every 30 min in a day with a given condition. We use the TACIFCA and FCM to analysis the power load pattern.

3.1 Clustering Results and Customer Distribution

Figure 1 shows the cluster results of TACIFCA. From Fig. 1 we can find that the 1st load pattern has significant complementary for the 2nd, 5th and 7th load patterns. The load character of the power system can be improved by adjusting their customers. This information is useful for the electricity company, for example, for the development of peak-valley electricity price policy.

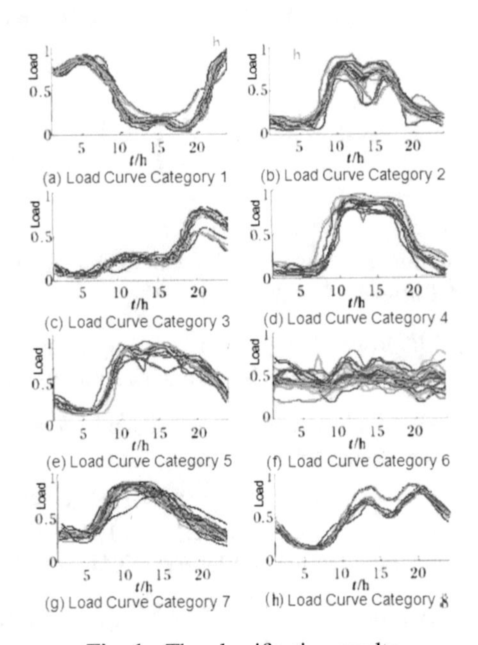

Fig. 1. The classification results

3.2 Comparison and Discussion

In order to compare the TACIFCA and FCM, we do a stability analysis. The final value of the objective function and iteration times are listed as Table 1.

From Table 1 it is easy to find that when the initial value make a change, the objective function in the traditional FCM method fluctuate greatly and the amplitude of

Table 1. The results of FCM and TACIFCA

Attributes	Algorithm	1	2	3	4	5	6	7	8	9	10
The objective function value	FCM	11.13	13.42	11.96	12.18	11.84	13.38	10.26	12.29	11.02	10.13
	TACIFCA	9.135	9.126	9.132	9.124	9.137	9.129	9.131	9.136	9.123	9.128
The number of iterations	FCM	35	38	29	26	29	33	30	28	31	37
	TACIFCA	18	18	18	19	18	18	18	18	16	18

variation is more than 30% and the iteration times also have a considerable change. However, the objective function value and the iteration times are basically stable in TACIFCA. So we can assume the TACIFCA is better than FCM for load pattern classification.

4 Conclusion

This paper presents a threshold method and a cloud based data weighting method to improve the traditional fuzzy clustering algorithms by reducing the affect of abnormal outliers, such as in-homogeneous and atypical homogeneous data, for the forming of pattern cluster. A case study based on real data demonstrates that the improved approach is much better than the traditional FCM.

Acknowledgement. This paper is supported by the Fundamental Research Funds for the Central Universities (2015MS51) and the National Natural Science Foundation of China (Grant No. 71671064). The author also thanks the faculty of mathematical department in Middle Tenseness State University and Professor Wu for their kindly help during my visiting period.

References

1. Wang, X.: Analysis of Modern Power System. Science and Technology Press, Beijing (2003)
2. Li, P., Li, X., Tang, W., et al.: Consumer choice for an industry in statistical synthesis method based load modeling. Autom. Electr. Power Syst. **29**(14), 34–38 (2005)
3. Gerbec, D., Gasperic, S., Smon, I., et al.: Determining the load profiles of consumers based on fuzzy logic and probability neural networks. IEEE Proc.: Gener. Transm. Distrib. **151**(3), 395–400 (2004)
4. Chicco, G., Ilie, I.S.: Support vector clustering of electrical load pattern data. IEEE Trans. Power Syst. **24**(3), 1619–1628 (2009)
5. Nagi, J., Yap, K.S., Tiong, S.K., et al.: Nontechnical loss detection for metered customers in power utility using support vector machines. IEEE Trans. Power Deliv. **25**(2), 1162–1171 (2010)
6. Meshram, R., Deorankar, A.V., Chatur, P.N.: Data mining application: classification of load pattern analysis for electricity customers. Int. J. Manag. IT Eng. **3**, 299 (2013)
7. Fidalgo, J.N., Matos, M.A., Ribeiro, L.: A new clustering algorithm for load profiling based on billing data. Electr. Power Syst. Res. **82**, 27–33 (2012)

8. Kim, I.-J., Lee, S.-K.: Short-term electric load forecasting using the real time weather information & electric power pattern analysis. Trans. Korean Inst. Electr. Eng. **6**(65), 934–939 (2016)
9. Milanovic, J.V., Yamashita, K., Martinez Villanueva, S., et al.: International industry practice on power system load modeling. IEEE Trans. Power Syst. **28**(3), 3038–3046 (2013)
10. Labeeuw, W., Deconinck, G.: Residential electrical load model based on mixture model clustering and Markov models. IEEE Trans. Ind. Inform. **9**(3), 1561–1569 (2013)
11. Tomoda, M., Matsuki, J., Hayashi, Y.: Parameter estimation of dynamic load model in power system by using measured data. J. Int. Counc. Electr. Eng. **1**(2), 200–206 (2014)
12. Gu, Y., Zhang, S., Feng, J.: Multi-model fuzzy synthesis forecasting of electric power loads for larger consumers. Trans. China Electro-Tech. Soc. **30**(12), 110–115 (2015)
13. Chicco, G.: Overview and performance assessment of the clustering methods for electrical load pattern grouping. Energy **42**(1), 68–80 (2012)
14. Zhang, T., Zhang, G., Lu, J., et al.: A new index and classification approach for load pattern analysis of large electricity customers. IEEE Trans. Power Syst. **27**(1), 153–160 (2012)
15. Song, Y., Li, C., Qi, Z.: Extraction of power load patterns based on cloud model and fuzzy clustering. Power Syst. Technol. **32**(8), 3376–3383 (2014)
16. Ryu, S., Kim, H., Oh, D., et al.: Customer load pattern analysis using clustering techniques. J. Electr. Power Energy **2**(1), 61–69 (2016)
17. Li, D., Du, Y.: Artificial Intelligence with Uncertainty, 2nd edn, pp. 110–116. National Defend Industry Press, Beijing (2005)
18. Chicco, G., Napoli, R., Piglione, F., Scutariu, M., et al.: Load pattern based classification of electricity customers. IEEE Trans. Power Syst. **19**(2), 1232–1239 (2004)

The Centrality Analysis of the Fuzzy Technology Innovation Network

Li-ping Liao[1(✉)] and Guang-yu Zhang[2]

[1] School of Management, Guangdong Polytechnic Normal University, Room C2-13A, No. 94, TianRun Road, Tianhe District, Guangzhou 510635, China
liping1110@hotmail.com

[2] School of Management, Guangdong University of Technology, Guangzhou 510520, China
guangyu@gdut.edu.cn

Abstract. Centrality analysis is a concept of measure which is used to analyze an actor's centrality in a fuzzy social network, and it reflects an actor's position or advantage difference in a fuzzy social network. This paper is a discussion on fuzzy node centrality, fuzzy closeness centrality, fuzzy betweenness centrality, and fuzzy centralization. In the case, 37 factors which influencing enterprise technology innovation are taken as actors to set up a fuzzy network of technology innovation. Judge the importance of each factor to enterprise technology innovation by calculating the fuzzy node centrality and fuzzy closeness centrality, thus provides technical support to the management of enterprise technology innovation.

Keywords: Centrality · Fuzzy social network · Fuzzy nodes centrality
Fuzzy closeness centrality · Fuzzy betweenness centrality

1 Introduction

At the start, social network analysis focuses on some prominent network actors (such as "Star"). They reflect different grades and advantages between social network actors, which is an important property of social structure (Ling 2009).

Analysis of social network centrality is one of the most important and commonly used tools of concept, which is a measurement concept reflects the different positions and advantages between social network actors. A number of different methods have been formed about social network centrality measurements; some focus on local actors, some focus on the overall network, and the measurements for directional relationship and non-directional relationship are different.

In social network analysis, determining the importance of actors in a network has been studied for a long time (Wasserman and Faust 1994). Degree centrality (Shaw 1954; Nieminen 1974) focuses on the level of communication activity, identifying the centrality of a node with its degree. Closeness centrality (Beauchamp, 1965; Sabidussi, 1966) considers the sum of the geodesic distances between a given actor and the remaining as a decentralist measure in the sense that the lower this sum is, the greater

B.-Y. Cao and Y.-B. Zhong (Eds.): ICFIE 2017, AISC 872, pp. 149–165, 2019.
https://doi.org/10.1007/978-3-030-02777-3_14

the centrality is. Closeness centrality is, then, a measure of independence in the communications, in the relations or in the bargaining, and thus, it measures the possibility to communicate with many others depending on a minimum number of intermediaries. Betweenness centrality (Bavelas, 1948; Freeman, 1977) emphasizes the value of the communication control: the possibility to intermediate in the relation of others. Here, all possible geodesic paths between pairs of nodes are considered. The centrality of each actor is the number of such paths in which it lies.

Centrality analysis is used extensively in social and behavioral sciences, as well as in political science, management science, economics, biology, and so forth. Bonacich (1972, 1987) suggested another concept of centrality. He proposed to measure the centrality of different nodes using the eigenvector associated with the largest characteristic eigenvalue of the adjacent matrix. Costenbader and Valente (2003) studied the stability of centrality measures when networks are sampled. In addition, social network analysts have made a distinction between centrality and centralization, with the latter referring to the overall closeness rather than the relative importance between certain points.

There is little research about fuzzy social network until now. Fan et al. (2007) and discussed structural equivalence and regular equivalence in fuzzy social network, but the study of fuzzy centrality about fuzzy social network has not been considered formally in the literature. In this paper, we extend the notion of centrality and centralization to fuzzy social network. We apply the fuzzy centrality theory to a case which is called fuzzy technology innovation network. For the aspect of R&D strategy, we encourage that decision makers should pay more attention to the factors such as *Research and development capabilities*, *Human resources quality*, *Production capacity*, *Level of economic development*, *Degree of industrial clusters*, *Technology accumulation* in fuzzy technology innovation network.

The framework of this paper is as follows. In Sect. 2, we briefly review the centrality analysis of social network. In Sect. 3, we discuss the fuzzy centrality analysis of fuzzy social network. In Sect. 4, we provide a case to further illustrate the fuzzy centrality analysis of fuzzy technology innovation network. Finally conclusions appear in Sect. 5.

2 Centrality Analysis of Social Network

Social analysts generally classify centrality into node centrality, closeness centrality, and betweenness centrality. The following describes respectively node centrality, closeness centrality, betweenness centrality and their measurements (Ling 2009).

2.1 Node Centrality

2.1.1 Actor's Node Centrality

An actor's node centrality is represented by the number of node, which refers to the number of line connected to a node, which is represented as:

$$C_D(n_i) = d(n_i) \tag{1}$$

This equation is based on calculating the absolute number of node, the actor with the maximum number of node is the center. A second equation is based on calculating the relative number of node, which refers to the ratio between node degree and the total number of connected lines. It can be represented as:

$$C_D^{'}(n_i) = \frac{d(n_i)}{N-1} \tag{2}$$

In the above equation, N refers to network scale, the maximum degree of any node being: N – 1.

While measuring the degree of node centrality based on the number of node degree, the following two points should be observed:

- First, this kind of measurement is based primarily on direct relationship, without considering the indirect relationship.
- Second, when measuring an actor's node centrality, it is carried out without considering the issue of 'only one' center in the whole network, which will be touched upon in the following discussion of group centralization.

2.1.2 Group Centralization

Centralization is a concept reflects the concentration degree in the whole network or graph. It is an important measurement which is complementary to density. The measurement formula is put forward firstly by ***Linden. C. Freeman*** based on node centrality.

$$C_D^{'} = \frac{\sum_{i=1}^{n} (C_{D\max} - C_{Di})}{\max \sum_{i=1}^{n} (C_{D\max} - C_{Di})} \tag{3}$$

What this formula implies is that the total sum of difference between the most prominent node and other nodes is divided by the total sum of the maximum possible difference.

The above calculation is based on absolute centrality. If calculating based on relative centrality, we need to have it divided by N – 1.

$$C_D^{'} = \frac{\sum_{i=1}^{n} (C_{D\max} - C_{Di})}{N-1} \tag{4}$$

2.2 Closeness Centrality

Closeness centrality is the centrality measured based on the closeness and distance between the nodes in the network. The shorter the total distance is, the closer the network is. It also demonstrates the closeness between one actor and other actors. The major difference between closeness centrality and node centrality is that indirect relationship involved; it equals to overall centrality.

2.2.1 Actor's Closeness Centrality

Its measurement formula is:

$$C_C = \left[\sum_{i=1}^{n} d(n_i, n_j)\right]^{-1} \tag{5}$$

The formula implies that the centrality of a node equals the sum of geodesics calculated from one node to other nodes.

Looking for relative closeness centrality, the above formula must be divided by N – 1.

$$C_C' = \frac{N-1}{\sum_{i=1}^{n} d(n_i, n_j)} = (n-1)C_C \tag{6}$$

2.2.2 Group Closeness Centralization

This closeness is group closeness, which belongs to group centralization, its measurement method being similar to the above-mentioned centralization measurement. The formula is:

$$C_C = \frac{\sum_{i=1}^{n} (C'_{C\max} - C'_{Ci})}{(N-2)(N-1)}(2N-3) \tag{7}$$

$C'_{C\max}$ refers to the maximum relative closeness centrality.

2.3 Betweenness Centrality

Betweenness Centrality measures the extent to what an actor controls other actors. This kind of actor also serves as a bridge for communication.

2.3.1 Actor's Betweenness Centrality

If there are several geodesics from one node to X and Z, then the proportion which is calculated by adding the geodesics from node Y to node X and Z(past Y and connected with X and Z) divided by the overall geodesics between node X and node Z, is called betweenness proportion. What it measures is to what extent Y is in between X and Z.

According to probability theory, if g_{jk} represents the possible number of geodesics between nodes J and K, then the probability that all these geodesics will be equally chosen as communication path is $1/g_{jk}$. Suppose that $g_{jk}(n_i)$ is used to represent the number of geodesics between two actors inclusive of actor n_i, then the betweenness of actor n_i will be the probability sum of $g_{jk}(n_i)/g_{jk}$, its formula is:

$$C_B = \sum_{j<k} \frac{g_{jk}(n_i)}{g_{jk}} \tag{8}$$

The formula of its relative betweenness centrality (non – directional graph) is:

$$C_B' = \frac{C_B}{(N-1)(N-2)/2} \tag{9}$$

Its directional graph is:

$$C_B' = \frac{C_B}{(N-1)(N-2)} \tag{10}$$

The value varies between 0 and 1, with 0 implies that this node is not capable of controlling any other actor, while 1 implies that this node is in complete control of all the other actors, and it is in the central position of the network.

2.3.2 Group Betweenness Centralization

The group centralization, based on betweenness, concerns with comparison of the website in which difference between its website membership exists. The computing formula is similar to other group centrality formulas:

$$C_B = \frac{2\sum_{i=1}^{n}(C'_{B\max} - C'_{Bi})}{(N-1)^2(N-2)} \tag{11}$$

or

$$C_B = \frac{\sum_{i=1}^{n}(C'_{B\max} - C'_{Bi})}{(N-1)} \tag{12}$$

$C_{B\max}$ refers to the maximum betweenness centrality.

3 The Calculation Method of Centrality About Fuzzy Social Network

Suppose that there is a fuzzy relationship structure $\widetilde{G} = (V, \widetilde{e}_{ij})$, in which $V = \{v_1, v_2, \ldots v_n\}$ is a non - empty set of actors, $\widetilde{e}_{ij}$ is the fuzzy relationship in V. We call $\widetilde{G}$ fuzzy social network. The following is a discussion of centrality analysis for fuzzy social network.

3.1 Fuzzy Node Centrality

Fuzzy centrality analysis is one of the most important and commonly used tools of concept in the analysis of social network. This is a measurement concept reflects the different positions and advantages between different actors in a fuzzy social network. Generally, according to the local difference and global difference, centrality is classified into local fuzzy centrality and global fuzzy centrality. The former, also known as the local fuzzy point of centrality, what it reflects is the fuzzy centrality, or a person's dominant position in the fuzzy network situation. The greater the centrality is, that is, more associated with more people, the more they are in the central location. The latter refers to the distances between a node with others in the whole network. This reflects the closeness between nodes, which is measured by the shortest distance between different nodes. Fuzzy centralization refers to the overall closeness, rather than the relative importance of certain nodes.

3.1.1 Actor's Fuzzy Node Centrality

Fuzzy node centrality is represented by the number of fuzzy node that is the total membership degree of the lines which are connected with the node. Its formula is:

$$\underset{\sim}{C}_D(n_i) = \underset{\sim}{d}(n_i) \tag{13}$$

In the formula, $\underset{\sim}{d}(n_i)$ refers to the sum of n_i-connected fuzzy-relation membership degree.

In the formula above, the centrality degree is calculated according to absolute number, and the actor with the maximum number of fuzzy node is the center. Another method to calculate the fuzzy centrality is according to the relative number, which refers to the ratio between the fuzzy node degree and the total number of connected lines.

$$\underset{\sim}{C'}_D(n_i) = \frac{\underset{\sim}{d}(n_i)}{N-1} \tag{14}$$

In this formula, N refers to network scale, and the maximum centrality of any node in the network is N – 1.

3.1.2 Fuzzy Group Centralization

Fuzzy group centralization is a concept used to represent the concentration degree of a fuzzy social network, and it is an important measure supplementary to fuzzy density.

$$\underset{\sim}{C}'_D = \frac{\sum_{i=1}^{n}(\underset{\sim}{C}_{D\max} - \underset{\sim}{C}_{Di})}{\max\sum_{i=1}^{n}(\underset{\sim}{C}_{D\max} - \underset{\sim}{C}_{Di})} \tag{15}$$

What this formula implies is that the total sum of difference between the most prominent node and others is divided by the total sum of the maximum possible difference.

The formula above is calculated according to absolute centrality. If calculated according to relative centrality, it needs to be divided by N − 1 as follows.

$$\underset{\sim}{C}'_D = \frac{\sum_{i=1}^{n}(\underset{\sim}{C}_{D\max} - \underset{\sim}{C}_{Di})}{N-1} \tag{16}$$

3.2 Fuzzy Closeness Centrality

Fuzzy closeness centrality is a centrality which is measured according to the closeness and distance in the network. The major difference between closeness centrality and node centrality lies in the fact that indirect relationship is taken into consideration.

Due to the fact that fuzzy closeness centrality is measured by distance, several formulas of common distance based on fuzzy equivalence matrices are presented here:

- Euclidean distance:

$$d(n_i, n_j) = \sqrt{\sum_{i=1}^{n}(x_{ik} - x_{jk})^2 + (x_{ki} - x_{kj})^2},\ i \neq k,\ j \neq k \tag{17}$$

- Manhattan distance or city-block metrics:

$$d_{ij} = \sum_{k=1}^{n}\left|x_{ik} - x_{jk}\right| \tag{18}$$

- Chebychev metrics:

$$d_{ij} = \max_k \left|x_{ik} - x_{jk}\right| \tag{19}$$

- Minkowski metrics:

$$d_{ij} = \left(\sum_{k=1}^{n} |x_{ik} - x_{jk}|^{r} \right)^{1/r} \tag{20}$$

3.2.1 Actor's Fuzzy Closeness Centrality

Its measurement formula is:

$$\underset{\sim}{C}_{C} = \left[\sum_{i=1}^{n} \underset{\sim}{d}(n_i, n_j) \right]^{-1} \tag{21}$$

This formula implies that the fuzzy closeness centrality of a point is the total distance calculated from one node to others. $\underset{\sim}{d}(n_i, n_j)$ is the distance between n_i and n_j.

To calculate the relative fuzzy closeness centrality, the formula above needs to be divided by N – 1.

$$\underset{\sim}{C}'_{C} = \frac{N-1}{\sum_{i=1}^{n} \underset{\sim}{d}(n_i, n_j)} = (n-1)\underset{\sim}{C}_{C} \tag{22}$$

3.2.2 Fuzzy Group Closeness Centralization

This is the closeness viewed from the perspective of group, and it belongs to group centralization, its measurement method is similar to the above-mentioned fuzzy centralization measurement.

$$\underset{\sim}{C}_{C} = \frac{\sum_{i=1}^{n} (\underset{\sim}{C}'_{C\max} - \underset{\sim}{C}'_{Ci})}{(N-2)(N-1)} (2N-3) \tag{23}$$

In the formula, $\underset{\sim}{C}'_{C\max}$ refers to the maximum fuzzy relative closeness centrality.

3.3 Fuzzy Betweenness Centrality

What fuzzy betweenness centrality measures is to what extent an actor controls other actors, and this kind of actor also serves as a bridge for communication.

3.3.1 Actor's Fuzzy Betweenness Centrality

If there are several walks from node X to Z, then the proportion, which is calculated by adding the strength from node X to Z (past Y, and connected with X and Z) divided by the overall strength between node X and node Z, is called betweenness proportion. What it measures is to what extent Y is in between X and Z.

The measurement formula for fuzzy closeness centrality is:

$$\underset{\sim}{C}_B = \sum_{j<k} \frac{\underset{\sim}{g}_{jk}(n_i)}{\underset{\sim}{g}_{jk}} \tag{24}$$

Among which, $\underset{\sim}{g}_{jk}(n_i)$ is the strength sum between the two actors which include the actor n_i; $\underset{\sim}{g}_{jk}$ represents the strength sum between node j and node k.

Its relative fuzzy closeness centrality (non-directional graph) is

$$\underset{\sim}{C}'_B = \frac{\underset{\sim}{C}_B}{(N-1)(N-2)/2} \tag{25}$$

Its value varies from 0 and 1. 0 means that this node is not capable of controlling any other controller; 1 means that this node is in complete control of other actors, and it is in the central position in the network.

3.3.2 Fuzzy Group Betweenness Centralization

The group centralization based on fuzzy betweenness concerns about the comparison of fuzzy network in which difference between the network members exists. Its calculation formula is similar to other group centrality situation.

$$\underset{\sim}{C}_B = \frac{2\sum_{i=1}^{n}(\underset{\sim}{C}'_{B\max} - \underset{\sim}{C}_{Bi})}{(N-1)^2(N-2)} \tag{26}$$

or

$$\underset{\sim}{C}_B = \frac{\sum_{i=1}^{n}(\underset{\sim}{C}'_{B\max} - \underset{\sim}{C}'_{Bi})}{(N-1)} \tag{27}$$

In this formula, $\underset{\sim}{C}_{B\max}$ refers to the maximum fuzzy betweenness centrality.

4 Case Study

Firstly, clear that the impact factors of fuzzy technology innovation network are as follows:

X_1: Social infrastructure; X_2: Level of economic development; X_3: Human resources quality; X_4: Information infrastructure environment; X_5: Social and cultural environment; X_6: Tax policy; X_7: Monetary policy; X_8: Intellectual property rights policy; X_9: Innovation incentives; X_{10}: Social service system; X_{11}: Technology property rights transaction; X_{12}: The level of technology assessment; X_{13}: Market demand; X_{14}: Fair market; X_{15}: Intensity of market competition; X_{16}: Degree of credit market; X_{17}: Level of industrial technology; X_{18}: Degree of industrial clusters; X_{19}: Industrial technology support platform; X_{20}: Corporate human resources constitute; X_{21}: Liquidity; X_{22}: Fixed capital formation; X_{23}: Technology accumulation; X_{24}: Information resources and platforms; X_{25}: Enterprise products constitute; X_{26}: Organizational structure; X_{27}: Ability to obtain resources; X_{28}: Marketing capability; X_{29}: Research and development capabilities; $X_{30:}$ Production capacity; X_{31}: Integration of collaborative capabilities; X_{32}: Management decision-making capacity; X_{33}: Organization execution; X_{34}: Brand Competitiveness; X_{35}: Corporate culture; X_{36}: Innovative driving forces; X_{37}: Rapid response force.

In fact, if the size of sample is too large, it becomes a difficult task to distinguish the network structure. On the other hand, if the size is too small, the information of network is limited. Thus, it is not appropriate if the size is either too large or too small, and it's a study of particularly interesting and worthy. In this case, the size of the data set (37 actors) is suitable.

We converted a relation data of 37 impact factors into an adjacency matrix (see Table 1), and then used Net draw in UCINET to draw the fuzzy technology innovation network. Figure 1 (ties have values > 0) depicts a fuzzy technology innovation network of 37 actors. By roughly inspecting the structure of the fuzzy technology innovation network, it is clear that Fig. 1 is too crowded to read, so centrality analysis is made in Sect. 4.

Table 1. Adjacency matrix of fuzzy technology innovation network

	X_1	X_2	X_3	X_4	X_5	X_6	X_7	–	X_{30}	X_{31}	X_{32}	X_{33}	X_{34}	X_{35}	X_{36}	X_{37}
X_1	1	1	0.7	0.8	0.6	0.3	0.6	–	0.9	0.7	0.7	0.6	0.6	0.2	0.2	0.2
X_2	1	1	0.8	0.9	0.8	0.3	0.3	–	1	0.8	0.8	0.7	0.9	0.8	0.6	0.7
X_3	0.7	0.8	1	0.7	0.8	0	0	–	1	0.8	0.9	0.8	0.8	0.7	0.7	0.8
X_4	0.8	0.9	0.7	1	0.4	0	0	–	0.8	0.8	0.8	0.6	0.6	0.5	0.5	0.7
X_5	0.6	0.8	0.8	0.4	1	0	0	–	0.6	0.4	0.6	0.6	0.6	0.8	0.6	0.7
X_6	0.3	0.3	0	0	0	1	0.7	–	0.7	0.2	0.2	0.1	0.1	0.1	0.7	0.1
X_7	0.6	0.3	0	0	0	0.7	1	–	0.7	0.2	0.2	0.1	0.1	0.1	0.7	0.1
X_8	0.3	0.8	0.8	0.2	0.1	0	0.2		0.9	0.2	0.2	0.1	0.1	0.1	1	0.6
–	–	–	–	–	–	–	–	–	–	–	–	–	–	–	–	–
X_{30}	0.9	1	1	0.8	0.6	0.7	0.7	–	1	0.6	0.2	0.2	0.3	0.2	0.7	0.7
X_{31}	0.7	0.8	0.8	0.8	0.4	0.2	0.2	–	0.6	1	0.8	0.8	0.7	0.3	0.7	0.8
X_{32}	0.7	0.8	0.9	0.8	0.6	0.2	0.2	–	0.2	0.8	1	0.7	0.6	0.4	0.7	0.9
X_{33}	0.6	0.7	0.8	0.6	0.6	0.1	0.1	–	0.2	0.8	0.7	1	0.8	0.5	0.8	0.9
X_{34}	0.6	0.9	0.8	0.6	0.6	0.1	0.1	–	0.3	0.7	0.6	0.8	1	0.7	0.6	0.5
X_{35}	0.2	0.8	0.7	0.5	0.8	0.1	0.1	–	0.2	0.3	0.4	0.5	0.7	1	0.7	0.8
X_{36}	0.2	0.6	0.7	0.5	0.6	0.7	0.7	–	0.7	0.7	0.7	0.8	0.6	0.7	1	0.5
X_{37}	0.2	0.7	0.8	0.7	0.7	0.1	0.1	–	0.7	0.8	0.9	0.9	0.5	0.8	0.5	1

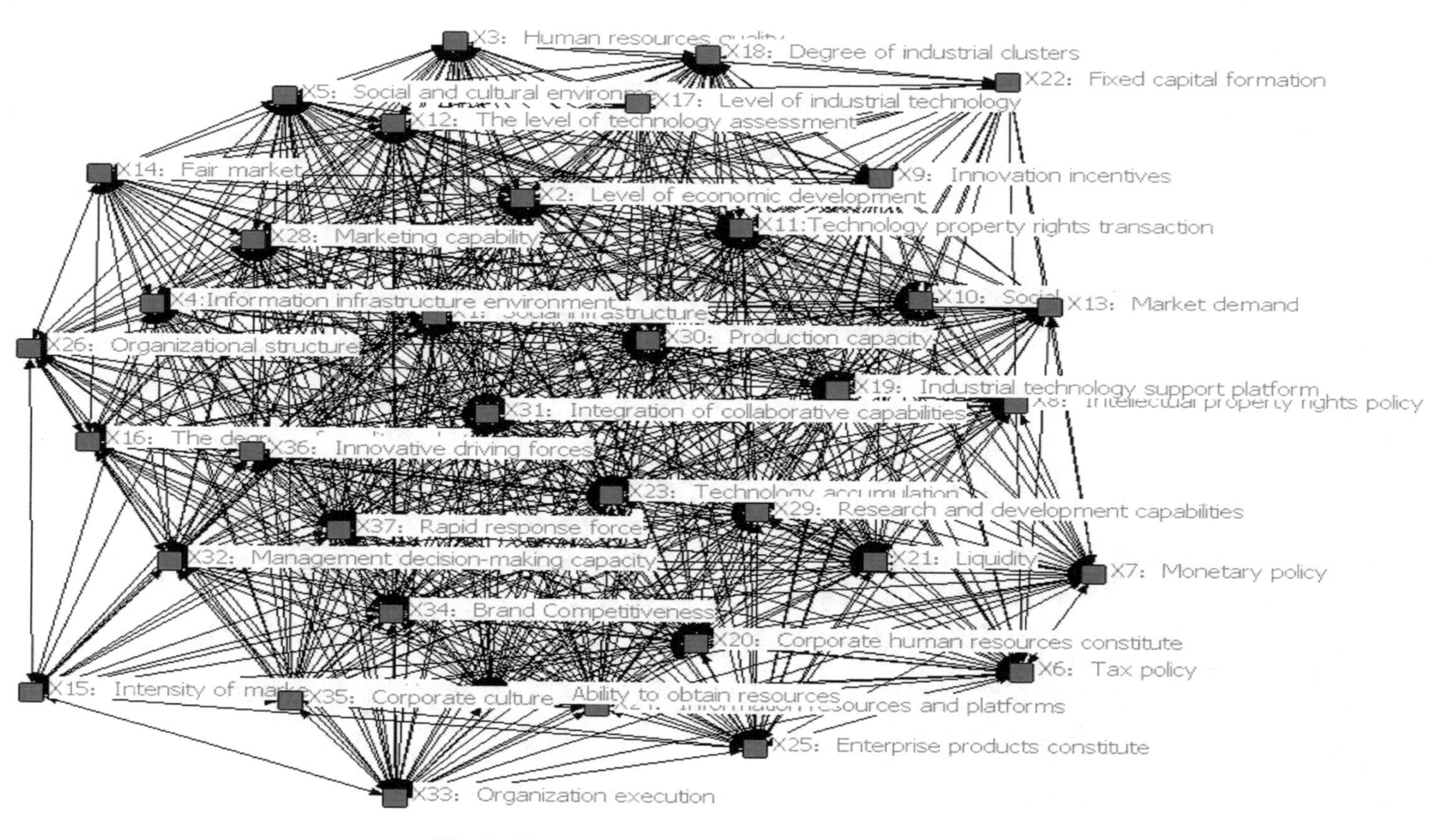

Fig. 1. Fuzzy technology innovation network of 37 actors

Table 2. Node centrality of each factor in fuzzy technology innovation network

NO.	Influencing factors	Fuzzy node centrality
1	v_{29}: Research and development capabilities	25.8
2	v_{3}: Human resources quality	23.24
3	v_{2}: Level of economic development	23.01
4	v_{30}: Production capacity	22.66
5	v_{17}: Level of industrial technology	22.36
6	v_{19}: Industrial technology support platform	21.76
7	v_{18}: Degree of industrial clusters	21.22
8	v_{27}: Ability to obtain resources	20.59
9	v_{37}: Rapid response force	20.460
10	v_{31}: Integration of collaborative capabilities	20.120
11	v_{36}: Innovative driving forces	20.090
12	v_{24}: Information resources and platforms	20.010
13	v_{23}: Technology accumulation	19.680
14	v_{34}: Brand Competitiveness	19.360
15	v_{20}: Corporate human resources constitute	19.320
16	v_{4}: Information infrastructure environment	19.200
17	v_{13}: Market demand	18.850
18	v_{5}: Social and cultural environment	18.840
19	v_{9}: Innovation incentives	18.280
20	v_{28}: Marketing capability	18.200
21	v_{33}: Organization execution	18.020
22	v_{32}: Management decision-making capacity	17.830
23	v_{11}: Technology property rights transaction	17.590
24	v_{10}: Social service system	17.340
25	v_{15}: Intensity of market competition	16.700
26	v_{26}: Organizational structure	16.520
27	v_{1}: Social infrastructure	16.460
28	v_{7}: Monetary policy	16.400
29	v_{35}: Corporate culture	16.380
30	v_{8}: Intellectual property rights policy	16.020
31	v_{6}: Tax policy	15.560
32	v_{25}: Enterprise products constitute	15.400
33	v_{16}: Degree of credit market	15.290
34	v_{14}: Fair market	15.140
35	v_{12}: The level of technology assessment	14.800
36	v_{21}: Liquidity	14.000
37	v_{22}: Fixed capital formation	12.160

4.1 Fuzzy Node Centrality of Technology Innovation Factors

Node centrality describes the direct impact that the factors which influencing enterprise technology innovation make on other factors. The important influential factors of technology innovation gotten from the analysis of node centrality are driven type factors, not only impact technological innovation directly themselves, but also have indirect influence through a lot of other relevant factors. According to formula (13), the fuzzy node centrality of each node is calculated, as shown in Table 2.

From Table 2 we can learn that v_{29} (Research and development capabilities) is the most important key fact influencing enterprise technology innovation from the respect of node centrality. Ranked by their importance, the order from NO. 2 to NO. 9 are as follows: v_3(Human resources quality), v_2 (Level of economic development), v_{30} (Production capacity), v_{17} (Level of industrial technology), v_{19} (Industrial technology support platform), v_{18} (Degree of industrial clusters), v_{27} (Ability to obtain resources), v_{37} (Rapid response force). All these factors can be divided into two levels: enterprise level and environment level. The enterprise level includes research and development capabilities, human resources quality, production capacity and rapid manufacturing force. Among them the factor research and development capabilities is the most important and direct, which relates the ability of enterprise technology innovation directly. Human resources quality is the most basic and the most essential factor for enterprise technology innovation, and it's the source of creativity. Production capacity is the guarantee for new technology, which reflects the hard resources condition of an enterprise. Rapid response force and ability to obtain resources is the reflection of the enterprises' sensitivity and adaptability to the environment, which is the motivation mechanism of technology innovation.

The environment level includes industrial environment and macro environment. There are three factors in industrial environment, level of industrial technology, industrial technology support platform and degree of industrial clusters included, which are of great significance to enterprise technology innovation. Among the factors concerning macro environment, level of economic development is the most relevant, which is the external impetus and guarantee for enterprise technology innovation.

4.2 Fuzzy Closeness Centrality of Technology Innovation Factors

Fuzzy node centrality reflects the influence degree of factors just from the respect of local nodes, i.e., the influence that one node and others relate directly; therefore, node closeness centrality must be calculated. Node closeness centrality considers all the factors, which is the effect from the most effective correlation between a certain node and the others in network. The factors influencing technology innovation gotten from the analysis of fuzzy closeness centrality demonstrates its comprehensive influence ability to others, namely the effective association degree to other factors. According to formula (21), the closeness centrality of each factor is calculated in Table 3.

From Table 3 we can learn that v_{29} (Research and development capabilities) is the most important factor among all the key factors influencing enterprise technology innovation from the view of closeness centrality. Ranked by their importance, the order from NO. 2 to NO. 9 are as follows: v_3 (Human resources quality), v_{18} (Degree of

Table 3. Closeness centrality of each fact in fuzzy technology innovation network

NO.	Influencing factors	Closeness centrality
1	v_{29}: Research and development capabilities	16.566
2	v_3: Human resources quality	14.304
3	v_{18}: Degree of industrial clusters	11.877
4	v_{33}: Organization execution	11.616
5	v_{30}: Production capacity	11.556
6	v_2: Level of economic development	11.15
7	v_4: Information infrastructure environment	11.134
8	v_{23}: Technology accumulation	11.004
9	v13: Market demand	10.927
10	v_5: Social and cultural environment	10.346
11	v_{19}: Industrial technology support platform	10.332
12	v_{32}: Management decision-making capacity	10.132
13	v_{11}: Technology property rights transaction	9.829
14	v_{12}: The level of technology assessment	9.586
15	v_{34}: Brand Competitiveness	9.348
16	v_{37}: Rapid response force	9.285
17	v_{15}: Intensity of market competition	9.075
18	v_{17}: Level of industrial technology	8.735
19	v_{14}: Fair market	8.530
20	v_{31}: Integration of collaborative capabilities	8.356
21	v_{20}: Corporate human resources constitute	8.237
22	v_{24}: Information resources and platforms	8.108
23	v_{10}: Social service system	8.051
24	v_1: Social infrastructure	8.029
25	v_7: Monetary policy	7.944
26	v_{25}: Enterprise products constitute	7.944
27	v_{27}: Ability to obtain resources	7.816
28	v_{28}: Marketing capability	7.784
29	v_8: Intellectual property rights policy	7.538
30	v_{35}: Corporate culture	7.407
31	v_6: Tax policy	7.047
32	v_{26}: Organizational structure	6.797
33	v_9: Innovation incentives	6.641
34	v_{36}: Innovative driving forces	6.619
35	v_{16}: The degree of credit market	6.263
36	v_{21}: Liquidity	6.189
37	v_{22}: Fixed capital formation	5.762

industrial clusters), v_{33} (Organization execution), v_{30} (Production capacity), v_2 (Level of economic development), v_4 (Information infrastructure environment), v_{23} (Technology accumulation), v_{13} (Market demand). All these factors can be divided into two levels: enterprise level and environment level. The enterprise level includes research and development capabilities, human resources quality, production capacity and organization execution. Among them, research and development capabilities influences enterprise technology innovation most directly. Human resources quality and organization execution are the soft factors influencing research and development capabilities while production capacity and technology accumulation are hard factors. The environment level includes industrial environment and macro environment. Degree of industrial clusters belongs to industrial environment while level of economic development, information infrastructure environment and market demand are included in macro environment.

4.3 Key Factors

The data in Table 2 is ordered by fuzzy node centrality, it is the reflection of the degree that a certain factor influences others directly, it's local influence. The data in Table 3 is ranked by fuzzy closeness centrality; it presents the effective correlation between a certain factor and others, which is a description of the comprehensive influence the certain factor makes on others. Reorder all the factors combined with the two aspects, the new comprehensive order is shown in Table 4.

Table 4. Rank of key factors' importance in influencing enterprise technology innovation

NO.	Factors	Rank of fuzzy node centrality	Rank of fuzzy closeness centrality	Key factors
1	v_{29}: Research and development capabilities	1	1	√
2	v_3: Human resources quality	2	2	√
3	v_{30}: Production capacity	4	5	√
4	v_2: Level of economic development	3	6	√
5	v_{18}: Degree of industrial clusters	7	3	√
6	v_{23}: Technology accumulation	9	8	√
7	v_{19}: Industrial technology support platform	6	11	
8	v_4: Information infrastructure environment	11	7	
9	v_{13}: Market demand	12	9	
10	v_{17}: Level of industrial technology	5	18	
11	v_5: Social and cultural environment	13	10	
12	v_{37}: Rapid response force	8	16	
13	v_{34}: Brand Competitiveness	10	15	

According to Table 4, there are 6 factors count not only from the aspect of node centrality but also closeness centrality, so they absolutely deserve much attention. They are v_{29} (Research and development capabilities), v_3 (Human resources quality), v_{30} (Production capacity), v_2 (Level of economic development), v_{18} (Degree of industrial clusters), v_{23} (Technology accumulation).

5 Conclusion and Discussion

In this paper, we extend the notion of centrality to the fuzzy social network, and propose actor's fuzzy node centrality, actor's fuzzy closeness centrality, actor's fuzzy betweenness centrality, fuzzy group centralization, fuzzy group closeness centralization and fuzzy group betweenness centralization. We apply the fuzzy centrality theory to a case which is called fuzzy technology innovation network. In this case, 37 factors that influencing enterprise technology innovation is taken as actors to set up the network of enterprise technology innovation. We calculate their fuzzy node centrality and fuzzy closeness centrality, determine the key factors influencing enterprise technology innovation. The results show that for the aspect of R&D strategy, we encourage that such factors as *Research and development capabilities*, *Human resources quality*, *Production capacity*, *Level of economic development*, *Degree of industrial clusters*, *Technology accumulation* should be focused more on by the decision makers in fuzzy technology innovation network.

All in all, analysis of fuzzy centrality is one of the most important and commonly used tools in fuzzy social network, which is the reflection of the central position and their advantages comparison of different actors (e.g.: factors influence technology innovation), and it contributes to the management decision of actors representing people, groups, organizations, enterprises, etc. The centrality and centralization analysis in fuzzy social network provide the theoretical foundation for further study of the fuzzy social network.

In fuzzy social network, links represent social relationships, for instance friendships, between actors. These relationships offer benefits in terms of favors, information, etc. Moreover, actors also benefit from indirect relationships. However, how to calculate the relationship between "actor of an actor"? How to calculate the fuzzy closeness centrality in fuzzy social network based on attenuation factor? Obviously, all these need further research.

Acknowledgement. Thanks to the support by Guangdong foundation of philosophy and Social Sciences, China (No. GD16XGL20) and Department of Education of Guangdong province, China (No. YQ2015107).

Recommender. Hong-Wei-Liu, Guangdong University of Technology Professor

References

Bavelas, A.: A mathematical model for small group structures. Hum. Organ. **7**, 16–30 (1948)
Beauchamp, M.A.: An improved index of centrality. Behav. Sci. **10**, 161–163 (1965)
Bonacich, P.: Factoring and weighting approaches to status scores and clique detection. J. Math. Soc. **2**, 113–120 (1972)
Bonacich, P.: Power and centrality: a family of measures. Am. J. Sociol. **92**, 1170–1182 (1987)
Costenbader, E., Valente, T.W.: The stability of centrality measures when networks are sampled. Soc. Netw. **25**, 283–307 (2003)
Freeman, L.C.: A set of measures of centrality based on betweenness. Sociometry **40**, 35–41 (1977)
Ling, J.: Social Network Analysis: Theory Methods and Applications. Beijing Normal University Press, Beijing (2009)
Del Pozo, M., Manuel, C., González-Arangüena, E., Owen, G.: Centrality in directed social networks. A game theoretic approach. Soc. Netw. **33**, 191–200 (2011)
Nieminen, J.: On the centrality in a directed graph. Soc. Sci. Res. **2**, 371–378 (1974)
Sabidussi, G.: The centrality index of a graph. Psychometrika **31**, 581–603 (1966)
Shaw, M.: Communication networks. In: Berkowitz, L. (ed.) Advances in Experimental Social Psychology, pp. 111–147. Academic Press, New York (1954)
Kang, S.M.: A note on measures of similarity based on centrality. Soc. Netw. **29**, 137–142 (2007)
Fan, T.F., Liau, C.J., Lin, T.Y.: Positional analysis in fuzzy social networks. In: Proceedings of the Third IEEE International Conference on Granular Computing, 2007, pp. 423–428 (2007)
Wasserman, S., Faust, K.: Social Network Analysis: Methods and applications. Cambridge University Press, Cambridge (1994)

Social Network User Feature Recognition Method Based on Weighted Graph and Fuzzy Set

Juan-li Zheng[1], Fu-yi Wei[1(✉)], and Xiao-feng Hu[2]

[1] College of Mathematics and Informatics, South China Agricultural University, Guangzhou, Guangdong 510642, China
weifuyi@scau.edu.cn

[2] Shenzhen Micro-An Computer Co., Ltd., Shenzhen, Guangdong 518036, China

Abstract. In this paper, we use the network model and graph theory techniques to introduce the concept of connected factor, and then we can assign the weight between all the friends in the "third degree range" and use it to quantify the user's influence on the structure of the speculated user. By using the fuzzy mathematics method, we can construct the user topic set and the frequency set, define the membership degree of the score, calculate the closeness of two users' topics, and then get the fuzzy influence from user to presumed user. By combining the network model and the fuzzy method, we can obtain a comprehensive formula for depicting the user's influence. Via using the users' influence vector and the user identification matrix, the score vector of the speculative user's identity is calculated, and a variety of identity features are obtained so that we can evaluated whether the user reaches the threshold.

Keywords: User identification · Closeness degree · Three-neighbor connection
Social network · Connected graph

1 Introduction

Social network has become an inseparable part of people's lives. Its rapid development has influenced human's thinking mode and changed human's way of life. Though the characteristics of social network users such as virtuality and anonymity bring convenience for people to propose their opinions, it also brings difficulties for us to identify the identification of users.

The identification of users in social networks is rather important. For example, user identification across networks can be applied to recommend users to interests on other platforms, the influential leaders of social networks can play a guiding role in the correct orientation of public opinion, and the identifying of the naval and zombie users of social networks plays an important role in the monitoring and management of the network.

Through the analysis of log records, a single feature-weighted Naive Bayesian classification algorithm is proposed by Liu et al. to identify the identity of users [1]. By simulating the voting behavior of human society, key users with high recognition and

B.-Y. Cao and Y.-B. Zhong (Eds.): ICFIE 2017, AISC 872, pp. 166–178, 2019.
https://doi.org/10.1007/978-3-030-02777-3_15

high coverage are identified by Wu et al. in [2]. A classification method based on two types of user names and microblog texts is proposed by Xue [3], via this method we can solve the problem of automatic identification of the individual attributes to microblog users. By calculating the sum of the weighted sum of the four eigenvalues of the average neighborhood, center, activity and heat of any two active users, Tian et al. proved that the users with larger link strength vector and influence vector than the influence threshold are the key users [4]. In [5], Hu et al. established the virtual mapping between the user's virtual identity and the real identity by using the user's geographical location information to infer the users' schools and work unit. By performing similarity calculation of nodes in the network and string similarity calculation, Sun et al. determined whether the user identity information matches [6]. Zhou et al. pointed out that we can use network awareness method to calculate user behavior characteristics through dynamic user connection and influence behavior [7]. Ye et al. proposed that we can identify the user-level cross-system user by calculating the similarity between the source user profile and the user profile to be matched [8]. In [9], Hu et al. established a method to identify identity features of full subgraphs and full subgraph identity feature recognition methods, which can effectively improve the instability of user identity features due to sparse social relationships. By using the frequent topics between restricted Boltzmann machines and users, the presented authors calculated the similarity between microblog accounts and similarity between users, and then we can assign the tendency of tendencies by using bipartite graphs to make recommendation for users [10]. Two of the presented authors showed that we can realize personalized recommendations that trade off between exactness and diversity by correcting weights based on tag usage and time effects [11].

The above methods mainly realizes the user group classification through the user's characteristic attributes, and they ignore the user's structured social relationship. In this paper, we use the three degrees of mutual relationship of powder network, combining the graph theory skills and fuzzy mathematics method to quantify the user's influence on the speculative users in the "third degree range" into the comprehensive weight formula of structural influence and fuzzy influence portrayed Using the user's influence vector and identity matrix, and we also obtain a method to identify the multiple identity features of recommended users. Our new result extends the corresponding result of [9], which used complete subgraphs and passing complete subgraphs to identity features by.

2 Basic Concepts

Let $G = \{V(G), E(G)\}$ be a simple graph, where $V(G)$ is the vertex set of G, and $E(G)$ is the edge set of G. A simple undirected graph with edges between any two vertices is called a complete graph, and denoted by K_n. By the definition: $E(K_n) = \binom{n}{2} = \frac{1}{2}n(n-1)$. The path from vertex x to vertex y in G is an alternating sequence of different vertices and edges from x to y, which is denoted by P(x, y). The number of edges in this alternating sequence of a path is called the length of the path. The shortest length of all paths connecting x and y in G is called the distance from x to y, and denoted as $d(x, y)$. The set of vertices, which are adjacent with u, are called the neighborhood of u, is denoted as $N(u)$. The vertex set consisting those vertices with the distance at most three from u will be referred as the three degrees of u. If $V_m \subseteq V$, then the induced subgraph by the vertices of V_m is denoted by $G(V_m)$.

Definition 1. If the vertex set V of a simple undirected graph G is divided into t disjointed classes $V_1, V_2, \cdots, V_t$ such that the following (i) and (ii) hold, then the subgraph $G(V_i)$ is called a connected branch of G, and t is called the connected branch number of G, where

(i) If $x, y \in V_i (1 \leq i \leq t)$, then there exists a path connecting x and y in G;
(ii) If $x \in V_i, y \in V_j (i \neq j)$, then there is not any path connecting x and y in G.

Definition 2. The graph $G_1 \wedge G_2$ is obtained from two graphs G_1 and G_2 by adding one edge between every vertex of G_1 and every vertex of G_2. Hereafter, $G_1 \wedge G_2$ is called the join of G_1 and G_2.

In 2011, Christakis showed that: "Six Degrees of Separation" spreading on social networks follows the rule of "The Degrees of Influence Rule" [12], that means, people's behaviors, attitudes and emotions are all influenced by three degrees of the social network Ripples. People can affect people within three separate sides each other, that is, one can affect one's own friends, friends can affect their friends and friends of friends' can also affect their friends. At the same time, man is also influenced by those people within three separate divisions. If more than three degrees apart, the influence will disappear. People within three degrees of separation have similarities in terms of age, occupation, hobbies, geography and so on. In a complex network structure, the impact between users is not only influenced by the friends whose distance is 1, but also by their mutual friends. More mutual friends and more stable structure will greatly enhance the users' influence. In this paper, we will introduce the concept of connected factors to study the impact of all users within three degrees on speculating users.

The influence of two adjacent users u_i and u_j (that is, satisfying $d(u_i, u_j) = 1$) is closely related to the structure of the induced subgraph formed by the vertex set of their common neighbors (that is, $N(u_i) \cap N(u_j)$). Let $|N(u_i) \cap N(u_j)| = m$. If the induced subgraph of vertex set (i.e., $N(u_i) \cap N(u_j)$) is a complete graph K_m, then the influence is maximum. While this induced subgraph is an empty graph (that is, $|E| = 0$), then the influence is minimum. When the order (namely, the number of vertices) of the corresponding graph G is fixed, the bigger size (i.e., the number of edges) of G, the greater influence. When the size of G is fixed, the fewer order, the greater influence. Now we introduce the concept of connected factor, which will be used to characterize the impact of relationship between users.

Definition 3. If $u_i u_j \in E(G)$ with $N(u_i) \cap N(u_j) = m$, then the induced subgraph with vertex set $N(u_i) \cap N(u_j)$ is denoted by $\tilde{G}_m$, and the real number $S(\tilde{G}_m)$ is called the connected factor of $\tilde{G}_m$, where

$$S(\tilde{G}_m) = \begin{cases} \frac{|E(\tilde{G}_m)|}{|V(\tilde{G}_m)|}, & \text{if } \tilde{G}_m \text{ is connected} \\ \begin{cases} \frac{m-1}{2m}, m \geq 2 \\ \frac{1}{8}, m = 1 \end{cases} & \text{if } \tilde{G}_m \text{ is empty (ie } |E_m| = 0) \end{cases},$$

Since the structure of the complete graph is the most stable, the mutual influence of its vertices are the largest. Let $S(K_m)$ be the connected factor of the complete graph K_m. Then, the connected factor of the complete graph is the largest, that is $S(\tilde{G}_m) \leq S(K_m)$. In what follows, we will use the connected factor to describe the mutual influence between these mutual friends.

3 Three Degrees-Connecting Network on the Speculation of the User's Structural Influence

In this section, we will study the effect of all users within the distance at most three on the speculation of users, and empowers users through the structure of mutual friends of two users. If the structure of mutual friends is compact, then the weight between two users is high. Otherwise, if the structure of mutual friends is loose, then the weight between two users is low.

3.1 The Weight of Edges Between the Users in the Three Degrees

For the speculative user u_c, the user sets whose distances from u_c are 1, 2 and 3 are called the first-neighbor, second-neighbor and third-neighbor set, respectively. From the definition, the three degrees of u_c is consisting of the first-neighbor, second-neighbor and third-neighbor set of u_c. As referred before, the impact on user u_c comes from its three degrees, so we only consider the three degrees of connection in this article.

It is presumed that the user u_c and its vertex set in the three degrees connection are denoted as $V = \{u_c, u_1, u_2, \cdots, u_n\}$. Now, we can construct a undirected graph, say G (V, E), from the mutual powder relationship between those vertices of V. Hereafter, the first-neighbor, second-neighbor and third-neighbor set of u_c will be denoted as $N(u_c), N_2(u_c), N_3(u_c)$, respectively. It is easy to see that $\{u_c\} \cup N(u_c) \cup N_2(u_c) \cup N_3(u_c) = V$. For convenience, we write $|N(u_c)| = p_1, |N_2(u_c)| = p_2, |N_3(u_c)| = p_3$ in the following. In what follows, we shall evaluate the weight of each edge within the three degrees-connecting network and we will consider the following three cases.

3.1.1 Weight of Edges Between the Speculative User u_c and Its First-Neighbor Set $N(u_c)$

For any vertex $u_x \in N(u_c)$, the impact of user u_x on the speculative user u_c can be evaluated by the following method.

(1) $N(u_x) \cap N(u_c) \neq \emptyset$

Let $N(u_x) \cap N(u_c) = \tilde{V}$, $|\tilde{V}| = q_1$, $G(\tilde{V}) = G(\tilde{V}_1) \cup G(\tilde{V}_2) \cup \cdots \cup G(\tilde{V}_{m_1-1}) \cup G(\tilde{V}_{m_1})$, where $G(\tilde{V}_{k_1})$ $(k_1 = 1, 2, \cdots, m_1 - 1)$ is a non-empty connected branch of G, that is $|E(G(\tilde{V}_{k_1}))| > 0$, $G(\tilde{V}_{m_1})$ is an empty graph consisting of h_1 isolated vertices, namely, $|E(G(\tilde{V}_{m_1}))| = 0$.

For any $\tilde{V}_{k_1} \subseteq \tilde{V}$, from Definition 3, we can conclude that the connected factor of the connected subgraph $G(\tilde{V}_{k_1})$ is $S(G(\tilde{V}_{k_1})) = \frac{|E(G(\tilde{V}_{k_1}))|}{|\tilde{V}_{k_1}|}$, where $k_1 = 1, 2, \cdots, m_1 - 1$.

Let $|\tilde{V}_{m_1}| = h_1$. Since the structure of the induced subgraph by the vertex set consisting of those vertices of $G(\tilde{V}_{m_1})$ and those vertices in the path $P(u_c, u_x)$ connecting u_c and u_x is the join graph $P(u_c, u_x) \wedge G(\tilde{V}_{m_1})$. Note that among these connected subgraphs with h_1 vertices, the tree with h_1 vertices has the worst connectivity, and hence the connected factor is $S(G(\tilde{V}_{h_1})) = \frac{h_1 - 1}{h_1}$. Its connectivity is better than that of an empty graph, so the connected factor of an empty graph $G(\tilde{V}_{m_1})$ connected to the path $P(u_x, u_c)$ should be less than $\frac{h_1 - 1}{h_1}$. Taking this into consideration, we define

$$S(G(\tilde{V}_{m_1})) = \begin{cases} \frac{h_1 - 1}{2h_1}, & h_1 \geq 2 \\ \frac{1}{8}, & h_1 = 1 \end{cases}.$$

Furthermore, if the users u_x and u_c are adjacent, then we define the weight between u_x and u_c as follows:

$\omega(u_x, u_c) = \frac{\sum_{k_1=1}^{m_1} S(G(\tilde{V}_{k_1}))}{S(K_{q_1})}$, where $S(K_{q_1}) = \frac{1}{2}$ if $q_1 = 1$, and

$\omega(u_x, u_c) = 1$, if $G(\tilde{V}) \cong K_{q_1}$ and $q_1 \neq 1$.

Note that when $G(\tilde{V}_{k_1})$ is a complete subgraph, it is just the case of the full degree subgraph of [9].

(2) $N(u_x) \cap N(u_c) = \emptyset$.

In this case, u_x and u_c have no common friends, and their influence is less than the worst connectivity (i.e., the connectivity of tree) when they have common friends, as the corresponding induced subgraph is an empty graph.

Let $\tilde{V}_0^{(1)} = \{u_x | N(u_x) \cap N(u_c) = \emptyset\}$, $|\tilde{V}_0^{(1)}| = l_1$, $|N(u_c)| = p_1$. For any $u_i \in N(u_c) / \tilde{V}_0^{(1)}$, let $V_s = N(u_c) / (\tilde{V}_0^{(1)} \cup \{u_i\})$ and $|V_s| = t_1$. Then, $t_1 = p_1 - 1 - l_1$.

Let $N(u_i) \cap N(u_c) = \tilde{V}_s \subseteq V_s$, $|\tilde{V}_s| = \tilde{t} \leq t_1$ and $G(\tilde{V}_s) = G(\tilde{V}_1) \cup G(\tilde{V}_2) \cup \cdots \cup G(\tilde{V}_{M_1-1}) \cup G(\tilde{V}_{M_1})$, where $G(\tilde{V}_{K_1})$ $(K_1 = 1, 2, \cdots, M_1 - 1)$ is a non-empty connected branch of G; $G(\tilde{V}_{M_1})$ is an empty graph consisting of m_1 isolated vertices. When $G(\tilde{V}_s)$ is an empty graph, then $m_1 = \tilde{t}$ and $\omega(u_i, u_c)$ is the smallest. Thus,

$$\omega(u_i, u_c) = \frac{\sum_{K_1=1}^{M_1} S(G(\tilde{V}_{K_1}))}{S(K_{\tilde{t}})} \geq \frac{\frac{\tilde{t}-1}{2\tilde{t}}}{S\left(K_{\frac{\tilde{t}}{t}}\right)} = \frac{1}{\tilde{t}} \geq \frac{1}{t_1}.$$

Now, from the former argument, if $u_x \in \tilde{V}_0^{(1)}$, then the weight between u_x and u_c is equal to $\omega(u_x, u_c) = \frac{1}{2t_1}$.

The weight of the edge $u_x u_c$ can be calculated as follows:

$$\omega(u_x, u_c) = \begin{cases} \frac{\sum_{k_1=1}^{m_1} S(G(\tilde{V}_{k_1}))}{S(K_{q_1})}, & N(u_x) \cap N(u_c) \neq \emptyset \\ \frac{1}{2t_1}, & N(u_x) \cap N(u_c) = \emptyset \end{cases}, \tag{1}$$

Actually, the weight of any edge between u_c and $N(u_c)$ can also deduced from the formula $\tilde{\omega}(u_x, u_c) = \frac{\omega(u_x, u_c)}{\sum_{u_x \in N(u_c)} \omega(u_x, u_c)}$.

3.1.2 Weight of Edges Between Vertex Sets $N_2(u_c)$ and $N(u_c)$

Let e be an edge between $N_2(u_c)$ and $N(u_c)$ such that $e = u_x u_j$, where $u_x \in N_2(u_c)$ and $u_j \in N(u_c)$. Then, the distance between u_x and u_c is two, that is, $d(u_x, u_c) = 2$.

For convenience, we suppose that $N(u_x) \cap N(u_c) = \{u_1, u_2, \cdots, u_l\}$ and $N(u_x) \cap N(u_j) = \tilde{V}^{(2)}$, where $|N(u_x) \cap N(u_c)| = 1$ and $\left|\tilde{V}^{(2)}\right| = q_2$, as shown in Fig. 1.

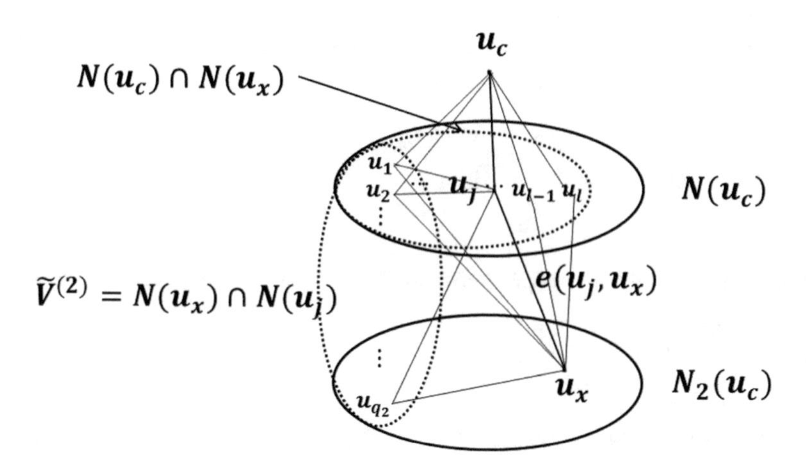

Fig. 1. The second-neighbors relationship of the edge $u_x u_j$ and related sets (Note: Some of the vertices and edges indicate the relationship between the second-neighbors of vertices in a friend's relationship.)

In this case, the weight of the edge $u_x u_j$ is determined by the structure of the induced subgraph $G(N(u_x) \cap N(u_j))$.

For any vertex $u_j \in \{u_1, u_2, \cdots, u_l\}$ with $u_c u_j \in E(G)$ and $u_j u_x \in E(G)$, we consider the following cases:

(1) $N(u_x) \cap N(u_j) \neq \emptyset$.

Suppose that $N(u_x) \cap N(u_j) = \tilde{V}^{(2)}$ and $G\left(\tilde{V}^{(2)}\right) = G(\tilde{V}_1^{(2)}) \cup G\left(\tilde{V}_2^{(2)}\right) \cup \cdots \cup G\left(\tilde{V}_{m_2-1}^{(2)}\right) \cup G\left(\tilde{V}_{m_2}^{(2)}\right)$, where $G\left(\tilde{V}_{k_2}^{(2)}\right)$ $(k_2 = 1, 2, \cdots, m_2 - 1)$ is a non-empty connected branch of G and $G\left(\tilde{V}_{m_2}^{(2)}\right)$ is an empty graph consisting of h_2 consisting vertices.

We can define the weight of the edge adjacent with u_x and u_j as follows:

$$\omega(u_x, u_j) = \frac{\sum_{k_2=1}^{m_2} S\left(G\left(\tilde{V}_{k_2}\right)\right)}{S\left(K_{q_2}\right)},$$

(2) $N(u_x) \cap N(u_j) = \emptyset$

Suppose that $\tilde{V}_0^{(2)} = \{u_x | N(u_x) \cap N(u_j) = \emptyset\}$, $\left|\tilde{V}_0^{(2)}\right| = l_2$, and $|N(u_j)| = p_2^{(j)}$. Let $V_s^{(2)} = N(u_j) / \left(\tilde{V}_0^{(2)} \cup \{u_i\}\right)$ and $\left|V_s^{(2)}\right| = t_2$. Then, $t_2 = p_2^{(j)} - 1 - l_2$.

Similarly as case (2) of u_c and its first-neighbor set $N(u_c)$, the weight of edge between u_x and u_j is defined as $\omega(u_x, u_j) = \frac{1}{2t_2}$.

By combining with (1) and (2), we can conclude that the weight of edge adjacent with u_x and u_j can be calculated as follows:

$$\omega(u_x, u_j) = \begin{cases} \frac{\sum_{k_2=1}^{m_2} S\left(G\left(\tilde{V}_{k_2}\right)\right)}{S\left(K_{q_2}\right)}, & N(u_x) \cap N(u_j) \neq \emptyset \\ \frac{1}{2t_2}, & N(u_x) \cap N(u_j) = \emptyset \end{cases},$$

Thus, the weight of edge between $N_2(u_c)$ and $N(u_c)$ can be written as $\tilde{\omega}(u_x, u_j) = \frac{\omega(u_x, u_j)}{\sum_{\substack{u_j \in N(u_c) \\ u_x \in N_2(u_c)}} \omega(u_x, u_j)}$.

3.1.3 Weights of Edges Between Vertex Sets $N_3(U_c)$ and $N_2(U_c)$

(1) $N(u_x) \cap N(v_i) \neq \emptyset$

For any vertex $u_x \in N_3(u_c)$, there must be a vertex $v_i \in N_2(u_c), u_j \in N(u_c)$ such that $u_x v_i, v_i u_j, u_j u_c \in E(G)$, that is $d(u_x, u_c) = 3$.

Let $N(u_x) \cap N(v_i) = \tilde{V}^{(3)}$ and $G\left(\tilde{V}^{(3)}\right) = G(\tilde{V}_1^{(3)}) \cup G\left(\tilde{V}_2^{(3)}\right) \cup \cdots \cup G\left(\tilde{V}_{m_3-1}^{(3)}\right) \cup G\left(\tilde{V}_{m_3}^{(3)}\right)$, where $G\left(\tilde{V}_{k_3}^{(3)}\right)$ $(k_3 = 1, 2, \cdots, m_3 - 1)$ is a non-empty connected branch of G and $G\left(\tilde{V}_{m_3}^{(3)}\right)$ is an empty graph consisting of h_3 isolated vertices.

In this case, we define the weight of edge adjacent with u_x and v_i as follows:

$$\omega(u_x, v_i) = \frac{\sum_{k_3=1}^{m_3} S(G(\tilde{V}_{k_3}))}{S(K_{q_3})},$$

(2) $N(u_x) \cap N(v_i) = \emptyset$

Similar to Case 2, for any $u_x \in N_3(u_c)$ and $v_i \in N(u_x) \cap N_2(u_c)$, let $|N(u_x) \cap N_2(u_c)| = r$ and $|N(v_i) \cap N(u_x)| = q_3$. For any $u_i \in N(v_i)/\tilde{V}_0^{(3)}$ with $\tilde{V}_0^{(3)} = \{u_x | N(u_x) \cap N(v_i) = \emptyset\}$, if $\left|\tilde{V}_0^{(3)}\right| = l_3$, $|N(v_i)| = p_3^{(i)}$ and $\tilde{V}_s^{(3)} = N(v_i)/\left(\tilde{V}_0^{(3)} \cup \{u_i\}\right)$, then $t_3 = p_3^{(i)} - 1 - l_3$.

Now, we define the weight of edge adjacent with u_x and v_i as follows:

$$\omega(u_x, v_i) = \begin{cases} \frac{\sum_{k_3=1}^{m_3} S(G(\tilde{V}_{k_3}))}{S(K_{q_3})}, & N(u_x) \cap N(v_i) \neq \emptyset \\ \frac{1}{2t_3}, & N(u_x) \cap N(v_i) = \emptyset \end{cases},$$

It is easy to see that the weight of edge between $N_3(u_c)$ and $N_2(u_c)$ can be also calculated as $\tilde{\omega}(u_x, v_i) = \frac{\omega(u_x, v_i)}{\sum_{\substack{v_i \in N_2(u_c) \\ u_x \in N_3(u_c)}} \omega(u_x, v_i)}$.

3.2 Three Degrees Users on the Structure Influence of the Speculative User

In this section, the weight of the vertices is used to characterize the influence of user u_x on the structure of the speculative user u_c. For any user $u_x \in V(G)/u_c$, we define $\omega(u_x)$ as the vertex weight of user u_x. We will construct the vertex weight for each vertex in the three-neighbor connection as follows:

(1) $u_x \in N(u_c)$

Here u_x is adjacent with u_c, let $\omega(u_x) = \tilde{\omega}(u_x, u_c)$. Then, $\tilde{\omega}(u_x)$ represents the influence of the first-neighbor user u_x on the structure of u_c.

(2) $u_x \in N_2(u_c)$

Suppose that $N(u_x) \cap N(u_c) = \{u_1, u_2, \cdots, u_l\}$, and $u_j \in \{u_1, u_2, \cdots, u_l\}$. The influence of the second-neighbor user u_x on the speculative user u_c is defined as: $\tilde{\omega}(u_j, u_c) \cdot \frac{1}{2}\tilde{\omega}(u_x, u_j)$. Since the influence of user u_x on the speculative user u_c through all path pairs with length two, the weight of the second-neighbor user u_x can be defined as:

$$\omega(u_x) = \sum_{j=1}^{l} \tilde{\omega}(u_j, u_c) \cdot \frac{1}{2}\tilde{\omega}(u_x, u_j) = \sum_{j=1}^{l} \omega(u_j) \cdot \frac{1}{2}\tilde{\omega}(u_x, u_j),$$

As the user u_x is a second-neighbor user of u_c, the weight of the edge in the second-neighbor relation in the formula is taken as one-half of the weight of the original weight. We make this choice according to the increasing of the path's length, as the longer the path, the less influence on the friend relationship strength.

(3) $u_x \in N_3(u_c)$

Let $N(u_x) \cap N_2(u_c) = \{v_1, v_2, \cdots, v_r\}$ and $|N(u_x) \cap N_2(u_c)| = r$. When $u_x \in N_3(u_c)$, then the distance between u_x and u_c is three, and hence the relation of vertices can be illustrated in Fig. 2.

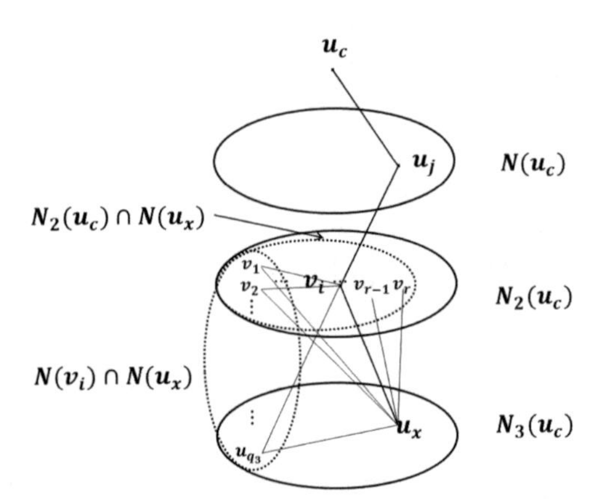

Fig. 2. Paths of length 3 $P(u_x, v_i, u_j, u_c)$ and related sets

For any $v_i \in N(u_x) \cap N_2(u_c)$, similar to case 2 of Sect. 3.2, we can define the weight of the users u_x as: $\omega(u_x) = \sum_{i=1}^{r} \omega(v_i) \cdot \frac{1}{3}\tilde{\omega}(u_x, v_i)$,

As user u_x is a third-neighbor user, the weight in the formula is taken as one-third of the weight of the original edge.

Now, for the speculative user u_c, we can conclude that the weight of user u_x can be defined as follows:

$$\omega(u_x) = \begin{cases} \tilde{\omega}(u_x, u_c), & u_x \in N(u_c) \\ \sum_{j=1}^{l} \omega(u_j) \cdot \frac{1}{2}\tilde{\omega}(u_x, u_j), & u_x \in N_2(u_c), u_j \in N(u_c) \\ \sum_{i=1}^{r} \omega(v_i) \cdot \frac{1}{3}\tilde{\omega}(u_x, v_i), & u_x \in N_3(u_c), v_i \in N_2(u_c) \end{cases}, \quad (2)$$

The weight $\omega(u_x)$ of the user u_x indicates its influence on the speculative user u_c is generated by network structure, which is referred as structure influence.

4 Topic Closeness to the Fuzzy Influence of Speculative User

Topics involved can be broadly divided into professional topics and personal interests of the topic. For example, users who focus on such specialized topics as machine learning and artificial intelligence can speculate that their identities are relevant. In order to more closely distinguish between close friends with common topics and quantitatively describing the frequency of topics, we can calculate the closeness of the topics by setting the user membership function on the topic scores.

Suppose that the set of users within three degrees of user u_c is $U = \{u_1, u_2, \cdots, u_n\}$ and the topic set of u_c is $X = \{x_1, x_2, \cdots, x_m\}$. For any $u_j \in U$, note that the frequency of the topic u_j to the topic u_c of the user can be set as $f_i^{(j)}$, where $f_i^{(j)} \neq 0$ if x_i is the topic of u_j. Otherwise, $f_i^{(j)} = 0$ if x_i is not the subject of u_j.

Suppose that $\mathcal{F}(X)$ is the fuzzy sets of the topic frequency. The membership function of u_j for the set of topic $X = \{x_1, x_2, \cdots, x_m\}$ is denoted as

$$A(u_j) = \left(f^{(j)}(x_1), f^{(j)}(x_2), \cdots, f^{(j)}(x_m)\right), \text{where } j = 1, 2, \cdots, n.$$

Then, $A(u_j) \in \mathcal{F}(X)$ can be expressed as:

$$A(u_j) = f^{(j)}(x_1)|x_1 + f^{(j)}(x_2)|x_2 + \cdots + f^{(j)}(x_m)|x_m,$$

According to the literature [13], if $A(u_j) \in \mathcal{F}(X)$, then we can define the closeness between users u_c and u_j as:

$$S(u_c, u_j) = \frac{\sum_{i=1}^{m}(A(u_c) \wedge A(u_j))}{\sum_{i=1}^{m}(A(u_c) \vee A(u_j))}, \quad (3)$$

where $m = |X|$, namely, m is the number of all topics.

The formula (3) can of the speculative user u_c and the user u_x within the three degrees connection be used to calculate the closeness $S(u_c, u_x)$. If w the speculative user u_c.

5 Identify the User's Identity

According to Eqs. (2) and (3), the structural influence and fuzzy influence are combined organically to obtain a comprehensive formula for the influence of u_i $(i = 1, 2, \cdots, n)$ on u_c within three degrees, that is, e define the fuzzy weight as $S(u_x) = S(u_c, u_x)$, then $S(u_x)$ indicates that the fuzzy influence of the user u_x on

$$\hat{\omega}(u_i) = \alpha\omega(u_i) + (1 - \alpha)S(u_i), \tag{4}$$

By formula (4), we get the user's influence vector of u_c by

$$W = (\hat{\omega}(u_1)\hat{\omega}(u_2) \cdots \hat{\omega}(u_n)),$$

Suppose that the known identities of users in the three degrees user set U are $D = \{d_1, d_2, \cdots, d_l\}$. For any $d_j \in D$, we define its recognition factor b_{ij} as follows: $b_{ij} = 1$ if and only if the user u_i has the identity feature d_j, that means, $b_{ij} = 0$ if and only if the user u_i does not have the identity feature d_j. Furthermore, we can construct the user ID matrix B as

$$B = \left(b_{ij}\right)_{n \times l'}$$

If we multiply the user's u_c influence vector and the user ID matrix, then we can get the user identity score vector, that is,

$$F(u_c) = WB = (f(d_1)f(d_2) \cdots f(d_l)),$$

Furthermore, we can construct the speculative user's identity feature vector as:

$$\tilde{F}(u_c) = (\tilde{f}(d_1)\ \tilde{f}(d_2) \cdots \tilde{f}(d_l)),$$

where $\tilde{f}\left(d_j\right) = \frac{f(d_j)}{\sum_{i=1}^{n} b_{ij}}$ and $j = 1, 2, \cdots, l$.

Now, we can apply the user's identity feature vector to analyze the identity of the user u_c. For instance, if $\tilde{f}\left(d_j\right) \geq 0.8$ $(j = 1, 2, \cdots, k)$ and $\tilde{f}\left(d_j\right) < 0.8$ $(j = k + 1, k + 2, \cdots, l)$, then we can predict the identity of the user u_c is $d_1, d_2, \cdots, d_k$. If $k = 1$, it indicates that u_c has only one identity d_1. If $k = 2$, it means that u_c has two identities d_1 and d_2 at the same time. If $k = 0$, all component values of the normalized scoring vectors are less than 0.8, which implies that the present information does not identify the identity of user u_c.

6 Conclusions

User identification in social networks is of great importance. This article calculates the impact of user friends within three degrees of influence on speculated users from two aspects to identify the identity characteristics of speculated users. Under the condition of the three degrees of connection principle, firstly, according to the connection factor of the connected subgraphs of two users, the edge weights between two users are transformed into the structural influence of user u_i by using graph theory skills. Secondly, the degree of closeness of the concerned topic is calculated, and the fuzzy influence of the user u_i is obtained. Combine the two influences organically and obtain the comprehensive influence of the u_i user within three degrees on the speculated user u_c. Through the user's u_c influence vector and the user identification matrix, the speculative user's identity feature score vector and the score over the threshold value of 0.8 are obtained, so as to obtain the conclusion that the speculative user possesses this identity feature.

Acknowledgments. This paper is supported by Guangdong Province joint training of graduate demonstration base personnel training project (South China Agricultural University, Shenzhen Microamount Computer Co., Ltd. joint training graduate demonstration base) and Training Program for Outstanding Young Teachers in University of Guangdong Province (No. YQ2015027).

Recommender: Fu-yi Wei, Professor. College of Mathematics and Informatics, South China Agricultural University.

References

1. Liu, L., Chen, X., Yin, X., Duan, Y., Zhao, L.: Network user identification based on feature weighted naive bayes classification. Comput. Appl. **31**(12), 3268–3270 (2011)
2. Wu, Z., Guo, Y., Chen, C.: Key user identification a Algorithm of online social network based on user relationship. J. Beijing Jiaotong Univ. **38**(05), 37–42 (2014)
3. Xue, Y.: Research on the method of microblogging user attributes recognition. Suzhou University (2015)
4. Tian, Z.: Research on key user identification methods in social networks based on influence. Jilin University (2015)
5. Hu, K., Liang, Y., Xu, H., Bi, X., Zuo, Y.: A social network user identity feature recognition method. Comput. Res. Dev. **53**(11), 2630–2644 (2016)
6. Sun, J., Wang, Z., Zhou, C., Wang, S., Zhan, H., Yu, Y., Chui, X.: Research on key technologies of individual identity based on multiple social media. Small Microcomput. Syst. **38**(02), 299–303 (2017)
7. Zhou, X., Wu, B., Jin, Q.: User role identification based on social behavior and networking analysis for information dissemination. Futur. Gener. Comput. Syst. (2017)
8. Ye, N., Zhao, Y., Bian, G., Li, J., He, J.: Pattern-independent social network user identification algorithm. J. Xi'an Jiaotong Univ. **47**(12), 19–25 (2013)
9. Hu, K., Liang, Y., Su, L., Xu, H., Fu, C.: Social network user feature recognition based on complete subgraphs. Pattern Recognit. Artif. Intell. **29**(08), 698–708 (2016)
10. Zheng, J., Wei, F., Hu, X.: User recommendation based on the perspective of weibo topic views. Math. Stat. Manag. **36**(01), 44–53 (2017)
11. Zheng, J., Wei, F.: Research on knowledge recommendation based on weighted directional tripartite graphic network structure. In: Chinese CSCW 2017 Proceedings of the 12th Chinese Conference on Computer Supported Cooperative Work and Social Computing, pp. 201–204 (2017)

12. Christakis, N.A., Fowler, J.H.: The surprising power of our social networks and how they shape our lives-how your friends' friends' friends affect everything you feel, think, and do. Back Bay Books, New York (2011)
13. Xue, Y., Niu, G., Zhao, H., Wang, Y., Qiao, F.: Intelligent diagnosis of hydro-turbine unit axis trajectory based on time-series fuzzy closeness characteristic and improved svm model. J. Irrig. Mach. Eng. 1–3 (2017)

A New Classifier Fusion Algorithm and Its Application in Practice

Zhuang-jian Mo and Sheng-quan Ma(✉)

School of Information Science and Technology, HaiNan Normal University, Haikou 571158, China
mashengquan@163.com

Abstract. In the paper, we introduce a new fuzzy set-valued integral algorithm based on the Sugeno fuzzy complex set-valued integral tool. In addition, we discuss its application in classifier fusion and give the operation steps using example. The results show that this new fuzzy complex set-valued integral classifier fusion algorithm has obvious superiority. Furthermore, adopting bi-direction membership evaluation and fuzzy reasoning are used to conduct final classification which can achieve better effect.

Keywords: Fuzzy complex set-valued function
Complex fuzzy set-valued measure · Sugeno fuzzy complex set-valued integral
Classifier fusion

1 Introduction

Fuzzy integral classifier is a new effective classifier training algorithm which emerge in the last ten years. It is especially suitable for two situations: the interaction between properties exist and classification results is not clear. Yao et al. [1] and Wang et al. [2] applied the fuzzy integral to the classification and achieve a better effect for the first time.

The study of classification has been over a century and a large number of classification algorithms have been developed. Because the classification problem in theory is an extremely complex function continuation problem, there does not exist an optimal algorithm which is suitable for different situation, and there are many classification algorithm appearing. Since most of learning search problem in classification is still NP-hard problem, and precise solution is still not possible, exploring various heuristic algorithms to approximately solving large-scale complex problem is still a meaningful work.

References [3–8] studied the complex fuzzy numeric function measure, complex fuzzy numeric measured function, and complex fuzzy numeric integral et al. The purpose of these studies make the theoretical methods can be applied preferably in the fields of classification technology and information fusion. References [9–11] in theory researched complex fuzzy-valued measure, complex fuzzy measurable function and complex fuzzy numerical integral, also offered Sugeno fuzzy complex numerical integral applied to the classification fusion technology. This paper will study the application of this Sugeno fuzzy set-valued integral in the classifier fusion.

Classification is a technology which construct a classification function or classification model (classification) according to the characteristics of the data sets. Its purpose

B.-Y. Cao and Y.-B. Zhong (Eds.): ICFIE 2017, AISC 872, pp. 179–188, 2019.
https://doi.org/10.1007/978-3-030-02777-3_16

map unknown category to a given a category. It is an important research field in data mining, machine learning, pattern recognition, it is also the core of knowledge processing. For example, Bayesian classification, neural network classification, linear classification, and the nearest classification are common classification methods.

With the emergence of fuzzy sets, fuzzy technology is combined with various classification algorithms to deal with the cognitive problems that classical classification algorithms can not resolve. Using fuzzy measure and fuzzy integral as classification tool is an effective method in practice.

2 Concepts of Sugeno Fuzzy Complex Set-Valued Complex Fuzzy Integral [9]

We suppose $\tilde{f}: Z \to F_0(K^+)$ is complex fuzzy set-vale complex fuzzy measure function belonging to the $(Z, \mathcal{F}(Z), \tilde{\mu})$, $\tilde{E} \in \mathcal{F}(Z)$

So, $(s)\int_{\tilde{E}} \tilde{f} d\tilde{\mu} = \underset{\alpha\in[0,\infty)}{sup} (\alpha \wedge \mathrm{Re}\,\tilde{\mu}[(\tilde{f}_1)_\alpha \cap \tilde{E}]) + \mathrm{i} \underset{\alpha\in[0,\infty)}{\sup} (\alpha \wedge \mathrm{Im}\,\tilde{\mu}[(\tilde{f}_2)_\alpha \cap \tilde{E}]$

$\tilde{f}$ is limited to $\tilde{E}$ which is Sugeno fuzzy complex set-valued fuzzy integral with respect to $\tilde{\mu}$. Abbreviation as S complex fuzzy integral.

Here $(\tilde{f}_1)_\alpha = \{x|\mathrm{Re}[\tilde{f}(x)] \geq \alpha\}$, $(\tilde{f}_2)_\alpha = \{x|\mathrm{Im}[\tilde{f}(x)] \geq \alpha\}$, $\alpha \in [0, \infty)$.

To simply, when X is limited set (that is $X = \{x_1, x_2, \ldots, x_n\}$), and the real part and imaginary part of $\tilde{\mu}$ are canonical fuzzy measure, complex fuzzy function is $\tilde{f}(x) = a_j = \mathrm{Re}a_j + i\mathrm{Im}a_j$.

In general, if $0 + i0 \leq a_1 \leq a_2 \leq \ldots \leq a_n \leq 1 + i, i = \sqrt{-1}$ (if the condition does not meet, X corner mark can be arranged to meet the expression), Sugeno fuzzy complex set-valued fuzzy integral can be simplified as

$$
\begin{aligned}
(s)\int_{\tilde{E}} \tilde{f} d\tilde{\mu} &= \underset{\alpha\in[0,\infty)}{sup} (\alpha \wedge \mathrm{Re}\,\tilde{\mu}[(\tilde{f}_1)_\alpha \cap \tilde{E}]) + i \underset{\alpha\in[0,\infty)}{\sup} (\alpha \wedge \mathrm{Im}\,\tilde{\mu}[(\tilde{f}_2)_\alpha \cap \tilde{E}] \\
&= \bigvee_{j=1}^{n} (\mathrm{Re}\, a_j \wedge \mathrm{Re}\,\tilde{\mu}\{x|\mathrm{Re}\tilde{f}(x) \geq a_j\}) + i \bigvee_{j=1}^{n} (\mathrm{Im}\, a_j \wedge \mathrm{Im}\,\tilde{\mu}\{x|\mathrm{Im}\tilde{f}(x) \geq \mathrm{Im}\, a_j\}) \\
&= \bigvee_{j=1}^{n} (\mathrm{Re}\, a_j \wedge \mathrm{Re}\,\tilde{\mu}(A_j)) + i \bigvee_{j=1}^{n} (\mathrm{Im}\, a_j \wedge \mathrm{Im}\,\tilde{\mu}(A'_j)),
\end{aligned}
$$

Here $A_j = \{x_1, x_2, \ldots, x_j\}$, $A'_j = \{x_j, x_{j+1}, \ldots, x_n\}$.

3 Basic Characters of Sugeno Fuzzy Complex Set-Valued Complex Fuzzy Integral [9]

To simply apply, the characters of S fuzzy complex set-valued fuzzy integral in reference [9] are concluded in the following. The demonstrate of these characters were described in references [4–8].

Theorem 1. If $\tilde{f}_1 \leq \tilde{f}_2$, $(s)\int_A \tilde{f}_1 d\tilde{\mu} \leq (s)\int_A \tilde{f}_2 d\tilde{\mu}$; $(c)\int \tilde{f}_1 d\tilde{\mu} \leq (c)\int \tilde{f}_2 d\tilde{\mu}$;

(1) If $\tilde{\mu}(A) = 0$, $(s)\int_A \tilde{f} d\tilde{\mu} = 0$;
(2) If N is a zero set and $\tilde{f}_1(x) = \tilde{f}_2(x)(\forall x \notin N)$, $(c)\int \tilde{f}_1 d\tilde{\mu} = (c)\int \tilde{f}_2 d\tilde{\mu}$;
(3) $(s)\int_A (\tilde{f}_1 \vee \tilde{f}_2) d\tilde{\mu} \geq (s)\int_A \tilde{f}_1 d\tilde{\mu} \vee (s)\int_A \tilde{f}_2 d\tilde{\mu}$;
(4) If $A \subseteq B$, $(s)\int_A \tilde{f} d\tilde{\mu} \leq (s)\int_B \tilde{f} d\tilde{\mu}$;
(5) $(s)\int_A \tilde{f} d\tilde{\mu} = (s)\int_X \tilde{f} \cdot \chi_A d\tilde{\mu}$; $(c)\int \chi_A d\tilde{\mu} = \tilde{\mu}(A)$;
(6) If c is a non-negative complex constant number, $(s)\int_A c d\tilde{\mu} = (s)\int_A c d\tilde{\mu} \wedge \tilde{\mu}(A)$;
(7) If a is a non-negative real number and b is a real number,

$$(c)\int_A (a\tilde{f} + b) d\tilde{\mu} = a(c)\int_A \tilde{f} d\tilde{\mu} + b\tilde{\mu}(X).$$

(8) If a is a real number, $(c)\int_A \tilde{f} d(a\tilde{\mu}) = a(c)\int_A \tilde{f} d\tilde{\mu}$.

4 Application of Sugeno Fuzzy Complex Set-Valued Integral in Classifier Fusion

The process of classic multiple classifier fusion is following. Firstly, individual classifier classify the unknown sample and obtain the output (that is decision profile DP). Then using fusion operator act on the DP and obtain a vector. Lastly each component corresponds to a class, and the maximum component corresponding class is classification results. This paper considers multi-classifier fusion based on this new S fuzzy complex set-valued integration.

The idea of resolve the classifier fusion is following. The first is to settle the fuzzy complex set-valued measure. The second is that regarding the fuzzy complex set-valued integral as the fusion operator for real and respectively fuse real and imaginary part (real part behalf the degree of membership corresponding to this class, imaginary part behalf non degree of membership corresponding to this class). The last is using proximity to solve the classification.

Because the fuzzy measure does not satisfy the add, for the practical application, we apply g_λ measure:

Definition 1. Supposing $(Z, \mathcal{F}(Z), \tilde{\mu})$ is measure space of fuzzy complex set-valued, and if $A, B \subset \mathcal{F}(Z)$, $A \cap B = \phi$, $\tilde{\mu}(A \cup B) = \tilde{\mu}(A) + \tilde{\mu}(B) + \lambda\tilde{\mu}(A)\tilde{\mu}(B)$, here $\lambda = \mathrm{Re}\,\lambda + i\,\mathrm{Im}\,\lambda$, $\mathrm{Re}\,\lambda, \mathrm{Im}\,\lambda \in (-1, 0) \bigcup (0, +\infty)$. We say that fuzzy complex set-valued measure $\tilde{\mu}$ is $\lambda-$ complex fuzzy set-valued measure. We still use g_λ as $\lambda-$ fuzzy complex set-valued measure.

In practice, X is finite. So suppose $X = \{x_1, x_2, \ldots, x_n\}$ is a finite set, $\tilde{f}$: $X \rightarrow [0, 1] \times [0, 1]$ is a complex function and $\tilde{f}(x_j) = a_j$ (complex value).

According to the convention, the complex values are sorted by the real part and imaginary part. The real part is large and the imaginary part is large, so the complex value is large. We can sort the complex value:

$$a_1 \geq a_2 \geq \ldots \geq a_n$$

$\tilde{\mu}$ is a fuzzy measure of g_λ. In practice, the real part $\mathrm{Re}\,\tilde{\mu}$ and the imaginary part $\mathrm{Im}\,\tilde{\mu}$ of $\tilde{\mu}$ can be limited in to some extent. For example, we can translated the "support degree" and "non support degree", "positive evaluation" and "negative evaluation", "important degree" and "unimportant degree". In general, this must be meet $0 < \mathrm{Re}\,\tilde{\mu} + \mathrm{Im}\,\tilde{\mu} \leq 1$ and $\mathrm{Im}\,\tilde{\mu}^j = 1 - \mathrm{Re}\,\tilde{\mu}^j$.

$\mathrm{Re}\tilde{\mu}$ can be computed according to the following equation. $\mathrm{Re}\,\tilde{\mu}(\{x_j\}) = \mathrm{Re}\,\tilde{\mu}^j$

$$\mathrm{Re}\,\tilde{\mu}(A_j) = \mathrm{Re}\,\tilde{\mu}^j + \mathrm{Re}\,\tilde{\mu}(A_{j-1}) + (\mathrm{Re}\,\lambda)\mathrm{Re}\,\tilde{\mu}^j \mathrm{Re}\,\tilde{\mu}(A_{j-1}), \quad 1 < j \leq n$$

Here $A_j = \{x_1, x_2, \ldots, x_j\}$

$\mathrm{Re}\,\lambda$ and $\mathrm{Im}\,\lambda$ can be obtained according to the following equation.

$$\mathrm{Re}\,\lambda + 1 = \prod_{j=1}^{n} (1 + \mathrm{Re}\,\lambda \mathrm{Re}\,\tilde{\mu}^j); \; \mathrm{Im}\,\lambda + 1 = \prod_{j=1}^{n} (1 + \mathrm{Im}\,\lambda\, \mathrm{Im}\,\tilde{\mu}^j)$$

The existence and uniqueness of the above $\mathrm{Re}\lambda$ and $\mathrm{Im}\lambda$ can be guarantee using the following theorem:

Theorem 2. For fixed real number set $\{g^j\}, 1 \leq j \leq m$, there is only real number $\lambda \in (-1, +\infty), \lambda \neq 0$ and it meet:

$$\lambda + 1 = \prod_{i=1}^{n} (1 + \lambda g^j),$$

$$\mathrm{Re}\,\mu^l = \frac{\mathrm{Re}\,n_{kk}^{(l)}}{\sum_{j=1}^{M} \mathrm{Re}\,n_{jk}^{(k)}}, k < M + 1,$$

$$\mathrm{Im}\,\mu^l = \frac{\mathrm{Im}\,n_{kk}^{(l)}}{\sum_{j=1}^{M} \mathrm{Im}\,n_{jk}^{(k)}}, k < M + 1$$

Here $Re\,n_{jk}^{(l)}$ represents output sample belonging to class j is judged as class k by sub classifier. $Im\,n_{jk}^{(l)}$ represents output sample belonging to non class j is judged as class k by sub classifier. When using complex fuzzy set-valued complex fuzzy measure compute complex fuzzy set-valued complex fuzzy integral, just needing to know the complex fuzzy set-valued complex fuzzy measure. The real part $\mathrm{Re}\mu^j$ of density values μ^j can be interpreted as important degree of x_j.

Accordingly, the Discrete formula of S complex fuzzy set-valued is following:

$$(s)\int_{\tilde{E}} \tilde{f} d\tilde{\mu} = \mathop{\vee}_{j=1}^{n} \left(\operatorname{Re} a_j \wedge \operatorname{Re} \tilde{\mu}(A_j)\right) + i \mathop{\vee}_{j=1}^{n} \left(\operatorname{Im} a_j \wedge \operatorname{Im} \tilde{\mu}(A_j^{'})\right) \tag{1}$$

Here $A_j = \{x_1, x_2, \ldots, x_j\}$, $A_j^{'} = \{x_j, x_{j+1}, \ldots, x_n\}$

Definition 2. Suppose (X, Λ, μ) be measure space of fuzzy complex set-valued. For all the $A, B \subset X, A \cap B = \phi$ such that $\mu(A \cup B) = \mu(A) + \mu(B) + \lambda\mu(A)\mu(B)$. Here $\lambda \in (-1, 0) \cup (0, +\infty)$. We can say that fuzzy complex set-valued measure degree μ is $\lambda-$ fuzzy complex set-valued measure. Using g_λ represent a $\lambda-$ fuzzy complex set-valued measure.

When X is algebra of $\sigma-$ and $g_\lambda(X) = 1$, g_λ is Sugeno fuzzy complex set-valued measure. Supposing $X = \{x_1, x_2, \ldots, x_n\}$ is a finite set, and $\bar{f}$: $X \to [0, 1] \times [0, 1]$ is a complex function and $\bar{f}(x_i) = a_i$ and $a_1 \geq a_2 \geq \ldots \geq a_n$, μ is a g_λ fuzzy complex set-valued measure.

$\operatorname{Im} \mu = 1 - \operatorname{Re} \mu$, the value of $\operatorname{Re} \mu$ can be computed by the $\operatorname{Re} \mu(\{x_i\}) = \operatorname{Re} \mu^i$

$$\operatorname{Re} \mu(A_i) = \operatorname{Re} \mu^i + \operatorname{Re} \mu(A_{i-1}) + (\operatorname{Re} \lambda)\operatorname{Re} \mu^i \operatorname{Re} \mu(A_{i-1}), 1 < i \leq n \tag{2}$$

Here $A_i = \{x_1, x_2, \ldots, x_i\}$. $\operatorname{Re} \lambda$ can be obtained by the following equation

$$\operatorname{Re} \lambda + 1 = \mathop{\Pi}_{i=1}^{n} \left(1 + \operatorname{Re} \lambda \operatorname{Re} \mu^i\right) \tag{3}$$

There are many ways to determine the fuzzy density (μ^i), and the simplest method is regarding recognition rate of each classifier as the fuzzy density value. In this paper, the complex fuzzy density of the real part and the imaginary part are respectively determined by the formulas proposed in the literature [1].

$$\operatorname{Re} \mu^k = \frac{\operatorname{Re} n_{jj}^{(k)}}{\sum_{i=1}^{M} \operatorname{Re} n_{ij}^{(k)}}, j < M + 1 \tag{4}$$

Here $Re\, n_{ij}^{(k)}$ represents output sample belonging to class i is judged as class j by sub classifier. $Im\, n_{ij}^{(k)}$ represents output sample belonging to non class i is judged as class j by sub classifier.

We use the above new Sugeno fuzzy complex set-valued integral as the fusion operator. The specific steps are following:

Step 1: According to Eqs. (4) and (3), computing μ^i and λ for each classifier, then computing fuzzy complex set-valued measure value according to Eq. (2).
Step 2: Computer integral value according to Eq. (1).

$$e_i(s) = (Re(e_i(s)), Im(e_i(s)))\ (i = 1, 2, \ldots, n), \text{here}$$

$$Re(e_i(s)) = \bigvee_{i=1}^{n} \left(Rea_i \wedge Re\,\mu\{x|Re\bar{f}(x_i) \geq Rea_i\}\right)$$

$$Im(e_i(s)) = \bigwedge_{i=1}^{n} \left(Ima_i \vee Im\mu\{x|Im\bar{f}(x_i) \leq Ima_i\}\right)$$

Step 3: Determining the ideal resolve $e^+(s) = \left(\max_i Re(e_i(s)), \min_i Im(e_i(s))\right)$.

Step 4: Computing the closeness of the Hamming $N(e^+(s), e_i(s))$,

$$N(e^+(s), e_i(s)) = 1 - \frac{1}{2}(|Re(e_i(s)) - Re(e^+(s))| + |Im(e_i(s)) - Im(e^+(s))|).$$

Step 5: Using $N(e^+(s), e_i(s))$ to obtained results, Classifying the sample X as the closest class.

Definition 3. Supposing there is c-classification, and $X = \{x_1, x_2, \cdots, x_n\}$ represent attribute set, and $f = (f(x_1), f(x_2), \cdots, f(x_n))$ is real function in X space, so each component of positive evaluation with each attribute is [0, 1], negative evaluation is $1 - f = (1 - f(x_1), 1 - f(x_2), \cdots, 1 - f(x_n))$. Let $a_i = \int_X f d\mu_i,\ b_i = \int_X (1 - f) d\mu_i$. Here μ_i is fuzzy complex set-valued measure value of class i. The output of integral classifier is two-dimension vector $\varphi^i(f) = ((a_1, b_1), (a_2, b_2), \cdots, (a_c, b_c))$. $k = 1, 2, \cdots, c$, (a_k, b_k) represent evaluation degree of positive and negative membership evaluation. Let $d_k = a_k - b_k,\ k = 1, 2, \cdots, c$, we can obtain the final classification results according to the principle of maximum membership degree.

If we regard two dimension vector as complex vector, the following definition can be obtained:

Definition 4. Supposing there is c-classification, and $X = \{x_1, x_2, \cdots, x_n\}$ represent attribute set, and $f = (f(x_1), f(x_2), \cdots, f(x_n))$ is real function in X space, and each component of positive evaluation with each attribute is [0,1], so the output of integral classifier is a group complex vector $\varphi^i(f) = ((a_1 + ib_1), (a_2 + ib_2), \cdots, (a_c + ib_c))$, here a_k, b_k. $k = 1, 2, \cdots, c$, $a_i \int_X f d\mu_i$, $b_i = \int_X (1 - f) d\mu_i$. The definition regard the output of attribute as complex function. Real part represent the positive evaluation for the attribute. Imaginary part represent the negative evaluation for the attribute. Then, using complex fuzzy set-value complex fuzzy integral of complex function can obtain membership evaluation of each class for sample. The obtained value is also complex. Real part is on behalf of positive membership evaluation. Imaginary part is on behalf of negative membership evaluation.

The above defined classification model which is still based on the integral, but it is the improvement of traditional integral classification. Because it join the negative membership evaluation, this can be entire describe the classification.

For example, the Table 1 is two class problem of four schemes.

Table 1. Two class problem of four schemes.

	Attribute x_1	Attribute x_2	Attribute x_3	Class
1	0.8	0.45	1	?
2	1	1	0.75	?
3	0.33	0.45	0.44	?
4	0.47	0.59	0.5	?

We can regard attribute valve as two-dimensional vector or plural which are shown in Table 2.

Table 2. Attribute value

	Attribute x_1	Attribute x_2	Attribute x_3	Class
1	0.8 + 0.2i	0.45 + 0.55i	1 + 0i	?
2	1 + 0i	1 + 0i	0.75 + 0.25i	?
3	0.33 + 0.67i	0.45 + 0.55i	0.44 + 0.56i	?
4	0.47 + 0.53i	0.59 + 0.41i	0.5 + 0.5i	?

In this article, fuzzy measure is shown in Table 3.

Table 3. Fuzzy measure

Set	Fuzzy measure μ_1	Fuzzy measure μ_2
$\emptyset$	0	0
$\{x_1\}$	0.1	0.2
$\{x_2\}$	0.5	0.3
$\{x_1, x_2\}$	0.7	0.8
$\{x_3\}$	0.2	0.6
$\{x_1, x_3\}$	0.3	0.9
$\{x_2, x_3\}$	0.9	0.7
$\{x_1, x_2, x_3\}$	1	1

Two treatment methods of membership evaluation will be introduced:

Because using $d_k = a_k - b_k$ describe membership evaluation which is rough, the following two methods can be adopted to improve the above evaluation method.

The first method:

Using optimistic and pessimistic factor, d_k can be expressed by the linear function of a_k and b_k. Let $d_k = \alpha a_k - (1 - \alpha)b_k$.

The second method:

Using positive and negative describe the evaluation. The specific steps are following.

(1) Establishing the fuzzy set of positive and negative membership evaluation;
(2) Establishing the fuzzy rules base of positive - negative and ultimate membership evaluation.
(3) Inputting the membership evaluation value of positive –negative, and using Matlab fuzzy logic tool to solve the problem.

The language variant fuzzy set of "high" "middle" "low" of positive and negative membership evaluation can be described:

If the positive membership evaluation is "high", let S fuzzy function is pos.high <0.5, 1>.

If the positive membership evaluation is "middle", let triangle fuzzy function is pos. mid<0, 0.5, 1>.

If the positive membership evaluation is "low", let opposite S fuzzy function is pos. low<0, 0.5>,

If the negative membership evaluation is "high", let S fuzzy function is, neg. high<0.5, 1>.

If the negative membership evaluation is "middle", let triangle fuzzy function is neg.mid<0, 0.5, 1>.

If the negative membership evaluation is "low", let opposite S fuzzy function is neg.low<0, 0.5>.

The language variant fuzzy set of "high" "little higher" "middle" "litter lower" "low" of composite membership evaluation can be described:

If composite membership evaluation is "high", let S fuzzy function is com. high<0.75, 1>.

If composite membership evaluation is "little higher", let triangle fuzzy function is com.lithigh<0.5, 0.75, 1>.

If composite membership evaluation is "middle", let triangle fuzzy function is com. mid<0.25, 0.5, 0.75>.

If composite membership evaluation is "litter lower", let triangle fuzzy function is com.litlow<0, 0.25, 0.5>.

If composite membership evaluation is "low", let opposite S fuzzy function is com. low<0, 0.25>.

According to experience of expert knowledge and estimator, we can give a few of fuzzy reasoning rules with double input and single output: positive membership evaluation: A, negative membership evaluation: B, comprehensive evaluation: C

(1) If A = "high",B = "low", then C = "high".
(2) If A = "high", B = "high", then C = "middle".
(3) If A = "high", B = "middle", then C="little higher".
(4) If A = "middle", B = "low", then C = "little higher".
(5) If A = "middle", B = "middle", then C = "middle".

(6) If A = “middle”, B = “high”, then C = “little lower”.
(7) If A = “low”, B = “low”, then C = “middle”.
(8) If A = “low”, B = “middle”, then C = “litter lower”.
(9) If A = “low”, B = “high”, then C = “low”.

According to the above nine rules and CRI algorithm of Mamdani to solve the fuzzy algorithm, the results of evaluation are shown in Table 4.

Table 4. The results of evaluation

	Class 1			Class 2		
	Positive	Negative	Synthesis	Positive	Negative	Synthesis
1	0.60	0.31	0.608	0.89	0.26	0.705
2	0.93	0.05	0.796	0.95	0.15	0.766
3	0.44	0.56	0.457	0.41	0.58	0.439
4	0.54	0.44	0.541	0.52	0.50	0.514

Mamdani implication operator is chosen and t-norm is minimum, then evaluations of each scheme can be obtained using Matlab tool. We know that the least of evaluation value, the highest of ranking. At last, the four schemes belong to class 2,1,1,1 respectively using improved fuzzy reasoning ultimate membership evaluation.

5 Conclusion

A lot of objective reality problem drives us to establish complex fuzzy set-valuedd function integral theory based on the fuzzy complex set-valuedd measure in term of mathematics. And, studying complex fuzzy set-valuedd integral in the application of classification, search engines, fuzzy system, earthquake system has important value. In this paper, using Sugeno complex fuzzy set-valued integral tool and bi-directional membership evaluation and fuzzy reasoning method to classify can achieve the classification effect.

Acknowledgements. This work was supported by the International Science and Technology Cooperation Program of China (No. **2012DFA11270**).

References

1. Yao, M.H., He, T.N.: Dynamic combination method of multiple classifiers. J. Zhejiang Univ. Technol. **30**(2), 156–159 (2002)
2. Wang, X.X.: Fuzzy measure and fuzzy integration and its application in classification technology. Science Press, Beijing (2008)
3. Zhang, G.: The convergence for a sequence of fuzzy integrals of fuzzy number-valued function on the fuzzy set. Fuzzy Sets Syst. **59**, 43–57 (1993)

4. Ma, S., Chen, F., Wang, Q.: Fuzzy complex-valued integral and its convergence. Fuzzy Eng. Oper. Res., AISC **147**, 265–273 (2012)
5. Zhao, Z., Ma, S.: Fuzzy complex-valued fuzzy measure base on fuzzy complex sets. Fuzzy Eng. Oper. Res., AISC **147**, 207–212 (2012)
6. Ma, S., Wang, Q.: The characters of the complex number-valued fuzzy measurable function. Adv. Intell. Soft Comput. **78**, 49–54 (2010)
7. Ma, S.Q., Zhao, H., Li, S.G.: The measured function and its property of complex fuzzy set-valued complex fuzzy measure space. Fuzzy Syst. Math. **28**(6), 73–79 (2014)
8. Ma, S.Q., Li, S.G., Zhao, H., Zhou, M.: Complex fuzzy set-valued complex fuzzy integrals and their properties. J. LanZhou Univ. (Nat. Sci.) **51**(1), 109–114 (2015)
9. Ma, S., Chen, F., Wang, Q., Zhao, Z.: Sugeno type fuzzy complex-value integral and its application in classification. Proc. Eng. **29**, 4140–4151 (2012)
10. Song, C., Ma, S.: An information fusion algorithm based on Sugeno fuzzy complex-valued integral. J. Comput. Inf. Syst. **7**, 2166–2171 (2011)
11. Ma, S., Chen, F., Wang, Q., Zhao, Z.: Sugeno type fuzzy complex-value integral and its application in classification. Proc. Eng. **29**, 4140–4151 (2012)

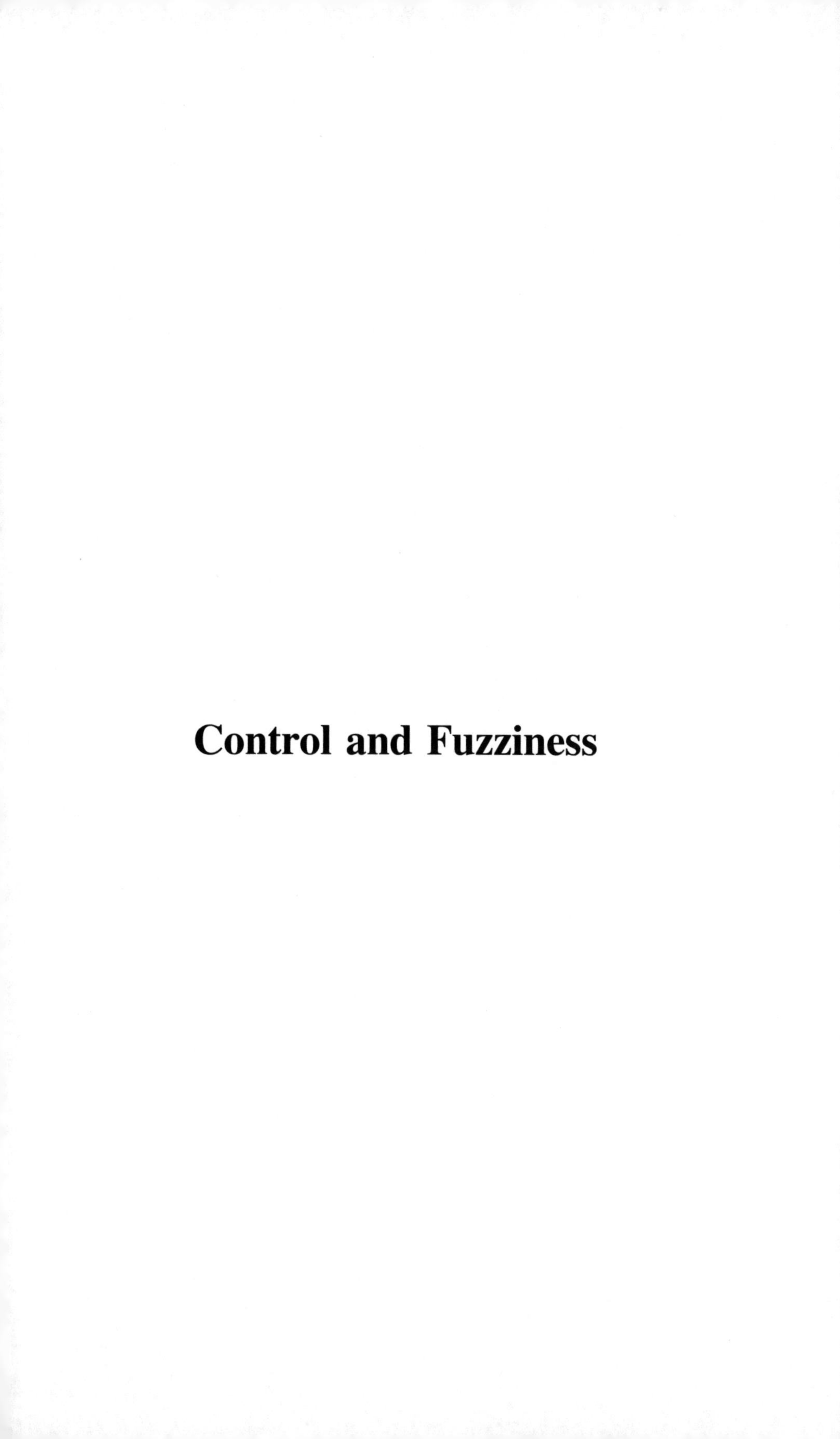

Control and Fuzziness

Some Results on Conditionally Subcancellative Triangular Subnorms

Gang Li[1(✉)], Zhenbo Li[2], and Hua-Wen Liu[3]

[1] Faculty of Science, Qilu University of Technology, Ji'nan 250353, China
sduligang@163.com
[2] School of Statistics, Shandong University of Finance and Economics, Ji'nan 250100, China
lizhenbo@126.com
[3] School of Mathematics, Shandong University, Ji'nan 250100, China
hw.liu@sdu.edu.cn

Abstract. Subcancellative triangular subnorms have been taken as aggregation functions and applied in multicriteria decision making. In this paper, the class of conditionally subcancellative triangular subnorms is introduced and its full characterization is given through a functional equation. Furthermore, we give the sufficient and necessary conditions under which such triangular subnorms become triangular norms.

Keywords: Triangular norms · Triangular subnorms
Conditional subcancellativity · Fuzzy negation

1 Introduction

The (left continuous) triangular norms have an important role in many domains such as fuzzy logic, fuzzy control, non-additive measures, fuzzy integrals and others. More details about triangular norms and their applications can be found in [1,6]. It is well known that continuous triangular norms were characterized as ordinal sums of Archimedean triangular norms. Furthermore, the basic stones for construction of triangular norms via the ordinal sum construction are triangular subnorms [5]. Therefore, triangular subnorms play an important role in the description of the structure of triangular norms. Many different classes of triangular subnorms have been discussed in [9,11]. Especially, there are some works [3,4,8,10] related with the class of conditionally cancellative triangular subnorms. In [3], Jayaram presented a complete characterization of such triangular subnorms M by a functional equation

$$I_M(x, M(x, y)) = \max(n_M(x), y) \tag{1}$$

for all $x, y \in [0, 1]$, where I_M is the residual implication induced from M which is given by, for all $x, y \in [0, 1]$,

$$I_M(x, y) = \sup\{t \in [0, 1] : M(x, t) \leq y\}, \tag{2}$$

B.-Y. Cao and Y.-B. Zhong (Eds.): ICFIE 2017, AISC 872, pp. 191–197, 2019.
https://doi.org/10.1007/978-3-030-02777-3_17

and n_M is the natural negation of M (see Definition 3 in Sect. 2).

Moreover, some other kinds of cancellativity about triangular subnorms was investigated in [8]. In this paper, we focus on the (conditionally) subcancellativity triangular subnorms [7]. A full characterization of this class of triangular subnorms is obtained through the functional Eq. (1).

2 Preliminaries

In this section, we summarize some of the essential results about triangular subnorms.

Definition 1 [6]. *A fuzzy negation is a function $N : [0,1] \to [0,1]$ that is non-increasing and such that $N(1) = 0, N(0) = 1$.*

Definition 2 [6]. *A triangular subnorm(t-subnorm for short) is a commutative, associative, non-decreasing function $M : [0,1]^2 \to [0,1]$ such that $M(x,y) \leq \min(x,y)$ for all $x, y \in [0,1]$. Furthermore, if 1 is the neutral element of M, then M is called a triangular norm (t-norm for short). To emphasize that a t-subnorm is not a t-norm, we will call it a proper t-subnorm.*

The associativity of t-subnorm allows us to extend each t-subnorm in a natural way to an $n-$ary operation in the usual way by induction.

Definition 3 [6]. *Let M be a t-subnorm.*

(i) M is said to satisfying the Conditional Cancellative Law if, for any $x, y, z \in [0,1]$,

$$M(x,y) = M(x,z) > 0 \Rightarrow y = z \tag{3}$$

Moreover, M is said to satisfying the Cancellative Law if, for any $x, y, z \in [0,1]$,

$$M(x,y) = M(x,z) \Rightarrow x = 0 \quad or \quad y = z \tag{4}$$

(ii) M is called subcancellative if $M(x,y) = M(x,z)$ implies that $y = z$, for every $x, y, z \in [0,1]$ s.t. $y, z \leq x$. Moreover, M is called conditionally subcancellative if $M(x,y) = M(x,z) > 0$ implies that $y = z$, for every $x, y, z \in [0,1]$ s.t. $y, z \leq x$.

(iii) M is said to be Archimedean, if for all x, y in open interval $]0,1[$ there exists a natural number $n \in \mathcal{N}$ such that $x_M^{(n)} < y$, where $x_M^{(1)} = x, x_M^{(n)} = M(\underbrace{x, ..., x}_{n})$ for $n > 1$.

(iv) An element $x \in]0,1[$ is a nilpotent element of M if there exists a natural number $n \in \mathcal{N}$ such that $x_M^{(n)} = 0$.

(v) A t-norm M is said to be nilpotent, if it is continuous and if each $x \in]0,1[$ is a nilpotent element.

(vi) The natural negation n_M of M is given by, for all $x \in [0,1]$,

$$n_M(x) = \sup\{t \in [0,1] : M(x,t) = 0\}. \tag{5}$$

Clearly, n_M is non-increasing and $n_M(0) = 1$, but need not to be fuzzy negation.

For the natural negation n_M, the following result holds.

Lemma 1 [2]. *Let M be any t-subnorm and n_M its natural negation. Then we have the following:*

(i) $M(x,y) = 0 \Rightarrow y \leq n_M(x)$.
(ii) $y < n_M(x) \Rightarrow M(x,y) = 0$.
(iii) If M is left continuous then $y = n_M(x)$.

In [3], a complete characterization of conditionally cancellative t-subnorm was shown.

Theorem 1 [3]. *Let M be any t-subnorm, not necessarily left continuous. Then the following are equivalent:*

(i) $I_M(x, M(x,y)) = \max(n_M(x), y)$ *for all* $x, y \in [0,1]$.
(ii) M is a conditionally cancellative t-subnorm.

More results on the conditionally cancellative t-subnorms can be founded in [3,4].

3 The Main Results

In this section, we focus on the class of (conditionally) subcancellative t-subnorms.

Lemma 2. *Let M be any t-subnorm. Then the following are equivalent:*

(i) M is a conditionally subcancellative t-subnorm.
(ii) If $M(x,y) = M(x,z) > 0, x,y,z \in [0,1], y,z < x$ *then* $y = z$.

Proof: $(i) \Rightarrow (ii)$: It is obvious.

$(ii) \Rightarrow (i)$ We only need to prove that $M(x,x) = M(x,z) > 0, x,z \in [0,1], z \leq x$ then $x = z$. On the contrary, suppose that $z < x$ and $M(x,x) = M(x,z) > 0$. Hence, $M(x,z) = M(x,t) > 0$ for all $t \in [z,x]$, a contradiction. □

Lemma 3. *Let M be any t-subnorm. Then the following statements hold:*

(i) If M is subcancellative then M is conditionally subcancellative.
(ii) If M is conditionally cancellative then M is conditionally subcancellative.
(iii) If M is Archimedean then M is cancellative if and only if M is subcancellative.

Proof: (i) and (ii) are obvious. The result in (iii) can be obtained from Proposition 6 in [8]. □

Remark 1: If M is Archimedean and conditionally subcancellative then M may be not conditionally cancellative. One example is presented here.

Example 1 [8]**:** Let $M : [0,1]^2 \to [0,1]$ be defined by

$$M(x,y) = \begin{cases} x+y-\frac{3}{2} & (x,y) \in]\frac{3}{4},\frac{7}{8}]^2 \cup]\frac{7}{8},1]^2, \\ \min(x,y)-\frac{5}{8} & \min(x,y) \in]\frac{3}{4},\frac{7}{8}] \text{ and } \max(x,y) \in]\frac{7}{8},1] \\ 0 & \text{otherwise} \end{cases} \quad (6)$$

It is obvious that M is an Archimedean and conditionally subcancellative t-subnorm. But, M is not conditionally cancellative. For example, $M(\frac{13}{16},1) = M(\frac{13}{16},\frac{15}{16}) = \frac{3}{16} > 0$.

Theorem 2. *Let M be any t-subnorm. Then the following are equivalent:*

(i) $I_M(x, M(x,y)) = \max(n_M(x), y)$ for all $x, y \in [0,1]$, s.t. $y < x$.
(ii) M is a conditionally subcancellative t-subnorm.

Proof: Let M be any t-subnorm. If M satisfies the equation: $I_M(x, M(x,y)) = \max(n_M(x), y)$ for all $x, y \in [0,1], y < x$. On the contrary, suppose that there exist $x, y, z \in [0,1], y, z \leq x$ such that $M(x,y) = M(x,z) > 0$ but $y < z$. Then we have that

$$I_M(x, M(x,y)) = I_M(x, M(x,z)) \geq z > y.$$

Since $M(x,y) > 0$, $y \geq n_M(x)$ and $\max(n_M(x), y) = y$, a contradiction. Hence, M is conditionally subcancellative.

Conversely, assume that M is conditionally subcancellative. Consider the arbitrary $x, y \in [0,1], y < x$. We divide the two cases in the following proof.

- $n_M(x) > y$. Then $M(x,y) = 0$ and $I_M(x, M(x,y)) = I_M(x,0) = n_M(x) = \max(n_M(x), y)$.
- $n_M(x) \leq y$. On one hand, if $M(x,y) = 0$ then $I_M(x, M(x,y)) = I_M(x,0) = n_M(x)$. Moreover, due to $M(x,y) = 0$, $y \leq n_M(x)$ and $\max(n_M(x), y) = n_M(x)$. On the other hand, if $M(x,y) > 0$ then $y \geq n_M(x)$ and $\max(n_M(x), y) = y$. On the contrary, suppose that $I_M(x, M(x,y)) = z > y$. Then $M(x,y) = M(x,w)$ for all $w \in [y, z[$. So, there exists $y_0 \in]y, \min(z,x)[$ such that $M(x,y) = M(x,y_0) > 0$, which is a contradiction with the fact that M is conditionally subcancellative. □

Remark 2: In (i) of Theorem 2, the restricted condition $y < x$ can not be replaced with $y \leq x$. One example is presented here.

Example 2: Let $M : [0,1]^2 \to [0,1]$ be defined by

$$M(x,y) = \begin{cases} 0 & x+y \leq 1, \\ \min(x,y) & \text{otherwise} \end{cases} \tag{7}$$

It is not difficult to show that M is a t-subnorm. Indeed, it is the nilpotent minimum t-norm. It is obvious that M is conditionally subcancellative. Taking $x = 0.6$, we have $I_M(x, M(x,x)) = I_M(x,x) = 1$, but, $\max(n_M(x), x) = 0.6$.

Lemma 4. *Let M be a conditionally subcancellative t-subnorm satisfying $M(1,1) = 1$. Then*

(i) If $M(1, y_0) = y_0$ for some $y_0 \in]0,1[$ then $M(1,y) = y$ for all $y \in [y_0, 1]$.
(ii) $M(1,y) = y$ for all $y \in]y^*, 1]$, where $y^* = \sup\{t : M(1,t) = 0\} = n_M(1)$.

Proof: (i) Let $y \in]y_0, 1[$. By the monotonicity and conditional subcancellativity of M, we have

$$0 < y_0 = M(1, y_0) < M(1,y) \leq y.$$

If $M(1,y) = y' < y$ then by the conditional subcancellativity of M, we have

$$M(1, M(1,y) = M(1,y') < M(1,y).$$

However, $M(M(1,1), y) = M(1,y)$, which a contradiction with the associativity of M. Hence, $M(1,y) = y$ for all $y \in [y_0, 1]$.

(ii) Let $y \in]y^*, 1]$. Clearly, $M(1,y) > 0$ and $M(1,y) \leq y$. The proof is similar to (i). □

Theorem 3. *Let M be any conditionally subcancellative t-subnorm. Then the following are equivalent:*

(i) M is a t-norm.
(ii) n_M is a fuzzy negation and $M(1,1) = 1$.

Proof: If n_M is a fuzzy negation then $y^* = 0$ in Lemma 4. The results holds. □

Remark 3: In (ii) of Theorem 3, the condition $M(1,1) = 1$ can not be omitted. For example, the t-subnorm $M(x,y) = \frac{1}{2}x \cdot y$ for $(x,y) \in [0,1]^2$ which is conditionally subcancellative and n_M is fuzzy negation. But M is not a t-norm.

Lemma 5. *Let M be any Archimedean and conditionally subcancellative t-subnorm, which is not subcancellative. Then every $x \in]0,1[$ is a nilpotent element of M.*

Proof: Note that there at least exists $x \in]0,1]$ such that $n_M(x) > 0$. On the contrary, suppose that $n_M(x) = 0$ for all $x \in]0,1]$ then it is obvious that $M(x,y) > 0$ for all $(x,y) \in]0,1]^2$. Hence, M is subcancellative, a contradiction.

So, there exist $x_0 \in]0,1]$ such that $n_M(x_0) > 0$. By Lemma 1, we have

$$M(x_0, x_1) = 0$$

for any $x_1 \in]0, n_M(x_0)[$. Due to the Archimedeanness of M, for any $x \in]0,1[$, there exist natural numbers $k, j \in \mathcal{N}$ such that $x_M^{(k)} < x_0$ and $x_M^{(j)} < x_1$. Hence,

$$x_M^{(k+j)} = M(x_M^{(k)}, x_M^{(j)}) \leq M(x_0, x_1) = 0,$$

i.e., x is a nilpotent element of M. □

Theorem 4. *Let M be any continuous, Archimedean and conditionally subcancellative t-subnorm. Then the following are equivalent:*

(i) M is a nilpotent t-norm or a cancellative proper t-subnorm.
(ii) n_M is a fuzzy negation.

Proof: $(i) \Rightarrow (ii)$: If M is a nilpotent t-norm then n_M is obviously a fuzzy negation. If M is a subcancellative t-subnorm then M is cancellative by Lemma 3. Hence, n_M is a fuzzy negation.

$(ii) \Rightarrow (i)$: Assume that n_M is a fuzzy negation. Hence, $M(1, x) > 0$ for all $x \in]0, 1]$.

- If $M(1,1) = 1$ then M is a t-norm by Theorem 3. Hence, M is nilpotent by Lemma 5.
- If $M(1,1) < 1$ then M is a proper t-subnorm. First, we prove that $M(1,x) < x$ for all $x \in]0,1]$. On the contrary, suppose that there exists $x_0 \in]0,1[$ such that $M(1, x_0) = x_0$. Since $M(1,1) < 1$, the limit of the sequence $(1_M^{(n)})_{n=1}^{\infty}$ equals to 0 by the Archimedean property of M. So, $x_0 = M(x_0, 1) = M(M(x_0, 1), 1) = M(x_0, 1_M^{(2)})$. By induction, we have $x_0 = M(x_0, 1_M^{(n)})$ for arbitrary natural number. Then $x_0 = 0$, a contradiction.

  Assume that M is conditionally subcancellativenot but not subcancellative. By the continuity of M, there exists the largest $x_0 \in]0,1[$ such that $M(x_0, y_0) = 0$ for some fixed $y_0 \in]0,1[$. By Theorem 2, for all $x \in [0,1[$, we have

  $$I_M(1, M(1,x)) = \max\{n_M(1), x\} = x$$

  i.e., $I_M(1, M(1,x)) = \sup\{t : M(1,t) \leq M(1,x)\} = x$. The partial function $M(1,\cdot) : [0,1] \to [0,1]$ is continuous and strictly increasing by the continuity of M. So, there exists $x_1 \in]0,1[$ such that $M(1, x_1) < x_0 < x_1$. We have

  $$M(M(1, x_1), y_0) = 0$$

  and

  $$M(1, M(x_1, y_0)) > 0,$$

  a contradiction with the associativity of M. Hence, M is a subcancellative proper t-subnorm. The result holds by Lemma 3. □

Remark 4: In [9], it is proved that each continuous, Archimedean, cancellative t-subnorm M can be represented in the form

$$M(x, y) = t^{-1}(t(x) + t(y))$$

where $t : [0, 1] \to [0, \infty]$ is a strictly decreasing continuous unbounded function.

4 Conclusion

In this paper, the class of conditionally subcancellative triangular subnorms has been full characterized through a functional equation. Furthermore, we have presented the sufficient and necessary conditions under which such triangular subnorms become triangular norms.

Acknowledgements. Thanks to the support by National Natural Science Foundation of China (No. 61403220 and No. 61573211).

References

1. Alsina, C., Frank, M.J., Schweizer, B.: Associative Functions. Triangular Norms and Copulas. World Scientific, New Jersey (2006)
2. Baczyński, M., Jayaram, B.: Fuzzy Implications. Studies in Fuzziness and Soft Computing, vol. 231. Springer, Berlin (2008)
3. Jayaram, B.: Solution to an open problem: a characterization of conditionally cancellative t-subnorms. Aequationes Mathematicae **84**, 235–244 (2012)
4. Jayaram, B.: T-subnorms with strong associated negations: some properties, Manucript in preparation
5. Jenei, S.: A note on the ordinal sum theorem and tis consequence for the construction of triangular norms. Fuzzy Sets Syst. **126**, 199–205 (2002)
6. Klement, E.P., Mesiar, R., Pap, E.: Triangular Norms. Kluwer Academic Publishers, Dordrecht (2000)
7. Kyselova, D.: Aggregation operators-based multicriteria decision making, Ph.D. thesis, STU, Bratislava (2007)
8. Maes, K.C., Mesiarová, A.: Cancellative prosperties for t-norms and t-subnorms. Inf. Sci. **179**, 1221–1233 (2009)
9. Mesiarová, A.: Continuous triangular subnorms. Fuzzy Sets Syst. **142**, 75–83 (2004)
10. Mesiarová, A.: Continuous additive generators of continuous, conditionally cancellative triangular subnorms. Inf. Sci. **339**, 53–63 (2016)
11. Ricci, R.G., Mesiar, R., Mesiarová, A.: Lipschitz continuity of triangular subnorms. Fuzzy Sets Syst. **240**, 51–65 (2014)

(a, b)-Roman Domination on Cacti

Yancai Zhao[1(✉)], H. Abdollahzadeh Ahangar[2], Zuhua Liao[3], and M. Chellali[4]

[1] Department of Basic Science, Wuxi City College of Vocational Technology, Jiangsu 214153, China
zhaoyc69@126.com

[2] Department of Mathematics, Babol Noshirvani University of Technology, Babol, I.R. of Iran
ha.ahangar@nit.ac.ir

[3] School of Science, Jiangnan University, Jiangsu 214122, China
liaozuhua57@163.com

[4] LAMDA-RO Laboratory, Department of Mathematics, University of Blida, B.P. 270, Blida, Algeria
m_chellali@yahoo.com

Abstract. Given two real numbers $b \geq a > 0$, an (a, b)-Roman dominating function on a graph $G = (V, E)$ is a function $f : V \rightarrow \{0, a, b\}$ satisfying the condition that every vertex v for which $f(v) = 0$ is adjacent to a vertex u for which $f(u) = b$. In the present paper, we design a linear-time algorithm to produce a minimum (a, b)-Roman dominating function for cacti, a superclass of trees and different from chordal graphs.

Keywords: Algorithm · Domination · Roman domination · Cacti

1 Introduction

Let $G = (V, E)$ be a simple graph. For a function $f\colon V \rightarrow \Re$, the *weight* of f is $w(f) = \sum_{v \in V} f(v)$. For $S \subseteq V$, let $f(S) = \sum_{v \in S} f(v)$, and so $w(f) = f(V)$. In [4], Cockayne, Dreyer, Hedetniemi, and Hedetniemi defined a *Roman dominating function* on G to be a function $f\colon V \rightarrow \{0, 1, 2\}$ satisfying the condition that every vertex v for which $f(v) = 0$ is adjacent to a vertex u for which $f(u) = 2$. The *Roman domination number*, denoted by $\gamma_R(G)$, is the minimum weight among all Roman dominating functions of G.

A generalization of Roman domination has been given by Liu and Chang in [11] as follows. For real numbers $b \geq a > 0$, an (a, b)-*Roman dominating function* on G is a function $f\colon V \rightarrow \{0, a, b\}$ satisfying the condition that every vertex v for which $f(v) = 0$ is adjacent to a vertex u for which $f(u) = b$. For $b = a = 1$, this is a dominating function; and for $b = 2$ and $a = 1$, this is a Roman dominating function. The (a, b)-*Roman domination number* of G is the minimum weight among all (a, b)-Roman dominating functions of G. It is worth mentioning that Liedloff et al. [10] showed that the Roman domination problem can be solved for interval graphs, cograph, AT-free graphs and graphs

B.-Y. Cao and Y.-B. Zhong (Eds.): ICFIE 2017, AISC 872, pp. 198–206, 2019.
https://doi.org/10.1007/978-3-030-02777-3_18

with a d-octopus. Moreover, for any fixed (a, b), the (a, b)-Roman domination problem has been shown in [11] to be NP-complete for bipartite and chordal graphs. Using the framework of linear programming and the strong elimination ordering as a tool, Liu and Chang [11] provided a linear-time algorithm for (a, b)-Roman domination problem with $2a \geq b \geq a > 0$ on strongly chordal graphs. For more results on Roman domination and its variations, see list of References [1–3,5,8,9].

In this paper, we continue the study on the algorithmic aspects of (a, b)-Roman domination. We design a linear-time algorithm to solve the (a, b)-Roman domination problem on cacti, by using a series of algorithms and procedures to deal with different graph structures in a cactus. Cacti is known to be a superclass of trees and different from strongly chordal graphs.

2 Definitions

For notation and graph theory terminology, we in general follow [6]. Let G be a graph with vertex set $V(G) = V$ of order $|V| = n(G)$ and size $|E(G)| = m(G)$. The *(open) neighborhood* $N(v)$ of a vertex v consists of the vertices adjacent to v and the *closed neighborhood* of v is $N[v] = N(v) \cup \{v\}$. The *degree* of a vertex $v \in V$ is $d_G(v) = |N(v)|$. If no ambiguity occurs, we simply write n, m, $N(v)$ and $d(v)$ in stead of $n(G)$,$m(G)$, $N_G(v)$ and $d_G(v)$, respectively. The *subgraph* of $G = (V, E)$ induced by $S \subseteq V$ is the graph $G[S]$ with vertex set S and edge set $\{uv \in E | u, v \in S\}$. A *clique* is a maximal set of vertices which induces a complete subgraph. The *deletion* of a subset S of vertices from G, denoted by $G - S$, is the graph $G[V \setminus S]$, and the deletion of a subset F of edges from G, denoted by $G - F$, is the spanning graph $(V, E \setminus F)$. For an element ε in G, we write $G - \varepsilon$ instead of $G - \{\varepsilon\}$.

The *length* of a path is the number of edges in the path. The *distance* $d(u, v)$ between two vertices u and v in G is the length of a shortest path from u to v. A *forest* is a graph without cycle. A *tree* is a connected forest. An *end vertex* or a *leaf* in a graph is a vertex with degree one. An *end edge* in a graph is an edge incident with a leaf. A vertex x is a *cut vertex* if deleting x increases the number of connected components. A *block* of G is a maximal connected subgraph without a cut vertex. If G has no cut vertex, G itself is a block. The intersection of two blocks contains at most one vertex and a vertex is a cut vertex if and only if it is the intersection of two or more blocks. A block B of G is called an *end block* if B contains at most one cut vertex of G. A *cactus* is a connected graph whose every block is either an edge or a cycle. Alternatively, a cactus is a connected graph in which two cycles have at most one vertex (cut vertex) in common. A cactus graph is a tree if all the blocks are edges. When an end block of G is a cycle, we call it an *end cycle*.

3 Linear-Time Algorithm for Cacti

To obtain an algorithm for cacti, we should first design algorithms or procedures to deal with different graph structures in a cactus.

3.1 Algorithms for Trees and Cycles

Given a tree T, we first root T at any vertex, say r. Then the *height* of T is the maximum distance from a vertex to r. Suppose the height of T is k. Let L_i $(0 \leq i \leq k)$ be the set of vertices of T which are at distance i from the root. We first number r with 1, then number in arbitrary order the vertices of L_1 with $2, 3, \ldots, |L_1| + 1$, and so on, till the vertices of L_k are numbered. We call this order *vertex array* of T, denoted $[1, 2, \ldots, n]$.

To find a minimum (a, b)-Roman dominating function of T, let T_i be the subtree with root v_i. We further introduce the following notes.

$\Gamma_i = \min\{w(f)|\ f$ is an (a,b)-Roman dominating function of $T_i\}$;
$\Gamma_i^0 = \min\{w(f)|\ f$ is an (a,b)-Roman dominating function of T_i and $f(v_i) = 0\}$;
$\Gamma_i^a = \min\{w(f)|\ f$ is an (a,b)-Roman dominating function of T_i and $f(v_i) = a\}$;
$\Gamma_i^b = \min\{w(f)|\ f$ is an (a,b)-Roman dominating function of T_i and $f(v_i) = b\}$.

The following lemma shows that the above four parameters can be computed recursively for T_i, and in particular, Γ_1 will be the (a, b) -Roman domination number of $T = T_1$.

Lemma 1. *Let T be a given rooted tree, and T_i be the subtree of T with root v_i, C_i be the set of children of v_i.*

If T_i consists of a single vertex, the root v_i, then $\Gamma_i = a, \Gamma_i^0 = \infty, \Gamma_i^a = a$, and $\Gamma_i^b = b$.

Otherwise, we have the following formulas.

$$\Gamma_i = \min\left\{\Gamma_i^0, \Gamma_i^a, \Gamma_i^b\right\} \tag{1}$$

$$\Gamma_i^0 = \min_{v_j \in C_i} \left\{\Gamma_j^b + \sum_{v_k \in C_i} \Gamma_k - \Gamma_j\right\} \tag{2}$$

$$\Gamma_i^a = a + \sum_{v_j \in C_i} \Gamma_j \tag{3}$$

$$\Gamma_i^b = b + \sum_{v_j \in C_i} \Gamma_j - a\left|\{v_j \in C_i|\ \Gamma_j^a = \Gamma_j\}\right| \tag{4}$$

Proof. If v_i is the only vertex in T_i, the correctness of the computed values of $\Gamma_i = a, \Gamma_i^0 = \infty, \Gamma_i^a = a$, and $\Gamma_i^b = b$ follows immediately. Assume T_i has more than one vertex.

The formula (1) is obvious. In a minimum (a,b)-Roman dominating function f of a subtree T_i, there are three possible values for the root v_i: $f(v_i) = 0$ or$f(v_i) = a$ or $f(v_i) = b$. If $f(v_i) = 0$, then there must be a child v_j of v_i such that $f(v_j) = b$, and thus the formula (2) is valid. If $f(v_i) = a$, then by the definition of (a,b)-Roman domination, the value a of v_i neither has any effect on nor has any requirement to the values of other vertices. Hence the formula (3).

Now consider the case that $f(v_i) = b$ and thus justify the formula (4) as follows. If $\Gamma_j^a = \Gamma_j$, suppose g is a Γ_j^a -function of T_j. Then the function f such that $f(v_i) = b$, $f(v_j) = 0$ and $f(x) = g(x)$ for each vertex $x \in V(T_j) \setminus \{v_j\}$ is clearly a Γ_i^b-function of T_i, which means that the value of v_j can decrease from a to 0. If $\Gamma_j^a > \Gamma_j = \Gamma_j^0$, suppose g is a Γ_j^0-function of T_j. Then the function f such that $f(v_i) = b$ and $f(x) = g(x)$ for each vertex $x \in V(T_i) \setminus \{v_i\}$ is clearly a Γ_i^b-function of T_i, which means that the value of v_j cannot decrease though v_j is adjacent to v_i for which $f(v_i) = b$. If $\Gamma_j^a > \Gamma_j = \Gamma_j^b$, suppose g is a Γ_j^b-function of T_j. Since $g(v_j) = b$, there must exist a child v_k of v_j such that $g(v_k) = 0$. This means that the value of v_j is determined by the value of its child v_k, and thus cannot decrease though v_j is adjacent to v_i for which $f(v_i) = b$.

Based on formulas (1)–(4), we have the following algorithm for the (a, b)-Roman domination problem in trees. The algorithm starts from the leaves of T and works inward. The status of a parent is updated when the statuses of its children are found.

Algorithm 1. RomDomTree (Determining the (a, b)-Roman domination number of a tree).
Input: a rooted tree T with its vertex array $[1, 2, \ldots, n]$.
Output: the (a, b)-Roman domination number of T.
for $i = n$ down to 1 **do**
  if there is only one vertex, the root v_i, or i is a leaf and $i \neq 1$ **then**
    $\Gamma_i^0 \leftarrow \infty; \Gamma_i^a \leftarrow a; \Gamma_i^b \leftarrow b; \Gamma_i \leftarrow a;$
  else
    $\Gamma_i^0 \leftarrow \min_{v_j \in C_i} \{\Gamma_j^b + \sum_{v_k \in C_i} \Gamma_k - \Gamma_j\};$
      {formula(2)};
    $\Gamma_i^a \leftarrow a + \sum_{v_j \in C_i} \Gamma_j;$
      {formula(3)};
    $\Gamma_i^b \leftarrow b + \sum_{v_j \in C_i} \Gamma_j - a \left|\{v_j \in C_i | \ \Gamma_j^a = \Gamma_j\}\right|;$
      {formula(4)};
    $\Gamma_i \leftarrow \min\{\Gamma_i^0, \Gamma_i^a, \Gamma_i^b\};$
      {formula(1)};
end.

The algorithm RomDomTree computes the four parameter $\Gamma_i, \Gamma_i^0, \Gamma_i^a$ and Γ_i^b for each vertex. To compute any parameter of a vertex v, the neighbors of v should be checked. So the total amount of time is $\sum_{v \in V} d(v) = O(m)$, where m is the edge number of the tree T. With a slight modification by using pointers, we cannot only find the (a, b)-Roman domination number of T but also the corresponding minimum (a, b)-Roman dominating function of T, which will be stated in detail by an example in the next section.

To obtain the desired algorithm for cacti, we give a slightly more general version of the (a, b)-Roman domination. If a vertex v with value 0 has a neighbor with value b, then we say that v is *satisfied*. Thus we use the following four labels $(0, N)$, $(0, S)$, a and b, which mean "valued 0 and not satisfied", "valued 0 and satisfied", "valued a" and "valued b", respectively. Given a graph $G = (V, E)$ with each vertex $v \in V$ having an arbitrary label $L(v) \in \{(0, N), (0, S), a, b\}$, an *optional* (a, b)*-Roman dominating function* of G is a function $f \colon V \rightarrow \{0, a, b\}$

satisfying conditions that (i) if $L(v) = b$ then $f(v) = b$, and (ii) if $L(v) = (0, N)$ then either $f(v) = 0$ and v is adjacent to a vertex u such that $f(u) = b$ (thus v is relabeled $(0, S)$) or $f(v) \neq 0$ (thus v is relabeled with its value). The minimum weight among all optional (a, b) -Roman dominating functions is called the *optional* (a, b)*-Roman domination number* of G.

It is easy to see that if the initial labels are all $(0, N)$, then the optional (a, b)-Roman domination problem turns to be the (a, b)-Roman domination problem. So an algorithm to find a minimum optional (a, b)-Roman dominating function suffices to find a minimum (a, b)-Roman dominating function.

The algorithm RomDomTree can be easily generalized to an algorithm, called OptRomDomTree, for producing a minimum optional (a, b)-Roman dominating function of a tree, which is stated as follows.

Algorithm 2. OptRomDomTree (producing a minimum optional (a, b)-Roman dominating function of a tree).
Input: a rooted tree T with its vertex array $[1, 2, \ldots, n]$, each vertex having a label $L(v) \in \{(0, N), (0, S), a, b\}$.
Output: a minimum optional (a, b)-Roman dominating function of T.
for $i = n$ down to 1 **do**
  if there is only one vertex, the root v_i, or v_i is a leaf and $i \neq 1$ **then**
    if $L(v_i) = (0, N)$ or a **then**
      $\Gamma_i^0 \leftarrow \infty; \Gamma_i^a \leftarrow a;\ \Gamma_i^b \leftarrow b;\ \Gamma_i \leftarrow a;$
    if $L(v_i) = (0, S)$ **then**
      $\Gamma_i^0 \leftarrow 0;\ \Gamma_i^a \leftarrow a;\ \Gamma_i^b \leftarrow b;\ \Gamma_i \leftarrow 0;$
    if $L(v_i) = b$ **then**
      $\Gamma_i^0 \leftarrow \infty; \Gamma_i^a \leftarrow \infty;\ \Gamma_i^b \leftarrow b;\ \Gamma_i \leftarrow b;$
  else
    if $L(v_i) \neq b$ **then**
      the same formulas as in algorithm RomDomTree;
    if $L(v_i) = b$ **then**
      $\Gamma_i^0 \leftarrow \infty;\ \Gamma_i^a \leftarrow \infty\ ;\ \Gamma_i^b \leftarrow b + \sum_{v_j \in C_i} \Gamma_j - a\left|\{v_j \in C_i |\ \Gamma_j^a = \Gamma_j\}\right|;$
  $\Gamma_i \leftarrow \min\{\Gamma_i^0, \Gamma_i^a, \Gamma_i^b\};$
end for
if $\Gamma_1 = \Gamma_1^b$ **then**
  $f =$ the function corresponding to Γ_1^b;
else
  if $\Gamma_1 = \Gamma_1^0$ **then**
    $f =$ the function corresponding to Γ_1^0;
  if $\Gamma_1 = \Gamma_1^a$ **then**
    $f =$ the function corresponding to Γ_1^a;

The purpose for OptRomDomTree to assign the root value b as possible is to minimize the weight of the whole cactus, which will be used in the procedure OptRomDomPendTree in the next section.

If a vertex has been valued (labeled) b, then its neighbors whose labels are $(0, N)$ or a can clearly be relabeled $(0, S)$ to minimize the weight of the whole graph. Thus we have the following procedure.

Release u.
let u be a vertex with $L(u) = b$;
for each vertex $v \in N[u]$ with $L(v) = (0, N)$ or a **do**
$L(v) \leftarrow (0, S)$;
end for

Based on algorithm OptRomDomTree, we have the following algorithm to find a minimum optional (a, b)-Roman dominating function of a cycle.

Algorithm 3. OptRomDomCYC (Finding a minimum optional (a, b)-Roman dominating function of a cycle).
Input: a cycle C with a $(0, S), (0, N), a, b$ assignment L.
Output: a minimum optional (a, b)-Roman dominating function of C.
if all vertices are labeled $(0, S)$ **then** stop;
else
  if there exists a vertex v such that $L(v) = b$ **then**
    Release v;
    OptRomDomTree$(C - v)$;
  else
    choose a vertex v such that $L(v) = (0, N)$ or a;
    $L(v) \leftarrow a$;
    f_v =a $\cup$ OptRomDomTree$(C - v)$;
    for each $v_i \in N[v]$ **do**
      $L(v_i) \leftarrow b$; Release v_i;
      $f_i = b \cup$ OptRomDomTree$(C - v_i)$;
    end for
  $f \leftarrow$ one of the functions f_v and f_i such that $w(f) = \min\{w(f_v), w(f_1), w(f_2), w(f_3)\}$;

The construction and correctness of OptRomDomCYC is straightforward, and the proof is omitted. For the time complexity, since OptRomDomTree is linear and OptRomDomCYC makes at most four calls to OptRomDomTree, it is clear that OptRomDomCYC is linear.

3.2 Algorithm on Cacti

To obtain an algorithm for finding a minimum (a, b)-Roman dominating function of a cactus, we also need the following procedures, to handle pending trees and end cycles in a cactus.

We call a tree T_x rooted at vertex x a *pending tree* on a cycle C, if $V(T_x) \cap V(C) = \{x\}$. We first give a procedure to handle a pending tree as follows.

Procedure. OptRomDomPendTree (G, C, T_x).
let T_x be a pending tree on cycle C with a vertex array $[1, 2, \ldots, n]$;
OptRomDomTree (T_x);

Obviously, OptRomDomPendTree (G, C, T_x) not only produces a minimum optional (a, b)-Roman dominating function of T_x, but also gives an optimal value to the root x (noting the method which assigning a value to the root in algorithm OptRomDomTree). We will use this procedure to deal with pending trees in a cactus.

For an end cycle in a cactus, we have the following procedure.

Procedure. OptRomDomEndCYC(G, C).
let C be an end cycle of G with a $(0, S), (0, N), a, b$ assignment L to C;
let x be the cut-vertex of G in C;
let u_1 and u_2 be the two neighbors of x in C;
$L(x) \leftarrow b$;
Release x in G;
let $f_2 = b \cup$ OptRomDomTree$(C - x)$;
set the labels back to their initial states;
$L(x) \leftarrow a$;
let $f_1 = a \cup$ OptRomDomTree$(C - x)$;
set the labels back to their initial states;
if $L(x) = (0, S)$, **then**
  let $f_S = 0 \cup$ OptRomDomTree$(C - x)$;
  let $m = \min\{w(f_2), w(f_1), w(f_S)\}$;
  if $w(f_2) = m$, **then** $f \leftarrow f_2$;
  else
    if $w(f_1) = m$, **then** $f \leftarrow f_1$;
    else, $f \leftarrow f_S$;
if $L(x) = (0, N)$ or a, **then**
  $L(u_1) \leftarrow b$; $L(x) \leftarrow (0, S)$;
  let $f_0 = 0 \cup$ OptRomDomTree$(C - x)$;
  set the labels back to their initial states;
  $L(u_2) \leftarrow b$; $L(x) \leftarrow (0, S)$;
  let $f_0' = 0 \cup$ OptRomDomTree$(C - x)$;
  let $m = \min\{w(f_2), w(f_1), w(f_0), w(f_0')$;
  if $w(f_2) = m$, **then** $f \leftarrow f_2$;
  else
    if $w(f_1) = m$, **then** $f \leftarrow f_1$;
    else, let f be one of f_0 and f_0' such that $w(f) = m$;
if $L(x) = b$, **then** $f \leftarrow f_2$;

The running time of procedure OptRomDomEndCYC(G, C) is clearly linear. The correctness of OptRomDomEndCYC(G, C) is followed by the fact that the function value of x is as large as possible among all minimum (a, b) -Roman dominating functions.

We are now ready to present our main algorithm, called RomDomCAC, to produce a minimum (a, b)-Roman dominating function of a cactus. Our algorithm takes algorithms OptRomDomTree, OptRomDomCYC and procedure OptRomDomEndCYC as subroutines.

Algorithm 4. RomDomCAC (Finding a minimum (a, b)-Roman dominating function of a cactus).
Input: a cactus K, with each vertex labeled $(0, N)$.
Output: a minimum (a, b)-Roman dominating function of K.
$K' \leftarrow K$;
if K' is a tree, **then** OptRomDomTree(K') and stop;
if K' is a cycle, **then** OptRomDomCYC(K') and stop;
else
  do while there exists a pending maximal induced subtree T_x
    OptRomDomPendTree (G, C, T_x);
    $K' \leftarrow K' - (V(T_x) \setminus \{x\})$;
  end while
  do while there is an end cycle C with a unique cut vertex x
    OptRomDomEndCYC (K', C);
    $K' \leftarrow K' - (V(C) \setminus \{x\})$
  end while

It is well-known that cacti can be recognized in linear-time [7]. In each step of algorithm RomDomCAC, it calls to one of OptRomDomTree, OptRomDomCYC, OptRomDomPendTree and RomDomEndCYC(G, C). Since each of these subroutine is linear, algorithm RomDomCAC is clearly linear.

Theorem 1. *Algorithm RomDomCAC produces a minimum (a, b)-Roman dominating function of a cactus in linear-time.*

4 An Example

We provide the following tree T for an example, which is simple but sufficient to show how the running result of Algorithm 1 corresponds to a minimum (a, b)-Roman dominating function of T.

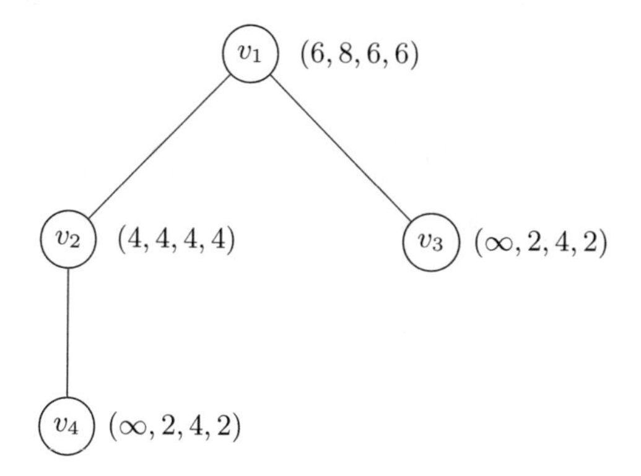

Fig. 1. A tree T.

Taking $a = 2$ and $b = 4$. The vector for each vertex v_i appearing in the figure is in the following notation $(\Gamma_i^0, \Gamma_i^2, \Gamma_i^4, \Gamma_i)$. Then by running Algorithm 1 on T, the vectors for the vertices of T are as shown in Fig. 1.

About how to get a function from these vectors, first notice that $\Gamma(T) = 6 = \Gamma_1^0 = \Gamma_1^4$. This means that the root v_1 can take an arbitrary value 0 or 4. Let $f(v_1) = 0$ for example. Notice formula (2) corresponding Γ_1^0. It holds for v_2. So $f(v_2) = 4$. Since $\Gamma_3^2 = \Gamma_3 = 2$, $f(v_3) = 2$. At last, v_4 is a leaf. Since $f(v_2) = 4$, we have $f(v_4) = 0$.

Acknowledgements. This paper is partially supported by the Natural Science Foundation of Jiangsu Province (No. BK20151117), and the second author was supported by the Babol Noshirvani University of Technology under research grant number BNUT/385001/97.

References

1. Abdollahzadeh Ahangar, H., Henning, M.A., Löwenstein, C., Zhao, Y., Samodivkin, V.: Signed Roman domination in graphs J. Comb. Optim. **27**(2), 241–255 (2014)
2. Atapour, M., Sheikholeslami, S.M., Volkmann, L.: Global Roman domination in trees. Graphs Comb. **31**(4), 813–825 (2015)
3. Chellali, M., Haynes, T.W., Hedetniemi, S.M., Hedetniemi, S.T., McRae, A.A.: A Roman domination chain. Graphs Comb. **32**(1), 79–92 (2016)
4. Cockayne, E.J., Dreyer Jr., P.A., Hedetniemi, S.M., Hedetniemi, S.T.: Roman domination in graphs. Discrete Math. **278**, 11–22 (2004)
5. Fernau, H.: Roman domination: a parameterized perspective. Int. J. Comput. Math. **85**, 25–38 (2008)
6. Haynes, T.W., Hedetniemi, S.T., Slater, P.J.: Fundamentals of Domination in Graphs. Marcel Dekker, New York (1998)
7. Hedetniemi, S., Laskar, R., Pfaff, J.: A linear algorithm for finding a minimum dominating set in a cactus. Discrete Appl. Math. **13**, 287–292 (1986)
8. Henning, M.A.: Defending the Roman Empire from multiple attacks. Discrete Math. **271**, 101–115 (2003)
9. Henning, M.A., Hedetniemi, S.T.: Defending the Roman Empire-a new strategy. Discrete Math. **266**, 239–251 (2003)
10. Liedloff, M., Kloks, T., Liu, J., Peng, S.L.: Efficient algorithms for Roman domination on some classes of graphs. Discrete Appl. Math. **156**, 3400–3415 (2008)
11. Liu, C.H., Chang, G.J.: Roman domination on strongly chordal graphs. J. Comb. Optim. **26**, 608–619 (2013)

The Semantic Information Method Compatible with Shannon, Popper, Fisher, and Zadeh's Thoughts

Chenguang Lu(✉)

College of Intelligence Engineering and Mathematics,
Liaoning Engineering and Technology University,
Fuxin, Liaoning 123000, China
survival99@gmail.com

Abstract. Popper and Fisher's hypothesis testing thoughts are very important. However, Shannon's information theory does not consider hypothesis testing. The combination of information theory and likelihood method is attracting more and more researchers' attention, especially when they solve Maximum Mutual Information (MMI) and Maximum Likelihood (ML). This paper introduces how we combine Shannon's information theory, likelihood method, and fuzzy sets theory to obtain the Semantic Information Method (SIM) for optimizing hypothesis testing better. First, we use the membership functions of fuzzy sets proposed by Zadeh as the truth functions of hypotheses; then, we use the truth functions to produce likelihood functions, and bring such likelihood functions into Kullback-Leibler and Shannon's information formulas to obtain the semantic information formulas. Conversely, the semantic information measure may be used to optimize the membership functions. The maximum semantic information criterion is equivalent to the ML criterion; however, it is compatible with Bayesian prediction, and hence can be used in cases where the prior probability distribution is changed. Letting the semantic channel and the Shannon channel mutually match and iterate, we can achieve MMI and ML for tests, estimations, and mixture models. This iterative algorithm is called Channels' Matching (CM) algorithm. Theoretical analyses and several examples show that the CM algorithm has fast speed, clear convergence reason, and wild potential applications. The further studies of the SIM related to the factor space and information value are discussed.

Keywords: Hypothesis testing · Shannon's theory · Maximum likelihood
Maximum mutual information · Membership function · Semantic information
Factor space · Information value

1 Introduction

Although Shannon's information theory [1] has achieved great successes, his information concept does not accord with daily usages of "information". For example, Shannon's (amount of) information is irrelevant to the truth and falsity of statements or predictions. Another problem is that the Shannon theory does not contain hypothesis

B.-Y. Cao and Y.-B. Zhong (Eds.): ICFIE 2017, AISC 872, pp. 207–222, 2019.
https://doi.org/10.1007/978-3-030-02777-3_19

testing thought, and therefore, we cannot use Shannon's mutual information as criterion to optimize tests, estimations, and predictions. Popper's theory of scientific advances [2] and Fisher's Likelihood Method (LM) [3] contain hypothesis testing thoughts. Popper initiates to use (semantic) information criterion to evaluate and optimize scientific hypotheses. His information concept is more accordant with daily usages. Yet, Popper did not provide proper semantic information formula. So far, what is used to resolve hypothesis testing problems is Fisher's likelihood method, which plays an important role in statistical inference and machine learning. Yet, it is unclear how the LM is related to semantic meaning and semantic information. Still, the LM is not compatible with Bayesian prediction well.

According Davidson's truth-conditional semantics [4], we can use the truth function of a hypothesis to represent its semantic meaning. According to Zadeh's fuzzy sets theory [5], a membership function is also a (fuzzy) truth function of a hypothesis.

There have been some iterative methods for Maximum Mutual Information (MMI) and Maximum Likelihood (ML), including the Newton method [6], EM algorithm [7], and minimax method [8]. Still, we want a better method.

Recently, we found that Lu's semantic information formulas [9–12] could combine information measures, likelihood functions, and membership functions better for hypothesis testing. Although Lu did not mention "likelihood" in his earlier studies, in fact, his "predicted probability distribution" is likelihood function. Using the concepts of likelihood and semantic channel, we can state Lu's Semantic Information Method (SIM) better. We also found that letting the semantic channel and the Shannon channel mutually match and iterate, we could achieve MMI and ML for tests, estimations, and mixture models conveniently.

In this paper, we first restate Lu's SIM in terms of likelihood and semantic channel. That is to use the fuzzy truth function to produce the likelihood function, and put the likelihood function into the Kullback-Laibler (KL) information formula and the Shannon mutual information formula to obtain sematic information formulas. Such a semantic information measure may contain Popper and Fisher's hypothesis testing thoughts. We shall show that new semantic information measure can be used to evaluate and optimize semantic communication, to improve the LM for variable sources, and to optimize the membership functions according to sampling distributions. Then, we simply introduce new iterative algorithm: Channels' Matching algorithm or the CM algorithm. For further studies, the information value related to portfolio and factor space are also simply discussed.

2 Semantic Channel, Semantic Communication Model, and Semantic Bayesian Prediction

2.1 Shannon Channel and Transition Probability Function

The semantic channel and the Shannon channel may mutually affect. First, we simply introduce the Shannon channel [1].

Let X be a discrete random variable representing a fact with alphabet $A = \{x_1, x_2, \ldots, x_m\}$, let Y be a discrete random variable representing a message with alphabet

$B = \{y_1, y_2, \ldots, y_n\}$, and let Z be a discrete random variable representing a observed condition with alphabet $C = \{z_1, z_2, \ldots, z_w\}$. A message sender chooses Y to predict X according to Z. For example, in weather forecasts, X is a rainfall, Y is a forecast such as "There will be light to moderate rain tomorrow", and Z is a set of meteorological data. In medical tests, X is an infected or uninfected person, Y is positive or negative (testing result), and Z is a laboratory datum or a set of laboratory data.

We use $P(X)$ to denote the probability distribution of X and call $P(X)$ the source, and we use $P(Y)$ to denote the probability distribution of Y and call $P(Y)$ the destination. We call $P(y_j|X)$ with certain y_j and variable X the transition probability function from X to y_j. Then a Shannon's channel is composed of a group of transition probability functions [1]:

$$P(Y|X) \Leftrightarrow \begin{bmatrix} P(y_1|x_1) & P(y_1|x_2) & \cdots & P(y_1|x_m) \\ P(y_2|x_1) & P(y_2|x_2) & \cdots & P(y_2|x_m) \\ \cdots & \cdots & \cdots & \cdots \\ P(y_n|x_1) & P(y_n|x_2) & \cdots & P(y_n|x_m) \end{bmatrix} \Leftrightarrow \begin{bmatrix} P(y_j|X) \\ P(y_j|X) \\ \cdots \\ P(y_n|X) \end{bmatrix} \quad (1)$$

The transition probability function has two properties:

(1) $P(y_j|X)$ is different from the conditional probability function $P(Y|x_i)$ or $P(X|y_j)$ in that whereas the latter is normalized, the former is not. In general, $\Sigma_i P(y_j|e_i) \neq 1$.
(2) $P(y_j|X)$ can be used to make Bayesian prediction to get the posterior probability distribution $P(X|y_j)$ of X. To use it by a coefficient k, the two predictions are equivalent, i.e.

$$\frac{P(X)kP(y_j|X)}{\sum_i P(x_i)kP(y_j|x_i)} = \frac{P(X)P(y_j|X)}{\sum_i P(x_i)P(y_j|x_i)} = P(X|y_j) \quad (2)$$

2.2 Semantic Channel and Semantic Communication Model

In terms of hypothesis testing, X is a sample point or a piece of evidence and Y is a hypothesis or a prediction. We need a sample sequence or a sampling distribution $P(X|.)$ to test a hypothesis to see how accurate the hypothesis is.

Let Θ be a random variable for a fuzzy set (defined by Zadeh [5]) and let θ_j be a value taken by Θ when $Y = y_j$. We also treat θ_j as a predictive model (or sub-model). A predicate $y_j(X)$ means "X is in θ_j" whose truth function is $T(\theta_j|X) \in [0, 1]$. Because $T(\theta_j|X)$ is constructed with some parameters, we may also treat θ_j as a set of model parameters.

In contrast to the popular likelihood method, we use sub-models $\theta_1, \theta_2, \ldots, \theta_n$ instead of one model θ or Θ, where a sub-model θ_j is defined by a truth function $T(\theta_j|X)$. The likelihood function $P(X|\theta_j)$ here is equivalent to $P(X|y_j, \theta)$ in popular likelihood method. A sample used to test y_j is a sub-sample or conditional sample. We use the sampling distribution $P(X)$ or $P(X|y_j)$ instead of the sample sequence $x(1)$, $x(2)$, …. to test a hypothesis. These changes will make the new method more flexible and more compatible with the Shannon information theory.

When $X = x_i$, $y_j(X)$ become $y_j(x_i)$, which is a proposition with truth value $T(\theta_j|x_i)$. We have the semantic channel:

$$T(\Theta|X) \Leftrightarrow \begin{bmatrix} T(\theta_1|x_1) & T(\theta_1|x_2) & \cdots & T(\theta_1|x_m) \\ T(\theta_2|x_1) & T(\theta_2|x_2) & \cdots & T(\theta_2|x_m) \\ \cdots & \cdots & \cdots & \cdots \\ T(\theta_n|x_1) & T(\theta_n|x_2) & \cdots & T(\theta_n|x_m) \end{bmatrix} \Leftrightarrow \begin{bmatrix} T(\theta_1|X) \\ T(\theta_2|X) \\ \cdots \\ T(\theta_n|X) \end{bmatrix} \quad (3)$$

This semantic channel can also be used for Bayesian prediction, i.e., semantic Bayesian prediction, to produce likelihood function:

$$P(X|\theta_j) = P(X)T(\theta_j|X)/T(\theta_j), \quad T(\theta_j) = \sum_i P(x_i)T(\theta_j|x_i) \quad (4)$$

where $T(\theta_j)$ may be called the logical probability of y_j. If $T(\theta_j|X) \propto P(y_j|X)$, then the semantic Bayesian prediction is equivalent to Bayesian prediction according to Eq. (2). Lu called this formula the set-Bayesian formula in 1991 [9] and put it into a semantic information measure. According to Dubois and Prade' paper [13], Thomas (1981) and Natvig (1983) proposed this formula earlier.

We can also consider that $T(\theta_j|X)$ is defined with normalized likelihood (function), i.e., $T(\theta_j|X) = kP(\theta_j|X)/P(\theta_j) = kP(X|\theta_j)/P(X)$, where k is a coefficient that makes the maximum of $T(\theta_j|X)$ be 1. With $P(X)$, $T(\theta_j|X)$ and $P(X|\theta_j)$ can ascertain each other.

Note that $T(\theta_j)$ is the logical probability of y_j, whereas $P(y_j)$ is the probability of choosing y_j. They are very different. $T(\Theta)$ is also not normalized, and generally there is $T(\theta_1)+T(\theta_2)\ldots+T(\theta_n) > 1$. Consider hypotheses y_1 = "There will be light rain", y_2 = "There will be moderate rain", and y_3 = "There will be light to moderate rain". According to their semantic meanings, $T(\theta_3) \approx T(\theta_1) + T(\theta_2)$; however, there may be $P(y_3) < P(y_1)$. Particularly, when y_j is a tautology, $T(\theta_j) = 1$ whereas $P(y_j)$ is almost 0. The $P(X|\theta_j)$ is a likelihood function and is also different from $P(X|y_j)$ which is a sampling distribution.

The semantic communication model is shown in Fig. 1.

A semantic channel is supported by a Shannon channel. For weather forecasts, the transition probability function $P(y_j|X)$ indicates the rule of choosing a forecast y_j. The rules used by different forecasters may be different and have more or fewer mistakes. Whereas, $T(\theta_j|X)$ indicates the semantic meaning of y_j that is understood by the audience. The semantic meaning is generally publicly defined and may also come from (or be affected by) the past rule of choosing y_j. To different people, the semantic meaning should be similar.

2.3 Is Likelihood Function or Truth Function Provided by the GPS's Positioning?

Consider the semantic meaning of the small circle (or the arrow) in the map on a GPS device. The circle tells where the position of the device is. A clock, a balance, or a

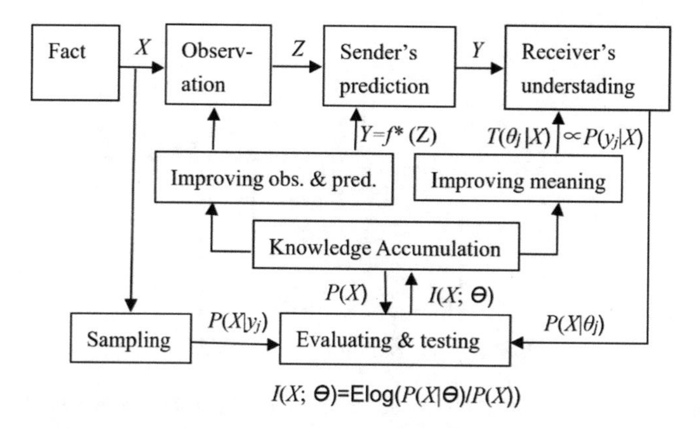

Fig. 1. The semantic communication model. Information comes from testing the semantic likelihood function $P(X|\boldsymbol{\theta}_j)$ by the sampling distribution $P(X|y_j)$.

thermometer is similar to a GPS device in that their actions may be abstracted as y_j = "$X \approx x_j$", $j = 1, 2, \ldots, n$. The Y with such a meaning may be called an unbiased estimate, and its transition probability functions $P(y_j|X)$ constitute a Shannon channel. This semantic channel may be expressed by

$$T(\theta_j|X) = \exp\left[-|X - x_j|^2/(2d^2)\right], \text{ j} = 1, 2, \ldots, n \tag{5}$$

where d is the standard deviation.

Consider a particular environment (in Fig. 2) where a GPS device is used in a car.

The positioning circle is on a building. The left side of the building is a highway and the right side is a road. We must determine the most possible position of the car. If we think that the circle provides a likelihood function, we should infer "The car is most possibly on the building". However, common sense would indicate that this conclusion is wrong. Alternatively, we can understand the semantic meaning of the circle by a transition probability function. However, the transition probability function is difficult to obtain, especially when the GPS has a systematical deviation. One may posit that we can use a guessed transition probability function and neglect its coefficient. This idea is a good one. In fact, the truth function in Eq. (5) is just such a function. With the truth function, we can obtain the likelihood function by the semantic Bayesian prediction:

$$P(X|\theta_j) = \frac{P(X)\exp[-(X - x_j)^2/(2d^2)]}{\sum_i P(X)\exp[-(X - x_j)^2/(2d^2)]} \tag{6}$$

This likelihood function accords with common sense and avoids conclusion "The car is most likely on the building". This example shows that a semantic channel is simpler and more understandable than the corresponding Shannon channel.

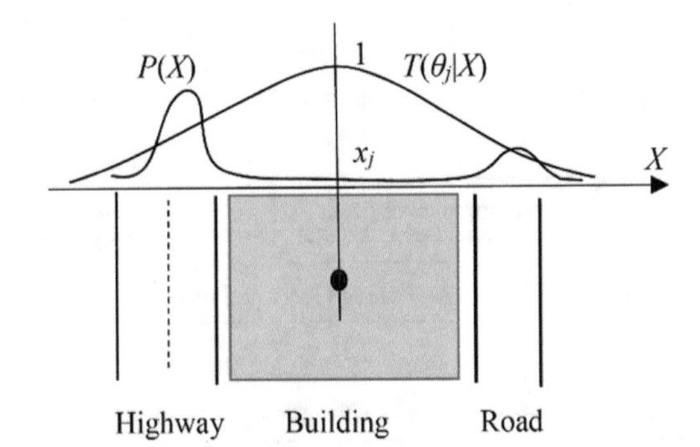

Fig. 2. The illustration of a GPS's positioning. When the prior distribution $P(X)$ is uneven and variable, using a truth function to make a semantic Bayesian prediction will be better than using a likelihood function to predict directly

3 Semantic Information Measure and the Optimization of the Semantic Channel

3.1 Semantic Information Measure Defined with Log Normalized Likelihood

In the Shannon information theory, there is only the statistical probability without the logical probability and likelihood (predicted probability). However, Lu defined semantic information measure by these three types of probabilities at the same time.

The (amount of) semantic information conveyed by y_j about x_i is defined as [10]:

$$I(x_i;\theta_j) = \log\frac{P(x_i|\theta_j)}{P(x_i)} = \log\frac{T(\theta_j|x_i)}{T(\theta_j)} \tag{7}$$

where semantic Bayesian prediction is used; it is assumed that the prior likelihood is equal to the prior probability distribution. For an unbiased estimation, its truth function and semantic information are illustrated in Fig. 3.

This formula contains Popper's thought [2] that the less the logical probability is, the more information there is if the hypothesis can survive tests; a tautology cannot be falsified and hence contains no information.

Bringing $T(\theta_j|X)$ in Eq. (5) into Eq. (7), we have

$$I(x_i;\theta_j) = \log(1/T(\theta_j)) - \left|X - x_j\right|^2/(2d^2) \tag{8}$$

where $\log(1/T(\theta_j))$ is the semantic information measure defined by Bar-Hillel and Carnap [14]. So, semantic information increases with either logical probability or deviation decreasing. The smaller deviation means that the hypothesis survives tests better.

Averaging $I(x_i;\theta_j)$, we obtain semantic (or generalized) Kullback-Leibler (KL) information (see [15] for the KL information or divergence):

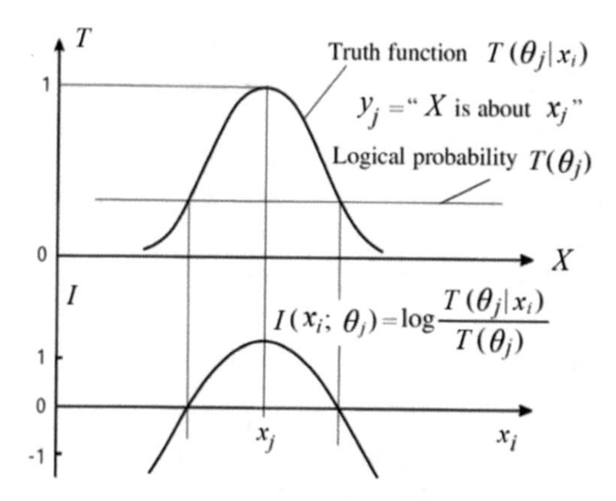

Fig. 3. Semantic information is defined with normalized likelihood. The less the logical probability is, the more information there is; the larger the deviation is, the less information there is; lastly, a wrong estimation may convey negative information.

$$I(X;\theta_j) = \sum_i P(x_i|y_j) \log \frac{P(x_i|\theta_j)}{P(x_i)} = \sum_i P(x_i|y_j) \log \frac{T(\theta_j|x_i)}{T(\theta_j)} \tag{9}$$

The statistical probability or frequency $P(x_i|y_j)$, $i = 1, 2, \ldots$, on the left of "log" above, represents a sampling distribution (note that a sample or sub-sample is also conditional) to test the hypothesis y_j or model θ_j. If $y_j = f(Z|Z \in C_j)$, then $P(X|y_j) = P(X|Z \in C_j) = P(X|C_j)$.

Although Akaike [16] revealed the relationship between likelihood and the KL divergence [15]. This relationship has attracted more attention in recent decades (see Cover and Thomas's text book ([17], Chap. 11.7)). In the following, we try to show that the relationship between the semantic information measure and likelihood is clearer.

Assume that the size of a sample used to test y_j is N_j, and the sample points come from independent and identically distributed random variables. Among N_j points, the number of x_i is N_{ij}. When N_j is infinite, $P(X|y_j) = N_{ij}/N_j$. Hence there is the following equation:

$$\log \prod_i \left[\frac{P(x_i|\theta_j)}{P(x_i)}\right]^{N_{ji}} = N_j \sum_i P(x_i|y_j) \log \frac{P(x_i|\theta_j)}{P(x_i)} = N_j I(X;\theta_j) \tag{10}$$

After averaging the above likelihood for different y_j, $j = 1, 2, \ldots, n$, we have

$$\begin{aligned} &\frac{1}{N}\sum_j \log \prod_i \left[\frac{P(x_i|\theta_j)}{P(x_i)}\right]^{N_{ji}} = \sum_j P(y_j) \sum_i P(x_i|y_j) \log \frac{P(x_i|\theta_j)}{P(x_i)} \\ &= \sum_i P(x_i) \sum_j P(y_j|x_i) \log \frac{T(\theta_j|x_i)}{T(\theta_j)} = I(X;\Theta) = H(X) - H(X|\Theta) \end{aligned} \tag{11}$$

where $N = N_1 + N_2 + \ldots + N_n$, $H(X)$ is the Shannon entropy of X, Θ is one of a group of models $(\theta_1, \theta_2, \ldots, \theta_n)$, $H(X|\Theta)$ is the generalized posterior entropy of X, and $I(X; \Theta)$ is the semantic mutual information. This equation shows that the ML criterion is equivalent to the maximum semantic mutual information criterion or the minimum generalized posterior entropy criterion. It is easy to find that when $P(X|\theta_j) = P(X|y_j)$ (for all j), the semantic mutual information $I(X; \Theta)$ will be equal to the Shannon mutual information $I(X; Y)$; the latter is the special case of the former.

3.2 The Optimization of Predictive Models or Semantic Channels

About how we get membership functions, we accept the statistical explanation of random sets [18]. However, to understand the evolution of membership functions, we may explain that membership functions evolves when they match the transition probability function, or say, semantic channels evolve when they match Shannon channels.

Optimizing a predictive model Θ is equivalent to optimizing a semantic Channel $T(\Theta|X)$. For given y_j, optimizing θ_j is equivalent to optimizing $T(\theta_j|X)$ by

$$T*(\theta_j|X) = \underset{T(\theta_j|X)}{\arg\max}\, I(X; \theta_j) \tag{12}$$

$I(X; \theta_j)$ can be written as the difference of two KL divergences:

$$I(X;\theta_j) = \sum_i P(x_i|y_j)\log\frac{P(x_i|y_j)}{P(x_i)} - \sum_i P(x_i|y_j)\log\frac{P(x_i|y_j)}{P(x_i|\theta_j)} \tag{13}$$

Because the KL divergence is greater than or equal to 0, when

$$P(X|\theta_j) = P(X|y_j) \tag{14}$$

$I(X; \theta_j)$ reaches its maximum and is equal to the KL information $I(X; y_j)$. Let the two sides be divided by $P(X)$; then

$$\frac{T(\theta_j|X)}{T(\theta_j)} = \frac{P(y_j|X)}{P(y_j)} \text{ and } T(\theta_j|X) \propto P(y_j|X) \tag{15}$$

Set the maximum of $T(\theta_j|X)$ to 1. Then we obtain

$$T*(\theta_j|X) = P(y_j|X)/P\left(y_j|x_j^*\right) \tag{16}$$

where x_j^* is the x_i that makes $P(y_j|x_j^*)$ be the maximum of $P(y_j|X)$. Generally, it is not easy to get $P(y_j|X)$. Yet, for given $P(X|y_j)$ and $P(X)$, it is easier to get $T(\theta_j|X)$ than to get $P(y_j|X)$ since from Eq. (16), we can obtain

$$T*(\theta_j|X) = [P(X|y_j)/P(X)]/[P\left(x_j^*|y_j\right)/P\left(x_j^*\right)] \tag{17}$$

The Eq. (12) fits parameter estimations with smaller samples, and Eqs. (16) and (17) fit non-parameter estimations with larger samples.

Similar to the Maximum-A-Posteriori (MAP) estimation, the above Maximum Semantic Information (MSI) estimation also uses the prior. The difference is that the MAP uses the prior of Y or Θ, whereas the MSI uses the prior of X. The MSI is more compatible with Bayesian prediction.

4 The CM Algorithm for Tests, Estimations, and Mixture Models

4.1 The Semantic Channel of a Medical Test

According to Eq. (11), we can obtain a new iterative algorithm, the CM algorithm, to achieve MMI and ML for uncertain Shannon channels.

For medical tests (see Fig. 4), $A = \{x_0, x_1\}$ where x_0 means no-infected person and x_1 means infected person, and $B = \{y_0, y_1\}$ where y_0 means test-negative and y_1 means test-positive.

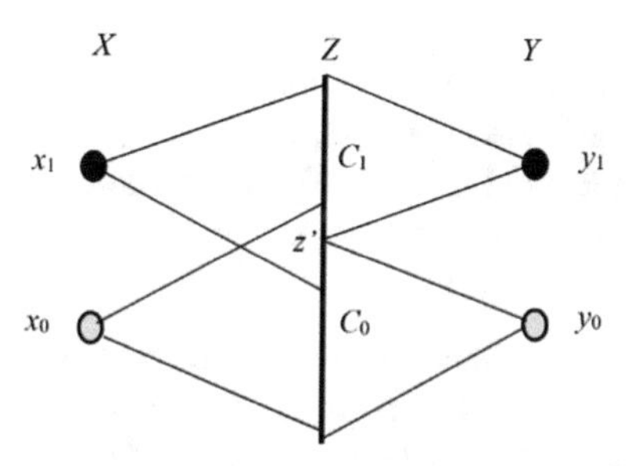

Fig. 4. A 2 × 2 Shannon nosy channel for tests. The channel changes with partition point z'.

In medical tests, the conditional probability in which the test-positive for an infected testee is called sensitivity, and the conditional probability in which the test-negative for an uninfected testee is called specificity [19]. The sensitivity and specificity form a Shannon channel as shown in Table 1.

Table 1. The sensitivity and specificity form a Shannon's channel $P(Y|X)$

Y	Infected x_1	Uninfected x_0
Test-positive y_1	$P(y_1|x_1)$ = sensitivity	$P(y_1|x_0)$ = 1-specificity
Test-negative y_0	$P(y_0|x_1)$ = 1-sensitivity	$P(y_0|x_0)$ = specificity

Assume that the no-confidence level of y_1 and y_0 are b_1' and b_0' respectively. Table 2 shows the semantic channel for a medical test.

Table 2. Two degrees of disbelief forms a semantic channel $T(\Theta|X)$

Y	Infected x_1	Uninfected x_0
Test-positive y_1	$T(\theta_1\|x_1) = 1$	$T(\theta_1\|x_0) = b_1$'
Test-negative y_0	$T(\theta_0\|x_1) = b_0$'	$T(\theta_0\|x_0) = 1$

According to Eq. (16), two optimized no-confidence levels are

$$b_1'* = P(y_1|x_0)/P(y_1|x_1); b_1'* = P(y_0|x_1)/P(y_0|x_0) \tag{18}$$

If we use popular likelihood method, when the source $P(X)$ is changed, the old likelihood function $P(X|\theta_j)$ will be improper. However, the above semantic channel is still proper for Bayesian prediction (see Eq. (4)) as well as the Shannon channel in Table 1. Therefore, the SIM can improve the popular likelihood method for variable sources.

4.2 Matching and Iterating

Matching I (Right-step): The semantic channel matches the Shannon channel. We keep the Shannon channel $P(Y|X)$ constant, and optimize the semantic channel $T(\Theta|X)$ (on the right of the log in Eq. (11)) so that $P(X|\theta_j)$ is equal or close to $P(X|y_j)$, or $T(\theta_j|X)$ is proportional or proximately proportional to $P(y_j|X)$ for all j,and hence $I(X; \Theta)$ reaches or approaches its maximum $I(X;Y)$.

Matching II (Left-step): The Shannon channel matches the semantic channel. While keeping the semantic channel $T(\Theta|X)$ constant, we change the Shannon channel $P(Y|X)$ (on the left of the log in Eq. (11)) to maximize $I(X; \Theta)$.

Iterating: The two channels mutually match in turn and iterate. The iterative convergence can be proved pictorially [20].

4.3 The Iterative Process for a Test

For the test as shown in Fig. 4, optimizing the Shannon channel is equivalent to optimizing the dividing point z'. When $Z > z$', we choose y_1; otherwise, we choose y_0.

As an example of the test, $Z \in C = \{1, 2, \ldots, 100\}$ and $P(Z|X)$ is a Gaussian distribution function:

$$P(Z|x_1) = K_1 \exp\left[-(Z - c_1)^2/\left(2d_1^2\right)\right],\ P(Z|x_0) = K_0 \exp[-(Z - c_0)^2/\left(2d_0^2\right)]$$

where K_1 and K_0 are normalizing constants. From $P(X)$ and $P(Z|X)$, we can obtain $P(X|Z)$. After setting the starting z', say z' = 50, as the input of the iteration, we perform the iteration as follows.

Right-Step: Calculate the following items in turn.

(1) Four transition probabilities $P(y_j|x_i)$ $(i,j = 0,1)$ for the Shannon channel:
(2) The b_1'* and b_0'* according to Eq. (18);
(3) $T(\theta_1) = P(x_1) + b_1$'*$P(x_0)$ and $T(\theta_0) = P(x_0) + b_0$'*$P(x_1)$;
(4) $I(X;\ \theta_1|Z)$ and $I(X;\ \theta_0|Z)$ for given Z (displaying as two curves):

$$I(X; \theta_j|z_k) = \sum_i P(x_i|z_k)I(x_i; \theta_j), \quad k = 1, 2, \ldots, 100; j = 0, 1 \tag{19}$$

Left-Step: Compare two information function curves $I(X;\ \theta_1|Z)$ and $I(X;\ \theta_0|Z)$ over Z to find their cross point. Use z under or over this point as new z'. If the new z' is the same as the last z' then let z^* = z' (z^*: optimal z') and quit the iteration; otherwise go to the Right-step. We may also use the following formula as classification function for new Shannon channel:

$$P(y_j|Z) = \lim_{s \to \infty} \frac{P(y_j)[\exp(I(X; \theta_j|Z))]^s}{\sum_{j'} P(y_{j'})[\exp(I(X; \theta_{j'}|Z))]^s}, \quad j = 1, 2, \ldots, n \tag{20}$$

4.4 Two Iterative Examples for Tests and Estimations

Iterative Example 1 (for a 2 × 2 Shannon Channel)
Input Data: $P(x_0) = 0.8$; $c_0 = 30$, $c_1 = 70$; $d_0 = 15$, $d_1 = 10$. The start point z' = 50.
Iterative Process: After the first Left-step, we get z' = 53; after the second Matching II, we get z' = 54; after the third Left-step, we get z^* = 54.
Iterative Example 2 (for a 3 × 3 Shannon channel)

This example is to examine a simplified estimation. The semantic channel is little complicated. The principle is the same as that for the test. A pair of good start points and a pair of bad start points are used to examine the reliability and speed of the iteration.

Input Data: $P(x_0) = 0.5$, $P(x_1) = 0.35$, and $P(x_2) = 0.15$; $c_0 = 20$, $c_1 = 50$, and $c_2 = 80$; $d_0 = 15$, $d_1 = 10$, and $d_2 = 10$.
Iterative Results:

(1) With the good start points: z_1' = 50 and z_2' = 60, the number of iterations is 4; z_1^* = 35 and z_2^* = 66.

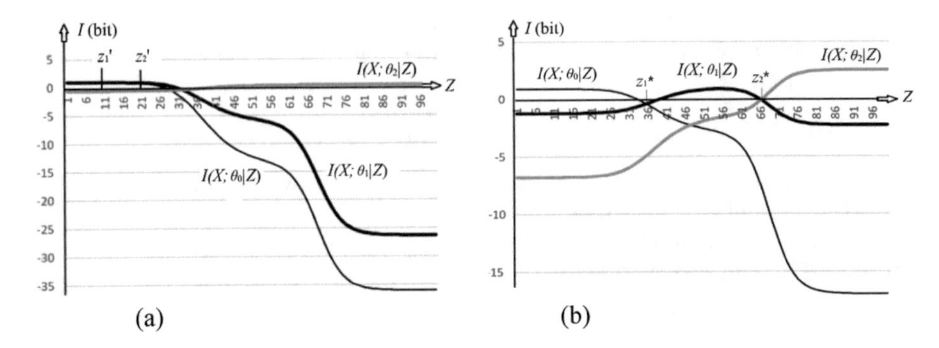

Fig. 5. The iteration with bad start points shows that the convergence is reliable. At the beginning (a), three information curves have small positive areas. At the end (b), three information curves have large positive areas so that $I(X; \Theta)$ reaches its maximum.

(2) With the bad start points: $z_1' = 9$ and $z_2' = 20$, the number of iterations is 11; $z_1^* = 35$ and $z_2^* = 66$ also. Figure 5 shows the information curves over Z before and after the iterative process.

4.5 Explaining the Evolution of Semantic Meaning

We may apply the CM algorithm to general predictions, such as weather forecasts. The difference is that the truth functions of predictions may be various. Then we can explain semantic evolution. A Shannon channel indicates a language usage, whereas a semantic channel indicates the comprehension of the audience. The Right-step is to let the comprehension match the usage, and the Left-step is to let the usage (including the observations and discoveries) match the comprehension. The mutual matching and iterating of two channels means that linguistic usage and comprehension mutually match and promote. Natural languages should have been evolving in this way.

4.6 The CM Algorithm for Mixture Models

A popular iterative algorithm for mixture models is the EM algorithm [7]. We can also use the CM algorithm to solve maximum likelihood mixture models or minimum relative entropy mixture models [21]. The convergence proof of the CM algorithm, without using Jensen's inequality, is clearer than that of the EM algorithm.

Table 3 shows an example [21]. Two Gussian distribution components with a group of real parameters produce the mixed distribution $P(X)$. Some guessed parameters (including $P(Y)$) are used to produce the mixed distribution $Q(X)$. The less the relative entropy or KL divergence $H(Q||P)$ is, the better the model is. For $H(Q||P) < 0.001$ bit, the number of iterations is 5.

In this example, Shannon mutual information with real parameters is less than that with start parameters. This example is a challenge to all authors who prove the standard EM algorithm convergent. For this example, maximizing likelihoods Q (in [7, 22]) or $Q + H(y)$ (in [23]) cannot be successful because Q or $Q + H(y)$ with true parameters

Table 3. Real and guessed model parameters and iterative results

	Real parameters in $P^*(X|Y)$ & $P^*(Y)$			Starting parameters; $H(Q||P) = 0.680$ bit			Parameters after 5 right-steps; $H(Q||P) = 0.00092$ bit		
	c	d	$P^*(Y)$	c	d	$P(Y)$	c	d	$P(Y)$
y_1	35	8	0.1	30	8	0.5	38	9.3	0.134
y_2	65	12	0.9	70	8	0.5	65.8	11.5	0.866

may be less than Q or $Q + H(y)$ with starting parameters. About how the CM algorithm solves this problem, see [21] for details.

5 Further Studies

5.1 Optimizing Membership Functions Under the Frame of Factor Space Theory

The factor space theory proposed by Wang [18] is a proper frame for knowledge representation and reasoning. Under this frame, many researchers have made meaningful results [24]. In these studies, the background distribution of objective facts in the factor space is not probability distribution, and the membership function is also not related to the probability distribution of facts. Now, we can use the probability distribution of facts in the factor space as the background distribution, by which we can set up the mutually matching relationship between the membership function and the probability distribution.

We use fuzzy color classification as example. The factor value of a color is a three primary color vector (r, g, b). The factor space R-G-B is a cubic with side length 1. The color vectors (0, 0, 0), (1, 0, 0), (1, 1, 0), (0, 1, 0), (0, 1, 1), (0, 0, 1), (1, 0, 1), and (1, 1, 1) represent typical black, red, yellow, green, cyan, blue, magenta, and white colors respectively. Assume the prior probability distribution of all possible colors in the R-G-B space is $P(\mathbf{X})$. For given y_1 = "red color", the distribution of $\mathbf{X}$ is $P(\mathbf{X}|y_1)$. If the sample is big enough, we can have continuous $P(\mathbf{X}| y_1)$ and $P(\mathbf{X})$. Using Eq. (17), we can obtain the optimized truth function $T^*(\theta_1|\mathbf{X})$ of y_1 = "$\mathbf{X}$ is red" or the membership function of fuzzy set θ_1. If the sample is not big enough, we may use Eq. (12) to obtain $T^*(\theta_1|\mathbf{X})$ with some parameters.

If we classify people into fuzzy sets {childhood}, {juvenile}, {youth}, {adult}, {middle-ager}… or classify weathers into {no rain}, {small rain}, {moderate rain}, {moderate to heavy rain}, {heavy rain}, … we may use similar method to obtain the membership function of each fuzzy set. This method does not require that these subsets form a partition or a fuzzy partition of A. That means we may allow fuzzy sets {adult} and {middle-ager}, or {moderate rain} and {moderate to heavy rain} (one may imply another), in A at the same time.

For a GPS device, main factors are the distances between the device and three satellites. Using this Semantic Information Method (SIM), we may eliminate the systematical deviation from other factors. Assume the Shannon channel of a GPS device is:

$$P(y_j|X) = K\exp[-|X - x_j - \Delta x|^2/(2d^2)],\ j = 1, 2, \ldots, n \tag{21}$$

where x_j denotes pointed position, y_j = "$X = x_j$", K is a constant, Δe is systematical deviation, and d denotes the precision. According Eq. (16), the corresponding semantic channel is

$$T(\theta_k|X) = \exp[-|X - x_k|^2/(2d^2)],\ k = 1, 2, \ldots, n \tag{22}$$

where $x_k = x_j + \Delta x$. About the applications of the SIM with the factor space theory, we need further studies.

5.2 Optimizing Predictions with Information Value as Criterion

The incremental entropy proposed by Lu [25] is a little different from that introduced by Cover and Thomas [17] (Chap. 16). The information for Lu's incremental entropy is semantic information. The incremental entropy is

$$U(\mathbf{X}|\theta_j) = \sum_{i=1}^{W} P(\mathbf{x}_i|\theta_j)\log R_i = \sum_{i=1}^{W} P(\mathbf{x}_i)\log\sum_{k=0}^{N} q_k R_{ik} \tag{23}$$

where x_i is a price vector of a portfolio; there are W possible price vectors. The $\boldsymbol{\theta}_j$ is the model parameters of prediction y_j. The number of securities in the portfolio is N. R_{ik} means the input-output ratio of the k-th security when $\mathbf{X} = \mathbf{x}_i$, and hence R_i is the input-output ratio of the portfolio as $\mathbf{X} = \mathbf{x}_i$. The q_k is investment ratio in the k-th security and q_0 means the ratio of safe asset or cash. U is the doubling rate of the portfolio.

If without the prediction y_j, $U = U(\mathbf{X})$, then there is the increment of U or information value brought by the information:

$$V(\mathbf{X};\theta_j) = \sum_{i}^{W} P(\mathbf{x}_i|\theta_j)\log[R_i(\mathbf{q}(y_j)*)/R_i(\mathbf{q}*)] \tag{24}$$

where $\mathbf{q}$* is the optimal vector of investment ratios without the prediction, and $\mathbf{q}(y_j)$* is the optimal vector based on the prediction.

In some cases, the information value criterion should be better than the information criterion. We need further studies for predictions with the information value criterion.

6 Conclusion

This paper restates Lu's semantic information method to clarify that his semantic information measure is defined with average log normalized likelihood, discusses the semantic channel and its optimization and evolution, and reveals that by letting the semantic channel and Shannon channel mutually match and iterate, we can achieve the maximum Shannon mutual information and maximum average log-likelihood for tests, estimations, and mixture models. Several iterative examples show that the CM algorithm has high speed, clear convergence reasons [20, 21], and wide potential applications[1].

The paper also concludes that the tight combination of Shannon information theory with likelihood method and fuzzy sets theory is necessary for hypothesis testing; with Lu's semantic information method, the combination is feasible.

Acknowledgement. The author thanks Professor Peizhuang Wang for his long term supports. Without his recent encouragement, the author wouldn't have continued researching to find the channels' matching algorithm.

References

1. Shannon, C.E.: A mathematical theory of communication. Bell Syst. Tech. J. **27**(3), 379–429 (1948). 623–656
2. Popper, K.: Conjectures and Refutations. Routledge, London/New York (1963/2005)
3. Fisher, R.A.: On the mathematical foundations of theoretical statistics. Philo. Trans. Roy. Soc. **222**, 309–368 (1922)
4. Davidson, D.: Truth and meaning. Synthese **17**, 304–323 (1967)
5. Zadeh, L.A.: Fuzzy sets. Inf. Control **8**(3), 338–353 (1965)
6. Kok, M., Dahlin, J., Schon, B., Wills, T.B.: A Newton-based maximum likelihood estimation in nonlinear state space models. IFAC-PapersOnLine **48**, 398–403 (2015)
7. Dempster, A.P., Laird, N.M., Rubin, D.B.: Maximum likelihood from incomplete data via the EM algorithm. J. R. Stat. Soc., Ser. B **39**, 1–38 (1977)
8. Barron, A., Roos, T., Watanabe, K.: Bayesian properties of normalized maximum likelihood and its fast computation. In: IEEE IT Symposium on Information Theory, pp. 1667–1671 (2014)
9. Lu, C.: B-fuzzy set algebra and a generalized cross-information equation. Fuzzy Syst. Math. (in Chin.) **5**(1), 76–80 (1991)
10. Lu, C.: A Generalized Information Theory (in Chinese). China Science and Technology University Press, Hefei (1993)
11. Lu, C.: Meanings of generalized entropy and generalized mutual information for coding. J. China Inst. Commun. (in Chin.) **15**(6), 37–44 (1994)
12. Lu, C.: A generalization of Shannon's information theory. Int. J. Gen. Syst. **28**(6), 453–490 (1999)

[1] More examples and the excel files for demonstrating the iterative processes can be found at http://survivor99.com/lcg/CM.html.

13. Dubois, D., Prade, H.: Fuzzy sets and probability: misunderstandings, bridges and gaps. In: Second IEEE International Conference on Fuzzy Systems, 28 March, 1 April (1993)
14. Bar-Hillel, Y., Carnap, R.: An outline of a theory of semantic information. Technical report No.247, Research Lab. of Electronics, MIT (1952)
15. Kullback, S., Leibler, R.: On information and sufficiency. Ann. Math. Stat. **22**, 79–86 (1952)
16. Akaike, H.: A new look at the statistical model identification. IEEE Trans. Autom. Control **19**, 716–723 (1974)
17. Cover, T.M., Thomas, J.A.: Elements of Information Theory, 2nd edn. Wiely, New York (2006)
18. Wang, P.W.: Fuzzy Sets and Random Sets Shadow (in Chinese). Beijing Normal University Press, Beijing (1985)
19. Thornbury, J.R., Fryback, D.G., Edwards, W.: Likelihood ratios as a measure of the diagnostic usefulness of excretory urogram information. Radiology **114**(3), 561–565 (1975)
20. Lu, C.: The Semantic Information Method for Maximum Mutual Information and Maximum Likelihood of Tests, Estimations, and Mixture Models. https://arxiv.org/abs/1706.07918, 24 June 2017
21. Lu, C.: Channels' matching algorithm for mixture models. In: Proceedings of International Conference on Intelligence Science, Shanghai, pp. 25–28, October 2017
22. Wu, C.F.J.: On the convergence properties of the EM algorithm. Ann. Stat. **11**(1), 95–103 (1983)
23. Neal, R., Hinton, G.: A view of the EM algorithm that justifies incremental, sparse, and other variants. In: Jordan, M.I. (ed) Learning in Graphical Models, pp. 355–368. MIT Press, Cambridge (1990)
24. Wang, P.Z.: Factor space and data science. J. Liaoning Tech. Univ. **34**(2), 273–280 (2015)
25. Lu, C.: Entropy Theory of Portfolio and Information Value (in Chinese). Science and Technology University Press, Hefei (1997)

Regression Analysis for Connection Number via Deviation Transmission

Wen-Ying Zhang(✉) and Bing-Jiang Zhang(✉)

School of Applied Science, Beijing Information Science and Technology University, Beijing 100192, China
zbj2013ch@163.com

Abstract. Connection number is one of the most important symbolic data, and a means for solving fuzzy problem as well. In this paper, the regression analysis for connection number via deviation transmission is proposed. Furthermore, a new similarity measure between connection numbers is defined based on the analysis of the characteristics of connection number. And, based upon this, evaluation indices of the regression analysis models are proposed. By calculation of numerical examples, the efficiency of this regression model can be proved.

Keywords: Set pair analysis · Symbolic data · Connection number Regression analysis · Deviation transmission

1 Introduction

Regression analysis is a statistical analysis method to determine the quantitative relationship between two or more variables, which are dependent on each other. According to the relation type between independent variables and dependent variables, regression analysis can be divided into linear regression analysis and nonlinear regression analysis. Based on the dependent relationship between variables established by observation data, regression analysis is one of application which is extremely extensive data analysis methods by analyzing the inherent law of data and being used for variable estimation, prediction and control. When the sample size of data is huge, the traditional regression analysis method is often difficult to grasp the intrinsic relationship of data attributes and can not get the implicit knowledge resources [1]. Facing with this problem, Diday first proposed a new data analysis method, i.e. Symbolic Data Analysis (SDA) in 1988 [2]. Its data table unit is no longer a general sense of quantitative or qualitative data and can be concepts, multi-valued sets, real domain of interval value or random variables with histogram described. Symbolic Data Analysis techniques overcome the disadvantages of the traditional methods of data analysis in some extent by compressing data. As the most common form of symbolic data, connection number has important significance for research.

In the real work of information collection and accumulation, the symbolic data analysis is especially suitable for the knowledge mining of large scale and complex data sets [3]. In recent years, the symbolic data analysis has made many achievements in theory method and application research [4–8]. Billard and Diday put forward the

B.-Y. Cao and Y.-B. Zhong (Eds.): ICFIE 2017, AISC 872, pp. 223–232, 2019.
https://doi.org/10.1007/978-3-030-02777-3_20

interval symbolic data regression analysis method using the covariance of interval data [9]. Guo et al. research interval principal component analysis based on the error theory [10]. Li et al., based on the deviation transmission theory, come up with the interval regression analysis method [11]. In fact, the concept of interval analysis is initially proposed from the angle of error analysis. This deviation analysis theory should be applied to set pair analysis theory system and connection numbers are formed by the data package.

Based on the deviation transmission theory this paper proposes a regression analysis method of connection number and analyze the inherent law of connection number. Set Pair Analysis [12] is a new uncertainty theory. Its core idea is that analyzes determinacy measure and uncertainty measure of objective things as a system, and overall handle mixed uncertainty problems caused by fuzzy, random, uncertain and intermediary. The main mathematical tool is connection numbers.

2 Binary Connection Number

The connection number, proposed by Chinese scholar Zhao in 1989, is a mathematical analysis tool in Set Pair Analysis. Its core idea can take determinacy and uncertainty as a system, describing transformation and connection between things from identical degree, differential degree and contrary degree [12]. The basic concept of Set Pair Analysis is set pair and connection number. A set pair is a pair of two sets with a certain relation, and its characteristic is depicted by connection number.

Definition 1 ([12]). Set Pair $Y = (F, G)$ is composed of two certain associated sets F and G. Under general circumstances, it's a equation:

$$U = A + Bi + Cj, \tag{1}$$

where describe identity degree, discrepancy degree, contrary degree and interconnection of the two sets. Namely A, B and C are called identity degree, discrepancy degree and contrary degree and $j = -1$, $i \in [-1, 1]$. Let $N = A + B + C$, and N is the total number of features, called connective norm. We let $u = U/N$, $a = A/N$, $b = B/N$, $c = C/N$, and then formula (1) can be rewritten as:

$$u = a + bi + cj, \tag{2}$$

where a, b, c is respectively identity degree, discrepancy degree and contrary degree of the two sets F and G. U and u are called ternary connection number.

If formulas (1) and (2) are rewritten as:

$$u = a + bi,\ U = A + Bi, \tag{3}$$

$$u = a + cj,\ U = A + Cj. \tag{4}$$

then U and u are called binary connection number.

Generally, we use binary connection number of formula (2). Given $i \in [-1, 1]$ in binary connection number [12], this paper consider it better express decision-making information. A and B are all natural numbers and $a, b \in [0, 1]$. This limits usable range. Therefore, this paper takes it generalized and gives Definition 2.

Definition 2. Suppose a and b arbitrary nonnegative real numbers, called $\beta = a + bi$ binary connection number, where a is identical number, b is uncertain degree (the maximum degree of uncertainty), bi is uncertain number, and according to specific circumstances, $i \in [0, 1]$. Sometimes i only plays the mark of uncertainty.

If there is no special note, this paper refers to binary connection number of this type.

Definition 3. Assuming $[x, y]$ $(0 \leq x \leq y \leq 1)$ nonnegative interval number, if a random variable, $r \in [x, y]$ is normally distributed $\{E, \sigma^2\}$, then $[x, y]$ is normal distribution interval number, named $\tilde{a} = \{E, \sigma\}$, where on the basis of 3σ principle of normal distribution, i.e. $P(r \in [x, y]) = 0.994$, formulas (5) and (6) define the population mean μ and the population standard deviation σ as follows:

$$\mu = (x + y)/2, \tag{5}$$

$$\sigma = (y - x)/6. \tag{6}$$

With regard to interval number $[x, y]$, it is rewritten the form of binary connection number, i.e.

$$\tilde{\beta} = x + (y - x)i, i \in [0, 1]. \tag{7}$$

According to formula (7), a binary connection number $\beta = a + bi$ can be rewritten as a interval number $\beta = [a, a + b]$.

Definition 4. Assuming $\beta = a + bi$ is binary connection number, called

$$E(\beta) = a + 0.5b \, (i = 0.5) \tag{8}$$

$$\sigma(\beta) = b/6 \tag{9}$$

expected value and standard deviation of β respectively.

Since the concept of connection number presented, it has been widely used in different branches of uncertainty inference, intelligent computing and group intelligent analysis, such as machine learning, decision analysis, knowledge acquisition and pattern matching, etc. Scholars, in the application of connection number to deal with these fuzzy problems, mostly focused on the theoretical basis of connection number [13, 14], transform approach between connection numbers and fuzzy sets [15, 16], the comparison and sorting method of connection number [17, 18], as well as the application of connection number [13, 19–21]. However, a few scholars make connection number regression analysis, predicting or controlling the process of a certain problem based on the relational expression of connection number.

In this paper, we put forward a regression analysis model of connection number based on the theory of propagation of deviation, and discussed evaluation index of regression analysis model.

3 Deviation Transfer Formula

Suppose direct measured values x1, x2, …, xn and indirect measured value y, they have continuous differentiable function, i.e.

$$y = f(x_1, x_2, \cdots, x_n). \tag{10}$$

then the realationship of deviation between y and x_1, x_2, …, x_n is called the deviation transmission. If standard deviations of random deviation of x_1, x_2, …, x_n are respectively $\sigma_{x_1}, \sigma_{x_2}, \cdots, \sigma_{x_n}$, then the general equation about propagation of deviation is

$$\sigma_y^2 = \sum_{i=1}^{n} (\frac{\partial f}{\partial x_i})^2 \sigma_{x_i}^2 + 2 \sum_{1 \le i \le j \le n} \frac{\partial f}{\partial x_i} \frac{\partial f}{\partial x_j} \rho_{ij} \sigma_{x_i} \sigma_{x_j}, \tag{11}$$

where ρ_{ij} is correlation coefficient of measured value x_i and x_j. Since $k_{x_i}\sigma_{x_i} = \delta_{x_i}$, where δ_{x_i} is single limit deviation, k_{x_i} is confidence coefficient of single limit deviation, then formula (11) can be rewritten

$$\delta_y^2 / k_y^2 = \sum_{i=1}^{n} (\frac{\partial f}{\partial x_i})^2 \delta_{x_i}^2 / k_{x_i}^2 + 2 \sum_{1 \le i \le j \le n} \frac{\partial f}{\partial x_i} \frac{\partial f}{\partial x_j} \rho_{ij} \delta_{x_i} \delta_{x_j} / k_{x_i} k_{x_j}. \tag{12}$$

When every single random deviation is normal distribution, the number of each single deviation is more, numerical value of single deviation is close and independent with each other, the total complex deviation is approximate the normal distribution, i.e. $k_y = k_{x_1} = k_{x_2} = \cdots = k_{x_n} = k$, $\rho_{ij} = 0$, and the complex limit deviation is

$$\delta_y = \sqrt{\sum_{i=1}^{n} (\frac{\partial f}{\partial x_i})^2 \delta_{x_i}^2}. \tag{13}$$

Formula (14) will be the main theory of regression analysis method for connection number via deviation transmission.

4 The Regression Model of Connection Number via Deviation Transmission

Assuming $p + 1$ connection numbers Y and X_j, $y_k = a_k + b_k i$ expresses the kth observed value of connection number of variable Y, $x_{kj} = c_{kj} + d_{kj}$ expresses the kth observed value of connection number of variable X_j, $k = 1, 2, \ldots, n$, $j = 1, 2, \ldots, p$. Therefore, the sample matrix is described by

$$[Y] = \begin{bmatrix} a_1 + b_1 i \\ a_2 + b_2 i \\ \vdots \\ a_n + b_n i \end{bmatrix}, \tag{14}$$

$$[X] = \begin{bmatrix} 1 & c_{11} + d_{11} i & \cdots & c_{1p} + d_{1p} i \\ 1 & c_{21} + d_{21} i & \cdots & c_{2p} + d_{2p} i \\ \vdots & \vdots & \ddots & \vdots \\ 1 & c_{n1} + d_{n1} i & \cdots & c_{np} + d_{np} i \end{bmatrix}. \tag{15}$$

The linear regression model between expected estimation $[Y]$ and $[X]$ is

$$[Y] = [X]\beta + \varepsilon. \tag{16}$$

So we obtain an regression equation as follows

$$[\hat{Y}] = [X]\hat{\beta}, \tag{17}$$

where expectation value and standard deviation of the kth observed value of connection number of dependent variable Y are severally $a_k + 0.5b_k$ $(i = 0.5)$ and $b_k/6$, and expectation value and standard deviation of the kth observed value of connection number of dependent variable X_j are severally $a_k + 0.5d_k$ $(i = 0.5)$ and $d_k/6$, $k = 1, 2, \ldots, n$, $j = 1, 2, \ldots, p$. And then get the expected value vector about connection number variables Y and X_j, i.e.

$$Y^c = \begin{bmatrix} a_1 + 0.5b_1 \\ a_2 + 0.5b_2 \\ \vdots \\ a_n + 0.5b_n \end{bmatrix}, X_j^c = \begin{bmatrix} c_{1j} + 0.5d_{1j} \\ c_{2j} + 0.5d_{2j} \\ \vdots \\ c_{nj} + 0.5d_{nj} \end{bmatrix}. \tag{18}$$

Consider the general linear regression model of variable Y^c and X_j^c,

$$Y^c = \beta_0 + \beta_1 X_1^c + \beta_2 X_2^c + \cdots + \beta_p X_p^c + \varepsilon. \tag{19}$$

Via the least squares method, obtain an regression equation

$$\hat{Y}^c = \hat{\beta}_0 + \hat{\beta}_1 X_1^c + \hat{\beta}_2 X_2^c + \cdots + \hat{\beta}_p X_p^c. \tag{20}$$

Consider deviation transmission Eq. (13), the limit deviation of $\hat{y}_i^c$, that is to say, the mean of the estimate value of total standard deviation is

$$\hat{y}_k^r = \sqrt{\sum_{j=1}^{p} \left(\frac{\partial \hat{Y}^c}{\partial X_j^c}\right)^2 (d_{kj}/6)^2} = \sqrt{\sum_{j=1}^{p} (\hat{\beta}_j)^2 (d_{kj}/6)^2}. \tag{21}$$

Then we can gain connection number of a dependent variable, namely

$$\hat{y}_k = \hat{y}_k^c - 3\hat{y}_k^r + 6\hat{y}_k^r i. \tag{22}$$

5 Evaluation of a Regression Model

After a series of analysis and deduction, we get the regression analysis model about connection numbers. It can be used to prediction variable of connection number. The smaller the error between the actual value and the predictive value of a dependent variable, the more effective the model is to a certain extent.

For the traditional regression analysis, the evaluation index of the frequently-used regression model has the sample mean of the dependent variable, the sample standard deviation of the dependent variable, the square sum of residuals, the standard deviation of the regression and the F statistic et al. With regard to connection number, the closeness degree of the actual value and predictive value can also be understood the similarity between the actual value and predictive value. Therefore, it can be to evaluate errors of a regression analysis model by using the similarity measure between connection numbers. This paper puts forward a new similarity measure between connection numbers to evaluate errors of the regression analysis model.

Definition 5. Assuming two binary connection numbers $x = x_1 + x_2 i$, $y = y_1 + y_2 i$, where $i \in [0, 1]$, $0 \leq x_1 + x_2 \leq 1$, $0 \leq y_1 + y_2 \leq 1$, then the similarity measure of two binary connection numbers is defined as follows:

$$M(x, y) = 1 - \frac{J + D}{2} \tag{23}$$

where

$$J = 1 - \frac{x_1 y_1 + x_2 y_2}{\sqrt{x_1^2 + x_2^2}\sqrt{y_1^2 + y_2^2}},$$

$$D = \frac{|x_1 - y_1| + |x_2 - y_2|}{2}.$$

Theorem 1. For formula (23), there exist facts as follow:

(1) Bounded: $0 \leq M(x, y) \leq 1$;
(2) $M(x, y) = 1$, if and only if $x = y$;
(3) $M(x, y) = 0$, if and only if $x = 0 + i$, $y = 1 + 0i$;
(4) $M(x, y) = M(y, x)$, for arbitrary x and y.

It seems from the definition that the bigger the similarity measured value, meaning two binary connection numbers more closer, the smaller error with each other. Due to formula (23), we can define evaluation indexes of the regression analysis model of a connection number.

Definition 6. Suppose the kth observed value and calculated values of connection numbers are $y_k = y_{k1} + y_{k2}i$ and $\hat{y}_k = \hat{y}_{k1} + \hat{y}_{k2}i$, $k = 1, 2, \ldots, n$, respectively, the similarity measure of two binary connection numbers is

$$M(Y,\hat{Y}) = \frac{1}{n}\sum_{k=1}^{n}(1 - \frac{J_k + D_k}{2}), \tag{24}$$

where

$$J_k = 1 - \frac{\hat{y}_{k1}y_{k1} + \hat{y}_{k2}y_{k2}}{\sqrt{\hat{y}_{k1}^2 + \hat{y}_{k2}^2}\sqrt{y_{k1}^2 + y_{k2}^2}},$$

$$D_k = \frac{|\hat{y}_{k1} - y_{k1}| + |\hat{y}_{k2} - y_{k2}|}{2}.$$

6 Illustrative Example

In consideration of proposing regression analysis method and evaluation index of regression model between connection numbers via deviation transmission, this paper takes some experimental data to verify analysis. Table 1 gives 20 groups connection

Table 1. Experimental data

Number	X_1	X_2	X_3	X_4	Y
1	$0.80 + 0.20i$	$0.60 + 0.35i$	$0.90 + 0.10i$	$0.95 + 0.05i$	$0.85 + 0.15i$
2	$0.90 + 0.10i$	$0.80 + 0.20i$	$0.90 + 0.10i$	$0.90 + 0.10i$	$0.85 + 0.15i$
3	$0.90 + 0.10i$	$0.80 + 0.20i$	$0.85 + 0.15i$	$1.00 + 0.00i$	$0.88 + 0.12i$
4	$0.90 + 0.10i$	$0.80 + 0.20i$	$0.94 + 0.06i$	$0.95 + 0.05i$	$0.85 + 0.15i$
5	$0.75 + 0.25i$	$0.70 + 0.30i$	$0.99 + 0.01i$	$0.95 + 0.05i$	$0.85 + 0.15i$
6	$0.97 + 0.03i$	$0.75 + 0.25i$	$0.85 + 0.15i$	$0.97 + 0.03i$	$0.85 + 0.15i$
7	$0.95 + 0.05i$	$0.95 + 0.05i$	$0.87 + 0.13i$	$100 + 0.00i$	$0.95 + 0.05i$
8	$0.90 + 0.10i$	$0.75 + 0.25i$	$0.90 + 0.10i$	$0.95 + 0.05i$	$0.90 + 0.10i$
9	$0.90 + 0.10i$	$0.83 + 0.17i$	$0.90 + 0.10i$	$0.95 + 0.05i$	$0.85 + 0.15i$
10	$0.85 + 0.15i$	$0.75 + 0.25i$	$0.90 + 0.10i$	$0.95 + 0.05i$	$0.85 + 0.15i$
11	$0.90 + 0.10i$	$0.95 + 0.05i$	$0.95 + 0.05i$	$1.00 + 0.00i$	$1.00 + 0.00i$
12	$1.00 + 0.00i$	$0.60 + 0.35i$	$0.90 + 0.10i$	$1.00 + 0.00i$	$0.88 + 0.12i$
13	$0.80 + 0.20i$	$0.65 + 0.35i$	$0.76 + 0.24i$	1.00 + 0.00i	$0.80 + 0.20i$
14	$0.95 + 0.05i$	$0.80 + 0.20i$	$0.95 + 0.05i$	$0.95 + 0.05i$	$0.93 + 0.07i$
15	$0.97 + 0.03i$	$0.88 + 0.12i$	$0.97 + 0.03i$	$0.97 + 0.03i$	$0.95 + 0.05i$
16	$0.88 + 0.12i$	$0.80 + 0.20i$	$0.98 + 0.02i$	0.98 + 0.02i	$0.92 + 0.08i$
17	$0.85 + 0.15i$	$0.70 + 0.30i$	$0.95 + 0.05i$	$0.98 + 0.02i$	$0.85 + 0.15i$
18	$0.90 + 0.10i$	$0.75 + 0.25i$	$0.90 + 0.10i$	$0.90 + 0.10i$	$0.88 + 0.12i$
19	$1.00 + 0.00i$	$0.85 + 0.15i$	$0.85 + 0.15i$	$0.98 + 0.02i$	$0.93 + 0.07i$
20	$0.95 + 0.05i$	$0.80 + 0.20i$	$0.94 + 0.06i$	$0.95 + 0.05i$	$0.90 + 0.10i$

Table 2. Expected values and mean square deviation of connection number

Number	X_1		X_2		X_3		X_4		Y	
	E	σ	E	σ	E	σ	E	σ	E	σ
1	0.90	0.033	0.83	0.058	0.95	0.017	0.98	0.008	0.93	0.025
2	0.95	0.017	0.90	0.033	0.95	0.017	0.95	0.017	0.93	0.025
3	0.95	0.017	0.90	0.033	0.93	0.025	1.00	0.000	0.94	0.020
4	0.95	0.017	0.90	0.033	0.97	0.010	0.98	0.008	0.93	0.025
5	0.88	0.042	0.85	0.050	0.99	0.002	0.98	0.008	0.93	0.025
6	0.99	0.005	0.88	0.042	0.93	0.025	0.99	0.005	0.93	0.025
7	0.98	0.008	0.98	0.008	0.94	0.023	1.00	0.00	0.98	0.008
8	0.95	0.017	0.88	0.042	0.95	0.017	0.98	0.008	0.95	0.017
9	0.95	0.017	0.92	0.028	0.95	0.017	0.98	0.008	0.93	0.025
10	0.93	0.025	0.88	0.042	0.95	0.017	0.98	0.008	0.93	0.025
11	0.95	0.017	0.98	0.008	0.98	0.008	1.00	0.000	1.00	0.000
12	1.00	0.000	0.83	0.058	0.95	0.017	1.00	0.000	0.94	0.020
13	0.90	0.033	0.83	0.058	0.88	0.040	1.00	0.000	0.90	0.033
14	0.98	0.008	0.90	0.033	0.98	0.008	0.98	0.008	0.97	0.012
15	0.99	0.005	0.94	0.02	0.99	0.005	0.99	0.005	0.98	0.008
16	0.94	0.020	0.90	0.033	0.99	0.003	0.99	0.003	0.96	0.013
17	0.93	0.025	0.85	0.050	0.98	0.008	0.99	0.003	0.93	0.025
18	0.95	0.017	0.88	0.042	0.95	0.017	0.95	0.017	0.94	0.020
19	1.00	0.00	0.93	0.025	0.93	0.025	0.99	0.003	0.97	0.012
20	0.98	0.008	0.90	0.033	0.97	0.010	0.98	0.008	0.95	0.017

number of experimental data, four independent variables X_1, X_2, X_3, X_4 and dependent variable Y.

Calculate expectation value and standard deviation of every connection number as centre and error of connection numbers. See in Table 2.

From formula (19), by using least square method and expectation values of X_1^c, X_2^c, X_3^c, X_4^c and Y^c (Table 2), we get regression equation as follows:

$$\hat{Y}^c = -0.2002 + 0.1769X_1^c + 0.3085X_2^c + 0.3070X_3^c + 0.4151X_4^c. \tag{25}$$

From formula (21), standard deviation (Table 2) and regression coefficient of regression Eq. (25), we have mean standard deviation of every dependent variable. Further, take use of formula (22) to gain dependent variable values of estimated connection numbers $\hat{Y}$. The results see in Table 3.

For the mean estimate values of dependent variable connection numbers obtained in Table 3 and dependent variable connection numbers presented in Table 1, their fitting degree should be further inspected. The above two groups' data substitute into formula (24) and the total similarity is 0.9845. This value is rather high as evaluation index of regression analysis method of connection numbers, and it explains the good estimate and forecast of the regression analysis model. The model has practical application value.

Table 3. Experimental results

Number	$\hat{y}_k^r$	$\hat{y}_k^c$	$\hat{y}_k$
1	0.8501	0.1196	$0.8501 + 0.1196i$
2	0.8903	0.0824	$0.8903 + 0.0824i$
3	0.9051	0.0790	$0.9051 + 0.0790i$
4	0.9131	0.0699	$0.9131 + 0.0699i$
5	0.8747	0.1047	$0.8747 + 0.1047i$
6	0.8914	0.0908	$0.8914 + 0.0908i$
7	0.9534	0.0437	$0.9534 + 0.0437i$
8	0.8905	0.0874	$0.8905 + 0.0874i$
9	0.9132	0.0666	$0.9132 + 0.0666i$
10	0.8850	0.0896	$0.8850 + 0.0896i$
11	0.9691	0.0280	$0.9691 + 0.0280i$
12	0.8818	0.1123	$0.8818 + 0.1123i$
13	0.8311	0.1354	$0.8311 + 0.1354i$
14	0.9202	0.0675	$0.9202 + 0.0675i$
15	0.9552	0.0405	$0.9552 + 0.0405i$
16	0.9256	0.0661	$0.9256 + 0.0661i$
17	0.8870	0.0979	$0.8870 + 0.0979i$
18	0.8766	0.0945	$0.8766 + 0.0945i$
19	0.9241	0.0658	$0.9241 + 0.0658i$
20	0.9183	0.0682	$0.9183 + 0.0682i$

7 Conclusion

A new method for connection numbers of regression analysis is proposed, based on the theory of error propagation, which is significant for the prediction and control problems in the fuzzy field with connection numbers. This paper also defined a new evaluation index of the connection number regression model.

The regression analysis method and the evaluation index are verified and analyzed by the numerical case. The results show that the regression analysis method is effective, and the evaluation index of the connection number regression model is appropriate and reasonable.

Acknowledgements. The research was supported by the National Natural Science Foundation of China (Grant No. 60972115) and the Graduate Student Science and Technology Innovation Project of Beijing Information Science and Technology University (Grant No. 5111623908).

References

1. Hu, Y., Wang, H.: A new data mining method based on huge data its application. J. Beijing Univ. Aeronaut. Astronaut. (Soc. Sci. Ed.) **17**(2), 40–44 (2004)
2. Bock, H.H., Diday, E.: Analysis of Symbolic Data. Springer, New York (2000)

3. Belson, W.A.: Prediction on the principle of biological classification. Appl. Stat. **8**, 65–67 (1995)
4. Billard, L., Diday, E.: From the statistics of data to the statistics of knowledge: symbolic data analysis. J. Am. Stat. Assoc. **98**(462), 470–487 (2003)
5. Billard, L., Diday, E.: Symbolic Data Analysis: Conceptual Statistics and Data Mining. Wiley, Chichester (2006)
6. Guo, J.P., Li, W.H.: Integrated evaluation of listed companies by factor analysis for symbolic data. In: AMIGE 2008 Proceedings, pp. 196–199 (2008)
7. Carvalho, F.D.A.T., Lechevallier, Y.: Partitional clustering algorithms for symbolic interval data based on adaptive distances. Pattern Recognit. **42**(7), 1223–1236 (2009)
8. Carvalho, F.A.T. Brito, P., Bock, H.H.: Dynamical clustering for symbolic quantitative data. In: Proceedings of International Conference on New Trends in Computational Statistics with Biomedical Applications, Osaka, pp. 203–215 (2001)
9. Billard, L., Diday, E.: Regression analysis for interval-valued data. In: Kiers, H.A.L., Rassoon, J.P., Groenen, P.J.F. (eds.) Data Analysis, Classification, and Related Methods, pp. 369–374. Springer, Berlin (2000)
10. Guo, J., Li, W.: Principal component analysis based on error theory and its application. Appl. Statics Manag. **26**(4), 636–640 (2007)
11. Li, W., Guo, J.: Methodology and application of regression analysis of interval-type symbolic data. J. Manag. Sci. China **13**(4), 38–43 (2010)
12. Zhao, K.Q.: Set Pair Analysis and Its Preliminary Application. Zhejiang Science and Technology Press, Hangzhou (2000)
13. Zhao, K.Q.: The theoretical basis and basis algorithm of binary connection A+Bi and its application in AI. CAAI Trans. Intell. Syst. **3**(6), 476–486 (2008)
14. Hiang, D.C., Zhao, K.Q., Lu, Y.Z., Hong, N.: The fundamental operation of arithmetic on connection number a+bi+cj and its application. Mech. Electr. Eng. Mag. **17**(3), 81–84 (2000)
15. Liu, X.M.: Multiple attribute decision making and its application based on the connection number with triangular fuzzy numbers. J. Huaiyin Inst. Technol. **17**(5), 30–33 (2008)
16. Wu, W.X.: The stochastic instinctive fuzzy decision based on the uncertainty analysis of correlate number. Fuzzy Syst. Math. **27**(6), 118–124 (2013)
17. Feng, X.S.: Expression and order of relative number. J. Qingdao Educ. Coll. **14**(3), 53–55 (2001)
18. Wang, X.F., Wang, J.Q., Yang, X.J.: Group decision making approaches based on binary connection number with incomplete information. J. Ind. Eng./Eng. Manag. **28**(1), 201–208 (2014)
19. Shi, N.G.: Application of connection number matter-element model in comprehensive evaluation. Comput. Eng. Des. **28**(13), 3192–3194 (2007)
20. Huang, D.C., Zhao, K.Q., Lu, Y.Z.: Fundamental operation of arithmetic on connection number a+bi and its application in network planning. J. Zhejiang Univ. Technol. **28**(3), 190–194 (2000)
21. Zhang, J.F., Chen, N., Zhang, L.: Application of SPA in safety assessment of coal mine. Ind. Saf. Environ. Prot. **37**(8), 1–3 (2011)

Research on Remote Control and Management Based on "4G Network" in Modern and High Efficiency Agriculture

Wei Tong(✉), Xi Feng, and Xian Jun Chen

Network Engineering College,
Haikou College of Economics, Haikou 570203, Hainan, China

Abstract. Based on the advantages of high-speed and high-quality transmission data in 4G network, either of audio, video images or flexible networking, a remote control and management research based on "4G network" in modern high-efficiency agriculture is proposed, which will lead for a modern way of automated remote monitoring and efficient management.

Keywords: 4G mobile communication technology · Efficient agriculture
Remote control · Wireless transmission

1 Introduction

With the coverage and popularization of 4G networks, at present, the three major domestic communication operators have invested a large amount of manpower and resources, and the new generation of high rate and high quality wireless mobile communication technologies represented by 4G has been put into full application. It can be upgraded smoothly on the basis of 3G, flexible networking and very suitable for a wide range of long-distance monitoring for its good quality of real-time video capture result. By studying on the remote control and management technology of "4G network" in modern and efficient agriculture [1], agricultural management has been advanced to the modern way of remote automatic monitoring and efficient management from traditional model. Therefore, the combination with the 4G mobile communication technology and the remote control and management of modern high efficiency agriculture is optimum solution and has good practical value and application prospect, which will surely become a new important research topic in mobile communication of agricultural management and monitoring [2]. Research in this paper is focused on the following areas:

(1) 4G mobile communication technology of video capture and compression used in high efficiency agriculture;
(2) Center monitoring and dispatching management platform;
(3) Data transmission of wireless data transmission.

B.-Y. Cao and Y.-B. Zhong (Eds.): ICFIE 2017, AISC 872, pp. 233–241, 2019.
https://doi.org/10.1007/978-3-030-02777-3_21

2 Remote Control and Management Monitoring Technology Based on 4G

In the technology integrated 4G mobile communication technology to agriculture environmental monitoring technology, the real-time picture and video data of remote scattered farmland will be transmitted to the agricultural information management staff of efficient agricultural planting, also to the managers and other mobile devices through 4G network. Real-time information of crop growth, weather forecast and pest etc. will be taken timely by the researchers, which could adopt effective and appropriate control measures, of highly practical. As shown in Fig. 1 4G network topology.

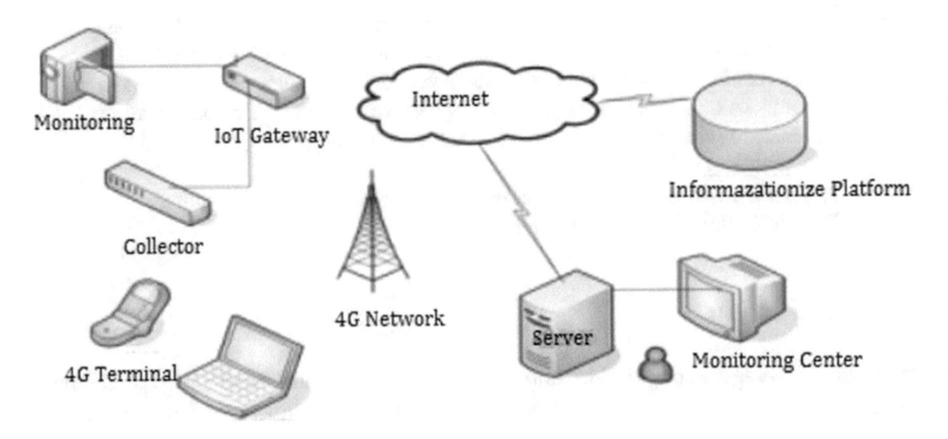

Fig. 1. 4G network topology diagram

2.1 4G Mobile Communication Technology

4G technology [3] includes two kinds of formats: TDD and FDD. Since the fast transmission rate, strong signal, capable of fast transmission of large data services, high quality audio, video and image services, etc., the latest testing results of 4G wireless Download Speed can reach more than 100 Mbps. Compared to 3G network, 4G can provide better technical support for real-time transmission of wireless big data in more practical areas such as long-distance image, video and audio services, wireless Internet access and electronic farming for efficient agriculture. It can also be utilized in remote locations not covered of the current cable networks. Therefore, it has incomparable superiority over 3G.

2.2 Embedded Technology

Embedded system is a processor board to which the control program has been stored in the ROM embedded. Due to its small size, high reliability, powerful functionality, flexibility and many other advantages, it has beed widely used in modern industrial control, agricultural monitoring, remote education, national defense science and technology, IoT and other modern technology integration fields. Embedded system consists of multiple sensor acquisition unit, 4G network unit, GPS/BeiDou positioning unit and others. The embedded system processing unit and transmitting unit are used to collect information of agricultural environment accurately. The collected video, sensor data and geographic location are transferred to the backstage server for processing and analysis through the 4G high-speed mobile communication network.

2.3 Sensor Technology

Sensor Technology [8] features data acquisition through a variety of sensors with accuracy, diversity of access, ease of data analysis, parallel processing of multiplexed sensor data, temperature for efficient agricultural environmental monitoring, humidity, light, CO2 concentration and other environmental information collection. The project network is designed on the basis of ZigBee’s wireless sensor. Its bottom is set up with wireless sensor networks for monitoring the status of efficient agricultural environment, remote controlling and acquisition of relevant data. The current ZigBee chip has become so commercialized of high degree to provide low-cost and low-power ZigBee chip very easy to implement. The middle layer is designed of wireless convergence layer gateway (4G-ZigBee gateway), whose role is the connection and conversion of ZigBee network and 4G network. Top layer is to be as a design of transferring 4G network to Internet network [11, 12].

2.4 Wireless Networking Technology

In this project, through the 4G router, ZigBee network data will be fused and processed. The router also converse the protocol which helps effective connection to the 4G network and WAN. The user can use the wireless terminal device or PC, through WiFi, 4G, the Internet and other a variety of ways to access to the site in order for real-time environmental control, which will greatly improve work efficiency, and ultimately sent to the remote client via the Internet.

2.5 Video Surveillance and Acquisition and Compression Technology

The technology mainly relates to the real-time collection and compression of high-definition video images in the cultivated planting area and sends them to the remote terminal through the monitoring platform. Agricultural staff could check related video information of remote farming planting area in real-time through the 4G network.

2.6 Database Technologies

In the construction of the monitoring platform, to implement of the appropriate database technology in order to efficiently store and access the information is the critical point. ZigBee sensor collects a variety of agricultural information in the backstage database, which make it convenient to facilitate real time crop growth for the observation, data acquisition and accurate analysis. The professional and technical backstage personnel can call on video capture and data at any time to obtain the indicators of the crop, and following diagnose and analyze. It makes online guidance by experts through real-time connection possible.

3 Advantages in the Rural Information Service in "4G Network"

With the acceleration of the construction of Hainan International Tourism Island, the network coverage of 4G mobile communication network has been rapidly developed and optimized. And it has now basically covered all the regions in the province. The majority of agricultural workers using 4G mobile terminals anywhere in the network coverage could fast access to rural information platform and be with huge control of real-time information such as crop growth or pest prevention of vegetable greenhouses via video, multimedia, etc. They can not only keep abreast of the latest agricultural technology, but also publish agricultural products information on information platform.

3.1 Big Data Transmission in 4G Mobile Communication Network

For a long time, the network for peasant, rural and agricultural informationize services based upon narrow band network business, has restricted the development of agriculture. A variety of video services are in urgent need in rural areas used to provide knowledge, information and tips in order to increase agricultural income. The introduction of 4G technology meet the needs of data transmission of fast, high quality audio and video image signals compatible with 3G and WLAN, with the result of a wide range of coverage in remote areas, to demonstrate the real-time monitoring of crop requirements to farmers on high-definition screen.

3.2 Wide Coverage and Easy Deployment

Based on 4G communications technology, self positioning search function provides a good network condition for remote monitoring and field monitoring in the process of information transmission. Mobile 4G base stations are over 5000 in Hainan, and in 2016 Mobile 4G Service will cover all the towns and villages in the province. Rural 4G mobile phone users are becoming more and more with the network signals more penetrating and better coverage. Especially in remote areas, the 4G network is easy to deploy, not requirement of network cabling and antenna placement. Surveillance systems can be installed outdoors in rural areas, including reservoirs, alpine zones and unmanned areas.

3.3 Integration of Wireless Sensor Network Platforms and 4G

Combined with ZigBee technology, embedded monitoring host technology and the wireless sensor network platform [9], the 4G mobile communication technology is developed to integrate to possibilitize long-distance data transmission, analysis and storage through the wireless sensor network platform and further implement through the 4G mobile communication network. Long-distance transmitted to the information of center database, a variety of agricultural information in the backstage database can be accessed by technical staff through screen in real-time, after the calculation, analysis, contrast, timing classification storage, and remote management.

3.4 High Communication Quality and Low Cost

With the popularization of 4G network, to download and transmit data of pictures and videos bidirectionally has been accomplished. In particular, it brings more convenient exchange and cooperation to rural agricultural technology exchange. The utilization of high bandwidth and high-quality transmission of 4G network, farmer handheld terminal watch video or video call and get high quality of wired video surveillance, faster and more convenient.

Due to gentle evolution to FDD-LTE in 4G network, the problem of compatibility with 3G communication has been solved so that more users can easily upgrade to the 4G communications with lower construction costs. Lower network line rental costs in 4G provited by operators vigorous promotion of a variety of packages, is more suitable for rural areas, providing farmers with convenient, fast, affordable, cheap information services.

4 Hardware Design in Remote Control and Management Monitoring System

The hardware structure of remote control and management and monitoring mainly includes ZigBee wireless communication module, sensor module, power management module and 4G router [4]. The collection node of the system is a information collection terminal such as data, picture, video, etc. The controller module of ZigBee connects relevant sensors, GPS, RFID and other modules through I2C bus and RS232 serial port, respectively, and collects various data in the node through ZigBee wireless network Unified format packaged is sent to the 4G router gateway, and then through the 4G router ZigBee network for data fusion and processing [10], and protocol conversion. Through the 4G network and WAN as effective connection, the user can use any wireless terminal device or PC, through WiFi, 4G, Internet and other means to access the site. Structure as shown in Fig. 2-4G remote control and management design.

5 Software Design of Monitoring System

In the construction of monitoring system platform, the appropriate database technology is needed to implement efficiently store information and access remote, data storage and display, and remote management. The data sending and receiving module completes the data sending and receiving with 4G communication module, adopting the technology of SQL Server database, socket network to receive data, multi-thread and queue [5, 6]. Functional structure is shown in Fig. 3-Remote monitoring system management software block diagram.

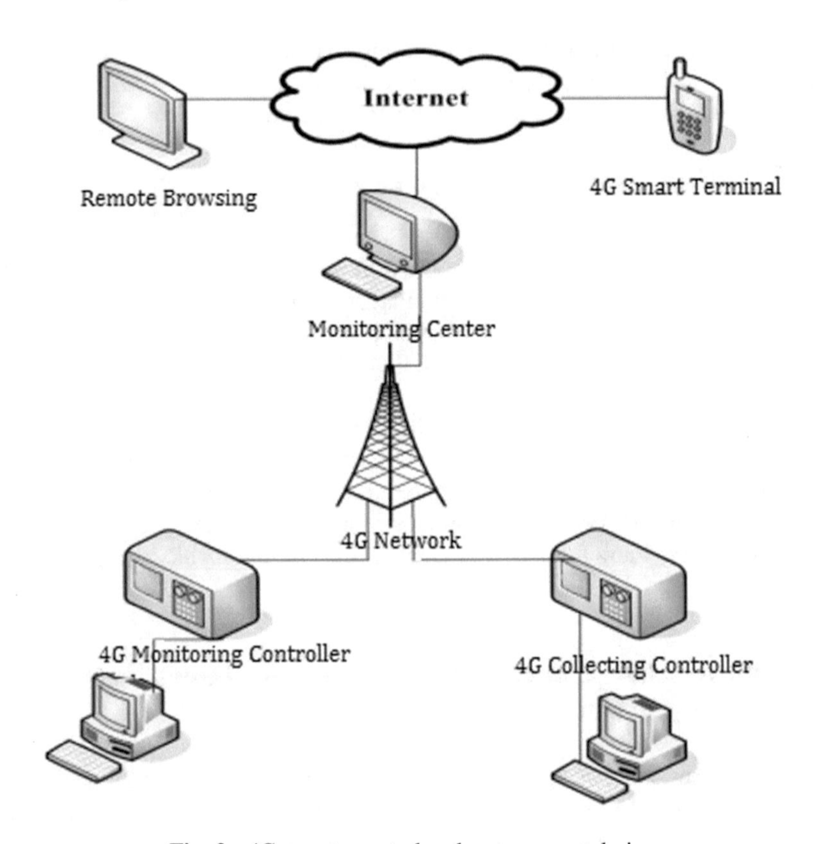

Fig. 2. 4G remote control and management design

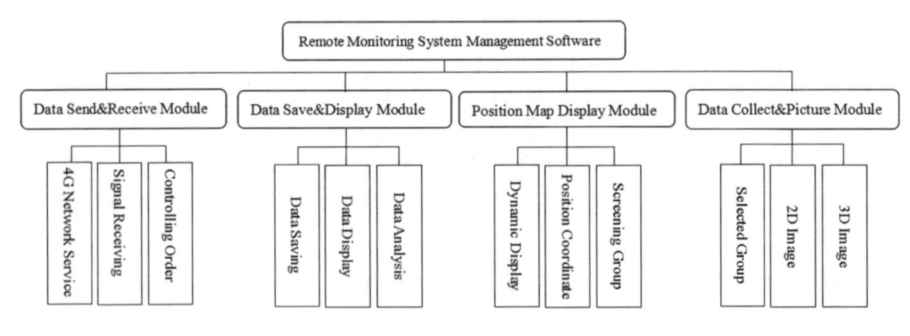

Fig. 3. Remote monitoring system management software block diagram

6 System Real-Time Communication Verification Test

6.1 SERED Algorithm to Reduce the Packet Loss Rate

Remote monitoring based on 4G network includes many places in outdoor areas, such as reservoirs, alpine zones and depopulated zone, so packet loss rate will be increased.

In order to make remote monitoring data transmission more smooth and reduce packet loss rate, the practical real-time long distance transmission of data is proposed to introduce a new SERED algorithm to reduce the packet loss rate [7].

In the SERED algorithm, when calculating the packet discards rate P, we consider combining the average queue length of the instantaneous queue length, so that the discard probability conforms to the current congestion degree and the trend of future congestion changes.

$$\dot{\min}' = \frac{\max_{th} - \min_{th}}{2} + \dot{\min}_{th} \quad (1)$$

In the SERED algorithm, the probability of packet loss is designed as an exponential function with a minimum of 0 and a maximum of 1. Taking into account the calculation is convenient, the definition of load factor:

$$Load = \frac{avg_g}{\min'} \times \frac{qlen}{\min'} \quad (2)$$

Packet discard probability according to the following formula:

$$p(k) = \begin{cases} 1 - e^{1-load} & Load \geq 1 \\ 0 & Load < 1 \end{cases} \quad (3)$$

While in the instantaneous queue decline stage, only the average queue is particularly high, and the decline is relatively small instantaneous queue will cause a high probability of packet loss; and in the rising phase of queue, if the average small queue, packet loss The probability is related to the increased size of the transient queue, so that it can both absorb the unexpected traffic and reflect the overload in time. As long as the average queue is elevated, it will cause a high probability of packet loss. Through simulation experiments, SERED algorithm can reduce the average queue length, improve throughput, and to a certain extent reduce the packet loss rate.

6.2 Communication Distance Test

The testing time is at 9:00–10:30 on June 10, 2017. The location is an open are in front of the Information Engineering College Building in the Haikou University of Economics. The test uses the collector to send the test data and the monitoring center to receive data. The test distance of 15, 25, 50, 100, 200, 500 m, and the maximum communication distance of transmission can reach 500 m. If the related hardware equipment will be upgraded, it is believed that the transmission distance can reach more than 1000–2000 m, which can meet and facilitate the monitoring of many places outdoors in rural areas, including reservoirs, alpine areas and unmanned areas.

6.3 Real-Time Communication Effectiveness Tests

Test communication effectiveness, through multiple hours of continuous testing, using collectors, 4G mobile terminals, remote monitoring system management software, where the collector and 4G mobile terminal distance of 50 m. The result is as Table 1 remote control and monitoring systems, dropped packets and the packet loss rate. The experiments show that the network packet loss rate is of less than 0.8%, compared to after taking the processing loss, packet loss rate could be lowered to 0.5%, meeting system design requirements.

Table 1. Control and monitoring system packet loss and packet loss rate

Sending terminal	Receiving terminal	Sending package number	Receiving package number	Loss package number	Loss package number	Permissible range
Collector	4G terminal	600	595	5	0.8%	<1%
Collector	Remote management system	600	595	5	0.8%	<1%

7 Conclusion

Through the implementation and promotion of "4G network" remote control and management and monitoring technology in rural areas, in rural areas in a variety of agricultural real time information, farmers can more quickly handheld terminals remotely monitor the growth of crops, modern and efficient agricultural management Monitoring provides a more efficient and timely program to optimize the management of efficient agriculture and production efficiency, which provides a good opportunity for the management and operation of modern and efficient agricultural bases.

Recommender: Sheng-quan Ma, School of Information Science and Technology, HaiNan Normal University, Haikou 571158, China, Professor.

Acknowledgements. This work is supported by the grant of Science Research Foundation of Haikou College of Economics(NO: hjkz13-07), the grant of Science Research Foundation of Haikou College of Economics (NO: hjkz16-02).

References

1. Han, H.F.: Agricultural environmental information remote monitoring and management system designs. Chinese Academy of Agricultural Sciences, Beijing (2009)
2. Yu, X.Q., Putt, W., Ting, K.: Farmland irrigation remote monitoring system is based on wireless sensor network. J. Irrig. Mech. Eng. **31**(1), 66–80 (2013)
3. Xu, J.: On 4G mobile communication technology. Cable TV Technol. (2014)

4. Wang, X.M., Wei, Y., Chan, L.M., Zheng, L.H., Cheng, Y.Q.: Integrated 3S, Soil ZigBee and RFID sampling remote intelligent management system. J. Agric. Eng. **33**, 143–149 (2017)
5. Sheng, P., Yang, G., Li, P.P., et al.: Facilities intelligent measurement and control system based ZigBee and 3G technology, agriculture. J. Agric. Mech. **43**(12), 229–233 (2012)
6. Zhang, M., Long, J., Han, Y.: Design of the remote monitoring system for greenhouse environment based on ZigBee and Internet. J. Agric. Eng. **29**(1), 171–176 (2013)
7. Feng, O.Y.: A modified RED algorithm to adjust the packet loss rate. J. Nanning Polytech. **17**(1) (2012)
8. Deng, X.L., Li, H., Che, Y.: Farmland information wirelesses sensor network based on ZigBee and PDA. J. Agric. Eng. **26**, 103–108 (2010)
9. Zhang, R.R., Zhao, C.J., Chen, L.P.: Farmland information acquisition wireless sensor network node design. J. Agric. Eng. **25**(11), 213–218 (2009)
10. Ya, P.J.: Development of farmland data acquisition device based on embedded system. Heilongjiang August First Land Reclamation University, Daqing (2009)
11. Shen, L.-T., Guan, S., Meng, Y.J.: Zig-Bee and Baidu Maps API application in the field information collection system. Chin. Agric. Mech. **4**(242), 184–188 (2012)
12. Yang, L., Wei, Y.: Manages farmland based networking environment was monitored wireless sensor networks. Chin. Agric. Mech. **1**(29), 97–99 (2015)

Extension of Fuzzy Set and System

A Study of Intuitionistic Fuzzy Rough Sets

Yun Bai(✉), Lan Shu, and Bang-sheng Yao

University of Electronic Science and Technology, Chengdu 611731, China
570535106@qq.com

Abstract. Axiomatic systems for rough sets are vital to rough sets theory and its application. First, two operators for both binary order number on fuzzy lattice and intuitionistic fuzzy sets are defined in this paper, and relevant properties are briefly discussed. Second, the intuitionistic fuzzy rough sets theory has been taken axiomatize treatment, and researches correlative conclusion and equivalence conditions.

Keywords: Intuitionistic fuzzy sets · Rough sets · Axioms

1 Introduction

The Rough sets (RS) theory, which is a useful mathematical tools to deal with incomplete and uncertain knowledge, was proposed by Pawlak. Recently, the theory has been widely successfully applied to machine learning, process control, pattern recognition, data mining and other fields [1]. The upper and lower approximation set satisfies many properties [2], which is the core concept of rough sets theory. The intuitionistic fuzzy sets (IFS) is another way to address with incomplete and imprecise information [3, 4], on reservation, on the basis of fuzzy set membership - adds a new attribute parameters of membership functions, its mathematical description more in line with the objective world the essence of the fuzzy object, decision analysis and pattern recognition in recent years is widely applied in areas such as [5, 6]. Although rough sets and intuitionistic fuzzy sets are used to handle imprecise knowledge, the study of them is very limited. Axiomatic method is one of the important contents of rough set theory, and have made many research results [8], the idea is assumed has been in existence for some axiom sets condition of approximation operator, and then search matching the binary relation. There are two reasons for the intuitionistic fuzzy rough sets (IFRS) axiomatization extended the fuzzy rough set of the method. First, the intuitionistic fuzzy rough sets be degenerated into fuzzy rough set. Secondly, fuzzy rough set theory is widely used, and theory is the prerequisite of application.

2 Preliminaries

In this section, we review some definitions in the properties of intuitionistic fuzzy sets and its correlation theorem.

B.-Y. Cao and Y.-B. Zhong (Eds.): ICFIE 2017, AISC 872, pp. 245–253, 2019.
https://doi.org/10.1007/978-3-030-02777-3_22

Definition 2.1 ([4]). Let U be a non-empty universe set, an intuitionistic fuzzy A is denoted on $U, A = \{\langle x, \mu_A(x), \gamma_A(x)\rangle | x \in U\}$, where for all $x \in U, 0 \leq \mu_A(x) + \gamma_A(x) \leq 1$, if we have $\mu_A(x) = \alpha, \gamma_A(x) = \beta$ for all $\alpha, \beta \in [0,1]$, $\alpha + \beta \leq 1$, then the intuitionistic fuzzy set is denoted $\langle \widehat{\alpha, \beta} \rangle$.

We know that an intuitionistic fuzzy relation (IFR) R on U is an intuitionistic fuzzy subset of $U \times U$, namely, R is given by $R = \{\langle (x,y), \mu_R(x,y), \gamma_R(x,y)\rangle | (x,y) \in U \times U\}$, where $\mu_R : U \times U \to [0,1]$ and $\gamma_R : U \times U \to [0,1]$ satisfy the condition $0 \leq \mu_R(x,y) + \gamma_R(x,y) \leq 1$ for all $(x,y) \in U \times U$. We denote the family of all IFR on U by $IFR(U \times U)$ [9].

Definition 2.2 ([8]). Let sets $A, B \in IFS(U)$, two operators $\oplus$ and $\otimes$ be denoted,

$$A \oplus B = \langle [\mu_A, \mu_B], (\gamma_A, \gamma_B) \rangle,$$

$$A \otimes B = \langle (\mu_A, \mu_B), [\gamma_A, \gamma_B] \rangle.$$

Where $[\mu_A, \mu_B] = \underset{x \in U}{\wedge} [\mu_A(x) \vee \mu_B(x)]$, $(\gamma_A, \gamma_B) = \underset{x \in U}{\vee} [\gamma_A(x) \wedge \gamma_B(x)]$,

$$(\mu_A, \mu_B) = \underset{x \in U}{\vee} [\mu_A(x) \wedge \mu_B(x)], [\gamma_A, \gamma_B] = \underset{x \in U}{\wedge} [\gamma_A(x) \vee \gamma_B(x)]$$

By Definition 2.2, the following properties be satisfied for all $A, B, C \in IFS(U)$,

(1) $A \oplus B = B \oplus A$, $A \otimes B = B \otimes A$,
(2) If $A \oplus B = A \oplus C$, then $B = C$,
(3) If $A \otimes B = A \otimes C$, then $B = C$,
(4) $(A \vee B) \oplus C = (A \oplus C) \vee (B \oplus C)$, $(A \wedge B) \otimes C = (A \otimes C) \wedge (B \otimes C)$.

Definition 2.3 ([9]). Let $R \in IFR(U \times U)$, we say that R is,

(1) Reflexive if $\mu_R(x,y) = 1$ and $\gamma_R(x,y) = 0$ for all $x \in U$.
(2) Symmetric if $\mu_R(x,y) = \mu_R(y,x)$ and $\gamma_R(x,y) = \gamma_R(y,x)$ for all $(x,y) \in U \times U$.
(3) Transitive if $R \geq R^{\vee}_{\wedge} \circ^{\wedge}_{\vee} R$, namely, for all $(x,z) \in U \times U$, $\mu_R(x,z) \geq \underset{y \in U}{\vee} [\mu_R(x,y) \wedge \mu_R(y,z)]$ and $\gamma_R(x,z) \leq \underset{y \in U}{\wedge} [\gamma_R(x,y) \vee \gamma_R(y,z)]$.

If R is reflexive, symmetric and transitive, then R is an equivalence relation.

Definition 2.4 ([8]). Let U be a nonempty universe of discourse, suppose $A \in IFS(U)$ and $R \in IFR(U \times U)$, the pair (U, R) is called an IF approximation space. The upper and lower approximations of IFS A w.r.t. (U, R) denote $\overline{R}A$ and $\underline{R}A$ respectively,

$$\overline{R}A = \{\langle x, \mu_{\overline{R}A}(x), \gamma_{\overline{R}A}(x)\rangle | x \in U\},$$

$$\underline{R}A = \left\{\langle x, \mu_{\underline{R}A}(x), \gamma_{\underline{R}A}(x)\rangle | x \in U\right\}.$$

Where $\mu_{\overline{R}A}(x) = \bigvee_{y \in U}[\mu_R(x,y) \wedge \mu_A(y)]$, $\gamma_{\overline{R}A}(x) = \bigwedge_{y \in U}[\gamma_R(x,y) \vee \gamma_A(y)]$,

$$\mu_{\underline{R}A}(x) = \bigwedge_{y \in U}[\gamma_R(x,y) \vee \mu_A(y)],\ \gamma_{\underline{R}A}(x) = \bigvee_{y \in U}[\mu_R(x,y) \wedge \gamma_A(y)].$$

By Definition 2.4 we can know that for any $A, B \in IF(U)$, $R \in IFR(U \times U)$. The upper and lower approximation sets of A and B satisfy:

(1) $\overline{R}(A \cap B) \subseteq \overline{R}A \cap \overline{R}B, \underline{R}(A \cup B) \supseteq \underline{R}A \cup \underline{R}B$;
(2) $\underline{R}(\sim A) = \sim(\overline{R}A), \overline{R}(\sim A) = \sim(\underline{R}A)$;
(3) $\underline{R}(A \cup \widehat{\langle \alpha, \beta \rangle}) = \underline{R}(A) \cup \widehat{\langle \alpha, \beta \rangle}, \overline{R}(A \cap \widehat{\langle \alpha, \beta \rangle}) = \overline{R}(A) \cap \widehat{\langle \alpha, \beta \rangle}$.

Theorem 2.1 ([8]). Let $A, B \in IFS(U)$, suppose $L: IFS(U) \to IFS(U)$ is an unary operator on $IFS(U)$ and satisfy $A \otimes L(B) = L(A) \otimes B$, the conditions hold:

(1) $L(A \vee \widehat{\langle \alpha, \beta \rangle}) = L(A) \vee \widehat{\langle \alpha, \beta \rangle}$,
(2) $L(A \wedge B) = L(A) \wedge L(B)$.

Theorem 2.2 ([8]). Let $A, B \in IFS(U)$, suppose $L: IFS(U) \to IFS(U)$ is an unary operator on $IFS(U)$ and satisfy $A \otimes L(B) = L(A) \otimes B$, therefore there exists $R \in IFR(U \times U)$ satisfy $L(A) = \underline{R}(A)$.

Theorem 2.3 ([8]). Let $A, B \in IFS(U)$, suppose $H: IFS(U) \to IFS(U)$ is an unary operator on $IFS(U)$ and satisfy $A \oplus H(B) = H(A) \oplus B$, the conditions hold,

(1) $H(A \wedge \widehat{\langle \alpha, \beta \rangle}) = H(A) \wedge \widehat{\langle \alpha, \beta \rangle}$,
(2) $H(A \vee B) = H(A) \vee H(B)$.

Theorem 2.4 ([8]). Let $A, B \in IFS(U)$, suppose $H: IFS(U) \to IFS(U)$ is an unary operator on $IFS(U)$ and satisfy $A \oplus H(B) = H(A) \oplus B$, therefore there exists $R \in IFR(U \times U)$ satisfy $H(A) = \overline{R}(A)$.

Definition 2.5 ([10, 11]). Let $L^* = \{(a,b) | a, b \in [0,1], a + b \leq 1\}$ be a fuzzy lattice (FL), the operation in L^* is as following:

(1) $(a_1, b_1) \prec (a_2, b_2) \Leftrightarrow a_1 \leq a_2, b_1 \geq b_2$,
(2) $(a_1, b_1) \vee (a_2, b_2) = (a_1 \vee a_2, b_1 \wedge b_2), (a_1, b_1) \wedge (a_2, b_2) = (a_1 \wedge a_2, b_1 \vee b_2)$,
(3) $(a, b)^c = (b, a)$.
(4) $\bigvee_{i \in I}(a_i, b_i) = (\bigvee_{i \in I} a_i, \bigwedge_{i \in I} b_i), \bigwedge_{i \in I}(a_i, b_i) = (\bigwedge_{i \in I} a_i, \bigvee_{i \in I} b_i)$,
(5) $\overline{1} = (1, 0), \overline{0} = (0, 1)$.

Where $(L^*, \vee, \wedge, c, \overline{1}, \overline{0})$ denote FL. In this paper, binary number $(a, b) \in L^*$ is called binary number on FL.

3 Axiomatization of Intuitionistic Fuzzy Rough Sets

In this section, firstly, binary number on FL, two operations of intuitionistic fuzzy and unary operator on IFS are defined. Secondly, the intuitionistic fuzzy rough sets Based on FL is axiomatic.

Definition 3.1. Let $A \in IFS(U)$, $\lambda = (a, b)$ and $\lambda \in L^*$, $\lambda \wedge A, \lambda \vee A$ be defined,

$$\lambda \wedge A = \{\langle x, a \wedge \mu_A(x), b \vee \gamma_A(x)\rangle | x \in U\},$$

$$\lambda \vee A = \{\langle x, a \vee \mu_A(x), b \wedge \gamma_A(x)\rangle | x \in U\}.$$

where $L^* = \{(a, b) | a + b \leq 1, a, b \in [0, 1]\}$ is fuzzy lattice.

Let $A, B \in IFS(U)$, suppose $\varphi, \psi : IFS(U) \to IFS(U)$ are unary operator, where $\varphi = (\varphi_\mu, \varphi_\gamma)$, $\psi = (\psi_\mu, \psi_\gamma)$, obviously, $\varphi_\mu, \psi_\mu : f(U) \to f(U); \varphi_\gamma, \psi_\gamma : \bar{f}(U) \to \bar{f}(U)$ where $f(U)$ is member functions on U and $\bar{f}(U)$ is non-member function on U. for all $A \in IFS(U)$, $\varphi(A) = (\varphi_\mu(\mu_A), \varphi_\gamma(\gamma_A))$, $\psi(A) = (\psi_\mu(\mu_A), \psi_\gamma(\gamma_A))$, $\varphi_\mu(\mu_A) = \mu_{\varphi(A)}$, $\varphi_\gamma(\gamma_A) = \gamma_{\varphi(A)}$, $\psi_\mu(\mu_A) = \mu_{\psi(A)}$ and $\psi_\gamma(\gamma_A) = \gamma_{\psi(A)}$.

Theorem 3.1. Let U be an arbitrary universal set, suppose that $\varphi :$ IFS$(U) \to$ IFS(U) is an unary operator which satisfy $\varphi(\cap_{i \in I}(\lambda_i \vee A_i)) = \cap_{i \in I}(\lambda_i \vee \varphi(A_i))$ for any index I and $i \in I, A_i \in IFS(U)$, where $\varphi_\mu(\cap_{i \in I}(\lambda_i \vee A_i)) = \cap_{i \in I}(a_i \vee \varphi_\mu(A_i))$, $\varphi_\gamma(\cap_{i \in I}(\lambda_i \vee A_i)) = \cup_{i \in I}(b_i \wedge \varphi_\gamma(A_i))$, the conclusions hold:

(1) R is reflexive $\Leftrightarrow \underline{R}A \subseteq A$ for all $A \in IFS(U)$;
(2) R is symmetric $\Leftrightarrow \underline{RR}A \subseteq \underline{R}A$ for all $A \in IFS(U)$;
(3) R is transitive $\Leftrightarrow A \otimes \varphi(B) = B \otimes \varphi(A)$, for all $A, B \in IFS(U)$.

Proof: First step is to prove the existence of binary intuitionistic fuzzy relation on U make satisfy $\varphi(A) = \underline{R}A$ for all $A \in IFS(U)$. Second step is that the conclusion (1), (2), (3) was established. (We prove that when x is a constant, y equals to z.)

Step 1: for all $\mu_A, \gamma_A \in FS(U)$, μ_A and γ_A are denoted,

$$\mu_A = \cup_{y \in U}\left(\mu_{1_y} \cap \hat{\mu_A}(y)\right) = \cap_{y \in U}\left(\mu_{1_{\{y\}'}} \cup \hat{\mu_A}(y)\right);$$

$$\gamma_A = \cup_{y \in U}\left(\gamma_{1_y} \cap \hat{\gamma_A}(y)\right) = \cap_{y \in U}\left(\gamma_{1_{\{y\}'}} \cup \hat{\gamma_A}(y)\right).$$

Where $1_{\{y\}'} = 1_{U - \{y\}} = \{\langle x, \mu_{\{y\}'}(x), \gamma_{\{y\}'}(x)\rangle | x \in U\}$, $1_y = \{\langle x, \mu_{1_y}(x), \gamma_{1_y}(x)\rangle | x \in U\}$

$$\mu_{1_y}(x) = \begin{cases} 1 & x = y \\ 0 & x \neq y \end{cases}, \gamma_{1_y}(x) = \begin{cases} 0 & x = y \\ 1 & x \neq y \end{cases}, \mu_{1_{\{y\}'}}(x) = \begin{cases} 1 & x \neq y \\ 0 & x = y \end{cases}, \gamma_{1_{\{y\}'}}(x) = \begin{cases} 0 & x \neq y \\ 1 & x = y \end{cases}.$$

The $\mu_R(x,y) = \varphi_\gamma(\gamma_{1_y})(x), \gamma_R(x,y) = \varphi_\mu(\mu_{1_{\{y\}'}})(x)$, then,

$$\begin{aligned}
\mu_{\varphi(A)}(x) &= \varphi_\mu(\mu_A)(x) \\
&= \varphi_\mu(\cap_{y\in I}(\mu_A(y) \vee \mu_{1_{\{y\}'}}(x))) \\
&= \bigwedge_{y\in I}(\mu_A(y) \vee \varphi_\mu(\mu_{1_{\{y\}'}})(x)) \\
&= \bigwedge_{y\in I}(\mu_A(y) \vee \gamma_R(x,y)) \\
&= \mu_{\underline{R}A}(x)
\end{aligned}$$

$$\begin{aligned}
\gamma_{\varphi(A)}(x) &= \varphi_\gamma(\gamma_A)(x) \\
&= \varphi_\gamma(\cap_{y\in I}(\gamma_A \overset{\wedge}{(y)} \vee \gamma_{1_{\{y\}}}(x))) \\
&= \bigvee_{y\in I}(\gamma_A \overset{\wedge}{(y)} \wedge \varphi_\gamma(\gamma_{1_{\{y\}}})(x)) \\
&= \bigvee_{y\in I}(\gamma_A(y) \wedge \mu_R(x,y)) \\
&= \gamma_{\underline{R}A}(x)
\end{aligned}$$

So exist $R \in IFR(U \times U)$ make satisfy $\varphi(A) = \underline{R}A, \forall A \in IFS(U)$.

Step 2:

(1) It is similarly proved by Theorem 2.5 (1).
(2) It is similarly proved by Theorem 2.5 (3).
(3) For one thing, we can know that $\varphi(A) = \underline{R}A$ by step 1, $A \otimes \varphi(B) = A \otimes \underline{R}B = B \otimes \underline{R}A = B \otimes \varphi(A)$. Where $A \otimes \underline{R}B = B \otimes \underline{R}A$ is proved by as following:

$$\begin{aligned}
(\mu_A, \mu_{\underline{R}B}) &= \bigvee_{x\in U}[\mu_A(x) \wedge \mu_{\underline{R}B}(x)] \\
&= \bigvee_{x\in U}[\mu_A(x) \wedge [\bigvee_{y\in U}[\gamma_R(x,y) \wedge \mu_B(y)]] \\
&= \bigvee_{x\in U}\bigvee_{y\in U}[\mu_A(x) \wedge \gamma_R(x,y) \wedge \mu_B(y)] \\
&= \bigvee_{y\in U}[\bigvee_{x\in U}[\mu_A(x) \wedge \gamma_R(y,x)] \wedge \mu_B(y)] \\
&= \bigvee_{y\in U}[\mu_{\underline{R}A}(y) \wedge \mu_B(y)] \\
&= (\mu_B, \mu_{\underline{R}A})
\end{aligned}$$

$$
\begin{aligned}
[\gamma_A, \gamma_{\underline{R}B}] &= \bigwedge_{x \in U} [\gamma_A(x) \vee \gamma_{\underline{R}B}(x)] \\
&= \bigwedge_{x \in U} [\gamma_A(x) \vee [\bigwedge_{y \in U} [\mu_R(x, y) \vee \gamma_B(y)]] \\
&= \bigwedge_{x \in U} \bigwedge_{y \in U} [\gamma_A(x) \vee \mu_R(y, x) \vee \gamma_B(x)] \\
&= \bigwedge_{y \in U} [\bigwedge_{x \in U} [\mu_A(x) \vee \gamma_R(y, x)] \vee \mu_B(y)] \\
&= \bigwedge_{y \in U} [\mu_{\underline{R}A}(y) \vee \mu_B(y)] \\
&= [\gamma_B, \gamma_{\underline{R}A}]
\end{aligned}
$$

For another thing, let $A = 1_x, B = 1_y$, because $A \otimes \varphi(B) = B \otimes \varphi(A)$, $A \otimes \varphi(B) = A \otimes \underline{R}B = ((\mu_A, \mu_{\underline{R}B}), \quad [\gamma_A, \gamma_{\underline{R}B}] = ((\mu_B, \mu_{\underline{R}A}), [\gamma_B, \gamma_{\underline{R}B}]) = B \otimes \underline{R}A = B \otimes \varphi(A)$ and $(\mu_A, \mu_{\underline{R}B}) = (\mu_B, \mu_{\underline{R}A}), [\gamma_A, \gamma_{\underline{R}B}] = [\gamma_B, \gamma_{\underline{R}A}]$.

$$
\begin{aligned}
(\mu_A, \mu_{\underline{R}B}) &= \bigvee_{x' \in U} [\mu_A(x') \wedge \mu_{\underline{R}B}(x')] \\
&= \bigvee_{x' \in U} [\mu_A(x') \wedge [\bigvee_{y' \in U} [\gamma_R(x', y') \wedge \mu_B(y')]]] \\
&= \bigvee_{x' \in U} [\mu_A(x') \wedge \gamma_R(x', y)] \\
&= \gamma_R(x, y)
\end{aligned}
$$

$$
\begin{aligned}
(\mu_B, \mu_{\underline{R}A}) &= \bigvee_{y' \in U} [\mu_B(y') \wedge \mu_{\underline{R}A}(y')] \\
&= \bigvee_{y' \in U} [\mu_B(y') \wedge [\bigvee_{x' \in U} [\gamma_R(y', x') \wedge \mu_A(x')]]] \\
&= \bigvee_{y' \in U} [\mu_A(y') \wedge \gamma_R(y', x)] \\
&= \gamma_R(y, x)
\end{aligned}
$$

So $\gamma_R(x, y) = \gamma_R(y, x)$, by the similar way, we can have $\mu_R(x, y) = \mu_R(y, x)$, so R is symmetric relation.

Theorem 3.2. Let U be an arbitrary universal set, suppose that $\varphi : \mathrm{IFS}(U) \to \mathrm{IFS}(U)$ is an unary operator satisfy the following conditions for all $A, B \in IFS(U)$.

(1) $\varphi(A) \subseteq A$;
(2) $A \otimes \varphi(B) = B \otimes \varphi(A)$;
(3) $\varphi(A) \subseteq \varphi(\varphi(A))$.

Then existence $R \in IFR(U \times U)$ satisfy $\varphi(A) = \underline{R}A$.

Proof: for any $B \in IFS(U)$

We have

$$\begin{aligned} B \otimes \varphi(\cap_{i\in I}(\lambda_i \vee A_i)) &= (\cap_{i\in I}(\lambda_i \vee A_i)) \otimes \varphi(B) \\ &= \bigwedge_{i\in I}((\lambda_i \vee A_i) \otimes \varphi(B)) \\ &= \bigwedge_{i\in I}((\lambda_i \otimes \varphi(B)) \vee (A_i \otimes \varphi(B))) \\ &= \bigwedge_{i\in I}((\varphi(\lambda_i) \otimes B) \vee (\varphi(A_i) \otimes B)) \\ &= \bigwedge_{i\in I}((\varphi(\lambda_i) \otimes B) \vee (\varphi(A_i) \otimes B) \\ &= \cap_{i\in I}(\lambda_i \vee \varphi(A_i)) \otimes B \end{aligned}$$

By Definition 2.5 (3), we can know that $\varphi(\cap_{i\in I}(\lambda_i \vee A_i)) = \cap_{i\in I}(\lambda_i \vee \varphi(A_i))$, therefore there exists $R \in IFR(U \times U)$ satisfy $\varphi(A) = \underline{R}A$ by Theorem 3.1.

Theorem 3.3. Let U be an arbitrary universal set, suppose that $\varphi : \mathrm{I}FS(U) \to \mathrm{I}FS(U)$ is an unary operator. The following propositions are equivalent for all $A, B \in IFS(U)$.

(1) $\varphi(\cap_{i\in I}(\lambda_i \vee A_i)) = \cap_{i\in I}(\lambda_i \vee \varphi(A_i))$, for all $\lambda_i = (a_i, b_i) \in L^*$;

(2) $\varphi(A \cup \widehat{\langle \alpha, \beta \rangle}) = \varphi(A) \cup \widehat{\langle \alpha, \beta \rangle}, \varphi(A \cap B) = \varphi(A) \cap \varphi(B)$.

Proof: (1) $\Rightarrow$ (2) by Theorem 3.1, we have existence $R \in IFR(U \times U)$ satisfy $\varphi(A) = \underline{R}A$ for all $A \in IFS(U)$. By Definition 2.5 (3), we have $\varphi(A \cup \widehat{\langle \alpha, \beta \rangle}) = \underline{R}(A \cup \widehat{\langle \alpha, \beta \rangle}) = \underline{R}(A) \cup \widehat{\langle \alpha, \beta \rangle} = \varphi(A) \cup \widehat{\langle \alpha, \beta \rangle}$.

(2) $\Rightarrow$ (1) because by Theorem 2.1 and Theorem 2.2, we have existence $R \in IFR(U \times U)$ satisfy $\varphi(A) = \underline{R}A$ for all $A \in IFS(U)$, and $\varphi(\cap_{i\in I}(\lambda_i \vee A_i)) = \underline{R}(\cap_{i\in I}(\lambda_i \vee A_i)) = \cap_{i\in I}\underline{R}(\lambda_i \vee A_i) = \cap_{i\in I}(\lambda_i \vee \underline{R}A_i) = \cap_{i\in I}(\lambda_i \vee \varphi(A_i))$.

The second equal: for any index I, $i \in I$, $A_i \in IFS(U)$,

$$\begin{aligned} \mu_{\underline{R}(\cap_{i\in I}A_i)}(x) &= \bigwedge_{y\in U}[\mu_R(x,y) \vee \mu_{\cap_{i\in I}A_i}(x)] \\ &= \bigwedge_{i\in I}\bigwedge_{y\in U}[\mu_R(x,y) \vee \mu_{A_i}(x)] \\ &= \bigwedge_{i\in I}\mu_{\underline{R}A_i}(x) \end{aligned}$$

$$\begin{aligned} \gamma_{\underline{R}(\cap_{i\in I}A_i)}(x) &= \bigvee_{y\in U}[\gamma_R(x,y) \wedge \gamma_{\cap_{i\in I}A_i}(x)] \\ &= \bigvee_{i\in I}\bigvee_{y\in U}[\gamma_R(x,y) \wedge \gamma_{A_i}(x)] \\ &= \bigvee_{i\in I}\gamma_{\underline{R}A_i}(x) \end{aligned}$$

so $\underline{R}(\cap_{i\in I}A_i) = \cap_{i\in I}\underline{R}(A_i)$.

The third equal:

$$\begin{aligned}\mu_{\underline{R}(\lambda_i \vee A_i)}(x) &= \bigwedge_{y\in U}[\gamma_R(x,y) \vee \mu_{\lambda_i \vee A_i}(y)] \\ &= \bigwedge_{y\in U}[\gamma_R(x,y) \vee [a_i \vee \mu_{A_i}(y)]] \\ &= a_i \vee \mu_{\underline{R}A_i}(x)\end{aligned}$$

By similar way, we can get $\gamma_{\underline{R}(\lambda_i \vee A_i)}(x) = b_i \wedge \gamma_{\underline{R}A_i}(x)$, then $\underline{R}(\lambda_i \vee A_i) = \lambda_i \vee \underline{R}(A_i)$, therefore $\cap_{i\in I}\underline{R}(\lambda_i \vee A_i) = \cap_{i\in I}(\lambda_i \vee \underline{R}A_i)$.

In summary, the above propositions are equivalent.

Theorem 3.4. Let U be an arbitrary universal set, suppose that $\psi : IFS(U) \rightarrow IFS(U)$ is an unary operator that satisfy $\psi(\cup_{i\in I}(\lambda_i \wedge A_i)) = \cup_{i\in I}(\lambda_i \wedge \psi(A_i))$ for any index I and $i \in I, A_i \in IFS(U)$, $\lambda_i = (a_i, b_i) \in L^*$, where $\psi_\mu(\cup_{i\in I}(\lambda_i \wedge A_i)) = \cup_{i\in I}(a_i \wedge \psi_\mu(A_i))$ and $\psi_\gamma(\cup_{i\in I}(\lambda_i \wedge A_i)) = \cap_{i\in I}(b_i \vee \psi_\gamma(A_i))$. The following conclusions are established:

(1) R is reflexive $\Leftrightarrow \overline{R}A \subseteq A$ for all $A \in IFS(U)$;
(2) R is symmetric $\Leftrightarrow \overline{R}\overline{R}A \supseteq \overline{R}A$ for all $A \in IFS(U)$;
(3) R is transitive $\Leftrightarrow A \oplus \psi(B) = B \oplus \psi(A)$, for all $A, B \in IFS(U)$.

Proof: It is similarly proved by Theorem 3.1.

Theorem 3.5. Let U be an arbitrary universal set, suppose that $\psi : IFS(U) \rightarrow IFS(U)$ is an unary operator satisfy the following conditions for all $A, B \in IFS(U)$:

(1) $\psi(A) \supseteq A$;
(2) $A \oplus \psi(B) = B \oplus \psi(A)$;
(3) $\psi(A) \supseteq \psi(\psi(A))$.

Then existence $R \in IFR(U \times U)$ satisfy $\psi(A) = \overline{R}A$.

Proof It is similarly proved by Theorem 3.2.

Theorem 3.6. Let U be an arbitrary universal set, suppose that $\psi : IFS(U) \rightarrow IFS(U)$ is an unary operator. The following propositions are equivalent for all $A, B \in IFS(U)$.

(1) $\psi(\cup_{i\in I}(\lambda_i \wedge A_i)) = \cup_{i\in I}(\lambda_i \wedge \psi(A_i))$, for all $\lambda_i = (a_i, b_i) \in L^*$;

$$\psi(A \cap \widehat{\langle \alpha, \beta \rangle}) = \psi(A) \cap \widehat{\langle \alpha, \beta \rangle}, \psi(A \cap B) = \psi(A) \cap \psi(B).$$

Proof: It is similarly proved by Theorem 3.3.

4 Conclusion

In this paper, we present the theory of IFRS based on FL. Firstly, we discuss the relevant conclusions and equivalent conditions in the intuitionistic fuzzy approximate space, and this theory also applies to fuzzy approximate space. Secondly, the axiomatic theory can be extended to two types of fuzzy approximate space.

References

1. Pawlak, Z., Grzymala-Bausse, J., Slowinski, R., et al.: Rough sets. Commun. Soft ACM **38** (11), 89–95 (1995)
2. Liu, G.L.: The axiomatization of the rough set upper approximation operations. Fundam. Inform. **69**, 331–342 (2006)
3. Atanassov, K.: More on intuitionistic fuzzy sets. Fuzzy Sets Syst. **33**, 37–45 (1989)
4. Atanassov, K.: Intuitionistic fuzzy sets. Fuzzy Sets Syst. **20**, 87–96 (1986)
5. Hu, B.Q.: Generalized interval-valued fuzzy variable precision rough sets determined by fuzzy logical operators. Int. J. Gen. Syst. **44**, 447–486 (2015)
6. Luo, S.Q., Xu, W.H., et al.: Rough Atanassov's intuitionistic fuzzy sets model over two universes and its applications. Sci. World J. 243-265 (2014)
7. Zeng, T.G., Zhang, X.X.: Variable precision intuitionistic fuzzy rough sets model and its application. Int. J. Mach. Learn. Cybern. **5**, 263–280 (2014)
8. Yang, Y., Zhu, X.Z., Li, L.: Axiomatization of intuitionistic fuzzy rough sets. J. Hefei Univ. Technol. **33**(04), 590–592 (2010)
9. Zhou, L., Wu, W.Z.: On generalized intuitionistic fuzzy rough approximation operators. Inf. Sci. **178**, 2448–2465 (2008)
10. Borzooei, R.A., Dvurečenskij, A., Zahiri, O.: L-ordered and L-lattice ordered groups. Inf. Sci. **314**, 118–134 (2015)
11. Zhang, C., Zhang, S., Liu, Z.X.: Extension principle on intuitionistic fuzzy set. J. Beijing Norm. Univ. (Nat. Sci.) (25), 106–113 (2011)

On the Characterizations of L-fuzzy Rough Sets Based on Fuzzy Lattices

Kai Hu(✉), Qiu Jin, and Ling-qiang Li

Department of Mathematics, Liaocheng University, Liaocheng 252059, People's Republic of China
hukai80@163.com

Abstract. In this paper, we use one axiom to characterize a series of upper (lower) approximations of L-fuzzy rough sets based on fuzzy lattices. The most interesting result of this paper is the investigation of two operators associated with an abstract operator defined on the fuzzy powerset of a universal set. These two operators will paly an essential role in this study. This work can be regarded as a continuation and generalization of Liu's work in 2013. In which, he characterized the upper approximations of rough sets and fuzzy rough sets by one axiom.

Keywords: Base · Subbase · Enriched L-topologies L-fuzzy rough set · Residuated lattice

1 Introduction

As a technique to information processing, rough set theory was proposed by Pawlak [21–23]. The classical rough set theory is based on equivalence relations, but in some situations, equivalence relations are not suitable for coping with granularity. Thus classical rough set theory has been extended to general relation-based rough sets [7,11,34–37], covering-based rough sets [15,17,37,39,40] and more general fuzzy and lattice-valued fuzzy rough sets [1,2,4–6,8,9,13,14,16,18–20,24–27,31,33].

Axiomatic approaches are important for the conceptual understanding of rough set theory. The axiomatic approach considers abstract upper and lower approximations subject to certain axioms as the primitive notions, and seeks for conditions restraining the axioms to guarantee the existence of a binary relation R (resp., a covering $\mathcal{C}$) on X such that the abstract upper and the abstract lower approximations can be derived from R (resp., $\mathcal{C}$) in the usual way. The research work of the axiomatic approach has been carried out by many authors in the study of rough set [3,12,17,36,38,39]. The most interesting axiomatic studies on rough sets were reported by Yao [36], who extended axiomatic approach to rough set algebras constructed from arbitrary binary relations. The study of the axiomatic approach have also been extended to approximation operators in fuzzy or lattice-valued fuzzy environment. Mi and Zhang [19] and Wu and

B.-Y. Cao and Y.-B. Zhong (Eds.): ICFIE 2017, AISC 872, pp. 254–266, 2019.
https://doi.org/10.1007/978-3-030-02777-3_23

Zhang [33] studied axioms for approximation operators in the context of fuzzy sets. Morsi and Yakout [20] investigated a set of axioms on fuzzy rough sets, but their work was restricted to fuzzy T-rough sets defined by fuzzy T-similarity relations. Thiele [28–30] investigated axiomatic characterizations of fuzzy rough approximation operators and rough fuzzy approximation operators within modal logic. Moreover, Liu [13] extended the axiomatic approach to generalized rough sets over fuzzy lattices. In [27], She and Wang discussed the axiomatic system of residuated lattice-valued fuzzy rough sets proposed by Radzikowska and Kerre [26]. However, almost all of the above axiomatic systems consist of more than one axiom.

Recently, invoked by the similarity between the upper approximation operators and closure operators and that closure operators can be described by a single axiom [10], Liu [16] used one axiom to describe the upper approximations generated by general binary relation and some special binary relations. In his study, the notion of inner product and the inverse operator of an abstract operator defined on the power set of a universal set, play an essential role. **However, in the study of the fuzzy upper approximations, Liu [16] did not define the inverse operator of an abstract operator defined on the fuzzy power set of a universal set. This leads some interesting characterizations on the fuzzy upper approximations are not available.** In addition, Liu also considered the single axiomatic characterizations on the (fuzzy) lower approximations. **However, in his study, only the (fuzzy) lower approximations generated by the (fuzzy) similar relations are discussed. For other cases, such as the general (fuzzy) reflexive, (fuzzy) symmetric, and (fuzzy) transitive relations have not been studied**. The aim of the present paper is to investigate and to solve these questions.

In this paper, we aim to present the single axiomatic characterizations on the theory of generalized rough sets over fuzzy lattice [13]. The contents are arranged as follows. In Sect. 2, we recall foundations of L-fuzzy sets and generalized rough sets over fuzzy lattice (L-fuzzy rough sets for short). In Sect. 3, we recall and generalize the notions of inner (out) products of L-fuzzy sets. Then we define two operators associated with each abstract operator f, which is defined on the L-fuzzy power set on a universal set. In Sects. 4 and 5, we use the notion of out (inner) products and the newly defined abstract operators to describe the lower and upper approximations generated by general L-fuzzy relations and some special L-fuzzy relations, respectively. The paper is completed with some concluding remarks.

2 Preliminaries

2.1 L-fuzzy Sets

In this paper, if not otherwise specified, $L = (L, \wedge, \vee, ', 0, 1)$ is always a fuzzy lattice [32], i.e., a completely distributive lattice with an order reversing involution $' : L \to L$. Let X be a set of objects called the universal set. We call a function $A : X \to L$ as an L-fuzzy set in X. We use L^X to denote the set of all

L-fuzzy sets in X and call it the L-fuzzy power set on X. Under the pointwise order, the set L^X also becomes a fuzzy lattice. For any $A, B, A_t (t \in T) \in L^X$ and for any $\alpha \in L$, we denote by $A \wedge B$, $A \vee B$, $\bigvee_{t \in T} A_t$, $\bigwedge_{t \in T} A_t$ and A' the L-fuzzy sets defined by

$$(A \wedge B)(x) = A(x) \wedge B(x), (A \vee B)(x) = A(x) \vee B(x),$$
$$(\bigvee_{i \in I} A_t)(x) = \bigvee_{t \in T} A_t(x), (\bigwedge_{t \in T} A_t)(x) = \bigwedge_{t \in T} A_t(x), A'(x) = (A(x))'.$$

Also, we make no difference between a constant function and its value since no confusion will arise. For $A \subseteq X$, let 1_A denote the characteristic function of A.

2.2 *L*-fuzzy Rough Sets

Let X be a universal set. Then a function $R : X \times X \longrightarrow L$ is said to be an L-fuzzy relation on X. R is called reflexive if $R(x, x) = 1$ for all $x \in X$; R is said to be symmetric if $R(x, y) = R(y, x)$ for all $x, y \in X$; R is said to be transitive if $R(x, y) \wedge R(y, z) \leq R(x, z)$ for all $x, y, z \in X$. R is refereed to as a similarity L-fuzzy relation if R is reflexive, symmetric and transitive. For each $x \in X$, the L-fuzzy set $R(-, x)$ defined by $R(-, x)(y) = R(x, y)$ is said to be the left R-neighborhood of x. For an L-fuzzy relation R on X, we use R^{-1} to denote its inverse relation (i.e., $R^{-1}(x, y) = R(y, x)$).

Definition 1.1 (Liu [13]). Let R be an L-fuzzy relation on a universal set X. For each $A \in L^X$, we associate two L-fuzzy sets $\underline{R}A$, $\overline{R}A$:

$$\forall x \in X, \underline{R}A(x) = \bigwedge_{y \in X} ((R(x, y))' \vee A(y)), \quad \overline{R}A(x) = \bigvee_{y \in X} (R(x, y) \wedge A(y)).$$

$\underline{R}A$ and $\overline{R}A$ are refereed to as the upper and lower approximations of A, respectively. The pair $(\underline{R}A, \overline{R}A)$ is said bo be a generalized rough set of A over L(L-fuzzy rough set for short). $\overline{R}, \underline{R}$ are referred to as the upper and lower L-fuzzy rough approximation operators, respectively.

The following lemma collects some basic properties of the upper and lower L-fuzzy rough approximation operators.

Lemma 1.2 (Liu [13]). Let R be an L-fuzzy relation on a universal set X. Then for all $A, B \in L^X$,$\{A_t\}_{t \in T} \subseteq L^X$, $\alpha \in L$ and $x, y \in X$ we have

(U1) $\overline{R}(\alpha \wedge A) = \alpha \wedge \overline{R}A$; (L1) $\underline{R}(\alpha \vee A) = \alpha \vee \underline{R}A$;
(U2) $A \leq B \Rightarrow \overline{R}A \leq \overline{R}B$; (L2) $A \leq B \Rightarrow \underline{R}A \leq \underline{R}B$;
(U3) $\overline{R}(\bigvee_{t \in T} A_t) = \bigvee_{t \in T} \overline{R}A_t$; (L3) $\underline{R}(\bigwedge_{t \in T} A_t) = \bigwedge_{t \in T} \underline{R}(A_t)$;
(U4) $\overline{R}(1_{\{x\}}) = R(-, x)$; (L4) $\underline{R}(1_{X - \{x\}}) = (R(-, x))'$;
(UL) $(\overline{R}(A'))' = \underline{R}A$, $(\underline{R}(A'))' = \overline{R}A$;
R is reflexive$\Leftrightarrow$(U5) $A \leq \overline{R}A \Leftrightarrow$(L5) $\underline{R}A \leq A$;
R is symmetric$\Leftrightarrow$(U6) $\overline{R}(1_{\{x\}})(y) = \overline{R}(1_{\{y\}})(x) \Leftrightarrow$(L6) $\underline{R}(1_{\{X - x\}})(y) = \underline{R}(1_{\{X - y\}})(x)$;
R is transitive$\Leftrightarrow$(U7) $\overline{RR}A \leq \overline{R}A \Leftrightarrow$(L7) $\underline{R}A \leq \underline{RR}A$.

3 Inner (out) Products and the Inverse Operators of Abstract Operators

In this section, we shall recall and generalize the notion of inner (out) product, and then define two inverse operators for each abstract operator $f : L^X \to L^X$. As we will see, both of them play an essential role in the following study.

Definition 2.1. Let A, B be L-fuzzy sets on X. Then

(1) the inner product of A, B (Liu [13]), denoted by (A, B), is defined by $(A, B) = \bigvee_{x \in X} (A(x) \wedge B(x))$;

(2) the out product of A, B (Liu [14] for $L = [0, 1]$), denoted by $[A, B]$, is defined by $[A, B] = \bigwedge_{x \in X} (A(x) \vee B(x))$.

When the lattice L being a complete residuated lattice, the notion of inner product also appeared in [2] under the name of related degree of two L-fuzzy sets. And this notion has been used to study more general rough sets than those generated by L-fuzzy relations. When $L = \{0, 1\}$, it is easily seen that $[A, B] = 1$ means $A \cup B = X$, however, $(A, B) = 1$ means $A \cap B \neq \emptyset$.

The following lemma collects the basic properties of inner (out) products used in this paper. Most of them have appeared in [13,14] except the last two.

Lemma 2.2. Let $A, B, C, A_t (t \in T) \in L^X, \alpha \in L$. Then

(N1) $(A, B) = (B, A)$, (O1) $[A, B] = [B, A]$,
(N2) if $B \leq C$, then $(A, B) \leq (A, C)$, (O2) if $B \leq C$, then $[A, B] \leq [A, C]$,
(N3) $(\bigvee_{t \in T} A_t, A) = \bigvee_{t \in T} (A_t, A)$, (O3) $[\bigwedge_{t \in T} A_t, A] = \bigwedge_{t \in T} [A_t, A]$,
(N4) $(\alpha \wedge A, B) = \alpha \wedge (A, B)$, (O4) $[\alpha \vee A, B] = \alpha \vee [A, B]$,
(N5) if $(A, C) \leq (B, C)$ for each $C \in L^X$ then $A \leq B$,
(O5) if $[A, C] \leq [B, C]$ for each $C \in L^X$ then $A \leq B$,
(N6) if $(A, C) = (B, C)$ for each $C \in L^X$ then $A = B$,
(O6) if $[A, C] = [B, C]$ for each $C \in L^X$ then $A = B$,
(N7) if R is an L-fuzzy relation on X then $(A, \overline{R}B) = (B, \overline{R^{-1}}A)$,
(O7) if R is an L-fuzzy relation on X then $[A, \underline{R}B] = [B, \underline{R^{-1}}A]$.

Proof. We prove only (N7). Indeed,

$$\begin{aligned}(A, \overline{R}B) &= \bigvee_{x \in X} (A(x) \wedge \overline{R}B(x)) = \bigvee_{x \in X} (A(x) \wedge \bigvee_{y \in X} (R(x, y) \wedge B(y))) \\ &= \bigvee_{x, y \in X} (A(x) \wedge R(x, y) \wedge B(y)) = \bigvee_{y \in X} (B(y) \wedge \bigvee_{x \in X} (R^{-1}(y, x) \wedge A(x))) \\ &= \bigvee_{y \in X} (B(y) \wedge \overline{R^{-1}}A(y)) = (B, \overline{R^{-1}}A).\end{aligned}$$

In [16], for each operator $f : P(X) \to P(X)$, where $P(X)$ denotes the power set on X, a operator $f^{-1} : P(X) \to P(X)$ is defined and then is used to study the upper approximations generated by (reflexive) binary relations. We observe that f^{-1} can be redefined by the notion of inner products. This observation leads the following definition.

Definition 2.3. Let $f : L^X \to L^X$ be an arbitrary operator. Then the associated operator $f_u^{-1} : L^X \to L^X$ defined by

$$f_u^{-1}(A)(x) = (f(1_{\{x\}}), A), A \in L^X, x \in X,$$

is called the upper inverse operator of f. The associated operator $f_l^{-1} : L^X \to L^X$ defined by

$$f_l^{-1}(A)(x) = [f(1_{X-\{x\}}), A], A \in L^X, x \in X,$$

is called the lower inverse operator of f.

Remark 2.4. As to our knowledge, the operator f_l^{-1} is defined for the first time. The following discussion shows that f_u^{-1} can be regarded as a generalization of f^{-1} in [16]. Indeed, when $L = \{0, 1\}$, the L-fuzzy sets reduce to the characteristic functions. Then $x \in A \Leftrightarrow A(x) = 1$. Now, for each $A \in L^X, x \in X$,

$$x \in f_u^{-1}(A) \Leftrightarrow f_u^{-1}(A)(x) = 1 \Leftrightarrow (f(1_{\{x\}}), A) = \bigvee_{y \in X} (f(1_{\{x\}})(y) \wedge A(y)) = 1$$
$$\Leftrightarrow \exists y \in X, y \in f(\{x\}) \cap A \Leftrightarrow f(\{x\}) \cap A \neq \emptyset,$$

i.e., $f_u^{-1}(A) = \{x \in X | f(\{x\}) \cap A \neq \emptyset\}$, which coincides with the definition of f^{-1} in [16].

In the following we always assume that $f : L^X \to L^X$ be an arbitrary operator.

Lemma 2.5. Let R be an L-fuzzy relation on X.

(1) If $f = \overline{R}$ then $f_u^{-1} = \overline{R^{-1}}$, (2) if $f = \underline{R}$ then $f_l^{-1} = \underline{R^{-1}}$.

Proof.(1) For each $A \in L^X, x \in X$ we have

$$f_u^{-1}(A)(x) = (f(1_{\{x\}}), A) = (\overline{R}(1_{\{x\}}), A) = \bigvee_{y \in X} (\overline{R}(1_{\{x\}})(y) \wedge A(y))$$
$$\overset{(U4)}{=} \bigvee_{y \in X} (R(y, x) \wedge A(y)) = \bigvee_{y \in X} (R^{-1}(x, y) \wedge A(y)) = \overline{R^{-1}}A(x).$$

It follows that $f_u^{-1}(A) = \overline{R^{-1}}A$ then $f_u^{-1} = \overline{R^{-1}}$ by the arbitrariness of A.
(2) For each $A \in L^X, x \in X$ we have

$$f_l^{-1}(A)(x) = [f(1_{X-\{x\}}), A] = [\underline{R}(1_{X-\{x\}}), A] = \bigwedge_{y \in X} (\underline{R}(1_{X-\{x\}})(y) \vee A(y))$$
$$\overset{(L4)}{=} \bigwedge_{y \in X} ((R(y, x))' \vee A(y)) = \bigwedge_{y \in X} ((R^{-1}(x, y))' \vee A(y)) = \underline{R^{-1}}A(x).$$

It follows that $f_l^{-1}(A) = \underline{R^{-1}}A$ then $f_l^{-1} = \underline{R^{-1}}$ by the arbitrariness of A.

4 A Single Axiomatic Characterization on the Lower Approximations

In this section, we shall use one axiom to describe the lower approximations generated by some special L-fuzzy relations. As we will see that the notion of out products and the operators f_l^{-1} paly an important role.

Theorem 3.1. There exists a unique ***L*-fuzzy relation** R on X such that $f = \underline{R}$ if and only if f satisfies
(LA) $[A, f(B)] = [B, f_l^{-1}(A)]$ for any $A, B \in L^X$.

Proof. Let $f = \underline{R}$. By Lemma ?? (2) we have $f_l^{-1} = \underline{R^{-1}}$ and then (LA) follows by (O7). Conversely, let f satisfy (LA). We prove below (LA)⇒ (LA′):
$f(\bigwedge_{t\in T}(\alpha_t \vee A_t)) = \bigwedge_{t\in T}(\alpha_t \vee f(A_t))$ for any given index set T, $\alpha_t \in L$ and $A_t \in L^X$.
Indeed, for each $B \in L^X$ we have

$$\begin{aligned}
[B, f(\bigwedge_{t\in T}(\alpha_t \vee A_t)] &\overset{(LA)}{=} [\bigwedge_{t\in T}(\alpha_t \vee A_t), f(B)] \overset{(O3)}{=} \bigwedge_{t\in T}[\alpha_t \vee A_t, f(B)] \\
&\overset{(O4)}{=} \bigwedge_{t\in T}(\alpha_t \vee [A_t, f(B)]) \overset{(LA)}{=} \bigwedge_{t\in T}(\alpha_t \vee [B, f(A_t)]) \\
&\overset{(O4)}{=} \bigwedge_{t\in T}[B, \alpha_t \vee f(A_t)] \overset{(O3)}{=} [B, \bigwedge_{t\in T}(\alpha_t \vee f(A_t))].
\end{aligned}$$

By (O6) we obtain (LA′). We use f to construct an L-fuzzy relation R as follows: $R(x, y) = (f(1_{X-\{y\}})(x))'$, i.e., $R(-, y) = (f(1_{X-\{y\}}))'$. For each $A \in L^X$, it is easily seen that $A = \bigwedge_{y\in X}(A(y) \vee 1_{X-\{y\}})$. Then

$$\begin{aligned}
f(A) = f(\bigwedge_{y\in X}(A(y) \vee 1_{X-\{y\}})) &\overset{(LA')}{=} \bigwedge_{y\in X}(A(y) \vee f(1_{X-\{y\}})) \\
&= \bigwedge_{y\in X}(A(y) \vee (R(-, y))') = \underline{R}A.
\end{aligned}$$

So, $f = \underline{R}$. The uniqueness of R is obvious.

Theorem 3.2. There exists a unique **reflexive *L*-fuzzy relation** R on X such that $f = \underline{R}$ if and only if f satisfies
(LR) $[A, B \wedge f_l^{-1}(B)] = [B, f(A)]$ for any $A, B \in L^X$.

Proof. Let R be a reflexive L-fuzzy relation on X and $f = \underline{R}$. Then R^{-1} is also reflexive. By Lemma ?? (2) we have $f_l^{-1} = \underline{R^{-1}}$. Thus for any $A, B \in L^X$ we have

$$[A, B \wedge f_l^{-1}(B)] = [A, B \wedge \underline{R^{-1}}B] \overset{(L5)}{=} [A, \underline{R^{-1}}B] \overset{(O7)}{=} [B, \underline{R}A] = [B, f(A)],$$

i.e., the axiom (LR) holds.

Conversely, let f satisfy (LR). Then

(1) (LR)$\Rightarrow$ (LR1): $f(A) \leq A$ for any $A \in L^X$. Indeed, for each $C \in L^X$

$$[C, A] \overset{(O1)}{=} [A, C] \overset{(O2)}{\geq} [A, C \wedge f_l^{-1}(C)] \overset{(LR)}{=} [C, f(A)].$$

It follows by (O5) that $f(A) \leq A$.

(2) (LR1)$\Rightarrow$ (LR2): $A \geq f_l^{-1}(A)$ for any $A \in L^X$. In fact, for any $x \in X$

$$f_l^{-1}(A)(x) = [f(1_{X-\{x\}}), A] \overset{(LR1),(O2)}{\leq} [1_{X-\{x\}}, A] = A(x).$$

It follows that $A \geq f_l^{-1}(A)$ by the arbitrariness of x.

(3) (LR)$\Rightarrow$ (LA). For any $A, B \in L^X$

$$[A, f(B)] \overset{(LR)}{=} [B, A \wedge f_l^{-1}(A)] \overset{(LR2)}{=} [B, f_l^{-1}(A)].$$

Thus the axiom (LA) holds. It follows by Theorem 3.1 that there exists a unique L-fuzzy relation R on X such that $f = \underline{R}$. The reflexivity of R holds for (LR1) and (L5).

Theorem 3.3. There exists a unique **symmetric** $\boldsymbol{L}$**-fuzzy relation** R on X such that $f = \underline{R}$ if and only if f satisfies

(LS) $[A, f(B)] = [B, f(A)]$ for any $A, B \in L^X$.

Proof. Let R be a symmetric L-fuzzy relation on X such that $f = \underline{R}$. Then $R = R^{-1}$ and for any $A, B \in L^X$, we have

$$[A, f(B)] = [A, \underline{R}B] \overset{(O7)}{=} [B, \underline{R^{-1}}A] = [B, \underline{R}A] = [B, f(A)],$$

i.e., the axiom (LS) holds.

Conversely, let f satisfy (LS). Then we construct an L-fuzzy relation R on X as $R(x, y) = (f(1_{X-\{y\}})(x))'$. Similar to Theorem 3.1, one can prove (LS)$\Rightarrow$(LA$'$) and then $f = \underline{R}$ follows. To prove R is symmetric we need check $R(x, y) = R(y, x)$ for any $x, y \in X$. Indeed, it follows by

$$\begin{aligned}(R(x, y))' = f(1_{X-\{y\}})(x) = [1_{X-\{x\}}, f(1_{X-\{y\}})] &\overset{(LS)}{=} [1_{X-\{y\}}, f(1_{X-\{x\}})] \\ = f(1_{X-\{x\}})(y) = (R(y, x))'.&\end{aligned}$$

Theorem 3.4. There exists a unique **transitive** $\boldsymbol{L}$**-fuzzy relation** R on X such that $f = \underline{R}$ if and only if f satisfies

(LT) $f(\bigwedge_{t \in T}(\alpha_t \vee A_t)) = (\bigwedge_{t \in T}(\alpha_t \vee f(A_t))) \wedge (\bigwedge_{t \in T}(\alpha_t \vee ff(A_t)))$ for any given index set T, $\alpha_t \in L$ and $A_t \in L^X$.

Proof. Let R be a transitive L-fuzzy relation on X such that $f = \underline{R}$. By Theorem 3.1 we get that f satisfies (LA$'$) and then (LT) follows by (LA$'$) and (L7).

Conversely, let f satisfy (LT). Then (LT)$\Rightarrow$(LT1): $f(A) \leq ff(A)$ for any $A \in L^X$. Indeed, taking $A_t = A, \alpha_t = 0$, $\alpha_s = 1 (s \neq t)$ in (LT) we obtain $ff(A) \wedge f(A) = f(A)$ and then (LT1) follows. Using (LT1) in (LT) we obtain (LA$'$). By Theorem 3.1 there exists a unique L-fuzzy relation R on X such that $f = \underline{R}$. The transitivity of R holds for (LT1) and (L7).

Theorem 3.5. There exists a unique **reflexive and symmetric L-fuzzy relation** R on X such that $f = \underline{R}$ if and only if f satisfies
(LRS) $[A, B \wedge f(B)] = [B, f(A)]$ for any $A, B \in L^X$.

Proof. Let R be a reflexive and symmetric L-fuzzy relation on X such that $f = \underline{R}$. Then the axiom (LRS) follows by $[A, B \wedge f(B)] \overset{(L5)}{=} [A, f(B)] \overset{(LS)}{=} [B, f(A)]$.

Conversely, let f satisfy (LRS). Then (LRS)$\Rightarrow$(LS). In fact, for each $A, B \in L^X$ we have $[A, B \wedge f(B)] \overset{(O2)}{=} [A, B] \wedge [A, f(B)] \overset{(LRS)}{=} [B, f(A)]$. It follows $[B, A] \overset{(O1)}{=} [A, B] \geq [B, f(A)]$ and then by (O5) we get (LR1). Using (LR1) in (LRS) we get (LS). By Theorem 3.3 there exists a unique symmetric L-fuzzy relation R on X such that $f = \underline{R}$. The reflexivity of R holds for (LR1) and (L5).

Theorem 3.6. There exists a unique **symmetric and transitive L-fuzzy relation** R on X such that $f = \underline{R}$ if and only if f satisfies
(LST) $[A, f(B) \wedge ff(B)] = [B, f(A)]$ for any $A, B \in L^X$.

Proof. Let R be a symmetric and transitive L-fuzzy relation on X such that $f = \underline{R}$.

Then (LST) follows by $[A, f(B) \wedge ff(B)] \overset{(L7)}{=} [A, f(B)] \overset{(LS)}{=} [B, f(A)]$.

Conversely, let f satisfy (LST). Then (LST)$\Rightarrow$(LS). Indeed, for any $A, B \in L^X$,

$$[A, f(B) \wedge ff(B)] \overset{(O2)}{=} [A, f(B)] \wedge [A, ff(B)] \overset{(LST)}{=} [B, f(A)].$$

It follows that $[A, f(B)] \geq [B, f(A)]$, exchanging A, B in $[A, f(B)] \geq [B, f(A)]$ we have $[B, f(A)] \geq [A, f(B)]$. Thus $[A, f(B)] = [B, f(A)]$, i.e., (LS) holds. By Theorem 3.3 there exists a unique symmetric L-fuzzy relation R on X such that $f = \underline{R}$. In addition, by

$$[A, f(B)] \wedge [A, ff(B)] \overset{(LST)}{=} [B, f(A)] \overset{(LS)}{=} [A, f(B)]$$

and (O6) we have $ff(B) \geq f(B)$. This means that R is transitive.

Theorem 3.7. There exists a unique **reflexive and transitive L-fuzzy relation** R on X such that $f = \underline{R}$ if and only if f satisfies (LRT)
$(\bigwedge_{t \in T} (\alpha_t \vee A_t)) \wedge (\bigwedge_{t \in T} (\alpha_t \vee f(A_t))) \wedge (\alpha \vee ff(A)) = (f(\bigwedge_{t \in T} (\alpha_t \vee A_t))) \wedge f(\alpha \vee A)$
for any given index set T, $\alpha_t \in L$, $A_t \in L^X$ and $\alpha \in L, A \in L^X$.

Proof. Let R be a reflexive and transitive L-fuzzy relation on X such that $f = \underline{R}$. Then by (L5) and (L7) we have $ff(A) = f(A)$ for any $A \in L^X$. Then (LRT) follows by (LR1), (LA′) and (L1).

Conversely, let f satisfy (LRT).

(1) Taking $\alpha_t \equiv 1, \alpha = 1$ in (LRT) we get $f(1) = 1$.
(2) Taking $\alpha_t \equiv 1, \alpha = 0$ in (LRT) we get $ff(A) = f(A)$.
(3) Taking $\alpha = 1, \alpha_t = 0, A_t = B$ and $\alpha_s = 1, s \neq t$ in (LRT) we get $B \wedge f(B) = f(B)$ then $B \geq f(B)$, i.e., the condition (LR1) hods.

By (1)–(3), taking $\alpha = 1$ in (LRT) we get (LA′). From the proof of Theorem 3.1 we conclude that there exists a unique L-fuzzy relation R on X such that $f = \underline{R}$. The transitivity and reflexivity of R follows by (2) and (3), respectively.

Theorem 3.8. There exists a unique **similar L-fuzzy relation** R on X such that $f = \underline{R}$ if and only if f satisfies
(LE) $[B, f(A)] = [A, B \wedge f(B) \wedge ff(B)]$ for any $A, B \in L^X$.

Proof. Let R be a similar L-fuzzy relation on X such that $f = \underline{R}$. Then (LE) follows by (LR1), (LT1) and (LS).

Conversely, let f satisfy (LE). Then

$$[B, f(A)] \overset{\text{(LE)}}{=} [A, B \wedge f(B) \wedge ff(B)] \overset{\text{(O3)}}{=} [A, B] \wedge [A, f(B)] \wedge [A, ff(B)].$$

So, $[B, A] = [A, B] \geq [B, f(A)]$, which means (LR1) holds by (O5). Therefore,

$$[B, f(A)] \overset{\text{(LE)}}{=} [A, B \wedge f(B) \wedge ff(B)] \overset{\text{(LR1)}}{=} [A, f(B) \wedge ff(B)].$$

That means the axiom (LST) holds. Thus by Theorem 3.6 we have that there exists a unique symmetric and transitive L-fuzzy relation R on X such that $f = \underline{R}$. The reflexivity of R follows by (LR1).

5 A Single Axiomatic Characterization on the Upper Approximations

In this section, we shall use one axiom to describe the upper approximations generated by some special L-fuzzy relations. As we will see that the notion of inner products and the operators f_u^{-1} paly an important role.

Theorem 4.1. There exists a unique **L-fuzzy relation** R on X such that $f = \overline{R}$ if and only if f satisfies
(UA) $(A, f(B)) = (B, f_u^{-1}(A))$ for any $A, B \in L^X$.

Proof. It is similar to Theorem 3.1.

The axiom (UA) appeared in [16] for the crisp case. The fuzzy case did not appear in Liu's series of paper, because he did not defined the operator f_u^{-1}. Alternatively, Liu [13] used the axiom
(UA′): $f(\bigvee_{t \in T} (\alpha_t \wedge A_t)) = \bigvee_{t \in T} (\alpha_t \wedge f(A_t))$ for any given index set T, $\alpha_t \in L$ and $A_t \in L^X$.

The following theorem shows that (UA)⇔ (UA′).

Theorem 4.2. (UA)⇔ (UA′).

Proof. (UA)$\Rightarrow$ (UA$'$). Indeed, for each $B \in L^X$ we have

$$\begin{aligned}(B, f(\bigvee_{t\in T}(\alpha_t \wedge A_t)) &\overset{(UA)}{=} (\bigvee_{t\in T}(\alpha_t \wedge A_t), f(B)) \overset{(N3)}{=} \bigvee_{t\in T}(\alpha_t \wedge A_t, f(B)) \\ &\overset{(N4)}{=} \bigvee_{t\in T}(\alpha_t \wedge (A_t, f(B))) \overset{(UA)}{=} \bigvee_{t\in T}(\alpha_t \wedge (B, f(A_t))) \\ &\overset{(N4)}{=} \bigvee_{t\in T}(B, \alpha_t \wedge f(A_t)) \overset{(N3)}{=} (B, \bigvee_{t\in T}(\alpha_t \wedge f(A_t))).\end{aligned}$$

By (N6) we obtain (UA$'$).

(UA$'$)$\Rightarrow$ (UA). For each $B \in L^X$, it is easily seen that $B = \bigvee_{x\in X}(B(x) \wedge 1_{\{x\}})$. Then

$$\begin{aligned}(A, f(B)) &= (A, f(\bigvee_{x\in X}(B(x) \wedge 1_{\{x\}}))) \overset{(UA')}{=} (A, \bigvee_{x\in X}(B(x) \wedge f(1_{\{x\}}))) \\ &\overset{(N3)}{=} \bigvee_{x\in X}(A, B(x) \wedge f(1_{\{x\}})) \overset{(N4)}{=} \bigvee_{x\in X}(B(x) \wedge (A, f(1_{\{x\}}))) \\ &\overset{(N1)}{=} \bigvee_{x\in X}(B(x) \wedge (f(1_{\{x\}}), A)) = \bigvee_{x\in X}(B(x) \wedge f_u^{-1}(A)(x)) = (B, f_u^{-1}(A)).\end{aligned}$$

Thus the axiom (UA) holds.

Theorem 4.3. There exists a unique **reflexive L-fuzzy relation** R on X such that $f = \overline{R}$ if and only if f satisfies

(UR) $(A, f(B)) = (B, A \vee f_u^{-1}(A))$ for any $A, B \in L^X$.

Proof. It is similar to Theorem ??.

The axiom (UR) appeared in [16] for the crisp case. The fuzzy case did not appear in Liu's series of paper, because he did not defined the operator f_u^{-1}. Alternatively, Liu [16] used the axiom

(UR$'$) $f(\bigvee_{t\in T}(\alpha_t \wedge A_t)) = (\bigvee_{t\in T}(\alpha_t \wedge A_t)) \vee (\bigvee_{t\in T}(\alpha_t \wedge f(A_t)))$ for any given index set T, $\alpha_t \in L$ and $A_t \in L^X$.

The following theorem shows that (UR)$\Leftrightarrow$ (UR$'$).

Theorem 4.4. (UR)$\Leftrightarrow$ (UR$'$).

Proof. (UR)$\Rightarrow$ (UR$'$).

(1) (UR)$\Rightarrow$ (UR1): $A \leq f(A)$ for any $A \in L^X$. Indeed, for each $C \in L^X$,

$$(C, A) \overset{(N1)}{=} (A, C) \overset{(N2)}{\leq} (A, C \vee f_u^{-1}(C)) \overset{(UR)}{=} (C, f(A)).$$

It follows by (N5) that $A \leq f(A)$.

(2) (UR1)$\Rightarrow$ (UR2): $A \leq f_u^{-1}(A)$ for any $A \in L^X$. In fact, for any $x \in X$

$$f_u^{-1}(A)(x) = (f(1_{\{x\}}), A) \overset{(UR1)}{\geq} (1_{\{x\}}, A) = A(x).$$

It follows that $A \leq f_u^{-1}(A)$ by the arbitrariness of x.

(3) (UR)⇒ (UA). For any $A, B \in L^X$

$$(A, f(B)) \overset{(UR)}{=} (B, A \vee f_u^{-1}(A)) \overset{(UR2)}{=} (B, f_u^{-1}(A)).$$

Now, we can prove (UR)⇒ (UR′).

$$(\bigvee_{t\in T}(\alpha_t \wedge A_t)) \vee (\bigvee_{t\in T}(\alpha_t \wedge f(A_t))) \overset{(UR1)}{=} \bigvee_{t\in T}(\alpha_t \wedge f(A_t)) \overset{(UA')}{=} f(\bigvee_{t\in T}(\alpha_t \wedge A_t)).$$

(UR′)⇒ (UR). For $A \in L^X$, taking $A_t = A, \alpha_t = 1, \alpha_s = 0, s \neq t$ in (UR′) we have $A \vee f(A) = f(A)$, so $A \leq f(A)$, i.e., (UR1) holds and so (UR2) holds. By using $A_t \leq f(A_t)$ in (UR′) we obtain (UA′) and then (UA). Therefore

$$(A, f(B)) \overset{(UA)}{=} (B, f_u^{-1}(A)) \overset{(UR2)}{=} (B, A \vee f_u^{-1}(A)),$$

i.e., the axiom (UR) holds.

Remark 4.5. We can prove similarly that (LA)⇔(LA′) and (LR)⇔(LR′): $f(\bigwedge_{t\in T}(\alpha_t \vee A_t)) = (\bigwedge_{t\in T}(\alpha_t \vee A_t)) \wedge (\bigwedge_{t\in T}(\alpha_t \vee f(A_t)))$ for any given index set T, $\alpha_t \in L$ and $A_t \in L^X$.

Theorem 4.6 (Liu [13]). There exists a unique **symmetric *L*-fuzzy relation** R on X such that $f = \overline{R}$ if and only if f satisfies
(US) $(A, f(B)) = (B, f(A))$ for any $A, B \in L^X$.

Theorem 4.7 (Liu [16] for $L = [0, 1]$). There exists a unique **transitive *L*-fuzzy relation** R on X such that $f = \overline{R}$ if and only if f satisfies
(UT) $f(\bigvee_{t\in T}(\alpha_t \wedge A_t)) = (\bigvee_{t\in T}(\alpha_t \wedge f(A_t))) \vee (\bigvee_{t\in T}(\alpha_t \wedge ff(A_t)))$ for any given index set T, $\alpha_t \in L$ and $A_t \in L^X$.

Theorem 4.8. There exists a unique **reflexive and symmetric *L*-fuzzy relation** R on X such that $f = \overline{R}$ if and only if f satisfies
(URS) $(A, B \vee f(B)) = (B, f(A))$ for any $A, B \in L^X$.

Proof. It is similar to Theorem 3.5.

Theorem 4.9. There exists a unique **symmetric and transitive *L*-fuzzy relation** R on X such that $f = \overline{R}$ if and only if f satisfies
(UST) $(A, f(B) \vee ff(B)) = (B, f(A))$ for any $A, B \in L^X$.

Proof. It is similar to Theorem 3.6.

Remark 4.11. When $L = \{0, 1\}$, the axiom (URT) splits into two axioms:
$f(0) = 0$ and $(\bigvee_{t\in T} A_t) \vee (\bigvee_{t\in T} f(A_t)) \vee ff(A) = f(\bigvee_{t\in T} A_t) \vee f(A)$.
It is easily seen that these two axioms are equivalent the axiom (RT):
$(\bigcup_{t\in T} A_t) \cup (\bigcup_{t\in T} f(A_t)) \cup ff(A) = f(A \cup \bigcup_{t\in T} A_t) - f(0)$.
Thus (URT) can be regarded as a generalization of (RT) in Liu [16].

Theorem 4.11 (Liu [16] for $L = [0, 1]$). There exists a unique **similar *L*-fuzzy relation** R on X such that $f = \overline{R}$ if and only if f satisfies
(UE) $(B, f(A)) = (A, B \vee f(B) \vee ff(B))$ for any $A, B \in L^X$.

6 Conclusion

Recently, Liu [16] used one axiom to describe the approximations of rough sets and fuzzy rough sets. In contrast to his nearly perfect study on the crisp case, some interesting results have not been generalized to the fuzzy case. In this paper, we investigated and solved these regards under a more general lattice context. The most interesting results are the operators f_l^{-1}, f_u^{-1} and the resulted characterizations on the upper and lower approximations of L-fuzzy rough sets.

Acknowledgements. Thanks to the support by National Natural Science Foundation of China (No. 11501278 and No. 11471152) and Shandong Provincial Natural Science Foundation, China (ZR2014AQ011),and Project Science Foundation of Liaocheng University.

References

1. Chen, D.G., Kwong, S., He, Q., et al.: Geometrical interpretation and applications of membership functions with fuzzy rough sets. Fuzzy Sets Syst. **193**, 122–135 (2012)
2. Chen, X., Li, Q.: Construction of rough approximations in fuzzy setting. Fuzzy Sets Syst. **158**, 2641–2653 (2007)
3. Comer, S.: An algebraic approach to the approximation of information. Fundam. Inform. **14**, 492–502 (1991)
4. Dai, J.H., Tian, H.W.: Fuzzy rough set model for set-valued data. Fuzzy Sets Syst. **229**, 54–68 (2013)
5. Dubois, D., Prade, H.: Rough fuzzy sets and fuzzy rough sets. Int. J. Gen. Syst. **17**, 191–208 (1990)
6. Dubois, D., Prade, H.: Putting fuzzy sets and rough sets together. In: Slowinski, R. (ed.) Intelligent Decision Support, pp. 203–232. Kluwer Academic, Dordrecht (2004)
7. Guan, L., Wang, G.: Generalized approximations defined by non-equivalence relations. Inf. Sci. **193**, 163–179 (2012)
8. Hao, J., Li, Q.G.: The relationship between L-fuzzy rough set and L-topology. Fuzzy Sets and Systems **178**, 74–8 (2011)
9. Jin, Q., Li, L.Q.: On the second type of L-fuzzy covering rough sets. Inf. Int. Interdiscip. J. **16**(2), 1101–1106 (2013)
10. Kelley, J.L.: General Topology. Graduate Texts in Mathematics, vol. 27, Springer (1955)
11. Li, Y., Tang, J., Chin, K., Luo, X., Han, Y.: Rough set-based approach for modeling relationship measures in product planning. Inf. Sci. **193**, 199–217 (2012)
12. Lin, T.Y., Liu, Q.: Rough approximate operators: axiomatic rough set theory. In: Ziarko, W. (ed.) Rough Sets, Fuzzy Sets and Knowledge Discovery, pp. 256–260. Springer, Berlin (1994)
13. Liu, G.: Generalized rough set over fuzzy lattices. Inf. Sci. **178**, 1651–1662 (2008)
14. Liu, G.: Axiomatic systems for rough sets and fuzzy rough sets. Int. J. Approx. Reason. **48**, 857–867 (2008)
15. Liu, G., Sai, Y.: A comparison of two types of rough sets induced by coverings. Int. J. Approx. Reason. **50**, 521–528 (2009)

16. Liu, G.: Using one axiom to characterize rough set and fuzzy rough set approximations. Inf. Sci. **223**, 285–296 (2013)
17. Liu, G.: The relationship among different covering approximations. Inf. Sci. **250**, 178–183 (2013)
18. Ma, Z.M., Hu, B.Q.: Topological and lattice structures of L-fuzzy rough sets determined by lower and upper sets. Inf. Sci. **218**, 194–204 (2013)
19. Mi, J., Zhang, W.: An axiomatic characterization of a fuzzy generalization of rough sets. Inf. Sci. **160**, 235–249 (2004)
20. Morsi, N.N., Yakout, M.M.: Axiomatics for fuzzy rough set. Fuzzy Sets Syst. **100**, 327–342 (1998)
21. Pawlak, Z.: Rough sets. Int. J. Comput. Inf. Sci. **11**, 341–356 (1982)
22. Pawlak, Z.: Rough Sets-Theoretical Aspects of Reasoning About Data. Kluwer Academic Publishers, Boston (1991)
23. Pawlak, Z., Skowron, A.: Rough sets: some extensions. Inf. Sci. **177**, 28–40 (2007)
24. Pei, D.W.: A generalized model of fuzzy rough sets. Int. J. Gen. Syst. **34**, 603–613 (2005)
25. Qin, K., Pei, Z., Yang, J.L.: Approximation operators on complete completely distributive lattices. Inf. Sci. **247**, 123–130 (2013)
26. Radzikowska, A.M., Kerre, E.E.: Fuzzy rough sets based on residuated lattices. In: Transactions on Rough Sets II. LNCS, vol. 3135, pp. 278–296 (2004)
27. She, Y.H., Wang, G.J.: An axiomatic approach of fuzzy rough sets based on residuated lattices. Comput. Math. Appl. **58**, 189–201 (2009)
28. Thiele, H.: On axiomatic characterization of fuzzy approximation operators. I, the fuzzy rough set based case. In: Conference Proceedings RSCTC 2000, Banff Park Lodge, Bariff, Canada, 19 October, pp. 239–247 (2000)
29. Thiele, H.: On axiomatic characterization of fuzzy approximation operators II, The rough fuzzy set based case. In: Proceedings of 31st IEEE International Symposium on Multiple-Valued Logic, pp. 330–335 (2001)
30. Thiele, H.: On axiomatic characterization of fuzzy approximation operators III, The fuzzy diamond and fuzzy box cases. In: The 10th IEEE International Conference on Fuzzy Systems, vol. 2, pp. 1148–1151 (2001)
31. Tiwari, S.P., Srivastava, A.K.: Fuzzy rough sets, fuzzy preorders and fuzzy topologies. Fuzzy Sets Syst. **210**, 63–68 (2013)
32. Wang, G.J.: Theory of L-fuzzy Topological Space. Shaanxi Normal University Press, Xi'an (1988). (in Chinese)
33. Wu, W., Zhang, W.: Constructive and axiomatic approaches of fuzzy approximation operators. Inf. Sci. **159**, 233–254 (2004)
34. Yao, Y.: Two views of the theory of rough sets in finite universes. Int. J. Approx. Reason. **15**, 291–317 (1996)
35. Yao, Y.: Relational interpretations of neighborhood operators and rough set approximation operators. Inf. Sci. **111**, 239–259 (1998)
36. Yao, Y.: Constructive and algebraic methods of the theory of rough sets. Inf. Sci. **109**, 21–47 (1998)
37. Yao, Y., Yao, B.: Covering based rough set approximations. Inf. Sci. **200**, 91–107 (2012)
38. Zhang, Y., Li, J., Wu, W.: On axiomatic characterizations of three pairs of covering based approximation operators. Inf. Sci. **180**, 274–287 (2010)
39. Zhang, Y., Luo, M.: On minimization of axiom sets characterizing covering-based approximation operators. Inf. Sci. **181**, 3032–3042 (2011)
40. Zhu, W.: Topological approaches to covering rough sets. Inf. Sci. **177**, 1499–1508 (2007)

Trigonometric Wavelet Methods for the Plane Elasticity Problem

De-jun Peng and Shen Youjian(✉)

Department of Mathematics and Statistics, Hainan Normal University, Haikou 571158, People's Republic of China
28043015@qq.com

Abstract. In this paper, we apply interpolatory Hermite-type trigonometric wavelet to investigate the numerical solution of the natural boundary integral equation of plane elasticity problem in the outer domain of a disk by Galerkin method. In our fast algorithm, the computational formulae of the stiffness matrix yield simple close-form and the stiffness matrix is sparse. The error estimates of the approximate solution are given and the test examples are presented in the end.

Keywords: Plane elasticity problem
Natural boundary integral equation · Wavelet-Galerkin method
Hermite trigonometric wavelet

1 Introduction

Plane elasticity problems include plane strain problem and plane stress problem which are widely applied in mechanics and engineering. Although these two kinds of problems have different mechanical background, they have a common property, i.e. the displacement argument depends only on two coordinate arguments and is free from the third coordinate argument. Consequently, the plane elasticity problem can be reduced as the boundary problem of Navier equation:

$$\mu \Delta \mathbf{u} + (\lambda + \mu)\mathrm{grad}\,\mathrm{div}\mathbf{u} = \mathbf{0} \tag{1}$$

where $\mathbf{u}$ represents the displacement vector function and λ, μ are Lamè constants.

Boundary element method is an important method to solve the boundary value problems of elliptic partial differential equations which can be reduced into equivalent boundary integral equations and further are discretized by the finite element methods as a linear algebraic equation systems. The different reduction ways may lead to different expressions and singularity of the boundary integral equations. The main advantage of the boundary element method is that the dimensionality of the problem is reduced by one, but the computational complexity of the stiffness matrices which are not sparse eliminates its advantage. In

This work is supported by the International Science and Technology Cooperation Foundation of China (Grant No. 2012DFA11270) and Natural Science Foundation of Hainan (Grant No. 117123).

B.-Y. Cao and Y.-B. Zhong (Eds.): ICFIE 2017, AISC 872, pp. 267–282, 2019.
https://doi.org/10.1007/978-3-030-02777-3_24

1923, Hadamard [1] introduced the concept of the finite part for divergent integral with high order singularity and discussed the corresponding hypersingular integral equations with the kernel of high-order singularity. In 1980's, Feng [2,3] and Yu [2,4] introduced a new boundary element method—-natural boundary element method. Their main idea is that the boundary value problem of partial differential equations, via Green function, Fourier series or complex analysis etc., can be converted into equivalent hypersingular boundary integral equations in Hadmard finite part sense, and then the corresponding equivalent variational problem can be solved by using some discrete techniques. In the last decade, the natural boundary element method has been efficiently used to solve some elliptic problems [4–7]. Besides the advantages of the boundary element method, the natural boundary element method has many advantages that have been stated in detail by Yu in [4]. These advantages conclude: (1) The energy functional of the original partial differential equation preserves unchanged. (2) The symmetry and the coerciveness of the bilinear form are preserved which result in the unique existence and stability of the solution of the natural boundary integral equations. (3) The stiffness matrix is both symmetric and positive definite, so that the approximate solution is still stabilized in the numerical computation. (4) Especially in the case of circular boundary, the stiffness matrix is circulant. This allows us only to calculate about one half row entries of the stiffness matrix, so that the computational complexity is reduced substantially by the natural boundary element method. However, the natural boundary element method has been limited to use since for a general domain the corresponding Green function can not be expressed in the explicit finite form.

In recent years, wavelet approximations have attracted much attention as a potentially efficient numerical technique for solving partial differential equations and integral equations [5,8–11,14,16]. In [12], Quak presented the Hermite multiresolution analysis for the space of square-integrable 2π-periodic functions, where two different type of trigonometric wavelet functions $\psi^0_{j,n}(\theta)$, $\psi^1_{j,n}(\theta)$ were constructed which represent different frequency at different levels j. Larger j is, higher frequency the wavelet functions represent. Quak's wavelet functions have the following interpolation properties:

$$\begin{aligned} &\psi^0_{j,n}(\theta_{j,k}) = \delta_{kn}, \psi^0_{j,n}{}'(\theta_{j,k}) = 0, \\ &\psi^1_{j,n}(\theta_{j,k}) = 0, \quad \psi^1_{j,n}{}'(\theta_{j,k}) = \delta_{k,n}, \end{aligned}$$

where

$$\theta_{j,n} = \frac{n\pi}{2^j},\ j \in N_0 = \{0\} \cup N, n = 0, 1, \cdots, 2^{j+1} - 1,$$

are dyadic points and δ_{kn} is the Kronecker's symbol. The trigonometric interpolants enable a completely explicit description of the decomposition and reconstruction coefficients.

For the plane elasticity problem in the outer domain of a disk, Yu [4] applied piecewise polynomial elements to discretize the natural boundary integral equations, where the computational formulae for the entries of the stiffness matrices are expressed as some convergent series. This results in larger numerical error since the series are truncated in the practical computations. In this paper, in order to improve Yu's results we utilize Quak's Hermite-type wavelet to discretize the natural boundary integral equation of the elasticity problem in the outer domain of a disk. As a result, the computational formulae of the stiffness matrix are of simple close-form and hence the stiffness matrix is very accurate and the approximate solution error is small in the practical computation. In the end we point out that despite the domain we consider is a outer disk, this is very special, our method is still useful to general domains since the natural boundary element method may couple naturally with the finite element method (see [4]).

The next section of the paper introduce the multiresolution analysis (MRA) of the interpolatory Hermite-type trigonometric wavelets. In Sect. 3, the natural boundary integral equation is solved by using wavelet-Galerkin method and some error estimates are presented. Finally, in Sect. 4 the numerical results of an example is displayed to illustrate our algorithm.

2 Hermite Interpolatory Trigonometric Wavelet

In this section we briefly state the Hermite interpolatory trigonometric wavelets and their basic properties that first introduced by Quak [12].

For $n \in \mathbb{N}$ ($\mathbb{N}$ denotes the set of all natural numbers), by T_n denotes the linear space of trigonometric polynomials of total degree not exceeding n. The Dirichlet kernel $D_n(\theta) \in T_n$ and its conjugate Dirichlet kernel $\tilde{D}_n(\theta) \in T_n$ are defined as

$$D_n(\theta) = \frac{1}{2} + \sum_{\ell=1}^{n} \cos \ell\theta = \begin{cases} \frac{\sin(n+\frac{1}{2})\theta}{2\sin\frac{\theta}{2}}, & \theta \notin 2\pi\mathbb{Z} \\ n+\frac{1}{2}, & \theta \in 2\pi\mathbb{Z} \end{cases}, \tag{1}$$

and

$$\tilde{D}_n(\theta) = \sum_{\ell=1}^{n} \sin \ell\theta = \begin{cases} \frac{\cos\frac{\theta}{2} - \cos(n+\frac{1}{2})\theta}{2\sin\frac{\theta}{2}}, & \theta \notin 2\pi\mathbb{Z} \\ 0, & \theta \in 2\pi\mathbb{Z} \end{cases}. \tag{2}$$

The equally spaced nodes on the interval $[0, 2\pi)$ with a dyadic step size are defined as

$$\theta_{j,n} = \frac{n\pi}{2^j}, \; j \in \mathbb{N}_0, n = 0, 1, \cdots, 2^{j+1}-1,$$

where $\mathbb{N}_0 = \{0\} \cup \mathbb{N}$.

Definition 1 (Scaling functions). *For $j \in \mathbb{N}_0$, two scaling functions $\phi^0_{j,0}(\theta)$, $\phi^1_{j,0}(\theta)$ are defined as*

$$\phi^0_{j,0}(\theta) = \frac{1}{2^{2j+1}} \sum_{\ell=1}^{2^{j+1}-1} D_\ell(\theta), \tag{3}$$

$$\phi^1_{j,0}(\theta) = \frac{1}{2^{2j+1}} \left(\tilde{D}_{2^{j+1}-1}(\theta) + \frac{1}{2}\sin 2^{j+1}\theta\right). \tag{4}$$

For $n = 1, 2, \cdots, 2^{j+1}-1$, define $\phi^i_{j,n}(\theta) = \phi^i_{j,0}(\theta - \theta_{j,n}), i = 0, 1$. Furthermore, let $\phi^i_{j,n}(\theta) = \phi^i_{j,n \bmod 2^{j+1}}(\theta)$ for $i = 0, 1$ and any $n \in \mathbb{N}$.

These functions are well known and have been studied in detail, e.g., in [18].

From the definition, it is no difficult to verify the following interpolatory properties of the scaling functions: for all $k, n = 0, 1, \cdots, 2^{j+1} - 1$, it holds that

$$\begin{aligned} \phi^0_{j,n}(\theta_{j,k}) &= \delta_{kn}, \ {\phi^0_{j,n}}'(\theta_{j,k}) = 0, \\ \phi^1_{j,n}(\theta_{j,k}) &= 0, \quad {\phi^1_{j,n}}'(\theta_{j,k}) = \delta_{k,n}. \end{aligned} \tag{5}$$

Define scaling function space V_j as

$$V_j = \text{span}\{\phi^0_{j,n}, \phi^1_{j,n} | \ n = 0, 1, \cdots, 2^{j+1} - 1\}, \tag{6}$$

then by the expressions of $\phi^0_{j,n}(\theta), \phi^1_{j,n}(\theta)$ and the interpolatory property (5), we can prove.

Theorem 1. *For $j \in \mathbb{N}_0$, we have*

$$V_j = \text{span}\{1, \cos\theta, \cdots, \cos(2^{j+1} - 1)\theta, \sin\theta, \cdots, \sin 2^{j+1}\theta\}. \tag{7}$$

Theorem 1 implies that $V_j \subset V_{j+1}$ for $j \in \mathbb{N}_0$. If we let $V_{-1} = \{0\}$, then

$$\text{clos}_{L^2}(\bigcup_{j=-1}^{\infty} V_j) = L^2[0, 2\pi], \qquad \bigcap_{j=-1}^{\infty} V_j = \{0\}.$$

Hence, $\{V_j\}_{j=-1}^{\infty}$ forms a multiresolution analysis (MRA) in $L^2[0, 2\pi]$.

Next we give the orthogonal complement W_j of V_j relative to V_{j+1}, i.e., the so-called wavelet space.

Definition 2 (Wavelet functions). *For $j \in \mathbb{N}_0$, two wavelet functions $\psi^0_{j,0}, \psi^1_{j,0}$ are defined by*

$$\psi^0_{j,0}(\theta) = \frac{1}{2^{j+1}} \cos 2^{j+1}\theta + \frac{1}{3 \cdot 2^{2j+1}} \sum_{\ell=2^{2j+1}+1}^{2^{j+2}-1} (3 \cdot 2^{j+1} - \ell) \cos \ell\theta, \tag{8}$$

$$\psi^1_{j,0}(\theta) = \frac{1}{3 \cdot 2^{2j+1}} \sum_{\ell=2^{2j+1}+1}^{2^{j+2}-1} \sin \ell\theta + \frac{1}{2^{2j+3}} \sin 2^{j+2}\theta. \tag{9}$$

For $j \in \mathbb{N}, n = 1, 2, \cdots, 2^{j+1} - 1$, define $\psi^i_{j,n}(\theta) = \psi^i_{j,0}(\theta - \theta_{j,n}), i = 0, 1$. Furthermore, let $\psi^i_{j,n}(\theta) = \psi^i_{j,n \bmod 2^{j+1}}(\theta)$ for $i = 0, 1$ and any $n \in \mathbb{N}$.

Definition 3 (Wavelet space). *For $j \in \mathbb{N}_0$, the wavelet spaces W_j are defined by*

$$W_j = \text{span}\{\psi^0_{j,n}(\theta), \psi^1_{j,n}(\theta) | n = 0, \cdots, 2^{j+1} - 1\}.$$

Corollary 1. *The space W_j is the orthogonal complement of V_j in V_{j+1}, i.e.,*

$$V_{j+1} = V_j \oplus W_j$$

and

$$W_j = \text{span}\{\cos 2^{j+1}\theta, \cdots, \cos(2^{j+2} - 1)\theta, \sin(2^{j+1} + 1)\theta, \cdots, \sin 2^{j+2}\theta\}.$$

The wavelet functions also have the similar interpolatory properties as the scaling functions.

Theorem 2. *For* $k, n = 0, 1, \cdots, 2^{j+1} - 1$, *it holds the following interpolatory properties:*

$$\begin{aligned} \psi^0_{j,n}(\theta_{j,k}) = \delta_{kn},\ {\psi^0_{j,n}}'(\theta_{j,k}) = 0, \\ \psi^1_{j,n}(\theta_{j,k}) = 0, \quad {\psi^1_{j,n}}'(\theta_{j,k}) = \delta_{k,n}. \end{aligned} \tag{10}$$

By taking first, second and third derivatives in the both sides of (1) and (2) with respect to θ respectively, we can get the following lemma that will be used in the next section. A very simple proof for it can be found in [13].

Lemma 1. *For* $j \in \mathbb{N}_0,\ n = 0, 1, \cdots, 2^{j+1} - 1$, *we have*

$$\sum_{\ell=1}^{2^{j+1}-1} \cos \ell\theta_{j,n} = \begin{cases} -1, & n \neq 0 \\ 2^{j+1} - 1, & n = 0 \end{cases}, \tag{11}$$

$$\sum_{\ell=1}^{2^{j+1}-1} \ell \cos \ell\theta_{j,n} = \begin{cases} -2^j, & n \neq 0 \\ 2^j(2^{j+1} - 1), & n = 0 \end{cases}, \tag{12}$$

$$\sum_{\ell=1}^{2^{j+1}-1} \ell^2 \cos \ell\theta_{j,n} = \begin{cases} -2^{2j+1} + 2^j \sin^{-2} \theta_{j+1,n}, & n \neq 0 \\ \frac{2^j}{3}(2^{j+1} - 1)(2^{j+2} - 1), & n = 0 \end{cases}, \tag{13}$$

$$\sum_{\ell=1}^{2^{j+1}-1} \ell^3 \cos \ell\theta_{j,n} = \begin{cases} -2^{3j+2} + 3 \cdot 2^{2j} \sin^{-2} \theta_{j+1,n}, & n \neq 0 \\ 2^{2j}(2^{j+1} - 1)^2, & n = 0 \end{cases}, \tag{14}$$

$$\sum_{\ell=1}^{2^{j+1}-1} \sin \ell\theta_{j,n} = 0,\ n \neq 0, \tag{15}$$

$$\sum_{\ell=1}^{2^{j+1}-1} \ell \sin \ell\theta_{j,n} = -2^j \text{ctg}\theta_{j+1,n},\ n \neq 0, \tag{16}$$

$$\sum_{\ell=1}^{2^{j+1}-1} \ell^2 \sin \ell\theta_{j,n} = -2^{2j+1} \text{ctg}\theta_{j+1,n},\ n \neq 0, \tag{17}$$

$$\sum_{\ell=1}^{2^{j+1}-1} \ell^3 \sin \ell\theta_{j,n} = -2^{j-1}(2^{2j+3} - 3)\text{ctg}\theta_{j+1,n} + 3 \cdot 2^{j-1} \text{ctg}^3\theta_{j+1,n},\ n \neq 0. \tag{18}$$

3 Wavelet-Galerkin Method

Let vector function $\mathbf{u} = (u_1(x_1, x_2), u_2(x_1, x_2))$, where $u_1(x_1, x_2)$ and $u_2(x_1, x_2)$ represent the displacement in the x_1 direction and x_2 direction respectively. Set

$$\varepsilon_{ij}(\mathbf{u}) = \varepsilon_{ji}(\mathbf{u}) = \frac{1}{2}(\frac{\partial u_i}{\partial x_j} + \frac{\partial u_j}{\partial x_i}),\ i = 1, 2,$$

$$\sigma_{ij}(\mathbf{u}) = \sigma_{ji}(\mathbf{u}) = \lambda \sum_{\ell=1}^{2} \varepsilon_{\ell\ell}(\mathbf{u})\delta_{ij} + 2\mu\varepsilon_{ij}(\mathbf{u}),\ i, j = 1, 2,$$

where ε_{ij} are strains, σ_{11}, σ_{22} are the direct stresses along the two principal directions of the x_1, x_2 axes, and σ_{12} is the shear stress, $\lambda > 0$, and $\mu > 0$ are Lamè constants.

Consider the second boundary value problem of the Navier equation in the outer domain of the disk B with radius R:

$$\begin{cases} \mu\Delta\mathbf{u} + (\lambda+\mu)\mathrm{grad}\,\mathrm{div}\mathbf{u} = \mathbf{0}, \text{ in } \mathbb{R}^2\backslash B, \\ \mathbf{t} = \mathbf{g}, \qquad\qquad\qquad\qquad \text{on } \Gamma = \partial B, \end{cases} \tag{1}$$

where $\mathbf{t} = (t_1, t_2), t_i = \sum_{j=1}^{2} \sigma_{ij}(\mathbf{u})n_j$ $(i = 1, 2)$, (n_1, n_2) is the unit exterior normal vector on the boundary Γ with respect to the domain $\mathbb{R}^2\backslash B$, $\mathbf{g} = (g_1, g_2)$ is the given vector function on Γ and satisfies the following compatible conditions:

$$\int_\Gamma g_i ds = 0, \qquad i = 1, 2. \tag{2}$$

For the problem (1), using Fourier series or complex analysis methods [4, Chap. IV], we can derive its natural boundary integral equation as

$$\begin{pmatrix} g_1(\theta) \\ g_2(\theta) \end{pmatrix} = \frac{1}{R}\mathcal{K}(\theta) * \begin{pmatrix} u_1(R,\theta) \\ u_2(R,\theta) \end{pmatrix}, \tag{3}$$

where

$$\mathcal{K}(\theta) = \begin{pmatrix} -\frac{ab}{2\pi(a+b)\sin^2\frac{\theta}{2}} & \frac{2b^2}{a+b}\delta'(\theta) \\ -\frac{2b^2}{a+b}\delta'(\theta) & -\frac{ab}{2\pi(a+b)\sin^2\frac{\theta}{2}} \end{pmatrix} \tag{4}$$

and where $\delta(\theta)$ is the Dirac function. It is obvious that the integral kernels of the natural integral operator $\mathcal{K}$ have the singularity of second order.

The Poisson formula of the problem (1) is

$$\begin{aligned} \begin{pmatrix} u_1(r,\theta) \\ u_2(r,\theta) \end{pmatrix} = \begin{pmatrix} \cos\theta & -\sin\theta \\ \sin\theta & \cos\theta \end{pmatrix} \Bigg\{ \begin{pmatrix} P_{rr} & P_{r\theta} \\ P_{\theta r} & P_{\theta\theta} \end{pmatrix} \\ * \begin{pmatrix} u_1(R,\theta)\cos\theta + u_2(R,\theta)\sin\theta \\ -u_1(R,\theta)\sin\theta + u_2(R,\theta)\cos\theta \end{pmatrix} \Bigg\}, \end{aligned} \tag{5}$$

where

$$\begin{aligned} P_{rr} &= \frac{[2br\cos\theta + (a-b)R](r^2-R^2)}{2\pi r(a+b)(R^2+r^2-2Rr\cos\theta)} \\ &\quad + \frac{(a-b)(r^2-R^2)(R^2\cos\theta - 2Rr + r^2\cos\theta)}{2\pi(a+b)(R^2+r^2-2Rr\cos\theta)^2}, \\ P_{r\theta} &= \frac{b(r^2-R^2)\sin\theta}{\pi(a+b)(R^2+r^2-2Rr\cos\theta)} + \frac{(a-b)(r^2-R^2)^2\sin\theta}{2\pi(a+b)(R^2+r^2-2Rr\cos\theta)^2}, \\ P_{\theta r} &= -\frac{a(r^2-R^2)\sin\theta}{\pi(a+b)(R^2+r^2-2Rr\cos\theta)} + \frac{(a-b)(r^2-R^2)^2\sin\theta}{2\pi(a+b)(R^2+r^2-2Rr\cos\theta)^2}, \end{aligned}$$

$$P_{\theta\theta} = \frac{[2ar\cos\theta + (b-a)R](r^2 - R^2)}{2\pi r(a+b)(R^2 + r^2 - 2Rr\cos\theta)} - \frac{(a-b)(r^2 - R^2)(R^2\cos\theta - 2Rr + r^2\cos\theta)}{2\pi(a+b)(R^2 + r^2 - 2Rr\cos\theta)^2}.$$

For the second boundary value problem of the Navier equation (1), we can get the solution via the Poisson formula (5) if we can find out the solution $\mathbf{u}_0(x)$ from the natural boundary integral equation (3). In the following, we use the Galerkin method to solve the natural boundary integral equation (3). Let

$$\begin{aligned}\hat{D}(\mathbf{u}_0, \mathbf{v}_0) &= \int_{\Gamma} \mathbf{v}_0 \cdot \frac{1}{R}\mathcal{K} * \mathbf{u}_0^T ds \\ &= \int_0^{2\pi} \mathbf{v}_0(\theta) \cdot \mathcal{K}(\theta) * \mathbf{u}_0(\theta)^T d\theta, \quad \mathbf{u}_0, \mathbf{v}_0 \in H^{1/2}(\Gamma)^2. \end{aligned} \tag{6}$$

Then the natural boundary integral equation (3) is equivalent to the following variational problem on the boundary Γ:

$$\begin{cases} \text{Find } \ \mathbf{u}_0 = (u_{10}(\theta), u_{20}(\theta)) \in H^{1/2}(\Gamma)^2, \text{s.t.} \\ \hat{D}(\mathbf{u}_0, \mathbf{v}_0) = \hat{F}(\mathbf{v}_0), \quad \forall \mathbf{v}_0 = (v_{r0}, v_{\theta 0}) \in H^{1/2}(\Gamma)^2, \end{cases} \tag{7}$$

where

$$\hat{F}(\mathbf{v}_0) = \int_{\Gamma} \mathbf{g} \cdot \mathbf{v}_0 ds = R\int_0^{2\pi} \mathbf{g}(\theta) \cdot \mathbf{v}_0(\theta) d\theta.$$

We state the following results for the variational problem (7) that can be found in detail in [4, Chap. IV].

Lemma 2. *$\hat{D}(\mathbf{u}_0, \mathbf{v}_0)$ is a symmetrical positive definite, V-elliptic, continuous bilinear form in the quotient space $H^{1/2}(\Gamma)^2/R(\Gamma)$, where*

$$R(\Gamma) = \{(C_1, C_2) | C_1, C_2 \in \mathbb{R}\}. \tag{8}$$

Theorem 3. *If the boundary load $\mathbf{g} \in H^{-1/2}(\Gamma)^2$ satisfies the compatible conditions (2) then the variational problem (7) has a unique solution in $H^{1/2}(\Gamma)^2/R(\Gamma)$ and the solution continuously depends on the given boundary load $\mathbf{g}$.*

Theorem 4. *Suppose boundary load $\mathbf{g} \in H^s(\Gamma)^2 (s \geq -\frac{1}{2})$ satisfies the compatible conditions (2). If $\mathbf{u}_0$ is the solution of the natural boundary integral equation (3), then $\mathbf{u}_0 \in H^{s+1}(\Gamma)^2$ and*

$$||\mathbf{u}_0||_{H^{s+1}(\Gamma)^2/R(\Gamma)} \leq \frac{(a+b)R}{\sqrt{2}b(a-b)}||\mathbf{g}||_{H^s(\Gamma)^2}.$$

Next, for notational convenience, we let

$$\begin{aligned}
&\psi^i_{-1,\ell}(\theta) = \phi^i_{0,\ell}(\theta),\ i = 1,2, \ell \in \mathbb{N}_0,\\
&d_\alpha = \frac{2ab}{a+b},\quad d_\beta = \frac{2b^2}{a+b},\\
&p_j = \begin{cases} 2, & j = -1 \\ 2^{j+1}, & j \in \mathbb{N}_0, \end{cases} \quad \mathbf{e}_i = \begin{cases} (1,0), & i = 1, \\ (0,1), & i = 2, \end{cases}\\
&s_{j\ell} = \operatorname{sgn}(\ell - p_j),\quad m_{j\ell} = (\ell - 1) \bmod p_j.
\end{aligned}$$

Now we approximate the space $H^{1/2}(\Gamma)$ by the space

$$V_n = V_0 \oplus W_0 \oplus W_1 \oplus \cdots \oplus W_n,$$

then the variational problem (7) is deduced as

$$\begin{cases} \text{find } \mathbf{u}_0 = (u_{10}(\theta), u_{20}(\theta)) \in V_n^2, \text{s.t.} \\ \hat{D}(\mathbf{u}_0, \mathbf{v}_0) = \hat{F}(\mathbf{v}_0),\ \forall \mathbf{v}_0 = (v_{r0}, v_{\theta 0}) \in V_n^2. \end{cases} \tag{9}$$

Since $\{\psi^i_{j\ell}(x) | j = -1, 0, \cdots, n,\ \ell = 0, 1, \cdots, p_j - 1,\ i = 1, 2\}$ is an orthonormal base of space V_n, the functions u_{10}, u_{20} may be approximated by the following finite sums respectively

$$u_{10} = \sum_{j=-1}^{n} \sum_{\ell=0}^{p_j - 1} (\alpha^0_{j,\ell} \psi^0_{j,\ell} + \alpha^1_{j,\ell} \psi^1_{j,\ell}), \tag{10}$$

$$u_{20} = \sum_{j=-1}^{n} \sum_{\ell=0}^{p_j - 1} (\beta^0_{j,\ell} \psi^0_{j,\ell} + \beta^1_{j,\ell} \psi^1_{j,\ell}), \tag{11}$$

where $\alpha^i_{j,\ell}$ and $\beta^i_{j,\ell}$ $(i = 1, 2)$ are expanding coefficients.

Substituting (10), (11) into (9) we obtain the following linear algebraic equation system

$$\begin{pmatrix} M^{-1,-1} & M^{-1,0} & \dots & M^{-1,n} \\ M^{0,-1} & M^{0,0} & \dots & M^{0,n} \\ \dots & \dots & & \dots\dots \\ M^{n,-1} & M^{n,0} & \dots & M^{n,n} \end{pmatrix} \begin{pmatrix} \gamma_{-1} \\ \gamma_0 \\ \vdots \\ \gamma_n \end{pmatrix} = \begin{pmatrix} F_{-1} \\ F_0 \\ \vdots \\ F_n \end{pmatrix}, \tag{12}$$

where $M^{j,j'}$ is $4p_j \times 4p_{j'}$ submatrix, γ_j and F_j are $4p_j$ column vectors. They may be expressed respectively as

$$M^{j,j'} = \begin{pmatrix} M^{j,j'}_{11} & M^{j,j'}_{12} \\ M^{j,j'}_{21} & M^{j,j'}_{22} \end{pmatrix},\ j, j' = -1, 0, \cdots, n, \tag{13}$$

$$[M^{j,j'}_{ii'}]_{kk'} = \hat{D}(\psi^{s_{j'k'}}_{j',m_{j'k'}} \mathbf{e}_{i'}, \psi^{s_{jk}}_{j,m_{jk}} \mathbf{e}_i), \tag{14}$$

$$i, i' = 1, 2,\ k = 1, \cdots, 2p_j,\ k' = 1, \cdots, 2p_{j'}, \tag{15}$$

$$\gamma_j = (\alpha_j, \beta_j)^T,\ j = -1, 0, \cdots, n, \tag{16}$$
$$\alpha_j = (\alpha^0_{j,0}, \cdots, \alpha^0_{j,p_j-1}, \alpha^1_{j,0}, \cdots, \alpha^1_{j,p_j-1}), \tag{17}$$
$$\beta_j = (\beta^0_{j,0}, \cdots, \beta^0_{j,p_j-1}, \beta^1_{j,0}, \cdots, \beta^1_{j,p_j-1}), \tag{18}$$
$$F_j = (b_j, c_j)^T,\ j = -1, 0, \cdots, n, \tag{19}$$
$$[b_j]_k = R\int_0^{2\pi} g_1(\theta)\psi^{s_{jk}}_{j,m_{jk}}(\theta)d\theta,\ k = 1, 2, \cdots, 2p_j, \tag{20}$$
$$[c_j]_k = R\int_0^{2\pi} g_2(\theta)\psi^{s_{jk}}_{j,m_{jk}}(\theta)d\theta,\ k = 1, 2, \cdots, 2p_j. \tag{21}$$

From the symmetry of the bilinear form $\hat{D}(\cdot,\cdot)$, we find that $[M^{j,j'}_{ii'}]_{kk'} = [M^{j',j}_{i'i}]_{k'k}$ which means that the stiffness matrix M, namely the coefficient matrix of the linear algebraic equation system, is symmetric. In addition, from the definition of the bilinear form $\hat{D}(\cdot,\cdot)$, we have

$$[M^{j,j'}_{22}]_{kk'} = \hat{D}((0, \psi^{s_{j'k'}}_{j',m_{j'k'}}), (0, \psi^{s_{jk}}_{j,m_{jk}})) \tag{22}$$
$$= \hat{D}((\psi^{s_{j'k'}}_{j',m_{j'k'}}, 0), (\psi^{s_{jk}}_{j,m_{jk}}, 0)) = [M^{j,j'}_{11}]_{kk'}, \tag{23}$$
$$[M^{j,j'}_{22}]_{kk'} = \hat{D}((\psi^{s_{j'k'}}_{j',m_{j'k'}}, 0), (0, \psi^{s_{jk}}_{j,m_{jk}})) \tag{24}$$
$$= -\hat{D}((0, \psi^{s_{j'k'}}_{j',m_{j'k'}}), (\psi^{s_{jk}}_{j,m_{jk}}, 0)) = -[M^{j,j'}_{12}]_{kk'}. \tag{25}$$

So that if set $A^{j,j'} = M^{j,j'}_{11}$, $B^{j,j'} = M^{j,j'}_{12}$, then we conclude that

Theorem 5. *The stiffness matrix M is symmetric and its block matrices may be expressed as*

$$M^{j,j'} = \begin{pmatrix} A^{j,j'} & B^{j,j'} \\ -B^{j,j'} & A^{j,j'} \end{pmatrix},\quad j, j' = -1, 0, \cdots, n,$$

where $A^{j,j'}, B^{j,j'}$ are $2p_j \times 2p_{j'}$ matrices and their elements may be expressed respectively as

$$[A^{j,j'}]_{kk'} = \hat{D}((\psi^{s_{j'k'}}_{j',m_{j'k'}}, 0), (\psi^{s_{jk}}_{j,m_{jk}}, 0)), \tag{26}$$
$$[B^{j,j'}]_{kk'} = \hat{D}((0, \psi^{s_{j'k'}}_{j',m_{j'k'}}), (\psi^{s_{jk}}_{j,m_{jk}}, 0)). \tag{27}$$

Since the stiffness matrix M is symmetrical we only compute the block matrices $M^{j,j'}(j' \geq j)$.

Using following important formula in the generalized function theory

$$\frac{1}{-4\sin^2\frac{\theta}{2}} = \sum_{\ell=1}^{\infty} \ell\cos\ell\theta \tag{28}$$

we get

$$[A^{j,j'}]_{kk'} = \hat{D}((\psi^{s_{j'k'}}_{j',m_{j'k'}},0),(\psi^{s_{jk}}_{j,m_{jk}},0))$$

$$= \frac{d_\alpha}{\pi}\int_0^{2\pi}\int_0^{2\pi}\frac{\psi^{s_{jk}}_{j,m_{jk}}(\theta)\psi^{s_{j'k'}}_{j',m_{j'k'}}(\theta')}{-4\sin^2\frac{\theta-\theta'}{2}}d\theta d\theta'$$

$$= \frac{d_\alpha}{\pi}\sum_{\ell=1}^{\infty}\ell\int_0^{2\pi}\int_0^{2\pi}\psi^{s_{jk}}_{j,m_{jk}}(\theta)\psi^{s_{j'k'}}_{j',m_{j'k'}}(\theta')\cos\ell(\theta-\theta')d\theta d\theta'$$

$$= \frac{d_\alpha}{\pi}\sum_{\ell=1}^{\infty}\ell[\int_0^{2\pi}\psi^{s_{jk}}_{j,m_{jk}}(\theta)\cos\ell\theta d\theta\int_0^{2\pi}\psi^{s_{j'k'}}_{j',m_{j'k'}}(\theta)\cos\ell(\theta)d\theta$$

$$+\int_0^{2\pi}\psi^{s_{jk}}_{j,m_{jk}}(\theta)\sin\ell\theta d\theta\int_0^{2\pi}\psi^{s_{j'k'}}_{j',m_{j'k'}}(\theta)\sin\ell(\theta)d\theta]. \tag{29}$$

In addition

$$[B^{j,j'}]_{kk'} = \hat{D}((0,\psi^{s_{j'k'}}_{j',m_{j'k'}}),(\psi^{s_{jk}}_{j,m_{jk}},0))$$

$$= d_\beta\int_0^{2\pi}\psi^{s_{jk}}_{j,m_{jk}}(\theta)\psi^{s_{j'k'}}_{j',m_{j'k'}}{}'(\theta)d\theta$$

$$= \frac{d_\beta}{\pi}\{\frac{1}{2}\int_0^{2\pi}\psi^{s_{jk}}_{j,m_{jk}}(\theta)d\theta\int_0^{2\pi}\psi^{s_{j'k'}}_{j',m_{j'k'}}{}'(\theta)d\theta$$

$$+\sum_{\ell=1}^{\infty}\ell[\int_0^{2\pi}\psi^{s_{jk}}_{j,m_{jk}}(\theta)\cos\ell\theta d\theta\int_0^{2\pi}\psi^{s_{j'k'}}_{j',m_{j'k'}}{}'(\theta)\cos\ell(\theta)d\theta$$

$$+\int_0^{2\pi}\psi^{s_{jk}}_{j,m_{jk}}(\theta)\sin\ell\theta d\theta\int_0^{2\pi}\psi^{s_{j'k'}}_{j',m_{j'k'}}{}'(\theta)\sin\ell(\theta)d\theta]\}. \tag{30}$$

By the orthogonality of trigonometric functions and (29), (30) it is easy to find that

$$M^{j,j'} = 0,\quad j,j' = -1,0,\cdots,n,\ j' \geq j+2.$$

For the matrices $M^{j,j}$ and $M^{j,j+1}$, we have the following theorems.

Theorem 6. *The matrix*

$$A^{j,j} = \begin{pmatrix} A^{j,j}_{11} & 0 \\ 0 & A^{j,j}_{22} \end{pmatrix},\ j = -1,0,\cdots,n,$$

where $A^{j,j}_{11}$ and $A^{j,j}_{22}$ are $p_j \times p_j$ matrices and

$$[A^{-1,-1}_{11}]_{kk'} = \frac{d_\alpha\pi}{4}(-1)^{k+k'}, \tag{31}$$

$$[A^{-1,-1}_{22}]_{kk'} = \frac{d_\alpha\pi}{8}[2(-1)^{k+k'}+1], \tag{32}$$

$$[A^{j,j}_{11}]_{kk'} = \begin{cases} \frac{d_\alpha\pi}{9p_j^2}(13p_j^2-3p_j-1), & k=k' \\ -\frac{d_\alpha\pi}{3p_j^2}(p_j+\sin^{-2}\theta_{j+1,k'-k}), & k\neq k', \end{cases} j\neq -1, \tag{33}$$

$$[A_{22}^{j,j}]_{kk'} = \begin{cases} \frac{d_\alpha \pi}{6p_j^3}(4p_j - 1), & k = k', \\ -\frac{d_\alpha \pi}{6p_j^3}, & k \neq k', \end{cases} \quad j \neq -1. \tag{34}$$

Proof. In the case of $j \neq -1$, we get from (29) and Lemma 1 that

$$\begin{aligned}
[A^{j,j}]_{k,p_j+k'} &= \frac{d_\alpha}{\pi} \sum_{\ell=p_j+1}^{2p_j-1} \ell[(\frac{2\pi}{3p_j^2}(3p_j - \ell)\cos \ell\theta_{j,k-1})(-\frac{2\pi}{3p_j^2}\sin \ell\theta_{j,k'-1}) \\
&+(\frac{2\pi}{3p_j^2}(3p_j - \ell)\sin \ell\theta_{j,k-1})(\frac{2\pi}{3p_j^2}\cos \ell\theta_{j,k'-1})] \\
&= -\frac{4d_\alpha \pi}{9p_j^4} \sum_{\ell=p_j+1}^{2p_j-1} \ell(3p_j - \ell)\sin \ell\theta_{j,k'-k} \\
&= -\frac{4d_\alpha \pi}{9p_j^4} \sum_{\ell=1}^{p_j-1} (p_j + \ell)(2p_j - \ell)\sin \ell\theta_{j,k'-k} = 0, \\
[A_{11}^{j,j}]_{kk'} &= [A^{j,j}]_{kk'} \\
&= \frac{d_\alpha}{\pi}\{p_j \cdot \frac{\pi}{p_j} \cdot \frac{\pi}{p_j} + \sum_{\ell=p_j+1}^{2p_j-1} \ell[\frac{4\pi^2}{9p_j^4}(3p_j - \ell)^2 \cos \ell\theta_{j,k-1}\cos \ell\theta_{j,k'-1} \\
&+\frac{4\pi^2}{9p_j^4}(3p_j - \ell)^2 \sin \ell\theta_{j,k-1}\sin \ell\theta_{j,k'-1}]\} \\
&= \frac{d_\alpha}{\pi}\{\frac{\pi}{p_j} + \frac{4\pi^2}{9p_j^4}\sum_{\ell=1}^{p_j-1}(p_j + \ell)(2p_j - \ell)^2 \cos \ell\theta_{j,k'-k}\} \\
&= \begin{cases} \frac{d_\alpha \pi}{9p_j^2}(13p_j^2 - 3p_j - 1), & k = k' \\ -\frac{d_\alpha \pi}{9p_j^2}(p_j + \sin^{-2}\theta_{j+1,k'-k}), & k \neq k', \end{cases}
\end{aligned}$$

and

$$\begin{aligned}
[A_{22}^{j,j}]_{kk'} &= [A^{j,j}]_{p_j+k,p_j+k'} \\
&= \frac{d_\alpha}{\pi}\{\sum_{\ell=p_j+1}^{2p_j-1} \ell[(\frac{2\pi}{3p_j^2})^2 \sin \ell\theta_{j,k-1}\sin \ell\theta_{j,k'-1} \\
&+(\frac{2\pi}{3p_j^2})^2 \cos \ell\theta_{j,k-1}\cos \ell\theta_{j,k'-1}] + (\frac{\pi}{2p_j^2})^2 \cdot 2p_j\} \\
&= \frac{d_\alpha}{\pi}\{\frac{4\pi^2}{9p_j^4}\sum_{\ell=1}^{p_j-1}(p_j + \ell)\cos \ell\theta_{j,k'-k} + \frac{\pi^2}{2p_j^3}\} \\
&= \begin{cases} \frac{d_\alpha \pi}{6p_j^3}(4p_j - 1), & k = k' \\ -\frac{d_\alpha \pi}{6p_j^3}, & k \neq k'. \end{cases}
\end{aligned}$$

These lead to the theorem. In the case of $j = -1$, the proof is similar.

Theorem 7. *The matrix*

$$B^{j,j} = \begin{pmatrix} B_{11}^{j,j} & B_{12}^{j,j} \\ B_{21}^{j,j} & B_{22}^{j,j} \end{pmatrix}, \ j = -1, 0, \cdots, n,$$

are anti-symmetric matrices, where $B_{ii'}^{j,j}$ $(i, i' = 1, 2)$ *are* $p_j \times p_j$ *matrices and*

$$[B_{11}^{-1,-1}]_{kk'} = 0, [B_{22}^{-1,-1}]_{kk'} = 0, \tag{35}$$

$$[B_{12}^{-1,-1}]_{kk'} = \frac{d_\beta \pi}{4}(-1)^{k+k'}, \tag{36}$$

$$[B_{11}^{j,j}]_{kk'} = \frac{d_\beta \pi}{9p_j^3}[(4p_j^2 + 3)\mathrm{ctg}\theta_{j+1,k'-k} + 3\mathrm{ctg}^3\theta_{j+1,k'-k}], k \neq k', j \neq -1 \tag{37}$$

$$[B_{22}^{j,j}]_{kk'} = -\frac{2d_\beta \pi}{9p_j^3}\mathrm{ctg}\theta_{j+1,k'-k}, \ k \neq k', j \neq -1, \tag{38}$$

$$[B_{12}^{j,j}]_{kk'} = \begin{cases} \frac{2d_\beta \pi}{27p_j^3}(p_j - 1)(13p_j - 1), & k = k' \\ -\frac{2d_\beta \pi}{9p_j^3}(4p_j + \sin^{-2}\theta_{j+1,k'-k}), & k \neq k', \end{cases} \ j \neq -1 \tag{39}$$

Proof. In the case of $j \neq -1$, we get from (30) and Lemma 1 that

$$\begin{aligned}
[B_{11}^{j,j}]_{kk'} &= [B^{j,j}]_{kk'} \\
&= \frac{d_\beta}{\pi} \sum_{\ell=p_j+1}^{2p_j-1} [\frac{4\pi^2}{9p_j^4}\ell(3p_j - \ell)^2 \sin \ell\theta_{j,k'-1} \cos \ell\theta_{j,k-1} \\
&\quad - \frac{4\pi^2}{9p_j^4}\ell(3p_j - \ell)^2 \cos \ell\theta_{j,k'-1} \sin \ell\theta_{j,k-1}]\} \\
&= \frac{4d_\beta \pi}{9p_j^4} \sum_{\ell=1}^{p_j-1} (p_j + \ell)(2p_j - \ell)^2 \sin \ell\theta_{j,k'-k}\} \\
&= \frac{d_\beta \pi}{9p_j^3}[(4p_j^2 + 3)\mathrm{ctg}\theta_{j+1,k'-k} + 3\mathrm{ctg}^3\theta_{j+1,k'-k}], k \neq k', \\
[B_{22}^{j,j}]_{kk'} &= [B^{j,j}]_{p_j+k,p_j+k'} \\
&= \frac{d_\beta}{\pi} \sum_{\ell=p_j+1}^{2p_j-1} [-\frac{4\pi^2}{9p_j^4}\ell \sin \ell\theta_{j,k-1} \cos \ell\theta_{j,k'-1} \\
&\quad + \frac{4\pi^2}{9p_j^4}\ell \cos \ell\theta_{j,k-1} \sin \ell\theta_{j,k'-1}] \\
&= \frac{4d_\beta \pi}{9p_j^4} \sum_{\ell=1}^{p_j-1} (p_j + \ell) \sin \ell\theta_{j,k'-k} \\
&= -\frac{2d_\beta \pi}{9p_j^3}\mathrm{ctg}\theta_{j+1,k'-k}, \ k \neq k'.
\end{aligned}$$

Similarly, we can prove the rest of the theorem.

Similar to the proof of Theorems 6 and 7 we can prove

Theorem 8. *The matrix* $A^{j,j+1} = 0$ *and the matrix*

$$B^{j,j+1} = \begin{pmatrix} 0 & 0 \\ B_{21}^{j,j+1} & 0 \end{pmatrix},$$

where $B_{21}^{j,j+1}$ *are* $p_j \times p_{j+1}$ *matrices and*

$$[B_{21}^{-1,0}]_{kk'} = -\frac{d_\beta \pi}{4}, \tag{40}$$

$$[B_{21}^{j,j+1}]_{kk'} = -\frac{d_\beta \pi}{2p_j^2}, \quad j \neq -1. \tag{41}$$

Since the variational problem (9) has a unique solution in $V_n^2/R(\Gamma)$, the linear algebraic system (12) has a unique solution $\gamma = (\gamma_{-1}, \gamma_0, \cdots, \gamma_n)$ in the sense of difference of one element in the set

$$\left\{ \gamma \left| \sum_{j=-1}^{n} \sum_{\ell=0}^{p_j-1} (\alpha_{j,\ell}^0 \psi_{j,\ell}^0 + \alpha_{j,\ell}^1 \psi_{j,\ell}^1, \beta_{j,\ell}^0 \psi_{j,\ell}^0 + \beta_{j,\ell}^1 \psi_{j,\ell}^1) \in R(\Gamma) \right. \right\}.$$

So that we must add two conditions to fix the solution as a result of containing two arbitrary constants in $R(\Gamma)$. On the other hand, we note that the first and second rows of the stiffness matrix M are linear dependent and so are the 5th and 6th rows. Hence we only need to replace the second and 5th rows of the matrix M by two adding equations in the computation.

In the end of this section, we state the following error estimates which can be proved as the same in our paper [13].

Theorem 9 (energy norm estimate). *Let* $\mathbf{u}_0 \in \overset{o}{C}^{k+1}(\Gamma)^2$ *be the solution of the variational problem (7) and* $\mathbf{u}_0^n$ *be the solution of the approximate variational problem (9). then*

$$||\mathbf{u}_0 - \mathbf{u}_0^n||_{\hat{D}} \leq C \cdot 2^{-(k+\frac{1}{2})n} E_{2^n}(\mathbf{u}_0^{(k+1)}), \tag{42}$$

where $E_n(f^{(q)})$ *is the best approximation to* $f^{(q)}$ *in* T_n *and* $||\cdot||_{\hat{D}}$ *is energy norm in the quotient space* $H^{1/2}(\Gamma)^2/R(\Gamma)$ *which is induced from bilinear form* $\hat{D}(\cdot,\cdot)$, *i.e.* $||\mathbf{u}||_{\hat{D}} = \hat{D}(\mathbf{u},\mathbf{u})^{1/2}$.

Theorem 10 (L^2 norm estimate). *If* $\mathbf{u}_0 \in \overset{o}{C}^{k+1}(\Gamma)^2$ *and*

$$(\mathbf{u}_0 - \mathbf{u}_0^n, \phi) = 0, \ \forall \phi \in R(\Gamma),$$

then

$$||\mathbf{u}_0 - \mathbf{u}_0^n||_{L^2(\Gamma)^2} \leq C \cdot 2^{-(k+\frac{1}{2})n} E_{2^n}(\mathbf{u}_0^{(k+1)}). \tag{43}$$

Theorem 11 **(L^∞ norm estimate).** *If* $\mathbf{u}_0 \in \overset{o}{C}{}^{k+1}(\Gamma)^2 (k \geq 1)$ *and*

$$(\mathbf{u}_0 - \mathbf{u}_0^n, \phi) = 0, \ \forall \phi \in R(\Gamma),$$

then

$$||\mathbf{u}_0 - \mathbf{u}_0^n||_\infty \leq C \cdot 2^{-kn} E_{2^n}(\mathbf{u}_0^{k+1}). \tag{44}$$

where constant C *is independent of* n.

4 Numerical Examples

In this section we present two numerical results to test our scheme discussed in Sect. 3. In the following examples, we set $\lambda = 1, \mu = 0.5$.

Example 1. Consider the plane elasticity problem in the outer domain of the unit disk:

$$\begin{cases} 0.5\Delta \boldsymbol{u} + 1.5\text{graddiv}\boldsymbol{u} = 0, & \text{in } \mathbb{R}^2 \backslash B \\ g_1(\theta) = 0.6\cos\theta, g_2(\theta) = 0.6\sin\theta & \text{on } \Gamma \end{cases}.$$

To make the solution unique we attach the conditions: $u_1(0) = 0, u_2(\frac{\pi}{2}) = 0$. The exact solution of this problem is

$$\begin{cases} u_1(r,\theta) = \frac{1}{r}(1.6\sin 2\theta\cos\theta - 0.4\cos 2\theta\sin\theta) - \frac{0.6}{r^3}\sin 3\theta \\ u_2(r,\theta) = \frac{1}{r}(1.6\sin 2\theta\sin\theta + 0.4\cos 2\theta\cos\theta) - \frac{0.6}{r^3}\cos 3\theta \end{cases}.$$

Its boundary value is

$$\begin{cases} u_1(1,\theta) = \sin\theta \\ u_2(1,\theta) = \cos\theta \end{cases}.$$

When $n = -1$, we calculate out the approximate boundary value is

$$\begin{cases} u_1^n(1,\theta) = \sin\theta - 5.551115 \times 10^{-17}\cos\theta\sin\theta \\ u_2^n(1,\theta) = \cos\theta \end{cases}.$$

Example 2. In the outer domain of the unit disk, let

$$\begin{cases} g_1(\theta) = 6\cos 2\theta + 24\cos 4\theta, \\ g_2(\theta) = -6\sin 2\theta + 24\sin 4\theta. \end{cases}$$

To make the solution unique we attach the conditions: $u_1(0) = 11, u_2(\frac{\pi}{2}) = 0$, then the exact solution of this problem is

$$\begin{cases} u_1(r,\theta) = \frac{1}{r^2}(5\cos 2\theta + 6\cos 4\theta) \\ u_2(r,\theta) = \frac{1}{r^2}(-5\sin 2\theta + 6\sin 4\theta) \end{cases}.$$

Its boundary value is

$$\begin{cases} u_1(1,\theta) = 5\cos 2\theta + 6\cos 4\theta \\ u_2(1,\theta) = -5\sin 2\theta + 6\sin 4\theta \end{cases}.$$

When $n = 1$, we calculate out the approximate boundary value is

$$\begin{cases} u_1^n(1,\theta) = 5\cos 2\theta + 6\cos 4\theta + 8.88178 \times 10^{-15} - 3.46945 \times 10^{-18} \sin\theta \\ \qquad -8.88178 \times 10^{-16} \cos 3\theta - 4.44089 \times 10^{-16} \cos 5\theta \\ \qquad -4.44089 \times 10^{-16} \cos 6\theta - 4.44089 \times 10^{-16} \cos 7\theta, \\ u_2^n(1,\theta) = -5\sin 2\theta + 6\sin 4\theta - 4.16334 \times 10^{-17} - 1.3877810^{-17} \cos\theta \\ \qquad -3.33067 \times 10^{-17} \cos 5\theta + 7.93016 \times 10^{-18} \cos 7\theta \\ \qquad +3.46945 \times 10^{-17} \sin 5\theta + 3.70074 \times 10^{-17} \sin 6\theta \\ \qquad +6.93889 \times 10^{-18} \sin 7\theta + 6.93889 \times 10^{-17} \sin 8\theta. \end{cases}$$

References

1. Hadamard, J.: Lectures on Cauchy's Problems in Linear Partial Differential Equations. Yale University Press, New Haven (1923)
2. Feng, K.: Finite element method and natural boundary reduction. In: Proceedings of the International Congress of Mathematicians, Warszawa, pp. 1439–1453 (1983)
3. Feng, K., Yu, D.: Cononical integral equations of elliptic boundary value problems and their numerical solutions. In: Feng, K., Lions, J.-L. (eds.) Proceedings of China-France Symposium on the Finite Element Method Science Press, Beijing, pp. 211–252 (1983)
4. Yu, D.: Mathematical Theory of Natural Boundary Element Methods. Science Press, Beijing (1993). (In Chinese)
5. Chen, W., Lin, W.: Hadamard singular integral equations and its Hermite wavelet. In: Li, Z., Wu, S., Yang, L. (eds.) Proceedings of the Fifth International Colloquium on Finite or Infinite Dimensional Complex Analysis, Beijing, China (1997)
6. Wu, J., Yu, D.: The natural integral equations of 3-D harmonic problems and their numerical solutions. Chinese J. Num. Math. Appl. **21**(1), 73–85 (1999)
7. Wu, J., Yu, D.: The overlapping domain decomposition method for harmonic equation over exterior three-dimensional domain. J. Comput. Math. **18**(1), 83–94 (2000)
8. Gilbert, R.P., Lin, W.: Wavelet solutions for time harmonic acoutic waves in a finite ocean. J. Comput. Acoust. **1**(1), 31–60 (1993)
9. Glowinski, R., Lawton, W., Ravachol, M., Tenenbaum, E.: Wavelets solution of linear and nonlinear elliptic, parabolic and hyperbolic problems in one space dimension. In: Glowinski, R., Lichnewsky, A. (eds.) Computing Methods in Applied Science and Engineering, pp. 55–120. SIAM, Philadelphia (1990)
10. Jaffard, S.: Wavelet methods for fast resolution of elliptic problems. SIAM J. Numer. Anal. **29**, 965–986 (1992)
11. Petersdorff, T.V., Schwab, C.: Wavelet approximations for first kind boundary integral equations on polygons. Numer. Math. **74**, 479–519 (1996)
12. Quak, E.: Trigonometric wavelets for Hermite interpolation. Math. Comput. **65**, 683–722 (1996)
13. Shen, Y., Lin, W.: The natural integral equations of plane elasticity problem and its wavelet methods, to appear
14. Shen, Z., Xu, Y.: Degenerate kernel schemes by wavelets for nonlinear integral equations on the real line. Appl. Anal. **59**(1–4), 163–184 (1996)
15. Timan, A.F.: Theory of Approximation of Functions of a Real Variable. Pergamon press, Oxford (1963)

16. Xu, J.C., Shann, W.C.: Galerkin-wavelet methods for two-point boundary value problems. Numer. Math. Anal. **24**, 246–262 (1993)
17. Xu, Y.: The generalized Marcinkiewicz-Zygmund inequality for trigonometric polynomials. J. Math. Anal. **161**, 447–456 (1991)
18. Zygmund, A.: Trigonometric Series. Cambridge University Press, Cambridge (1959)

Study on Destruction Transformation of Error System Structure

Li-ping Liao[1(✉)] and Kai-zhong Guo[2]

[1] School of Management, Guangdong Polytechnic Normal University, Room C2-13A, No. 94, TianRun Road, Tianhe District, Guangzhou 510635, China
liping1110@hotmail.com

[2] School of Management, Guangzhou Vocational College of Science and Technology, Guangzhou 510450, China
guokaizhong@163.com

Abstract. Firstly this paper analyzes the basic structures of error system, which can be broken down to six basic types including series, parallel, expansion & contraction, inclusion, feedback and others. Then gives the physical meaning of destruction, T_{hsw} (things destruction) → things that do not exist → nothing to be discussed, or there's no need to discuss the things or the things are killed, extinct, destroyed, disappeared, fired, sold, thrown, transferred, moved and so on. And next we describe the logic rules of destruction transformation by corresponding logic connectives, quantitatively represents the relationship between system structure and error under destruction transformation by using error function and structure function. Finally the transformation system is given.

Keywords: Error system · System structure · Destruction transformation Functions

1 Preface

When repairing machines or equipment, we usually remove the bad or broken parts, and then replace them with new parts, so as to make the machines or equipment operational. Therefore, tearing down the bad parts here is the method of destruction [1–3].

When studying the reliability or safety of a system, especially a large or complex one, we need to study how the reliability or safety of the system is affected when one or more elements in a system or one or more subsystems appear to have error or fault, namely the ways and rules for transfer and transformation, thus looking for ways to avoiding system error, or to eliminating error when appears [4–6]. For example, in order to eliminate some subsystem errors in a system, firstly we can decompose the system into subsystem set which including the error one, and then remove it and substitute a corresponding subsystem without error for it. Similarly, in the research of enterprise management, we need to study how to learn from a successful enterprise management system, and similarly copy it to another, for example, the chain's management [7–9]. However, due to the differences in cultural background, legal system, the success of the enterprise management system can't be copied completely. This is

B.-Y. Cao and Y.-B. Zhong (Eds.): ICFIE 2017, AISC 872, pp. 283–291, 2019.
https://doi.org/10.1007/978-3-030-02777-3_25

the decomposition, substitution, reduction, similarity, destruction and their inverse transformation etc. of enterprise management system [10].

Therefore, in order to study the change rule of system error, we should research the change rule of system error when its structure changes. This paper mainly discusses the destruction transformation of error system structure [11–13].

2 Basic Types of Error System Structure

2.1 Series

Series structure system CS (s_1, s_2 ...s_n), describing function of error CSC (s_1, s_2 ...s_n) (Fig. 1).

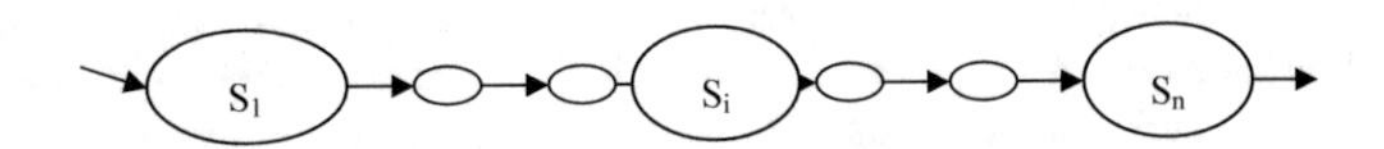

Fig. 1. Series structure of error system

2.2 Parallel

Parallel structure system BS (s_1, s_2 ...s_n), describing function of error BSC (s_1, s_2 ...s_n) (Fig. 2).

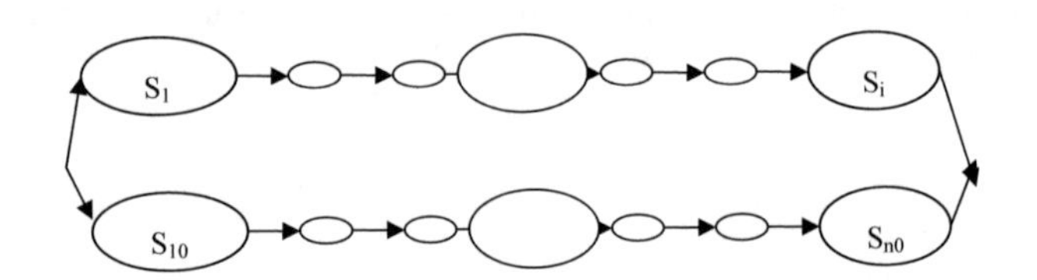

Fig. 2. Parallel structure of error system

2.3 Expansion and Contraction

(1) **Expansion Type**

(See Fig. 3)

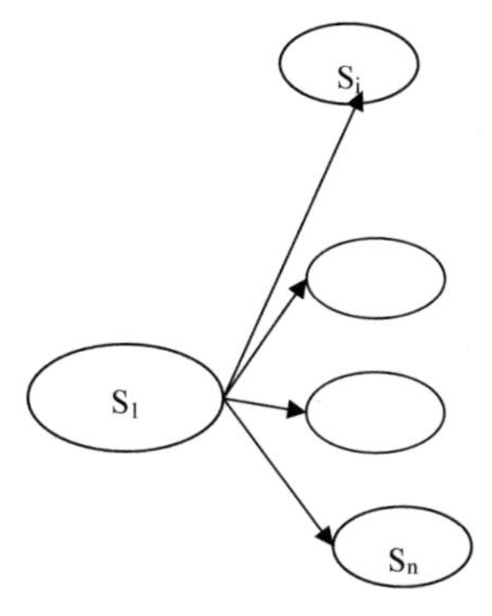

Fig. 3. Expansion type of expansion and contraction structure of error system

(2) **Contraction Type**

Expansion & contraction structure system KS (s_1, s_2 …s_n), describing function of error KSC (s_1, s_2 …s_n) (Fig. 4).

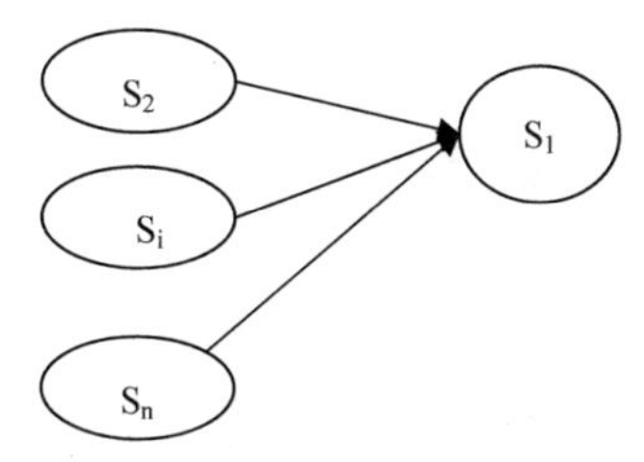

Fig. 4. Contraction type of expansion and contraction structure of error system

2.4 Inclusion Structure

(1) **Only Inclusion in Center**

(See Fig. 5)

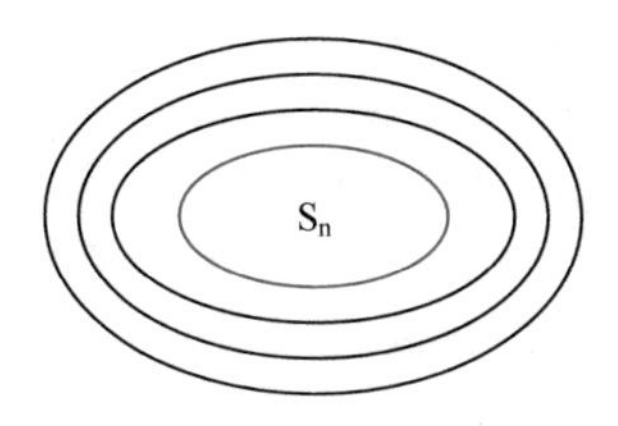

Fig. 5. Only inclusion in center structure of error system

(2) **Inclusion in Layers**

Inclusion structure system YS (s_1, s_2 …s_n), describing function of error YSC (s_1, s_2 … s_n) (Fig. 6).

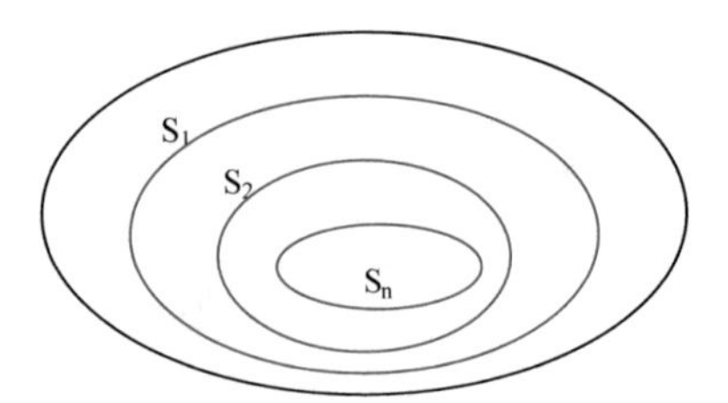

Fig. 6. Inclusion in layers structure of error system

2.5 Feedback Structure

Feedback structure system FS (s_1, s_2 …s_n), describing function of error FSC (s_1, s_2 … s_n) (Fig. 7).

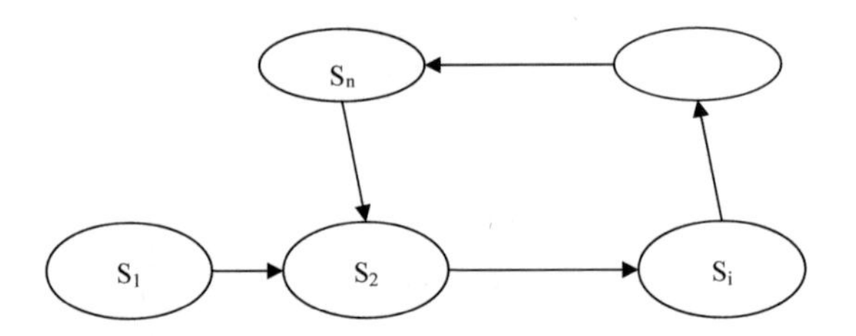

Fig. 7. Feedback structure of error system

3 Destruction Mode of Error System Structure Decomposition Transformation

Logic propositions representations of error system structure destruction transformation are as follows:

Set A((U, S(t), $\vec{p}$, T(t), L(t)), x(t) = f((u(t), $\vec{p}$), G(t))), G(t) is the rule for judging error based on U, for the error logic variables of G(t), if T_h(A((U, S(t), $\vec{p}$, T(t), L(t)), x(t) = f((u(t), $\vec{p}$), G(t)))) = A(((Φ, Φ, Φ, Φ, Φ), Φ = Φ(Φ, Φ),Φ)), then T_h is called in U, the error destruction transformation connective of the rule G(t) and A((U, S(t), $\vec{p}$, T(t), L(t)), x(t) = f((u(t), $\vec{p}$), G(t))) is denoted by T_h.

Set A((U, S(t), $\vec{p}$, T(t), L(t)),x(t) = f((u(t), $\vec{p}$), G(t))), G(t) is the rule for judging error based on U, for the error logic variables of G(t), if T_{hsw}(A((U, S(t), $\vec{p}$, T(t), L(t)), x(t) = f((u(t), $\vec{p}$), G(t)))) = A((U, Φ, $\vec{p}$, T(t), L(t)), x(t) = f((u(t), $\vec{p}$), G(t))), then T_{hsw} is called in U, the error destruction transformation connective of the rule G(t) and A((U, S (t), $\vec{p}$, T(t), L(t)), x(t) = ((u(t), $\vec{p}$), G(t))) is denoted by T_{hsw}.

The meaning of destruction: T_{hsw} (things destruction) → things doesn't exist → nothing to be discussed, or there's no need to discuss the things or the things are killed, extinct, destroyed, disappeared, fired, sold, thrown, transferred, moved and so on.

4 Several Typical Structures of Destruction Transformation of Error System

4.1 Series Structure Be Destroyed

(1) Logical Representation of Series Structure Destruction

$T_h\{A_i((V_i(t), CS(s_1, s_2 \ldots s_3)_{iu}(t), \vec{P}_{iu}, T_{iu}(t), L_{iu}(t)), x_i(t) = f_i(u_i(t), G_{iA}(t))), I = 1, 2, \ldots, m\} = \{B_i((V_i(t), \Phi, \vec{P}_{iv}, T_{iv}(t), L_{iv}(t)), y_i(t) = g_i(v_i(t), G_{iB}(t))), I = 1, 2, \ldots, n\}$,

(2) Formalized Representation of System

$$S=S(s_1, s_2 \ldots, CS(s_1, s_2 \ldots\ldots s_3)_{iu}(t), \ldots s_3) \longrightarrow$$

$$S(s_1, s_2 \ldots, \Phi, \ldots s_3)$$

(3) Representation of System Error Value

$$SC(s_1, s_2 \ldots, CS(s_1, s_2 \ldots\ldots s_3)_{iu}(t), \ldots s_3) \longrightarrow$$

$$SC(s_1, s_2 \ldots, \Phi, \ldots s_3)$$

4.2 Parallel Structure Be Destroyed

(1) Logical Representation of Parallel Structure Destruction

$T_h\{A_i((V_i(t), BS(s_1, s_2 \ldots s_3)_{iu}(t), \vec{P}_{iu}, T_{iu}(t), L_{iu}(t)), x_i(t) = f_i(u_i(t), G_{iA}(t))), I = 1, 2, \ldots, m\} = \{B_i((V_i(t), \Phi, \vec{P}_{iv}, T_{iv}(t), L_{iv}(t)), y_i(t) = g_i(v_i(t), G_{iB}(t))), I = 1, 2, \ldots, n\}$,

(2) **Formalized Representation of System**

$$S=S(s_1, s_2 \ldots, BS(s_1, s_2 \ldots\ldots s_3)_{iu}(t), \ldots s_3) \longrightarrow S(s_1, s_2 \ldots, \Phi, \ldots s_3)$$

(3) **Representation of System Error value**

$$SC(s_1, s_2 \ldots, BS(s_1, s_2 \ldots\ldots s_3)_{iu}(t), \ldots s_3) \longrightarrow SC(s_1, s_2 \ldots, \Phi, \ldots s_3)$$

4.3 Expansion and Contraction Structure Be Destroyed

(1) **Logical Representation of Expansion and Contraction Structure Destruction**

$T_h\{A_i((V_i(t), KS(s_1, s_2 \ldots s_3)_{iu}(t), \vec{P}_{iu}, T_{iu}(t), L_{iu}(t)), x_i(t) = f_i(u_i(t), G_{iA}(t))), I = 1, 2, \ldots, m\} = \{B_i((V_i(t), \Phi, \vec{P}_{iv}, T_{iv}(t), L_{iv}(t)), y_i(t) = g_i(v_i(t), G_{iB}(t))), I = 1, 2, \ldots, n\}$,

(2) **Formalized Representation of System**

$$S=S(s_1, s_2 \ldots, KS(s_1, s_2 \ldots\ldots s_3)_{iu}(t), \ldots s_3) \longrightarrow S(s1, s2 \ldots, \Phi, \ldots s3)$$

(3) **Representation of System Error Value**

$$SC(s_1, s_2 \ldots, KS(s_1, s_2 \ldots\ldots s_3)_{iu}(t), \ldots s3) \longrightarrow SC(s_1, s_2 \ldots, \Phi, \ldots s_3)$$

4.4 Inclusion Structure Be Destroyed

(1) **Logical Representation of Contains Structure Destruction**

$T_h\{A_i((V_i(t), YS(s1, s2 \ldots s3)_{iu}(t), \vec{P}_{iu}, T_{iu}(t), L_{iu}(t)), x_i(t) = f_i(u_i(t), G_{iA}(t))), I = 1, 2, \ldots, m\} = \{B_i((V_i(t), \Phi, \vec{P}_{iv}, T_{iv}(t), L_{iv}(t)), y_i(t) = g_i(v_i(t), G_{iB}(t))), I = 1, 2, \ldots, n\}$,

(2) **Formalized Representation of System**

$$S=S(s_1, s_2 \ldots, YS(s_1, s_2 \ldots\ldots s_3)_{iu}(t), \ldots s_3) \longrightarrow S(s1, s2 \ldots, \Phi, \ldots s3)$$

(3) **Representation of System Error Value**

$$SC(s_1, s_2 \ldots, YS(s_1, s_2 \ldots\ldots s_3)_{iu}(t), \ldots s_3) \longrightarrow SC(s_1, s_2 \ldots, \Phi, \ldots s_3)$$

4.5 Feedback Structure Be Destroyed

(1) **Logical Representation of Feedback Structure Destruction**

$T_h\{A_i((V_i(t), FS(s_1, s_2 \ldots s_3)_{iu}(t), \vec{P}_{iu}, T_{iu}(t), L_{iu}(t)), x_i(t) = f_i(u_i(t), G_{iA}(t))), I = 1, 2, \ldots, m\} = \{B_i((V_i(t), \Phi, \vec{P}_{iv}, T_{iv}(t), L_{iv}(t)), y_i(t) = g_i(v_i(t), G_{iB}(t))), I = 1, 2, \ldots, n\}$,

(2) **Formalized Representation of System**

$$S=S(s_1, s_2 \ldots, FS(s_1, s_2 \ldots\ldots s_3)_{iu}(t), \ldots s_3) \longrightarrow S(s_1, s_2 \ldots, \Phi, \ldots s_3)$$

(3) **Representation of System Error Value**

$$SC(s_1, s_2 \ldots, FS(s_1, s_2 \ldots\ldots s_3)_{iu}(t), \ldots s_3) \longrightarrow SC(s_1, s_2 \ldots, \Phi, \ldots s_3)$$

5 Application Examples

System S is consisted of three production lines in a plant, total output $S(GY_0) = S$ (process $S1(GY_{01}(a_{01}, b_{01}))$, process $S_2(GY_{02}(a_{02}, b_{02}))$, …, process $S_n(GY_{0n}(a_{0n}, b_{0n})))$, as Fig. 8 shows.

Supposing that due to the change of the market, for the function of total output S (GY_0) of the production system S, excess production is almost equal to the output of a production line, and then the decision-maker can achieve his goals by destruction transformation. He can remove a series production line, i.e. change production system S into production system S_1, which is shown in Fig. 9:

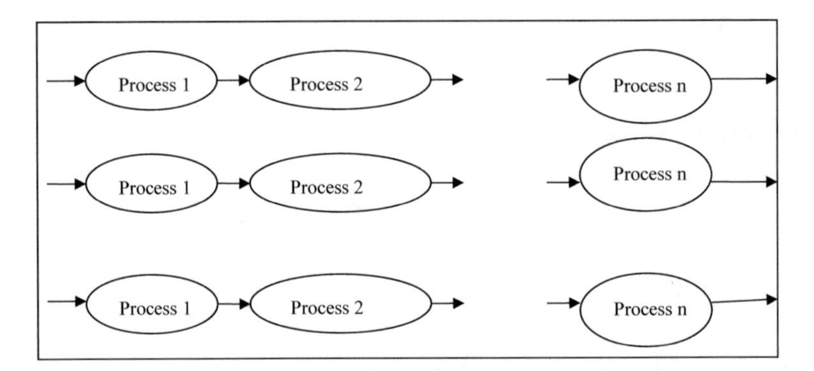

Fig. 8. Production system S of a plant

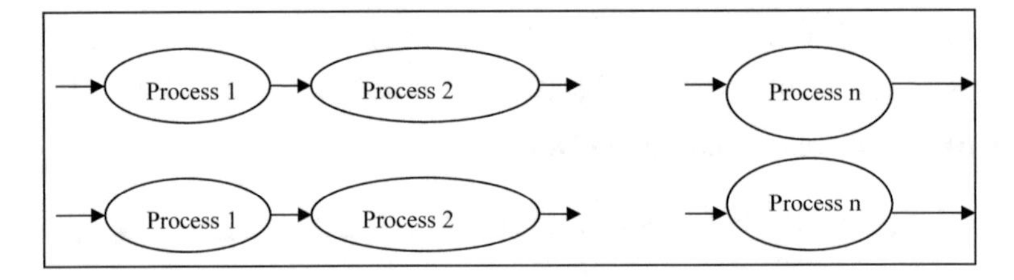

Fig. 9. Production system S_1 of a plant

6 Conclusion

This paper mainly studies the basic structures of error system structure which is composed of series, parallel, expansion & contraction, inclusion and feedback. Based on what we discussed above we can find that the basic system structures are almost consistent with the static structures of error system, based on which, for such types of error system, the logical performance of the demonstration of five basic structure systems, the formalized performance and the performance of its error value [14–16]. Furthermore, the sample of its application and its destruction transformation are given respectively. Therefore, we believe that part of the error in the real system can be avoided or eliminated via its destruction transformation [17, 18].

Acknowledgements. Thanks to the support by Guangdong foundation of philosophy and Social Sciences, China (No. GD16XGL20) and Department of Education of Guangdong Province, China (No. YQ2015107).

Recommender: Xiang-sheng Xie, Guangdong University of Technology, Professor.

References

1. Guo, K., Liu, S.: Introduction for the theory of error-eliminating. Adv. Model. Anal. A: Gen. Math. Comput. Tools **39**(2), 39–66 (2002)
2. Guo, K., Huang, J.: The research on systematic mechanism of "1% error can result in 100% error" and the ways of eliminating errors. J. Guangdong Univ. Technol. **25**(2), 1–5 (2008)
3. Guo, K., Zhang, S.: Introduction for Error-Eliminating. Press of South China University of Technology, Guangzhou (1995)
4. Guo, K., Zhang, S.: Theory and Method for Distinguishing the Error of Enterprise Fixed Assets Investment Decision. Press of South China University of Technology, Guangzhou (1995)
5. Liu, Y., Guo, K.: Large Complicated System Conflict and Wrong Theory and Theory and Method of Error. Press of South China University of Technology, Guangzhou (2000)
6. Guo, K., Zhang, S.: Error Theory. Press of Central South China University of Technology, Changsha (2001)
7. Shengyong, C.: Risk Prevention and Trap Avoidance. Enterprise Management Press, Beijing (1998)
8. Guo, K., Liu, S.: Research on laws of security risk-error logic system with critical point. Model. Meas. Control. D **22**(1), 1–10 (2001)
9. Wang, C.: Introduction for Fuzzy Mathematics. Press of Beijing University of Technology, Beijing (1998)
10. Liu, S., Guo, K.: Exploration and application of redundancy system in decision making-relation between fuzzy error logic increase transformation word and connotative model implication word. Adv. Model. Anal. A Gen. Math. Comput. Tools **39**(4), 17–29 (2002)
11. Liu, S., Guo, K.: Substantial change of decision-making environment-mutation of fuzzy error system. Adv. Model. Anal. A Gen. Math. Comput. Tools **39**(4), 29–39 (2002)
12. Barwise, J.: Handbook of Mathematical Logic. North-Holland Publishing Company, Amsterdam (1977)
13. Guo, K., Xiong, H.: Research on fuzzy error-logic destruction transforming word. Fuzzy Syst. Math. **20**(2), 34–39 (2006)
14. Xiong, H., Guo, K.: Research on fuzzy error-logic decomposition transformation word in domain. Fuzzy Syst. Math. **20**(1), 24–29 (2006)
15. Li, M., Guo, K.Z.: Research on decomposition of fuzzy error set. Adv. Model. Anal. A: Gen. Math. Comput. Tools **43**(26), 15–26 (2006)
16. Liu, H.B., Guo, K.Z.: Discussion about risc-control of investment: research on transformation system of fuzzy error set. Model. Meas. Control. D **27**(1), 63–70 (2006)
17. Liu, H.B., Guo, K.Z.: One-element fuzzy error-matrix. Model. Meas. Control. D **27**(2), 33–42 (2006)
18. Guo, K.: Error System. Press of Science, Beijing (2012)

Interval Type-2 Fuzzy System Based on NT Method and Its Probability Representation

Wei Zhou[1,2](✉), Hong-xing Li[1], and De-gang Wang[1]

[1] School of Control Science and Engineering, Dalian University of Technology, Dalian 116024, China
zhou_bnu@126.com
[2] School of Applied Mathematics, Beijing Normal University, Zhuhai 519087, China

Abstract. In this paper, we study the interval type-2 fuzzy system based on NT method. We establish an interval type-2 fuzzy system based on a set of input-output data and reveal the probability significance of the interval type-2 fuzzy system. We prove that the crisp output of the interval type-2 fuzzy system based on NT method can be regarded as the conditional mathematical expectation of certain random vectors. And we provide the expression of the joint probability density function and this interval type-2 fuzzy system.

Keywords: Type-2 fuzzy system · Probability density function NT method

1 Introduction

In 1975, Zadeh [1] first proposed the concept of type-2 fuzzy set based on the concept of an ordinary fuzzy set (type-1 fuzzy set). Since the membership grades of a type-2 fuzzy set are type-1 fuzzy subsets of [0,1], type-2 fuzzy sets are more suitable for modeling uncertainty than type-1 fuzzy sets [2].

Karnik and Mendel et al. [3] expanded type-1 fuzzy system to type-2 fuzzy system and have established a relatively complete theoretical system of type-2 fuzzy system. But because of the computational complexity of general type-2 fuzzy system, most of people research interval type-2 system which use interval type-2 fuzzy sets [4].

Since fuzzy systems and probability system both can model uncertainty, there may exist certain relationship between themselves. In [5] and [6], the issue of the probability representation of type-1 fuzzy system was first discussed and the probability significance of a type-1 fuzzy system was first revealed. Then some related researches have been developed. In [7], it was pointed out that the fuzzy system based on center-of-gravity method is a regression function. A method for constructing joint probability density function using non-singleton fuzzifier was proposed in [8]. The fuzzy trustworthiness system and probability representation based on two kinds of fuzzy implication was studied in [9].

In this paper, we study the probability representation of an interval type-2 fuzzy system based on NT type-reduction method. Section 2 establishes an interval type-2 fuzzy system based on a set of input-output data. In Sect. 3, we reveal the probability

B.-Y. Cao and Y.-B. Zhong (Eds.): ICFIE 2017, AISC 872, pp. 292–304, 2019.
https://doi.org/10.1007/978-3-030-02777-3_26

significance of such fuzzy system and provide the expression of probability density function. In Sect. 4, we draw conclusions.

2 The Mathematical Representation of Interval Type-2 Fuzzy System

For a single input and single output system S with uncertainty, a set of input-output data $IOD = \{(x_i, y_i)|i = 1, 2\cdots, n\}$ can be obtained by means of experiment, and then an interval type-2 fuzzy system $\bar{S}$ can be established as a approximator of S.

Suppose the input-output data set IOD satisfies:

$$a = x_1 < x_2 < \cdots < x_n = b, c = y_1 < y_2 < \cdots < y_n = d.$$

We can construct type-1 triangle fuzzy sets by these data first as follows,

$$A_1(x) = \begin{cases} \frac{x-x_2}{x_1-x_2} & x \in [x_1, x_2] \\ 0 & else \end{cases}, A_n(x) = \begin{cases} \frac{x-x_{n-1}}{x_n-x_{n-1}} & x \in [x_{n-1}, x_n] \\ 0 & else \end{cases},$$

$$B_1(y) = \begin{cases} \frac{y-y_2}{y_1-y_2} & y \in [y_1, y_2] \\ 0 & else \end{cases}, B_n(y) = \begin{cases} \frac{y-y_{n-1}}{y_n-y_{n-1}} & y \in [y_{n-1}, y_n] \\ 0 & else \end{cases},$$

for $i = 2, \cdots, n-1$

$$A_i(x) = \begin{cases} \frac{x-x_{i-1}}{x_i-x_{i-1}} & x \in [x_{1-1}, x_i) \\ \frac{x-x_{i+1}}{x_i-x_{i+1}} & x \in [x_i, x_{i+1}] \\ 0 & else \end{cases}, B_i(y) = \begin{cases} \frac{y-y_{i-1}}{y_i-y_{i-1}} & y \in [y_{i-1}, y_i) \\ \frac{y-y_{i+1}}{y_i-y_{i+1}} & y \in [y_i, y_{i+1}] \\ 0 & else \end{cases},$$

where $A_i(x_i) = 1$, $B_i(y_i) = 1$, $i = 1, 2\cdots, n$, and when $x \in [x_i, x_{i+1}]$, $y \in [y_i, y_{i+1}]$, $A_i(x) + A_{i+1}(x) = 1$, $B_i(x) + B_{i+1}(x) = 1$.

Considering the uncertainty of output data, the membership degree $B_i(y)$ is fuzzified to an interval $[\delta_i B_i(y), 1 \wedge (1 + \gamma_i)B_i(y)]$ and the output interval type-2 fuzzy set $\tilde{B}_i$ is obtained as follows,

$$\tilde{B}_i(y) = 1/[\delta_i B_i(y), 1 \wedge (1 + \gamma_i)B_i(y)], i = 1, 2, \cdots, n,$$

where $\delta_i, \gamma_i \in [0, 1], i = 1, 2, \cdots, n$. Meanwhile the input type-1 fuzzy set A_i can be regarded as a reduced type-2 fuzzy set $\tilde{A}_i$ as follows,

$$\tilde{A}_i(x) = 1/A_i(x),\ \tilde{A}_i(x)(u) = \begin{cases} 1, & u = A_i(x) \\ 0, & else \end{cases},\quad i = 1, 2, \cdots, n.$$

Then we obtain the type-2 fuzzy inference rules,

$$\text{If } x \text{ is } \tilde{A}_i \text{ then } y \text{ is } \tilde{B}_i,\ i = 1, \cdots, n,$$

and get the type-2 fuzzy relationships $\tilde{R}_i, i = 1, 2, \cdots, n$ and $\tilde{R} = \cup_{i=1}^{n} \tilde{R}_i$,

$$\tilde{R}_i(x, y) = \tilde{A}_i(x) \sqcap \tilde{B}_i(y) = 1/[\underline{R}_i(x, y), \bar{R}_i(x, y)],$$

$$\tilde{R}(x, y) = \sqcup_{i=1}^{n} \tilde{R}_i(x, y),$$

where $\underline{R}_i(x, y) = A_i(x) \wedge \delta_i B_i(y)$, $\bar{R}_i(x, y) = A_i(x) \wedge (1 + \gamma_i) B_i(y)$.

For the monotonicity of the input-output data, the total inference relationship $\tilde{R}$ can be revised as follows.

$$\tilde{R}(x, y) = \begin{cases} \tilde{R}_i(x, y) \sqcup \tilde{R}_{i+1}(x, y), & (x, y) \in [x_i, x_{i+1}) \times [y_i, y_{i+1}) \\ 1/0, & else \end{cases}.$$

An input $x' \in X = [a, b]$ of a type-2 fuzzy system can be singleton fuzzified into a type-2 fuzzy set $\tilde{A}'$,

$$\tilde{A}'(x) = \begin{cases} 1/1, & x = x' \\ 1/0, & x \neq x' \end{cases}$$

Compositing $\tilde{A}'$ and the total inference relationship $\tilde{R}$, the inference result $\tilde{B}'$ can be obtained as $\tilde{B}' = \tilde{A}' \circ \tilde{R}$.

When $x' \in [x_i, x_{i+1})$,

$$\begin{aligned} \tilde{B}'(y) &= \sqcup_{x \in X} \left(\tilde{A}'(x) \sqcap \tilde{R}(x, y) \right) \\ &= \tilde{R}\left(x', y\right) = \begin{cases} \tilde{R}_i(x, y) \sqcup \tilde{R}_{i+1}(x, y), & y \in [y_i, y_{i+1}) \\ 1/0, & else \end{cases} \\ &= 1/[\underline{B}'(y), \bar{B}'(y)] \end{aligned} \tag{1}$$

where,

$$\underline{B}'(y) = \begin{cases} \underline{R}_i(x', y) \vee \underline{R}_{i+1}(x', y) & y \in [y_i, y_{i+1}) \\ 0 & else \end{cases} \tag{2}$$

$$\bar{B}'(y) = \begin{cases} \bar{R}_i(x', y) \vee \bar{R}_{i+1}(x', y) & y \in [y_i, y_{i+1}) \\ 0 & else \end{cases} \tag{3}$$

It is obvious that the inference result $\tilde{B}'$ is an interval type-2 fuzzy set. Using NT method, the type reduction fuzzy set B'' can be obtained. And the crisp output y' can be obtained by using center-of-gravity method.

$$B''(y) = \frac{1}{2}\left(\underline{B}'(y) + \bar{B}'(y)\right) \tag{4}$$

$$y' = \frac{\int_Y yB''(y)dy}{\int_Y B''(y)dy} \tag{5}$$

As the notations of x', y' are just for clear presentation of the inference process, we obtain the function relationship as follows, by transforming x' to x and y' to $\bar{S}(x)$.

$$\bar{S}: X \to Y, x \to \bar{S}(x) = \frac{\int_Y yB''(y)dy}{\int_Y B''(y)dy}$$

So, we get the interval type-2 fuzzy system $\bar{S}$, as a approximator of the original system S.

3 The Probability Representation of Interval Type-2 Fuzzy System

As the probability significance of type-1 fuzzy system was first revealed in [5] and [6], we reveal the probability significance of the above interval type-2 fuzzy system in this section. Observing formula (1)–(5), we can find that $B''(y)$ is related to the input x', because $B''(y)$ is related to $B'(y)$ and $B'(y)$ is related to the input x'. So $B''(y)$ can be denoted as

$$B''\left(y|x = x'\right) \triangleq B''(y).$$

Considering the arbitrariness of the input $x' \in X$, we can denote $B''\left(y|x = x'\right)$ as a form of binary function $p: X \times Y \to \mathbb{R}$, $p(x, y) \triangleq B''\left(y|x = x'\right)$. And $p(x, y)$ can be expanded as a function on $\mathbb{R}^2$, denoted as $q(x, y) \triangleq p(x, y)I_{X \times Y}$, where $X = [a, b], Y = [c, d]$.

Let

$$H(2, n, \sqcap, \sqcup) \triangleq \int_{-\infty}^{+\infty} \int_{-\infty}^{+\infty} q(x, y)dydx.$$

Function $H(2, n, \sqcap, \sqcup)$ has parameters "$2, n, \sqcap, \sqcup$", and "2" indicates that the integrand $q(x, y)$ is binary function, "n" figures the amount of the rules, "$\sqcap, \sqcup$" are the operators which are used for generating and aggregating type-2 fuzzy inference relationships.

If $H(2, n, \sqcap, \sqcup) > 0$, let

$$f(x,y) \triangleq q(x,y)/H(2,n,\theta^*,\vee^*). \tag{6}$$

It is obvious that function $f(x,y)$ satisfies $\int_{-\infty}^{+\infty}\int_{-\infty}^{+\infty} f(x,y)dydx = 1$ and $f(x,y) \geq 0$, so $f(x,y)$ can be regarded as a joint probability density function of some random vector (ξ,η) and

$$\bar{S}(x) = \frac{\int_{-\infty}^{+\infty} yf(x,y)dy}{\int_{-\infty}^{+\infty} f(x,y)dy} \tag{7}$$

just is the conditional mathematical expectation $E(\eta|\xi = x)$.

Theorem 1. Given a single input single output interval type-2 fuzzy system denoted as above, if $\int_Y yp(x,y)dy < +\infty, 0 < \int_Y p(x,y)dy < \infty$, then there exists a probability space $(\Omega,\mathcal{F},P)$ and a random vector (ξ,η) on the probability space $(\Omega,\mathcal{F},P)$, which satisfies

$$E(\eta|\xi = x) = \bar{S}(x).$$

Theorem 1 indicates that the output of an interval type-2 fuzzy system based on NT method can be regarded as the conditional mathematical expectation of certain random vector. From the perspective of probability theory, the input variable and output variable of an uncertainty system can be regarded as two random variables depended on each other. We know that the conditional mathematical expectation is the least square approximation of the value of output variable, when the sample values of input variable are known. Hence an interval type-2 fuzzy system also is a stochastic system in nature.

Theorems 2 and 3 in the below provide the expression of the fuzzy system $\bar{S}(x)$ and the probability density function $f(x,y)$.

Theorem 2. Given a single input single output interval type-2 fuzzy system denoted as above. If $\delta_i = 2/3,\ \gamma_i = 1/3,\ i = 1,2,\cdots,n$, then the output of the fuzzy system

$$\bar{S}(x) = A_i^*(x)y_i + A_{i+1}^*(x)y_{i+1}, x \in [x_i, x_{i+1}),$$

where

$$A_i^*(x) = \begin{cases} -\dfrac{162A_i(x)^3 - 405A_i(x)^2 + 261A_i(x) - 199}{12(18A_i(x)^2 - 30A_i(x) + 35)} & x \in \left[x_i, x_i^{(1)}\right] \\ \dfrac{90A_i(x)^3 - 63A_i(x)^2 - 57A_i(x) - 13}{4(36A_i(x)^2 - 42A_i(x) - 11)} & x \in \left[x_i^{(1)}, x_i^{(3)}\right] \\ \dfrac{90A_i(x)^3 - 63A_i(x)^2 - 33A_i(x) - 25}{4(36A_i(x)^2 - 30A_i(x) - 17)} & x \in \left[x_i^{(3)}, x_i^{(2)}\right] \\ -\dfrac{162A_i(x)^3 - 297A_i(x)^2 + 9A_i(x) - 95}{12(18A_i(x)^2 - 6A_i(x) + 23)} & x \in \left[x_i^{(2)}, x_{i+1}\right] \end{cases}$$

$$A_{i+1}^*(x) = 1 - A_i^*(x),\ x_i^{(1)} = \frac{2x_i + x_{i+1}}{3}, x_i^{(2)} = \frac{x_i + 2x_{i+1}}{3},\ x_i^{(3)} = \frac{x_i + x_{i+1}}{2}.$$

Proof: From formulas (4), (5) and (7),

$$\bar{S}(x) = \frac{\int_{-\infty}^{+\infty} yf(x,y)dy}{\int_{-\infty}^{+\infty} f(x,y)dy} = \frac{\int_c^d yp(x,y)dy}{\int_c^d p(x,y)dy} = \frac{\int_c^d y\left(\underline{B}'(y) + \bar{B}'(y)\right)dy}{\int_c^d \left(\underline{B}'(y) + \bar{B}'(y)\right)dy}.$$

We compute $\int_c^d \underline{B}'(y)dy$ and $\int_c^d y\underline{B}'(y)dy$ first.
Let

$$x_i^{(1)} = \frac{2x_i + x_{i+1}}{3}, x_i^{(2)} = \frac{x_i + 2x_{i+1}}{3}, y_i^{(3)} = \frac{1}{2}(y_i + y_{i+1}),$$

$$y_i^{(1)} = \frac{3}{2}(y_i A_i(x) + y_{i+1}A_{i+1}(x)) - \frac{1}{2}y_{i+1}, y_i^{(2)} = y_i^{(1)} + \frac{1}{2}(y_{i+1} - y_i),$$

$$y_i^{(4)} = \frac{3}{2}(y_i A_{i+1}(x) + y_{i+1}A_i(x)) - \frac{1}{2}y_{i+1}, y_i^{(5)} = y_i^{(4)} + \frac{1}{2}(y_{i+1} - y_i).$$

Then

$$A_i\left(x_i^{(1)}\right) = \frac{2}{3}, A_i\left(x_i^{(2)}\right) = \frac{1}{3}, A_i(x) = \frac{2}{3}B_i\left(y_i^{(1)}\right), A_{i+1}(x) = \frac{2}{3}B_{i+1}\left(y_i^{(2)}\right),$$

$$B_i\left(y_i^{(3)}\right) = B_{i+1}\left(y_i^{(3)}\right), \frac{2}{3}B_i\left(y_i^{(4)}\right) = A_{i+1}(x), \frac{2}{3}B_{i+1}\left(y_i^{(5)}\right) = A_i(x).$$

So, we can know that

a. when $x \in \left[x_i, x_i^{(1)}\right]$

If $y \in \left[y_i, y_i^{(4)}\right)$, then $A_i(x) > \frac{2}{3} \geq \frac{2}{3}B_i(y) > A_{i+1}(x)$, $\underline{B}'(y) = \frac{2}{3}B_i(y)$;

If $y \in \left[y_i^{(4)}, y_{(i+1)}\right]$, then $A_i(x) > \frac{2}{3} > A_{i+1}(x) \geq \frac{2}{3}B_i(y)$, $A_{i+1}(x) < \frac{2}{3}B_{i+1}(y)$, $\underline{B}'(y) = A_{i+1}(x)$.

Hence,

$$\int_c^d \underline{B}'(y)dy = \int_{y_i}^{y_{i+1}} \underline{B}'(y)dy = \int_{y_i}^{y_i^{(4)}} \frac{2}{3}B_i(y)dy + \int_{y_i^{(4)}}^{y_{i+1}} A_{i+1}(x)dy$$
$$= \frac{\left(y_i^{(4)}\right)^2 - y_i^2 - 2y_{i+1}\left(y_i^{(4)} - y_i\right)}{3(y_i - y_{i+1})} + A_{i+1}(x)\left(y_{i+1} - y_i^{(4)}\right),$$

$$\int_c^d y\underline{B}'(y)dy = \frac{-2y_i^3 + 2\left(y_i^{(4)}\right)^3 - 3y_{i+1}\left(\left(y_i^{(4)}\right)^2 - y_i^2\right)}{9(y_i - y_{i+1})} + \frac{1}{2}A_{i+1}(x)\left(y_{i+1}^2 - \left(y_i^{(4)}\right)^2\right);$$

b. when $x \in \left[x_i^{(1)}, x_i^{(2)}\right]$,

If $y \in \left[y_i, y_i^{(1)}\right)$, then $\frac{2}{3}B_i(y) > A_i(x) > \frac{1}{3} > \frac{2}{3}B_{i+1}(y)$, $\underline{B}'(y) = A_i(x)$;

If $y \in \left[y_i^{(1)}, y^{(3)}\right)$, then $\frac{2}{3}B_{i+1}(y) < \frac{2}{3}B_i(y) \le A_i(x)$, $\underline{B}'(y) = \frac{2}{3}B_i(y)$;

If $\left[y_i^{(3)}, y^{(2)}\right)$, then $A_{i+1}(x) > \frac{2}{3}B_{i+1}(y) \ge \frac{2}{3}B_i(y)$, $\underline{B}'(y) = \frac{2}{3}B_{i+1}(y)$;

If $y \in [y_i^{(2)}, y_{i+1}]$, then $B_i(y) < \frac{1}{3} < A_{i+1}(x) \le \frac{2}{3}B_{i+1}(y)$, $\underline{B}'(y) = A_{i+1}(x)$.
Hence,

$$\int_c^d \underline{B}'(y)dy = \int_{y_i}^{y_i^{(1)}} A_i(x)dy + \int_{y_i^{(1)}}^{y_i^{(3)}} \frac{2}{3}B_i(y)dy + \int_{y_i^{(3)}}^{y_i^{(2)}} \frac{2}{3}B_{i+1}(y)dy + \int_{y_i^{(2)}}^{y_{i+1}} A_{i+1}(x)dy$$

$$= A_i(x)\left(y_i^{(1)} - y_i - y_{i+1} + y_i^{(2)}\right) - (y_i^{(2)} + 2y_i^{(3)})/3 + y_{i+1}$$

$$+ \frac{2\left(y_i^{(3)}\right)^2 - \left(y_i^{(1)}\right)^2 - \left(y_i^{(2)}\right)^2 - 2y_{i+1}\left(2y_i^{(3)} - y_i^{(1)} - y_i^{(2)}\right)}{3(y_i - y_{i+1})},$$

$$\int_c^d y\underline{B}'(y)dy = \frac{1}{2}A_i(x)\left(\left(y_i^{(1)}\right)^2 - y_i^2 - y_{i+1}^2 + \left(y_i^{(2)}\right)^2\right)$$

$$+ \frac{2\left(y_i^{(3)}\right)^3 - \left(y_i^{(1)}\right)^3 - \left(y_i^{(2)}\right)^2 - 3y_{i+1}\left(2\left(y_i^{(3)}\right)^2 - \left(y_i^{(1)}\right)^2 - \left(y_i^{(2)}\right)^2\right)}{9(y_i - y_{i+1})}$$

$$+ \frac{\left(y_i^{(2)}\right)^2 - \left(y_i^{(3)}\right)^2}{3} + \frac{(1 - A_i(x))y_{i+1}^2}{2};$$

c. when $x \in \left[x_i^{(2)}, x_{i+1}\right]$,

If $y \in \left[y_i, y_i^{(5)}\right)$, then $A_{i+1}(x) > \frac{2}{3} > A_i(x) \ge \frac{2}{3}B_{i+1}(y)$, $A_i(x) < \frac{2}{3}B_i(y)$, $\underline{B}'(y) = A_i(x)$;

If $y \in \left[y_i^{(5)}, y_{i+1}\right]$, then $A_{i+1}(x) > \frac{2}{3} \ge \frac{2}{3}B_{i+1}(y) > A_i(x)$, $\underline{B}'(y) = \frac{2}{3}B_{i+1}(y)$.

Hence,

$$\int_c^d \underline{B}'(y)dy = \int_{y_i}^{y_i^{(4)}} \frac{2}{3}B_i(y)dy + \int_{y_i^{(4)}}^{y_{i+1}} A_{i+1}(x)dy$$
$$= A_i(x)\left(y_i^{(5)} - y_i\right) + \frac{2y_{i+1} - 2y_i^{(5)}}{3}$$
$$- \frac{\left(y_i^{(2)}\right)^2 - \left(y_i^{(5)}\right)^2 + 2y_{i+1}\left(y_i^{(5)} - y_{i+1}\right)}{3(y_i - y_{i+1})},$$

$$\int_c^d y\underline{B}'(y)dy = \frac{1}{2}A_i(x)\left(\left(y_i^{(5)}\right)^2 - y_i^2\right)$$
$$- \frac{2\left(y_i^{(2)}\right)^3 - 2\left(y_i^{(5)}\right)^3 + 3y_i\left(\left(y_i^{(5)}\right)^2 - y_{i+1}^2\right)}{9(y_i - y_{i+1})};$$

From the above analysis, we can know the expression of $\underline{B}'(y)$ in addition.
$\underline{B}'(y)$

$$\underline{B}'(y) = \begin{cases} \frac{2}{3}B_i(y), & (x,y) \in \left[x_i, x_i^{(1)}\right) \times \left[y_i, y_i^{(4)}\right) \cup \left[x_i^{(1)}, x_i^{(2)}\right) \times \left[y_i^{(1)}, y^{(3)}\right) \\ A_i(x), & (x,y) \in \left[x_i^{(1)}, x_i^{(2)}\right) \times \left[y_i, y_i^{(1)}\right) \cup \left[x_i^{(2)}, x_{i+1}\right) \times \left[y_i, y_i^{(5)}\right) \\ A_{i+1}(x), & (x,y) \in \left[x_i, x_i^{(1)}\right) \times \left[y_i^{(4)}, y_{(i+1)}\right) \cup \left[x_i^{(1)}, x_i^{(2)}\right) \times \left[y_i^{(2)}, y_{i+1}\right) \\ \frac{2}{3}B_{i+1}(y), & (x,y) \in \left[x_i^{(1)}, x_i^{(2)}\right) \times \left[y_i^{(3)}, y^{(2)}\right) \cup \left[x_i^{(2)}, x_{i+1}\right) \times \left[y_i^{(5)}, y_{(i+1)}\right) \\ 0, & else \end{cases}$$

Then we compute $\int_c^d \bar{B}'(y)dy$ and $\int_c^d y\bar{B}'(y)dy$.

Let

$$x_i^{(3)} = \frac{x_i + x_{i+1}}{2},$$

$$y_i^{(6)} = \frac{3}{4}(y_iA_i(x) + y_{i+1}A_{i+1}(x)) + \frac{1}{4}y_{i+1},\ y_i^{(7)} = y_i^{(6)} + \frac{1}{4}(y_i - y_{i+1}),$$

$$y_i^{(8)} = \frac{3}{4}(y_iA_{i+1}(x) + y_{i+1}A_i(x)) + \frac{1}{4}y_{i+1},\ y_i^{(9)} = y_i^{(8)} + \frac{1}{4}(y_i - y_{i+1})$$

Then

$$A_i\left(x_i^{(3)}\right) = A_{i+1}\left(x_i^{(3)}\right), A_i(x) = \frac{4}{3}B_i\left(y_i^{(6)}\right), A_{i+1}(x) = \frac{4}{3}B_{i+1}\left(y_i^{(7)}\right),$$

$$\frac{4}{3}B_i\left(y_i^{(8)}\right) = A_{i+1}(x),\ \frac{4}{3}B_{i+1}\left(y_i^{(9)}\right) = A_i(x)$$

Similarly, we can know the expression of $\bar{B}'(y)$.

$$\bar{B}'(y)=\begin{cases}\frac{2}{3}B_i(y) & (x,y)\in\left[x_i,x_i^{(3)}\right)\times\left[y_i^{(6)},y_i^{(8)}\right)\\ A_i(x) & (x,y)\in\left[x_i,x_i^{(3)}\right)\times\left[y_i,y_i^{(6)}\right)\cup\left[x_i^{(3)},x_{i+1}\right)\times\left[y_i,y_i^{(9)}\right)\\ A_{i+1}(x) & (x,y)\in\left[x_i,x_i^{(3)}\right)\times\left[y_i^{(8)},y_{i+1}\right)\cup\left[x_i^{(3)},x_{i+1}\right)\times\left[y_i^{(7)},y_{i+1}\right)\\ \frac{2}{3}B_{i+1}(y) & (x,y)\in\left[x_i^{(3)},x_{i+1}\right)\times\left[y_i^{(9)},y_i^{(7)}\right)\\ 0 & else\end{cases}$$

So, we can know that

a. when $x\in\left[x_i,x_i^{(3)}\right]$,

$$\begin{aligned}\int_c^d\bar{B}'(y)dy&=\int_{y_i}^{y_i^{(6)}}A_i(x)dy+\int_{y_i^{(6)}}^{y_i^{(8)}}\frac{4}{3}B_i(y)dy+\int_{y_i^{(8)}}^{y_{i+1}}A_{i+1}(x)dy\\&=A_i(x)\left(y_i^{(6)}-y_i\right)+A_{i+1}(x)\left(y_{i+1}-y_i^{(8)}\right)\\&\quad+\frac{-2\left(y_i^{(6)}\right)^2+2\left(y_i^{(8)}\right)^2-4y_{i+1}(y_i^{(8)}-y_i^{(6)})}{3(y_i-y_{i+1})},\end{aligned}$$

$$\begin{aligned}\int_c^d y\bar{B}'(y)dy&=\frac{1}{2}A_i(x)\left(\left(y_i^{(6)}\right)^2-y_i^2\right)+\frac{A_{i+1}(x)\left(y_{i+1}^2-\left(y_i^{(8)}\right)^2\right)}{2}\\&\quad+\frac{-4\left(y_i^{(6)}\right)^3+4\left(y_i^{(8)}\right)^3-6y_{i+1}\left(\left(y_i^{(8)}\right)^2-\left(y_i^{(6)}\right)^2\right)}{9(y_i-y_{i+1})};\end{aligned}$$

b. when $x\in\left[x_i^{(3)},x_{i+1}\right]$,

$$\begin{aligned}\int_c^d\bar{B}'(y)dy&=\int_{y_i}^{y_i^{(9)}}A_i(x)dy+\int_{y_i^{(9)}}^{y_i^{(7)}}\frac{4}{3}B_{i+1}(y)dy+\int_{y_i^{(7)}}^{y_{i+1}}A_{i+1}(x)dy\\&=A_i(x)\left(y_i^{(9)}-y_i\right)\\&\quad+\frac{-2\left(y_i^{(7)}\right)^2+2\left(y_i^{(9)}\right)^2-4y_{i+1}\left(y_i^{(9)}-y_i^{(7)}\right)}{3(y_i-y_{i+1})}\\&\quad+\frac{y_i^{(7)}-4y_i^{(9)}}{3}+y_{i+1}-A_i(x)\left(y_{i+1}-y_i^{(7)}\right),\end{aligned}$$

$$\int_c^d y\bar{B}'(y)dy = \frac{1}{2}A_i(x)\left(\left(y_i^{(9)}\right)^2 - y_i^2\right)$$
$$+\frac{-4\left(y_i^{(7)}\right)^3 + 4\left(y_i^{(9)}\right)^3 - 6y_{i+1}\left(\left(y_i^{(9)}\right)^2 - \left(y_i^{(7)}\right)^2\right)}{9(y_i - y_{i+1})}$$
$$+\frac{6\left(y_i^{(7)}\right)^2 - 6\left(y_i^{(9)}\right)^2}{3} + \frac{A_{i+1}(x)\left(y_{i+1}^2 - \left(y_i^{(7)}\right)^2\right)}{2}.$$

Because of $x_i < x_i^{(1)} < x_i^{(3)} < x_i^{(2)} < x_{i+1}$, we can know that from the discussion above, when $x \in [x_i, x_{i+1})$,

$$\bar{S}(x) = \frac{\int_c^d y(\underline{B}'(y) + \bar{B}'(y))dy}{\int_c^d (\underline{B}'(y) + \bar{B}'(y))dy} = A_i^*(x)y_i + A_{i+1}^*(x)y_{i+1}$$

where

$$A_i^*(x) = \begin{cases} -\dfrac{162A_i(x)^3 - 405A_i(x)^2 + 261A_i(x) - 199}{12\left(18A_i(x)^2 - 30A_i(x) + 35\right)} & x \in \left[x_i, x_i^{(1)}\right] \\ \dfrac{90A_i(x)^3 - 63A_i(x)^2 - 57A_i(x) - 13}{4\left(36A_i(x)^2 - 42A_i(x) - 11\right)} & x \in \left[x_i^{(1)}, x_i^{(3)}\right] \\ \dfrac{90A_i(x)^3 - 63A_i(x)^2 - 33A_i(x) - 25}{4\left(36A_i(x)^2 - 30A_i(x) - 17\right)} & x \in \left[x_i^{(3)}, x_i^{(2)}\right] \\ -\dfrac{162A_i(x)^3 - 297A_i(x)^2 + 9A_i(x) - 95}{12\left(18A_i(x)^2 - 6A_i(x) + 23\right)} & x \in \left[x_i^{(2)}, x_{i+1}\right] \end{cases}$$

$$A_{i+1}^*(x) = 1 - A_i^*(x).$$

Theorem 3. Given a single input single output interval type-2 fuzzy system denoted as above. If $\delta_i = 2/3, \gamma_i = 1/3, i = 1, 2, \cdots, n,$ $L = \frac{137}{144}\sum_{i=1}^{n-1}(y_{i+1} - y_i)(x_{i+1} - x_i)$, then Probability density function

$$f(x, y) = \frac{\underline{B}'(y) + \bar{B}'(y)}{L}.$$

Proof: $f(x, y)$ is determined by formula (6)

$$f(x, y) = \frac{q(x, y)}{H(2, n, \sqcap, \sqcup)}.$$

When $(x, y) \in [x_i, x_{i+1}) \times [y_i, y_{i+1})$,

$$q(x,y) = p(x,y) = B''(y) = \frac{1}{2}\left(\underline{B}'(y) + \bar{B}'(y)\right),$$

$$\begin{aligned} H(2,n,\sqcap,\sqcup) &\triangleq \int_{-\infty}^{+\infty}\int_{-\infty}^{+\infty} q(x,y)dydx \\ &= \frac{1}{2}\sum_{i=1}^{n-1}\left(\int_{x_i}^{x_{i+1}}\int_{y_i}^{y_{i+1}} \underline{B}'(y)dydx + \int_{x_i}^{x_{i+1}}\int_{y_i}^{y_{i+1}} \bar{B}'(y)dydx\right) \end{aligned}$$

Where $\underline{B}'(y) = \underline{R}_i(x',y) \vee \underline{R}_{i+1}(x',y), \bar{B}'(y) = \bar{R}_i(x',y) \vee \bar{R}_{i+1}(x',y), \underline{R}_i(x,y) = A_i(x) \wedge \delta_i B_i(y), \bar{R}_i(x,y) = A_i(x) \wedge (1+\gamma_i)B_i(y)$.

From the analysis in Theorem 2, we can know

$$\underline{B}'(y) = \begin{cases} \frac{2}{3}B_i(y), & (x,y) \in \left[x_i, x_i^{(1)}\right) \times \left[y_i, y_i^{(4)}\right) \cup \left[x_i^{(1)}, x_i^{(2)}\right) \times \left[y_i^{(1)}, y^{(3)}\right) \\ A_i(x), & (x,y) \in \left[x_i^{(1)}, x_i^{(2)}\right) \times \left[y_i, y_i^{(1)}\right) \cup \left[x_i^{(2)}, x_{i+1}\right) \times \left[y_i, y_i^{(5)}\right) \\ A_{i+1}(x), & (x,y) \in \left[x_i, x_i^{(1)}\right) \times \left[y_i^{(4)}, y_{(i+1)}\right) \cup \left[x_i^{(1)}, x_i^{(2)}\right) \times \left[y_i^{(2)}, y_{i+1}\right) \\ \frac{2}{3}B_{i+1}(y), & (x,y) \in \left[x_i^{(1)}, x_i^{(2)}\right) \times \left[y_i^{(3)}, y^{(2)}\right) \cup \left[x_i^{(2)}, x_{i+1}\right) \times \left[y_i^{(5)}, y_{(i+1)}\right) \\ 0, & else \end{cases}$$

$$\bar{B}'(y) = \begin{cases} \frac{2}{3}B_i(y) & (x,y) \in \left[x_i, x_i^{(3)}\right) \times \left[y_i^{(6)}, y_i^{(8)}\right) \\ A_i(x) & (x,y) \in \left[x_i, x_i^{(3)}\right) \times \left[y_i, y_i^{(6)}\right) \cup \left[x_i^{(3)}, x_{i+1}\right) \times \left[y_i, y_i^{(9)}\right) \\ A_{i+1}(x) & (x,y) \in \left[x_i, x_i^{(3)}\right) \times \left[y_i^{(8)}, y_{i+1}\right) \cup \left[x_i^{(3)}, x_{i+1}\right) \times \left[y_i^{(7)}, y_{i+1}\right) \\ \frac{2}{3}B_{i+1}(y) & (x,y) \in \left[x_i^{(3)}, x_{i+1}\right) \times \left[y_i^{(9)}, y_i^{(7)}\right) \\ 0 & else \end{cases}$$

So, we can obtain

$$\int_{x_i}^{x_{i+1}}\int_{y_i}^{y_{i+1}} \underline{B}'(y)dxdy = \frac{7}{18}(y_{i+1} - y_i)(x_{i+1} - x_i)$$

$$\int_{x_i}^{x_{i+1}}\int_{y_i}^{y_{i+1}} \bar{B}'(y)dxdy = \frac{9}{16}(y_{i+1} - y_i)(x_{i+1} - x_i).$$

Hence,

$$H(2,n,\sqcap,\sqcup) = \frac{1}{2}\sum_{i=1}^{n-1}\left\{\int_{x_i}^{x_{i+1}}\int_{y_i}^{y_{i+1}} \underline{B}'(y)dydx + \int_{x_i}^{x_{i+1}}\int_{y_i}^{y_{i+1}} \bar{B}'(y)dydx\right\}$$

$$= \frac{137}{288}\sum_{i=1}^{n-1}(y_{i+1}-y_i)(x_{i+1}-x_i)$$

$$f(x,y) = \frac{q(x,y)}{H(2,n,\sqcap,\sqcup)} = \frac{\underline{B}'(y)+\bar{B}'(y)}{L}$$

where $L = \frac{137}{144}\sum_{i=1}^{n-1}(y_{i+1}-y_i)(x_{i+1}-x_i)$.

Note: (1) The above results are obtained in the suppose that $a = x_1 < x_2 < \cdots < x_n = b, c = y_1 < y_2 < \cdots < y_n = d$. And we can obtain the similar conclusion when $a = x_1 < x_2 < \cdots < x_n = b, c = y_1 > y_2 > \cdots > y_n = d$.

(2) In this paper, we assume $S(x)$ is a monotone function. For a non-monotonic function, we can split X into several small intervals to make sure $S(x)$ monotonous in each interval.

4 Conclusion

Based on a set of input-output data, we establish an interval type-2 fuzzy system, and reveal the probability significance of this fuzzy system. We prove that the output of this interval type-2 fuzzy system based on NT type-reduction method can be regarded as the conditional mathematical expectation of certain random vector.

From the perspective of probability theory, the input variable and output variable of an uncertainty system can be regarded as two random variables depended on each other. Hence, an interval type-2 fuzzy system also is a stochastic system in nature, and either the joint probability density function $f(x,y)$ of the input-output random vector can be researched using the probability statistics method and then the approximation function relationship of input and output of the system can be obtained, or the probability distribution of input and output random variables can be obtained based on fuzzy inference method.

Then we provide the expression of the probability density function and the fuzzy system when the parameters values are given.

Acknowledgements. This work is supported by the National Natural Science Foundation of China (61374118, 61773088) and the programme of Zhuhai Municipal Key Laboratory of Intelligent Control.

References

1. Zadeh, L.A.: The concept of a linguistic variable and its application to approximate reasoning-I. Inf. Sci. **8**(3), 199–249 (1975)
2. Mendel, J.M.: Uncertain Rule-Based Fuzzy Logic Systems: Introduction and New Directions. Prentice-Hall, Upper Saddle River (2001)
3. Karnik, N.N., Mendel, J.M.: Introduction to type-2 fuzzy logic systems. IEEE Int. Conf. Fuzzy Syst. Proc. **2**, 915–920 (1998)
4. Karnik, N.N., Mendel, J.M., Liang, Q.: Type-2 fuzzy logic systems. IEEE Trans. Fuzzy Syst. **7**(6), 643–658 (1999)
5. Li, H.-X.: The probability representation of fuzzy system. China Sci. E Ser. Inf. Sci. **36**(4), 373–397 (2006)
6. Li, H.-X.: The united theory of uncertainty systems. Chinese J. Eng. Math. **24**(1), 1–21 (2007)
7. Wang, Y.-T., Yuan, X.-H., Li, H.-X.: Probability density function and fuzzy system based on a set of input—output data. In: Engineering, pp. 3820–3824(2010)
8. Yuan, Y., Yuan, X., Li, H.: The probability distribution and fuzzy system based on bounded product implication. In: FSKD 2010, pp. 786–790 (2010)
9. Zhong, Y.-B., Liu, Z.-L., Yuan, X.-H.: A Fuzzy Trustworthiness System with Probability Presentation Based on Center-of-gravity Method. Ann. Data. Sci. **2**(3), 335–362 (2015)

Coordination of the Supply Chain with Nash Bargaining Fairness Concerns

Wen-Qian Liu[1], Guo-Hang Huang[1], Zhong-Ping Li[1], Cai-Min Wei[1](✉), and Yan Chen[2]

[1] Department of Mathematics, Shantou University, Shantou 515063, People's Republic of China
{16wqliu,16ghhuang,14zpli}@stu.edu.cn, cmwei@stu.cn

[2] Department of Natural Sciences, Shantou Polytechnic, Shantou 515078, People's Republic of China
cylxq331@126.com

Abstract. This paper investigates the contracts and the coordination of the supply chain when the retailer and the manufacturer are both concerned with fairness. We assume that the market demand is stochastic, where defective products are considered. Firstly, based on Nash bargaining fairness-concerned reference point, we obtain the optimal order quantity in both the centralized and decentralized decision-making cases, where we find that wholesale price contract cannot achieve the coordination of the supply chain. Secondly, we get the optimal wholesale price under both buy-back and revenue sharing contracts such that the coordination of the supply chain is achieved. Finally, numerical examples show the influence of the fairness-concerned behavioral preference of the channel members on the optimal order quantity, the optimal wholesale price, the maximum utility of the channel members and the whole supply chain. Some managerial insights are obtained.

Keywords: Nash bargaining · Coordination · Fairness concerns Buy-back contract · Revenue sharing contract

1 Introduction

With the rapid advances of scientific technology and the rapid development of the productive forces, the consumption level of customers is improved continuously. Competition among enterprisers is increasingly fierce with the great changes of the political, economic and social environment, leading to the increasing demand uncertainty greatly (e.g. [1]). Due to supply chain members are always primarily concerned with optimizing their own interests in real commercial activities, leading to the well-known problem of "double marginalization" (e.g. [2]). In order to change this behavior, many scholars have proposed various kinds of contracts to achieve the coordination of the supply chain and avoid the problem of "double marginalization", such as wholesale price contract, buy-back contract and revenue sharing contract and so on (e.g. [3]).

B.-Y. Cao and Y.-B. Zhong (Eds.): ICFIE 2017, AISC 872, pp. 305–319, 2019.
https://doi.org/10.1007/978-3-030-02777-3_27

While a large amount of research has paid attention to the coordination of the supply chain under fairness-neutrality, for example, Su [4] supposed that decision makers were fully rational in the traditional decision-making. However, the current study found that people were not only concerned with their own profits but also cared about the profit of the channel' other member. Many researchers have paid close attention to fairness, for example, Kahneman *et al.* [5] firstly found that profit-maximizing firms were using fairness to keep the individuals employees motivated and business moving forward when they were willing to resist unfair transactions and punish unfair firms at some cost to themselves, customers and staff were both concerned with fairness for price and salary respectively in the market transaction process. Loch *et al.* [6] found fairness concerns were incompatible with conventional theory, it violated the assumption that the humans were fully rational, many experiments and empirical sciences have demonstrated the existence of fairness-concerned behavior. Pavlov *et al.* [7] showed that they failed to achieve the coordination of the supply chain by experimental studies of coordinating contracts due to information asymmetry with regard to their preferences for fairness.

Some scholars have been studied the coordination of the supply chain under fairness concerns. For, example, Du *et al.* [8] investigated the effects of the fairness-concerned behavioral preference of the retailer on wholesale price contract, buy-back and revenue sharing contract, respectively, in a dyadic supply chain. Du *et al.* [9] investigated newsvendor problem in a dyadic supply chain, but they did not consider the effects of Nash bargaining fairness concerns on the coordination of the supply chain under buy-back and revenue sharing contracts. Wu *et al.* [10] studied the influence of inequity averse preferences on the performance of the supply chain in a two-party random-demand newsvendor setting. However, these scholars did not consider the influence of the fairness-concerned behavioral preference of the channel members on the coordination of the supply chain under buy-back and revenue sharing contracts, and they also did not consider the effects of the fairness-concerned behavioral preference of the channel members on the utility of the channel members and the whole supply chain, and the optimal wholesale price, the optimal order quantity. At the same time, they did not consider the existence of defective products and the defective products being tested were whether this was advantageous for the whole supply chain.

Based on the above studies and the existence of problem, we consider the influence of the fairness-concerned behavioral preference of the channel members on the optimal order quantity in the decentralized decision-making and centralized decision-making cases, the optimal wholesale price under buy-back and revenue sharing contracts, the coordination of the supply chain, and the maximum utility of the channel members and the whole supply chain, where the existence of defective products is considered and Nash bargaining game solution is introduced as fairness-concerned reference point. We discuss about the influence of defective products being tested on the benefit of the whole supply chain.

The rest of the paper is organized as follows. In Sect. 2, the basic model and assumptions are introduced. We study the coordination of the supply chain under wholesale price contract, buy-back contract and revenue sharing contract when the retailer and the manufacturer are both concerned with fairness in Sect. 3. In Sect. 4, numerical examples show the influence of fairness concerns on the decision of the channel members under three contracts. The final section gives conclusion.

2 The Basic Model

2.1 Assumptions and Notations

The paper considers a dyadic supply chain with one manufacturer and one retailer who are both concerned with fairness. Now, we adopt some assumptions in establishing our model:

Assumption 1. It is assumed that the existence of defective products is unavoidable in the dyadic supply chain. And the manufacturer is to test the quality of the product.

Assumption 2. Assume that the manufacturer produces only one type of product, and the market demand is stochastic in the one-cycle selling marketing.

Assumption 3. Nash bargaining solution is introduced as the fairness-concerned reference point. For simplicity, assume that decision makers are the same sensitivity for the same profit and loss. Without loss of generality, assume that the profit of each channel member is the linear ratio of that of the whole supply chain.

The parameters and notations with respect to this paper are given as follows:

p: the unit sale price;
ω: the unit wholesale price;
c: the cost of unit production;
q: the order quantity of retailer;
q^*: the optimal order quantity of the decentralized channel under fairness-neutrality (the retailer and the manufacturer are both not concerned with fairness);
q^o: the optimal order quantity of the centralized channel under fairness-neutrality;
q_f^*: the optimal order quantity of the decentralized channel under fairness concerns;
q_f^o: the optimal order quantity of the centralized channel under fairness concerns;
π_r, π_m and π: the expected profits of the retailer, the manufacturer and the whole supply chain, respectively;

θ_1: the defective products rate (the ratio of the number of defective products to the total number of the products);
θ_2: the probability of the defective products being tested;
c_I: the test cost of unit production;
c_{IF}: the internal failure cost (with regard to the defective products being tested, manufacturers have internal failure cost) of the unit defective product;
c_{EF}: the external failure cost (but not being tested and flow the terminal customer, manufacturers have external failure cost) of the unit defective product; where D denotes the stochastic market demand with the mean value μ, $f(x)$ and $F(x)$ are the corresponding probability density and the distribution function of D, and $F(x)$ is continuous, differentiable and strictly increasing, $F(0) = 0$, and $\overline{F}(x) = 1 - F(x)$; $S(q)$ is the expected sale quantity of the retailer, and $S(q) = \int_0^q \overline{F}(x)dx$; $I(q)$ is the inventory at the sale end of season, and $I(q) = q - S(q)$.

Before discussing the model, we need to illustrate fairness and utility. When the manufacturer and the retailer are both concerned with fairness, they care about their own profits as well as the gap to the fairness reference. Thus, they will pursue the maximization of utility μ_i, i.e.

$$\mu_i = \pi_i + \lambda_i(\pi_i - \overline{\pi}_i), \ i = r, m, \tag{1}$$

where μ_i accounts for the channel member's profit as well as his concern about fairness; $\overline{\pi}_i$ is the final stable distribution agreement of the two-side game, i.e. $\overline{\pi}_r + \overline{\pi}_m = \pi$ and $\pi_r + \pi_m = \pi$; λ_i denotes the fairness-concerned degree of the channel members, $\lambda_i > 0$. Note that if λ_i is equal to zero, the utility of channel members is equal to the profits of the channel members.

2.2 Model Analysis Under Fairness-Neutrality

In this section, assume that the inventory is positive at the sale end of season. If the channel members are all fairness-neutral, the expected profits of the retailer, the manufacturer and the whole supply chain are given respectively as follows:

$$\pi_r = [p - c_{EF}(1 - \theta_2)\theta_1]S(q) - \omega q, \tag{2}$$

$$\pi_m = (\omega - c - c_{IF}\theta_1\theta_2 - c_I)q, \tag{3}$$

$$\pi = [p - c_{EF}(1 - \theta_2)\theta_1]S(q) - (c + c_{IF}\theta_1\theta_2 + c_I)q. \tag{4}$$

Proposition 1. When the manufacturer and the retailer are both fairness-neutral, the optimal order quantity under the decentralized case q^* is subjected to $\overline{F}(q^*) = \frac{\omega}{p - c_{EF}(1-\theta_2)\theta_1}$ and the optimal order quantity under the centralized case q^o is subjected to $\overline{F}(q^o) = \frac{c + c_{IF}\theta_1\theta_2 + c_I}{p - c_{EF}(1-\theta_2)\theta_1}$. And $q^* < q^o$, the wholesale price contract cannot achieve channel coordination under fairness-neutrality.

Proof. Taking first-order derivative of π_r and π with respect to q, we obtain

$$\frac{\partial \pi_r}{\partial q} = [p - c_{EF}(1-\theta_2)\theta_1][1-F(q)] - \omega,$$
$$\frac{\partial \pi}{\partial q} = [p - c_{EF}(1-\theta_2)\theta_1][1-F(q)] - (c + c_{IF}\theta_1\theta_2 + c_I).$$

Because the second-order derivative of π_r and π with respect to q are all subjected to

$$\frac{\partial^2 \pi_r}{\partial q^2} = \frac{\partial^2 \pi}{\partial q^2} = -[p - c_{EF}(1-\theta_2)\theta_1]F'(q) < 0,$$

then π_r and π are all concave with respect to q.

Let $\partial \pi_r/\partial q = 0$, $\partial \pi/\partial q = 0$, we obtain q^* and q^o. Because $\omega > c + c_{IF}\theta_1\theta_2 + c_I$, so the optimal order quantity $q^* < q^o$, i.e. we get Proposition 1.

3 Coordination of the Supply Chain Under Fairness Concerns

In this section, when the retailer and the manufacturer are both concerned with fairness, we describe the utility function of the retailer and the manufacturer through introducing dependence of the fairness reference point.

Du [9] had a more detailed discussion about the case. According to Lemma 1 in [9], then $\overline{\pi}_i = \frac{1+\lambda_i}{2+\lambda_i+\lambda_{i-}}\pi$. Then, the utility function of the retailer, the manufacturer and the whole supply chain is given respectively as follows:

$$\mu_r = (1+\lambda_r)\pi_r - \frac{1+\lambda_r}{2+\lambda_r+\lambda_m}\lambda_r\pi, \tag{5}$$

$$\mu_m = (1+\lambda_m)\pi_m - \frac{1+\lambda_m}{2+\lambda_r+\lambda_m}\lambda_m\pi, \tag{6}$$

$$\mu = (\lambda_r - \lambda_m)\pi_r + \frac{2+2\lambda_m+\lambda_r\lambda_m-\lambda_r^2}{2+\lambda_r+\lambda_m}\pi. \tag{7}$$

Lemma 1. The size relationship of the profit and the utility of the whole supply chain is connected with the fairness-concerned degree of the channel members and the difference between the channel members' profits and the fairness-concerned reference, namely,

(1) For $\lambda_r = \lambda_m$, then $\pi = \mu$;
(2) For $\lambda_r > \lambda_m$, if $\pi_r > \overline{\pi}_r$, then $\pi > \mu$, otherwise, $\pi < \mu$;
(3) For $\lambda_m > \lambda_r$, if $\pi_m > \overline{\pi}_m$, then $\pi > \mu$, otherwise, $\pi < \mu$.

Proof. Due to $\mu = (\lambda_r - \lambda_m)\pi_r + \frac{2+2\lambda_m+\lambda_r\lambda_m-\lambda_r^2}{2+\lambda_r+\lambda_m}\pi$, when $\lambda_r = \lambda_s$, we get $\pi = \mu$; when $\lambda_i > \lambda_{i^-}$ and $\pi_i > \overline{\pi}_i$, we obtain $\pi_i > \frac{1+\lambda_i}{2+\lambda_i+\lambda_{i^-}}\pi$. By simplifying Eq. (7), we get the conclusions of Lemma 1.

Lemma 1 suggests the utility of the whole supply chain is related to the fairness-concerned degree of the channel members. Especially, the retailer is just concerned with fairness as the manufacturer, the utility of the whole supply chain will keep unchanged. The utility of the whole supply chain can be improved if and only if the sign of $\lambda_i - \lambda_{i^-}$ is consistent with the sign of $\pi_i - \overline{\pi}_i$.

3.1 Wholesale Price Contract

Wholesale price contract is the most traditional of all contracts, which results in the well-known problem of "double marginalization". Thus, the profit of the whole supply chain is always suboptimal, and wholesale price contract cannot achieve the coordination of the supply chain. Usually, wholesale price contract is as a reference whether or not other contracts are effective [11]. When the retailer and the manufacturer are both concerned with fairness, the utility function of the retailer and the whole supply chain is

$$\begin{aligned} u_r = (1+\lambda_r)\left\{\left[p - c_{EF}(1-\theta_2)\theta_1\right]S(q) - \omega q\right\} - \tfrac{1+\lambda_r}{2+\lambda_r+\lambda_m}\lambda_r \\ \cdot\left\{\left[p - c_{EF}(1-\theta_2)\theta_1\right]S(q) - \left(c + c_{IF}\theta_1\theta_2 + c_I\right)q\right\}, \end{aligned} \tag{8}$$

$$\begin{aligned} u = (\lambda_r - \lambda_m)\left\{\left[p - c_{EF}(1-\theta_2)\theta_1\right]S(q) - \omega q\right\} + \tfrac{2+2\lambda_m+\lambda_r\lambda_m-\lambda_r^2}{2+\lambda_r+\lambda_m} \\ \cdot\left\{\left[p - c_{EF}(1-\theta_2)\theta_1\right]S(q) - \left(c + c_{IF}\theta_1\theta_2 + c_I\right)q\right\}. \end{aligned} \tag{9}$$

Proposition 2. Utility function of the retailer μ_r is strictly concave with respect to q, and there exists a unique optimal order quantity q_f^* such that $\overline{F}\left(q_f^*\right) = \frac{(2+\lambda_r+\lambda_m)\omega - \lambda_r\left(c+c_{IF}\theta_1\theta_2+c_I\right)}{(2+\lambda_m)\left[p-c_{EF}(1-\theta_2)\theta_1\right]}$, which is smaller than the optimal order quantity of the decentralized channel under fairness-neutrality, i.e. $q_f^* < q^*$.

Proof. Taking the second derivative of Eq. (8), we have

$$\frac{\partial^2 u_r}{\partial q^2} = -\frac{2+2\lambda_r+\lambda_s+\lambda_r\lambda_s}{2+\lambda_r+\lambda_s}\left[p - c_{EF}(1-\theta_2)\theta_1\right]F'(q) < 0.$$

So that μ_r is strictly concave for all q, let $\partial u_r/\partial q = 0$, the conclusion can be obtained. Besides, we get $\overline{F}(q_f^*) > \overline{F}(q^*)$ with regard to $\omega > (c + c_{IF}\theta_1\theta_2 + c_I)$, i.e. $q_f^* < q^*$.

This suggests that "when the channel members are all concerned with fairness, the optimal order quantity under the centralized channel are more deviation from that under decentralized channel". Wholesale price contract cannot achieve channel coordination. What's more, the coordination of the supply chain under fairness concerns is more difficult to be achieved.

Proposition 3. Utility function of the supply chain μ is strictly concave for all q in the range of $\lambda_m < \sqrt{2+2\lambda_r}$, and there exists a unique optimal order quantity q_f^o such that

$$\bar{F}\left(q_f^o\right) = \frac{\left(\lambda_r - \lambda_m\right)\left(2+\lambda_r+\lambda_m\right)\omega + \left(2+2\lambda_m+\lambda_r\lambda_m-\lambda_r^2\right)\left(c+c_{IF}\theta_1\theta_2+c_I\right)}{\left(2+2\lambda_r-\lambda_m^2\right)\left[p-c_{EF}\left(1-\theta_2\right)\theta_1\right]}.$$

Proof. Taking the second derivative of Eq. (9), we have

$$\frac{\partial^2 u}{\partial q^2} = -\frac{2+2\lambda_r-\lambda_m^2}{2+\lambda_r+\lambda_m}[p-c_{EF}(1-\theta_2)\theta_1]F'(q) < 0. \tag{10}$$

So that μ is strictly concave for all q in the range of $\lambda_m < \sqrt{2+2\lambda_r}$. Obviously, let $\partial u/\partial q = 0$, we obtain the above conclusion.

Proposition 4. The optimal order quantity q_f^o, q^o, q_f^* and q^* satisfy the following size relationship:

(1) For $\lambda_r < \lambda_m$, if $\lambda_m < \sqrt{2+2\lambda_r}$, then $q_f^o > q^o > q^* > q_f^*$;
(2) For $\lambda_r > \lambda_m$, if $\lambda_m < \sqrt{2+2\lambda_r}$, then $q^o > q_f^o > q^* > q_f^*$;
(3) For $\lambda_r < \lambda_m$, if $\lambda_m > \sqrt{2+2\lambda_r}$, then $q^o > q^* > q_f^* > q_f^o$;
(4) For $\lambda_r > \lambda_m$, if $\lambda_m > \sqrt{2+2\lambda_r}$, then $q_f^o > q^o > q^* > q_f^*$.

Proof. From Proposition 2, we find that utility function u is strictly concave with respect to all q in the range of $\lambda_m < \sqrt{2+2\lambda_r}$, and

$$\frac{\partial u\left(q^o\right)}{\partial q} = \left(\lambda_r - \lambda_m\right)\left(c+c_{IF}\theta_1\theta_2+c_I-\omega\right).$$

(1) If $\lambda_r < \lambda_m$, then $\partial u(q^o)/\partial q > 0$, $q_f^o > q^o$, from Sect. 2.2, we get $q_f^o > q^o > q^* > q_f^*$;
(2) If $\lambda_r > \lambda_m$, then $\partial u(q^o)/\partial q < 0$, $q_f^o < q^o$, and $\bar{F}\left(q^*\right) = \frac{\omega}{p-c_{EF}(1-\theta_2)\theta_1}$,

$$\bar{F}\left(q_f^o\right) = \frac{\left(\lambda_r - \lambda_m\right)\left(2+\lambda_r+\lambda_m\right)\omega + \left(2+2\lambda_m+\lambda_r\lambda_m-\lambda_r^2\right)\left(c+c_{IF}\theta_1\theta_2+c_I\right)}{\left(2+2\lambda_r-\lambda_m^2\right)\left[p-c_{EF}\left(1-\theta_2\right)\theta_1\right]},$$

so that $\bar{F}\left(q_f^o\right) < \bar{F}\left(q^*\right)$, we get $q^o > q_f^o > q^* > q_f^*$.

From Eq. (10), we get

$$\frac{\partial^2 u}{\partial q^2} = \frac{2+2\lambda_r-\lambda_m^2}{2+\lambda_r+\lambda_m}\left[p-c_{EF}\left(1-\theta_2\right)\theta_1\right]\left(-F'\left(q\right)\right),$$

and $\omega > \left(c+c_{IF}\theta_1\theta_2+c_I\right)$, so that utility function μ is strictly convex with respect to all q in the range of $\lambda_m < \sqrt{2+2\lambda_r}$.

(3) If $\lambda_r < \lambda_m$, then $\partial u(q^o)/\partial q > 0$, $q_f^o < q^o$, $2+2\lambda_m+\lambda_r\lambda_m-\lambda_r^2 > 2+2\lambda_r > 0$, $\bar{F}\left(q_f^o\right) > \bar{F}\left(q^*\right)$, we get $q^o > q^* > q_f^o$. And $\bar{F}\left(q_f^o\right) - \bar{F}\left(q_f^*\right) > 0$, then $q_f^o < q_f^*$, namely, $q^o > q^* > q_f^* > q_f^o$;

(4) If $\lambda_r > \lambda_m$, then $\partial u(q^o)/\partial q < 0$, $q_f^o > q^0$. Similarly, we get $q_f^o > q^o > q^* > q_f^*$.

From Proposition 4, we find that wholesale price contract still cannot achieve the coordination of the supply chain under fairness concerns. What's more, the fairness-concerned behavioral preference of the channel members effects on the coordination status of the supply chain.

Proposition 5. When the expected sale quantity of the retailer satisfies $S(q) > \frac{c_{IF}}{c_{EF}} q = q'$ under fairness-neutrality, the probability of the defective products being tested θ_2 is the bigger and the expected profit of the whole supply chain π is the more; conversely, the bigger θ_2, the less π.

Proof. Because $\partial\pi/\partial\theta_2 = \theta_1 (c_{EF} S(q) - c_{IF} q)$, for $S(q) > \frac{c_{IF}}{c_{EF}} q$, then $\partial\pi/\partial\theta_2 > 0$. Conversely, we get $\partial\pi/\partial\theta_2 < 0$.

This suggests whether the increase of θ_2 improves the expected profit of the whole supply chain with relation to the expected sale quantity of the retailer. Generally speaking, the higher $S(q)$, the more π. When the expected sale quantity of the retailer is the more and products are in the sales season particularly, the probability of the defective products being tested is the bigger, the profit of the whole supply chain is the more; however, when products are in sales off-season, the defective products being not tested is beneficial to the whole supply chain. So we need to test the more products in the sales season.

Proposition 6. When the expected sale quantity of the retailer satisfies $S(q_f) > \frac{(2+2\lambda_m+\lambda_r\lambda_m-\lambda_r^2)}{(2+2\lambda_r-\lambda_m^2)} \cdot \frac{c_{IF}}{c_{EF}} q \triangleq q_f' \left(\lambda_m < \sqrt{2+2\lambda_r}\right)$ under fairness concerns, the probability of the defective products being tested θ_2 is the bigger, the utility of the whole supply chain μ is the more; conversely, the bigger θ_2, the less μ.

Particularly, it is worth nothing that Proposition 6 shows that the manufacturer is the more concerned with fairness, the sales volume q_f' is the more. It means that to satisfy the channel's utility requirement is more difficult. When the manufacturer is more concerned with fairness than the retailer, we get $q_f' > q'$, i.e. the expected sale quantity of the retailer under fairness concerns need to be more than that under fairness-neutrality, the defective products being tested is more advantageous for the supply chain.

3.2 Buy-Back Contract

Pasternack [12] was the first to identify that buy-back contract can coordinate the fairness-neutral supply chain in the newsvendor setting. The manufacturer charges a wholesale price ω_b and pays the retailer b for the salvage value of unit production with that contract at the sale end of season, and $b < \omega_b$. The expected profits function of the retailer and the manufacturer are given respectively as follows:

$$\pi_r(b) = [p - b - c_{EF}(1-\theta_2)\theta_1]S(q) + (b - \omega_b)q, \tag{11}$$

$$\pi_m(b) = (\omega_b - c - c_{IF}\theta_1\theta_2 - c_I - b)q + bS(q). \tag{12}$$

Proposition 7. When the retailer and the manufacturer are both concerned with fairness, buy-back contract can achieve the coordination of the supply chain. And coordination condition:

$$\begin{aligned}\omega_b^* = &\frac{b\{(2+2\lambda_m+\lambda_r\lambda_m-\lambda_r^2)(c+c_{IF}\theta_1\theta_2+c_I)-(2+2\lambda_r-\lambda_m^2)[p-c_{EF}(1-\theta_2)\theta_1]\}}{(\lambda_r\lambda_m-2\lambda_m-2)[p-c_{EF}(1-\theta_2)\theta_1]-b(\lambda_r-\lambda_m)(2+\lambda_m+\lambda_r)} \\ &-\frac{[(2+2\lambda_m)-\frac{\lambda_m\lambda_r}{2+\lambda_m+\lambda_r}][p-c_{EF}(1-\theta_2)\theta_1](c+c_{IF}\theta_1\theta_2+c_I)}{(\lambda_r\lambda_m-2\lambda_m-2)[p-c_{EF}(1-\theta_2)\theta_1]-b(\lambda_r-\lambda_m)(2+\lambda_m+\lambda_r)}.\end{aligned} \tag{13}$$

Proof. The expected utility function of the retailer is given under buy-back contract:

$$u_r(b) = (1+\lambda_r)\,\pi_r(b) - \frac{1+\lambda_r}{2+\lambda_r+\lambda_m}\lambda_r\pi \Rightarrow \frac{\partial^2 u_r(b)}{\partial q^2} = \frac{\partial^2 u_r}{\partial q^2} - b(1+\lambda_r) < 0,$$

so that utility function $u_r(b)$ is strictly concave with respect to all q, and let $\partial u_r(b)/\partial q = 0$, we get

$$\bar{F}(q_{bf}^*) = \frac{(2+\lambda_r+\lambda_m)(\omega-b) - \lambda_r(c+c_{IF}\theta_1\theta_2+c_I)}{(2+\lambda_m)[p-c_{EF}(1-\theta_2)\theta_1] - b(2+\lambda_r+\lambda_m)},$$

$$\bar{F}(q_f^o) = \frac{(\lambda_r-\lambda_m)(2+\lambda_r+\lambda_m)\omega + (2+2\lambda_m+\lambda_r\lambda_m-\lambda_r^2)(c+c_{IF}\theta_1\theta_2+c_I)}{(2+2\lambda_r-\lambda_m^2)[p-c_{EF}(1-\theta_2)\theta_1]}.$$

However, $\bar{F}(x)$ is continuous, differentiable and strictly decreasing. Then

$$\bar{F}(q_{bf}^*) = \bar{F}(q_f^o),$$

we get Eq. (13).

When the value of λ_r and λ_m are both equal to zero, in other words, the retailer and the manufacturer are both fairness-neutral, then

$$\omega_b^* = b + (c+c_{IF}\theta_1\theta_2+c_I) - b\frac{(c+c_{IF}\theta_1\theta_2+c_I)}{p-c_{EF}(1-\theta_2)\theta_1}.$$

This means that the coordination of the supply chain always can be achieved as long as the relationship between the wholesale price and buy-back price satisfies Eq. (13). At the same time, the conditions of coordination are related to the fairness-concerned behavioral preference of the channel members.

3.3 Revenue Sharing Contract

The manufacturer charges a wholesale price ω_r with revenue sharing contract, and the retailer shares the ϕ times of its revenue with the manufacturer, the retailer retains the $1-\phi$ times of its own revenue. The expected profits function of the retailer and the manufacturer are given respectively as follows:

$$\pi_r(\phi) = [(1-\phi)p - c_{EF}(1-\theta_2)\theta_1]S(q) - \omega q, \tag{14}$$

$$\pi_m(\phi) = (\omega - c - c_{IF}\theta_1\theta_2 - c_I)q + \phi pS(q). \tag{15}$$

Proposition 8. When the retailer and the manufacturer are both concerned with fairness, revenue sharing contract can achieve the coordination of the supply chain in the range of $0 < \phi < 1 - \frac{c_{EF}(1-\theta_2)\theta_1}{p}$. And coordination condition:

$$\omega_r^* = \frac{\phi p(2+2\lambda_m+\lambda_r\lambda_m-\lambda_r^2)(c+c_{IF}\theta_1\theta_2+c_I)}{(\lambda_r\lambda_m-2\lambda_m-2)[p-c_{EF}(1-\theta_2)\theta_1]-\phi p(\lambda_r-\lambda_m)(2+\lambda_m+\lambda_r)} - \frac{[(2+2\lambda_m)-\frac{\lambda_m\lambda_r}{2+\lambda_m+\lambda_r}][p-c_{EF}(1-\theta_2)\theta_1](c+c_{IF}\theta_1\theta_2+c_I)}{(\lambda_r\lambda_m-2\lambda_m-2)[p-c_{EF}(1-\theta_2)\theta_1]-\phi p(\lambda_r-\lambda_m)(2+\lambda_m+\lambda_r)}. \quad (16)$$

Proof. The expected utility function of the retailer is given under revenue sharing contract:

$$u_r(\phi) = (1+\lambda_r)\pi_r(\phi) - \frac{1+\lambda_r}{2+\lambda_r+\lambda_m}\lambda_r\pi,$$

$$\frac{\partial^2 u_r(\phi)}{\partial q^2} = \frac{(1+\lambda_r)(2+\lambda_m)}{2+\lambda_r+\lambda_m}\left[\frac{\partial^2\pi_r}{\partial q^2} - \phi p(-F'(q))\right] < 0\left(0 < \phi < 1 - \frac{c_{EF}(1-\theta_2)\theta_1}{p}\right).$$

So that utility function $u_r(\phi)$ is strictly concave with respect to all in the range of $0 < \phi < 1 - \frac{c_{EF}(1-\theta_2)\theta_1}{p}$, and let $\partial u_r(\phi)/\partial q = 0$, we obtain

$$\bar{F}(q_{rf}^*) = \frac{(2+\lambda_r+\lambda_m)(\omega-b) - \lambda_r(c+c_{IF}\theta_1\theta_2+c_I)}{(2+\lambda_m)[p-c_{EF}(1-\theta_2)\theta_1] - \phi p(2+\lambda_r+\lambda_m)}.$$

$$\bar{F}(q_f^o) = \frac{(\lambda_r-\lambda_m)(2+\lambda_r+\lambda_m)\omega + (2+2\lambda_m+\lambda_r\lambda_m-\lambda_r^2)(c+c_{IF}\theta_1\theta_2+c_I)}{(2+2\lambda_r-\lambda_m^2)[p-c_{EF}(1-\theta_2)\theta_1]},$$

however, $\bar{F}(x)$ is continuous, differentiable and strictly decreasing, then we have

$$\bar{F}(q_{rf}^*) = \bar{F}(q_f^o),$$

we obtain Eq. (16).

Especially, when the value of λ_r and λ_s are both equal to zero, in other words, the retailer and the manufacturer are both fairness-neutral, then

$$\omega_r^* = (c+c_{IF}\theta_1\theta_2+c_I) - \phi p\frac{(c+c_{IF}\theta_1\theta_2+c_I)}{p-c_{EF}(1-\theta_2)\theta_1}.$$

This means that the coordination of the supply chain always can be achieved when the relationship between the wholesale price and the ratio of revenue distribution satisfies Eq. (16). At the same time, the conditions of coordination are related to the fairness-concerned behavioral preference of the channel members.

4 Numerical Analysis

In this section, numerical examples are used to illustrate the effects of the fairness-concerned behavioral preference of the channel members on contracts and the coordination of the supply chain. Without loss of generality, we assume that the market demand follows normal distribution, i.e. $D \sim N(1000, 100^2)$, and the other parameters as follows: $p = 160, c = 40, c_I = 10, c_{EF} = 100, c_{IF} = 40, \theta_2 = 0.8, \theta_1 = 0.1$. We get the optimal order quantity of the whole supply chain $q^o = 1042$, the expected sale quantity of the retailer $S(q^o) = 966$. Suppose that parameter b is 30 under the buy-back contract; ϕ is 0.1 under the revenue sharing contract.

4.1 Effects of the Fairness-Concerned Behavioral Preference on the Optimal Order Quantity Under Wholesale Price Contract

This subsection illustrates how the fairness-concerned behavioral preference of the channel members influences the optimal order quantity in the decentralized and centralized decision-making cases. Here, we set $\omega = 100$, the values of λ_r, λ_m fall in the range of $[0, 1]$. Figure 1(a) and (b) show the optimal order quantity with respect to λ_r, λ_m in the decentralized and centralized channels, respectively. We see that the optimal order quantity in both the decentralized and centralized decision-making cases increases as λ_r increases and λ_m decreases. Note that the optimal order quantity in the centralized decision-making case is always higher than that in the decentralized decision-making case. Moreover, we find that as λ_m decreases, λ_r increases, the optimal order quantity in the decentralized decision-making case increases more quickly than that in the centralized decision-making case, which means that the coordination of the supply chain is more difficult to achieved under wholesale price contract.

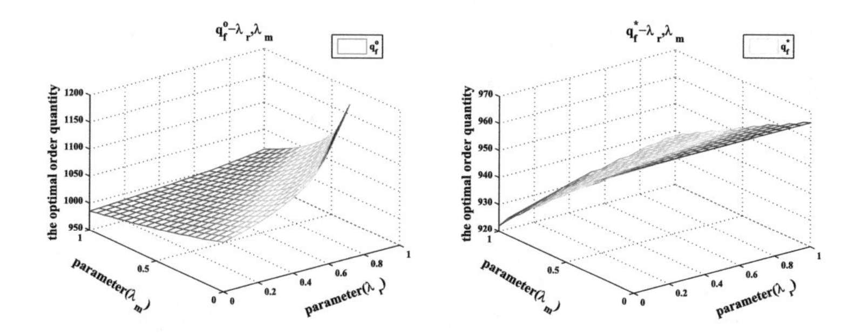

Fig. 1. Effects of λ_r, λ_m on the optimal order quantity.

4.2 Effects of the Fairness-Concerned Behavioral Preference on the Maximum Utility of the Channel Members Under Wholesale Price Contract

Furthermore, we explore how the fairness-concerned behavioral preference of the channel members influences the maximum utility of the retailer and the manufacturer. We set $\omega = 100$, the value of λ_r, λ_m vary from 0 to 1. Figure 2(a) and (b) show the effects of λ_r, λ_m on the maximum utility of the retailer and the manufacturer, respectively. We find that the maximum utility of the retailer increases as λ_r increases and λ_m decreases, the maximum utility of the manufacturer increases as λ_r decreases and λ_m increases. Apparently, when the fairness-concerned behavioral preference of the retailer is the lower, the influence of λ_m on the maximum utility of the retailer is the more obvious. Similarly, when λ_m is the lower, the influence of λ_r on the maximum utility of the manufacturer is also the more obvious.

The most interesting problem is whether the fairness-concerned behavioral preference of the channel members is profitable for the utility of the whole supply

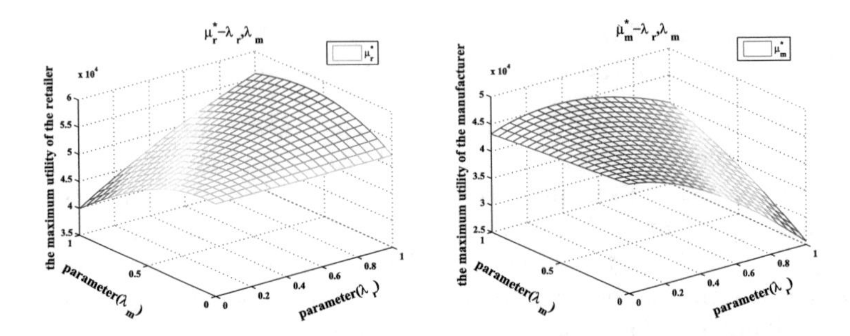

Fig. 2. Effects of λ_r, λ_m on the maximum utility of the retailer and the manufacturer.

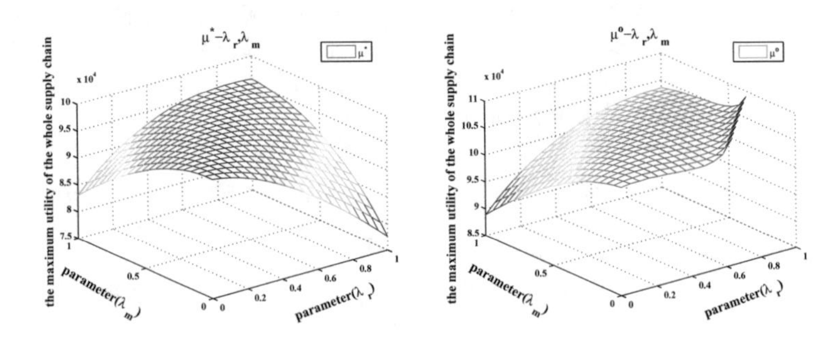

Fig. 3. Effects of λ_r, λ_m on the maximum utility of the whole supply chain.

chain. Then, we find how the fairness-concerned behavioral preference of the channel members influences the maximum utility of the whole supply chain. Set $\omega = 100$, the value of λ_r, λ_m fall in the range of $[0, 1]$. Figure 3(a) shows the effects of λ_r, λ_m on the maximum utility of the whole supply chain under the decentralized decision-making case. We see that the maximum utility of the whole supply chain is more as the manufacturer is just concerned with fairness as the retailer. Figure 3(b) shows the effects of λ_r, λ_m on the maximum utility of the whole supply chain under the centralized decision-making case. We find that the maximum utility of the whole supply chain is more as the retailer is more concerned with fairness than the manufacturer. The gap of the maximum utility of the whole supply chain between the centralized channel and the decentralized channel is the more, as the retailer are more concerned with fairness and the manufacturer are less concerned with fairness, so that the coordination of the supply chain is more difficult to be achieved.

4.3 Effects of the Fairness-Concerned Behavioral Preference on the Optimal Wholesale Price Under Buy-Back and Revenue Sharing Contracts

Then, we discuss how the fairness-concerned behavioral preference of the channel members influences the optimal wholesale price under buy-back and revenue sharing contracts. Similarly, we set $b = 30, \theta = 0.1$, the value of λ_r, λ_m vary from 0 to 1. Figure 4(a) and (b) show the effects of λ_r, λ_m on the wholesale price under buy-back and revenue sharing contracts, respectively. We see that the optimal wholesale price increases as λ_r, λ_m increase. Especially, the influence of λ_m on the optimal wholesale price under buy-back contract is more obvious; the influence of λ_r on the optimal wholesale price under revenue sharing contract is more obvious. Thus, the optimal wholesale price is the more under both buy-back and revenue sharing contracts as the retailer and the manufacturer are both the more concerned with fairness.

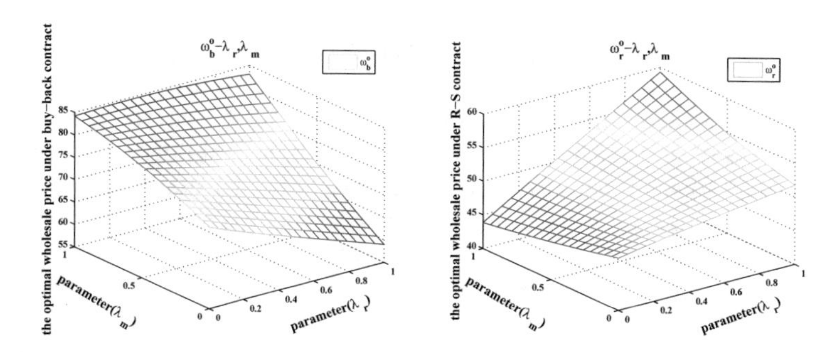

Fig. 4. Effects of λ_r, λ_m on the optimal wholesale price under buy-back and revenue sharing contracts.

4.4 Effects of the Fairness-Concerned Behavioral Preference on the Maximum Utility of the Whole Supply Chain Under Buy-Back and Revenue Sharing Contracts

In addition, we explore how the fairness-concerned behavioral preference of the channel members impacts on the maximum utility of the whole supply chain under buy-back and revenue sharing contracts. Set $b = 30, \theta = 0.1$, the value of λ_r, λ_m vary from 0 to 1. Figure 5(a) and (b) show the effects of λ_r, λ_m on the maximum utility of the whole supply chain under buy-back and revenue

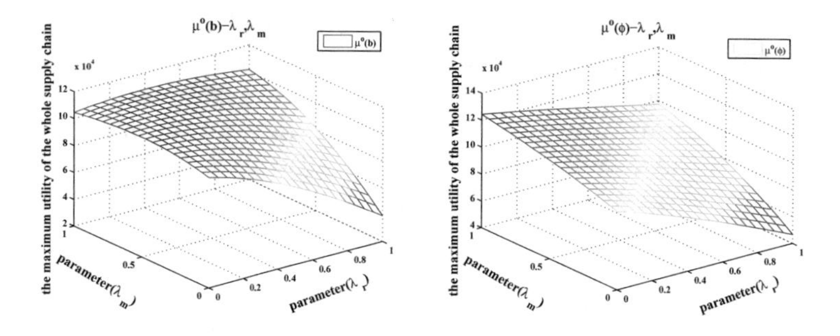

Fig. 5. Effects of λ_r, λ_m on the maximum utility of the whole supply chain.

sharing contracts, respectively. We see that the maximum utility of the whole supply chain is more as the manufacturer is more concerned with fairness than the retailer. Particularly, the influence of λ_m on the maximum utility of the whole supply chain is more obvious under both buy-back and revenue sharing contracts. Thus, the maximum utility of the whole supply chain is the more as the retailer are less concerned with fairness and the manufacturer are more concerned with fairness.

5 Conclusions

The paper studies the coordination of the supply chain by introducing Nash bargaining fairness concerns into the supply chain with a retailer and a manufacturer. We develop the utility model of the retailer and the manufacturer based on behavior of the newsvendor problem in a dyadic supply chain. Firstly, in wholesale price contract case, we obtain the optimal order quantity in the decentralized decision-making and centralized decision-making cases, respectively, and discuss that the fairness-concerned behavioral preference of the channel members impacts on the coordination of the supply chain and the utility of the whole supply chain. Secondly, we get the optimal wholesale price such that the coordination of the supply chain is achieved under buy-back revenue sharing contracts. Finally, some numerical experiments illustrate the effects of the fairness-concerned behavioral preference of the channel members on the optimal order quantity in both the decentralized and centralized decision-making cases, the maximum utility of the channel members under wholesale price contract, and the optimal wholesale price and the maximum utility of the whole supply chain under buy-back and revenue sharing contracts.

Through the above discussion, we find some interesting results as follows: when the retailer is the more concerned with fairness and the manufacturer is the less concerned with fairness, the optimal order quantity in both the decentralized and centralized decision-making cases is the more, at the same time the coordination of the supply chain is more difficult to be achieved under wholesale price contract. In wholesale price contract case, when the retailer is more concerned than fairness than the manufacturer, the maximum utility of the retailer is more; similarly, so it is with the maximum utility of the manufacturer; when the manufacturer is just concerned with fairness as the retailer, the maximum utility of the whole supply chain is more under the decentralized decision-making case. When the retailer and the manufacturer are both more concerned with fairness, the optimal wholesale price under both buy-back and revenue sharing contracts are more. When the manufacturer is the more concerned with fairness and the retailer is the less concerned with fairness, the maximum utility of the whole supply chain under both buy-back revenue sharing contracts.

In this paper, there are still some limitations as follows: on the one hand, it does not consider the influence of imperfect information on the coordination of the supply chain. Thus, we can further study that and make up the insufficiency of the existing research. On the other hand, this paper only investigates a simple dyadic supply chain. In the future, we can generate this study to research the effects of competition among multiple manufacturers or multiple retailers on coordination of the supply chain under fairness concerns.

Acknowledgements. Thanks to the support by Key Projects of National Social Science Foundation of China (Grant Nos. 16AGL010), Guangdong Natural Science Foundation of China (Grant Nos. C2017A030313005) and the General Foundation of Shantou Polytechnic of China (No. SZK2017Y01).

References

1. Qin, Y., Li, Y.: Supply chain coordination model under fairness concern of retailer and supplier. J. Inf. Comput. Sci. **11**(1), 211–217 (2014)
2. Ho, T.H., Zhang, J.J.: Designing pricing contracts for boundedly rational customers: does the framing of the fixed fee matter. Manag. Sci. **54**(4), 686–700 (2008)
3. Cachon, G.P., Lariiere, M.A.: Supply chain coordination with revenue-sharing contracts: strengths and limitations. Manag. Sci. **51**(1), 30–46 (2005)
4. Su, X.: Bounded rationality in newsvendor models. Manuf. Serv. Oper. Manag. **10**(4), 566–589 (2008)
5. Kahneman, D., Knetsch, J.L., Thaler, R.: Fairness as a constraint on profit seeking: entitlements in the market. Am. Econ. Rev. **76**(4), 728–741 (1986)
6. Loch, C.H., Wu, Y.: Social preferences and supply chain performance: an experimental study. Manag. Sci. **54**(11), 1321–1326 (2008)
7. Pavlov, V., Katok, E.: Fairness and coordination failures in supply chain contracts. Working Paper, University of Auckland (2011)
8. Du, S., Du, C., Liang, L., Liu, T.: Supply chain coordination considering fairness concerns. J. Manag. Sci. China **13**(4), 41–48 (2010)
9. Du, S., Nie, T., Chu, C., Yu, Y.: Newsvendor model for a dyadic supply chain with Nash bargaining fairness concerns. Int. J. Prod. Res. **52**(17), 5070–5085 (2014)
10. Wu, X., Niederhoff, J.A.: Fairness in selling to the newsvendor. Prod. Oper. Manag. **23**(11), 2002–2022 (2012)
11. Lariviere, M.A., Porteus, E.L.: Selling to the newsvendor: an analysis of price-only contracts. Manuf. Serv. Oper. Manag. **3**(4), 293–305 (2001)
12. Pasternack, B.A.: Optimal pricing and return policies for perishable commodities. Mark. Sci. **27**(1), 133–140 (2008)

OR

Normality and Shared Values of Meromorphic Functions with Differential Polynomial

Li-xia Cui and Wen-jun Yuan(✉)

School of Mathematics and Information Science, Guangzhou University, Guangzhou 510006, People's Republic of China
wjyuan1957@126.com

Abstract. In this paper, we discuss the normality and shared values of meromorphic functions with differential polynomial. We obtain the main result: Let $\mathcal{F}$ be a family of meromorphic functions in a domain D and k, q be two positive integers. Let $P(z,w) = w^q + a_{q-1}(z)w^{q-1} + \cdots + a_1(z)w$ and $H(f, f', \ldots, f^{(k)})$ be a differential polynomial with $\frac{\Gamma}{\gamma}|_H < k+1$. If $P(z, f^{(k)}) + H(f, f', \cdots, f^{(k)}) - 1$ has at most $q(k+1) - 1$ distinct zeros (ignoring multiplicity) for each function $f \in \mathcal{F}, f(z) \neq 0$, then $\mathcal{F}$ is normal in D. This result generalizes that of Chang [1].

Keywords: Meromorphic functions · Normal families
Uniformly convergence · Zeros · Differential polynomial

1 Introduction and Main Results

In this paper, we consider that the reader has already been familiar with Nevanlinna's Theory of reference mark and basic results.

A family $\mathcal{F}$ of meromorphic functions defined in a plane domain $D \subset \mathbf{C}$ is said to be normal in D, if each sequence $\{f_n\} \subset \mathcal{F}$ has a subsequence $\{f_{n_j}\}$ which converges spherically locally uniformly in D to a meromorphic function or ∞, see [5,10,11].

For the convenience of the reader, we make the following notation.

Let $q, k, m,$ be positive integers, $a_i(z)(i = 1, 2, \cdots, q-1)$, $b_j(z)(j = 1, 2, \cdots, m)$ be analytic in D and $n_0, n_1, \cdots,\ n_k$ be non-negative integers. Set

$$P(z,w) = w^q + a_{q-1}(z)w^{q-1} + \cdots + a_1(z)w,$$

$$M(f, f', \cdots, f^{(k)}) = f^{n_0}(f')^{n_1} \cdots (f^{(k)})^{n_k},$$

$$\gamma_M = n_0 + n_1 + n_2 + \cdots + n_k,$$

$$\Gamma_M = n_0 + 2n_1 + 3n_2 + \cdots + (k+1)n_k.$$

B.-Y. Cao and Y.-B. Zhong (Eds.): ICFIE 2017, AISC 872, pp. 323–333, 2019.
https://doi.org/10.1007/978-3-030-02777-3_28

$M(f, f', \cdots, f^{(k)})$ is called the differential monomial of f, γ_M is called the degree of $M(f, f', \cdots, f^{(k)})$ and Γ_M is called the weight of $M(f, f', \cdots, f^{(k)})$.

Let $M_1(f, f', \cdots, f^{(k)})$, $M_2(f, f', \cdots, f^{(k)}), \cdots, M_m(f, f', \cdots, f^{(k)})$ be differential monomials of f. Set

$$H(f, f', \cdots, f^{(k)}) = b_1(z)M_1(f, f', \cdots, f^{(k)}) + \cdots + b_m(z)M_m(f, f', \cdots, f^{(k)}),$$

$$\gamma_H = \max\{\gamma_{M_1}, \gamma_{M_2}, \cdots, \gamma_{M_m}\},$$

$$\Gamma_H = \max\{\Gamma_{M_1}, \Gamma_{M_2}, \cdots, \Gamma_{M_m}\}.$$

$H(f, f', \cdots, f^{(k)})$ is called the differential polynomial of f, γ_H is called the degree of $H(f, f', \cdots, f^{(k)})$ and Γ_H is called the weight of $H(f, f', \cdots, f^{(k)})$. Set

$$\frac{\Gamma}{\gamma}\Big|_H = \max\left\{\frac{\Gamma_{M_1}}{\gamma_{M_1}}, \frac{\Gamma_{M_2}}{\gamma_{M_2}}, \cdots, \frac{\Gamma_{M_m}}{\gamma_{M_m}}\right\}.$$

In 1959, Hayman [6] proved the following theorem.

Theorem 1.1. Let f be a nonconstant meromorphic function in $\mathbf{C}$ and k be a positive integer, then f or $f^{(k)} - 1$ has at least one zero. Moreover, if f is transcendental, then f or $f^{(k)} - 1$ has infinitely many zeros.

In 1967, Hayman [7] proposed the following famous conjecture which is proved by Gu [3] in 1979.

Theorem 1.2. Let $\mathcal{F}$ be a family of meromorphic functions in a domain D and k be a positive integer. If for each $f \in \mathcal{F}, f(z) \neq 0, f^{(k)}(z) \neq 1$, then $\mathcal{F}$ is normal in D.

In 2012, Chang [1] generalized Theorem 1.2 by allowing $f^{(k)}(z) - 1$ to has zeros but restricting it's numbers and proved the following theorem.

Theorem 1.3. Let $\mathcal{F}$ be a family of meromorphic functions in a domain D and k be a positive integer. If for each $f \in \mathcal{F}, f(z) \neq 0, f^{(k)}(z) - 1$ has at most k distinct zeros (ignoring multiplicity), then $\mathcal{F}$ is normal in D.

In 2000, Fang and Hong [2] considered a differential polynomial of f which omitted a value and proved the following theorem.

Theorem 1.4. Let $\mathcal{F}$ be a family of meromorphic functions in a domain D, $k, q(\geq 2)$ be two positive integers, and $H(f, f', \cdots, f^{(k)})$ be differential polynomial with $\frac{\Gamma}{\gamma}|_H < k + 1$. If for each $f \in \mathcal{F}$, the zeros of f are of multiplicity at least $k + 1$, and $(f^{(k)})^q + H(f, f', \cdots, f^{(k)}) \neq 1$, then $\mathcal{F}$ is normal in D.

In 2014, Yuan *et al.* [12] relaxed the condition in Theorem 1.4 and proved the following result.

Theorem 1.5. Let $\mathcal{F}$ be a family of meromorphic functions in a domain D and k be a positive integer. Let $h(z)$ be a polynomial with degree at least 2 and $H(f, f', \cdots, f^{(k)})$ be a differential polynomial with $\frac{\Gamma}{\gamma}|_H < k+1$. If $h(z) - 1$ has at least two distinct zeros, for each $f \in \mathcal{F}$, the zeros of f are of multiplicity at least $k+1$ and $h(f^{(k)}) + H(f, f', \cdots, f^{(k)}) - 1$ has at most one zero, then $\mathcal{F}$ is normal in D.

In this paper, we combine the differential polynomial of Fang and Yuan to study the problem, we extend Theorem 1.3 as follows.

Theorem 1.6. Let $\mathcal{F}$ be a family of meromorphic functions in a domain D and k, q be two positive integers. Let $P(z,w) = w^q + a_{q-1}(z)w^{q-1} + \cdots + a_1(z)w$ and $H(f, f', \ldots, f^{(k)})$ be a differential polynomial with $\frac{\Gamma}{\gamma}|_H < k+1$. If for each function $f \in \mathcal{F}, f(z) \neq 0, P(z, f^{(k)}) + H(f, f', \cdots, f^{(k)}) - 1$ has at most $q(k+1)-1$ distinct zeros (ignoring multiplicity) in D, then $\mathcal{F}$ is normal in D.

By the idea of shared values, in 2009, Shang [9] proved the following normality criterion.

Theorem 1.7. Let $\mathcal{F}$ be a family of meromorphic functions in a domain D, k, q be two positive integers and a, b be two non-zero complex numbers. Let $P(z,w) = w^q + a_{q-1}(z)w^{q-1} + \cdots + a_1(z)w$ and $H(f, f', \ldots, f^{(k)})$ be a differential polynomial with $\frac{\Gamma}{\gamma}|_H < k+1$. If for each function $f \in \mathcal{F}$, the zeros of f are of multiplicity at least $k+1$, the poles of f are of multiplicity at least 2 and $P(z, f^{(k)}) + H(f, f', \cdots, f^{(k)}) = a \Rightarrow f(z) = b$, then $\mathcal{F}$ is normal in D.

In this paper, we omit the condition of poles and extend Theorem 1.7 as follows.

Theorem 1.8. Let k, q be two positive integers and a, b be two non-zero complex numbers. And let $P(z,w) = w^q + a_{q-1}(z)w^{q-1} + \cdots + a_1(z)w$ and $H(f, f', \ldots, f^{(k)})$ be a differential polynomial with $\frac{\Gamma}{\gamma}|_H < k+1$. Let $\mathcal{F}$ be a family of meromorphic functions in a domain D. If for each function $f \in \mathcal{F}, f(z) \neq 0$, and $P(z, f^{(k)}) + H(f, f', \cdots, f^{(k)}) = a \Rightarrow f(z) = b$, then $\mathcal{F}$ is normal in D.

Example: Let $D = \{z : |z| < 1\}$ and $F : \{f_n(z) = \frac{1}{nz}, n = 1, 2, \cdots\}$, then

$$(f_n^{(k)}(z))^q + f_n^{(k)}(z) - 1 = \left(\frac{\frac{(-1)^k k!}{n}}{z^{k+1}}\right)^q + \frac{\frac{(-1)^k k!}{n}}{z^{k+1}} - 1$$
$$= \frac{\left(\frac{(-1)^k k!}{n}\right)^q + \left(\frac{(-1)^k k!}{n}\right) z^{(q-1)(k+1)} - z^{q(k+1)}}{z^{q(k+1)}}.$$

For each $f_n \in \mathcal{F}$, $(f_n^{(k)}(z))^q + f_n^{(k)}(z) - 1$ has $q(k+1)$ distinct zeros. However, $\mathcal{F}$ is not normal in D.

This example shows that $f(z) \neq 0$ and $P(z, f^{(k)}) + H(f, f', \cdots, f^{(k)}) - 1$ has at most $q(k+1)-1$ distinct zeros are necessary in Theorem 1.6.

2 Some Lemmas

For the proof of theorems, we require the following Lemmas.

Lemma 2.1 ([8,13]). Let $k \in \mathbf{N}_+$, let $\mathcal{F}$ be a family of meromorphic functions on the unit disc Δ, all of whose zeros have multiplicity at least k, and suppose that there exists $A \geq 1$ such that $|f^{(k)}(z)| \leq A$ whenever $f(z) = 0$. Then if $\mathcal{F}$ is not normal at z_0, there exist α, $0 \leq \alpha \leq k$ and

(a) functions $f_n \in \mathcal{F}$,

(b) points $z_n \in \Delta$, $z_n \to z_0$,
(c) positive numbers $\rho_n \to 0$,

such that $g_n(\xi) = \rho_n^{-\alpha} f_n(z_n + \rho_n \xi) \to g(\xi)$ locally uniformly with respect to the spherical metric, where $g(\xi)$ is a nonconstant meromorphic function on $\mathbf{C}$, all of whose zeros have multiplicity at least k, such that $g^{\#}(\xi) \leq g^{\#}(0) = kA + 1$. In particular, g has order at most 2.

Here, $g^{\#}(\xi) = \frac{|g'(\xi)|}{1+|g(\xi)|^2}$ is the spherical derivative of g.

Lemma 2.2 **([4]).** Let $f(z)$ be a non-polynomial meromorphic function in $|z| < R(\leq \infty)$, k be a positive integer. If $f(0) \neq 0, \infty$, $f^{(k)}(0) \neq 1$, $f^{(k+1)}(0) \neq 0$, and

$$(k+1)f^{(k+2)}(0)(f^{(k)}(0) - 1) - (k+2)(f^{(k+1)}(0))^2 \neq 0.$$

Then,

$$T(r, f) \leq (2 + \frac{1}{k})N(r, \frac{1}{f}) + (2 + \frac{2}{k})\overline{N}(r, \frac{1}{f^{(k)} - 1}) + S(r, f),$$

where $0 < r < R$.

Lemma 2.3 **([2]).** Let $f(z)$ be a meromorphic function with finite order in $\mathbf{C}$, k, q be two positive integers. If the zeros of $f(z)$ are of multiplicity at least $k+1$, $(f^{(k)}(z))^q \neq 1$, then $f(z)$ be a constant.

Lemma 2.4. Let $f(z)$ be a nonconstant rational function, k, q be two positive integers, and $h(z) = z^q + a_{q-1}z^{q-1} + \cdots + a_1 z$. If $f(z) \neq 0$, then $h(f^{(k)}(z)) - 1$ has at least $q(k+1)$ distinct zeros, ignoring multiplicity.

Proof: Since $f(z)$ is a nonconstant rational function, and $f(z) \neq 0$, we know that $f(z)$ is not a polynomial. Hence $f(z)$ has at least one pole. Set

$$f(z) = \frac{C_1}{\prod_{i=1}^{n}(z + z_i)^{p_i}}, \tag{2.1}$$

where C_1 is a nonzero constant. By (2.1) we have

$$f^{(k)}(z) = \frac{P(z)}{\prod_{i=1}^{n}(z + z_i)^{p_i + k}}, \tag{2.2}$$

where $P(z)$ is polynomial of degree $(n-1)k$. Hence,

$$(f^{(k)}(z))^j = \frac{P^j(z)}{\prod_{i=1}^{n}(z + z_i)^{j(p_i + k)}}, 1 \leq j \leq q. \tag{2.3}$$

Set

$$h(f^{(k)}(z)) = (f^{(k)}(z))^q + a_{q-1}(f^{(k)}(z))^{q-1} + \cdots + a_1 f^{(k)}(z)$$
$$= \frac{P^q(z)}{\prod_{i=1}^{n}(z+z_i)^{q(p_i+k)}} + \cdots + \frac{P(z)}{\prod_{i=1}^{n}(z+z_i)^{p_i+k}}. \tag{2.4}$$

From (2.3) and (2.4) we conclude that $h(f^{(k)}(z))$ is a proper rational expression. We write

$$h(f^{(k)}(z)) = 1 + \frac{C_2 \prod_{j=1}^{s}(z+w_j)^{l_j}}{\prod_{i=1}^{n}(z+z_i)^{q(p_i+k)}}, \tag{2.5}$$

where C_2 is a nonzero constant, n, s, p_i, l_j are positive integers, $w_j (1 \le j \le s)$ and $z_i (1 \le i \le n)$ are distinct complex numbers.

By (2.4) we have

$$h(f^{(k)}(z)) = \frac{Q(z)}{\prod_{i=1}^{n}(z+z_i)^{q(p_i+k)}}, \tag{2.6}$$

where $Q(z)$ is polynomial of degree $q(n-1)k$.

By (2.5) and (2.6) we get

$$\prod_{i=1}^{n}(z+z_i)^{q(p_i+k)} + C_2 \prod_{j=1}^{s}(z+w_j)^{l_j} = Q(z). \tag{2.7}$$

Thus by (2.7), it follows that $\sum_{j=1}^{s} l_j = \sum_{i=1}^{n} q(p_i + k) = q(\sum_{i=1}^{n} p_i + nk), C_2 = -1$. In (2.7), set $t = \frac{1}{z}$, then we have

$$\prod_{i=1}^{n}(1+z_i t)^{q(p_i+k)} - \prod_{j=1}^{s}(1+w_j t)^{l_j} = t^{q(p+k)} A(t), \tag{2.8}$$

where $A(t) = t^{q(n-1)k} Q(\frac{1}{t})$ is a polynomial of degree less than $q(n-1)k$ and $p = \sum_{i=1}^{n} p_i$.

Now by dividing $\prod_{j=1}^{s}(1+w_j t)^{l_j}$ of both sides of (2.8), we have

$$\frac{\prod_{i=1}^{n}(1+z_i t)^{q(p_i+k)}}{\prod_{j=1}^{s}(1+w_j t)^{l_j}} = 1 + \frac{t^{q(p+k)} A(t)}{\prod_{j=1}^{s}(1+w_j t)^{l_j}}. \tag{2.9}$$

Noting that for $t \to 0$,

$$\frac{t^{q(p+k)} A(t)}{\prod_{j=1}^{s}(1+w_j t)^{l_j}} = t^{q(p+k)}(a_0 + a_1 t + \cdots) = o(t^{q(p+k)+1}), \tag{2.10}$$

where $a_0 \neq 0$. By (2.7) and (2.8), for $t \to 0$, we have

$$\frac{\prod_{i=1}^{n}(1+z_i t)^{q(p_i+k)}}{\prod_{j=1}^{s}(1+w_j t)^{l_j}} = 1 + o(t^{q(p+k)+1}). \tag{2.11}$$

Thus by taking logarithmic of both sides of (2.11), we get

$$\log \prod_{i=1}^{n}(1+z_i t)^{q(p_i+k)} - \log \prod_{j=1}^{s}(1+w_j t)^{l_j} = \log(1 + o(t^{q(p+k)+1})).$$

Further, by taking derivatives of t, we have

$$\sum_{i=1}^{n} \frac{q(p_i+k)z_i}{1+z_i t} - \sum_{j=1}^{s} \frac{l_j w_j}{1+w_j t} = o(t^{q(p+k)}). \tag{2.12}$$

By (2.12), we write the power expansion of t:

$$\sum_{i=1}^{n} q(p_i+k) z_i^m t^m - \sum_{j=1}^{s} l_j w_j^m t^m = 0, \quad m = 0, 1, \ldots, q(p+k)-1. \tag{2.13}$$

Comparing the coefficients for t^m, we get

$$\sum_{i=1}^{n} q(p_i+k) z_i^m - \sum_{j=1}^{s} l_j w_j^m = 0, \quad m = 0, 1, \ldots, q(p+k). \tag{2.14}$$

Set $z_{n+j} = w_j,\ j = 1, 2, \ldots, s$, and

$$\sum_{i=1}^{n+s} [q(p_i+k) - l_j] z_i^m = 0.$$

We deduce that the system of linear equations

$$\sum_{i=1}^{n+s} x_i z_i^m = 0, \quad m = 0, 1, \ldots, q(p+k) \tag{2.15}$$

has a nonzero solution,

$$(x_1, \cdots, x_n, x_{n+1}, \cdots, x_{n+s}) = (q(p_1+k), \cdots, q(p_n+k), -l_1, \cdots, -l_s). \tag{2.16}$$

If $q(p+k) \geq n+s$, it follows from Cramers rule, the linear equations have unique solution, then the determinant $\det((z_i^m)_{(n+s)\times(n+s)})$ of the coefficients of equations (2.15) for $0 \leq m \leq n+s-1$ is equal to zero. However, noting that $\det((z_i^m)_{(n+s)\times(n+s)})$ is a Vandermonde determinant and all $z_i, i =$

$1, 2, \cdots, n+s$ are distinct complex numbers, therefore, $\det((z_i^m)_{(n+s)\times(n+s)}) \neq 0$. This is a contradiction with our assumption.

Thus, we deduce that $q(p+k) < n+s$. It follows that

$$\begin{aligned} s &> q(p+k) - n \\ &= qk + q - 1 + q(p-1) + n + 1 \\ &= qk + q - 1 + q\sum_{i=1}^{n}(p_i - 1) + (n-1)(q-1). \end{aligned}$$

Noting that $p_i \geq 1,\ q \geq 1,\ n \geq 1$, we have $q\sum_{i=1}^{n}(p_i - 1) \geq 0,\ (n-1)(q-1) \geq 0$.

Hence $s \geq q(k+1)$. Lemma 2.4 is proved.

3 Proofs of Theorems

Proof of Theorem 1.6: We show that $\mathcal{F}$ is normal in D. Otherwise, there exists at least one point $z_0 \in D$ such that $\mathcal{F}$ is not normal at z_0. Then by Lemma 2.1, there exist functions $f_n \in \mathcal{F}$, points $z_n \in D$, $z_n \to z_0$, positive numbers $\rho_n \to 0$, such that $g_n(\xi) = \rho_n^{-\alpha} f_n(z_n + \rho_n \xi)$ converges local uniformly with respect to the spherical metric to a nonconstant meromorphic function $g(\xi)$ on $\mathbf{C}$.

According to the hypothesis of the theorem, we can deduce $g(\xi) \neq 0$. For $\alpha = k$,

$$f_n^{(k)}(z_n + \rho_n \xi) = g_n^{(k)}(\xi).$$

It is easy to see that

$$\begin{aligned} &H(f_n, f_n', \cdots, f_n^{(k)})(z_n + \rho_n \xi) \\ &= \sum_{j=1}^{m} b_j(z_n + \rho_n \xi)\rho_n^{(k+1)\gamma_{M_j} - \Gamma_{M_j}} M_j(g_n, g_n', \cdots, g_n^{(k)})(\xi). \end{aligned}$$

Noting that all $b_j(z)(j = 1, 2, \ldots, m)$ are analytic on D and implies

$$|b_j(z_n + \rho_n \xi)| \leq M(\frac{1+r}{2}, b_j(z)) < \infty$$

for sufficiently large n, we infer from $\frac{\Gamma}{\gamma}|_H < k+1$ that

$$\sum_{j=1}^{m} b_j(z_n + \rho_n \xi)\rho_n^{(k+1)\gamma_{M_j} - \Gamma_{M_j}} M_j(g_n, g_n', \cdots, g_n^{(k)})(\xi)$$

converges uniformly to 0 on $\mathbf{C}$.

Thus we deduce that

$$P(z_n+\rho_n\xi, f_n^{(k)}(z_n+\rho_n\xi))+\sum_{j=1}^{m} b_j(z_n+\rho_n\xi)\rho_n^{(k+1)\gamma_{M_j}-\Gamma_{M_j}} M_j(g_n, g_n', \cdots, g_n^{(k)})(\xi)-1$$

converges local uniformly to $P(z_0, g^{(k)}(\xi)) - 1$.

By Hurwitz's Theorem, we have $P(z_0, g^{(k)}(\xi)) - 1 = 0$ or $P(z_0, g^{(k)}(\xi)) - 1 \neq 0$.

Noting that $P(z_0, 0) = 0$, we know that q zeros (counting multiplicity), say $d_i \neq 0, i = 1, 2, \cdots, q$, of $P(z_0, w) - 1$ are zero-free. Hence, we have

$$P(z_0, g^{(k)}(\xi)) - 1 = (g^{(k)} - d_1)(g^{(k)} - d_2)\cdots(g^{(k)} - d_q).$$

If $P(z_0, g^{(k)}(\xi)) - 1 \neq 0$, that is $(g^{(k)} - d_1)(g^{(k)} - d_2)\cdots(g^{(k)} - d_q) \neq 0$, by Lemma 2.3, $g^{(k)} \neq d_1$ and noting that $g \neq 0$, then we deduce $g(\xi)$ is a constant, a contradiction.

Hence $P(z_0, g^{(k)}(\xi)) - 1 = 0$. We consider two cases.

Case 1. $P(z_0, g^{(k)}(\xi)) - 1 \equiv 0$.

By Picard's Theorem, we get

$$g^{(k)}(\xi) \equiv C,$$

and then $g(\xi)$ is a polynomial with degree at most k. Noting that $g(\xi) \neq 0$, thus, $g(\xi)$ is a constant. This is impossible.

Case 2. $P(z_0, g^{(k)}(\xi)) - 1 \not\equiv 0$.

Now, we claim that $P(z_0, g^{(k)}(\xi)) - 1$ has at most $q(k+1) - 1$ distinct zeros.

Suppose that $P(z_0, g^{(k)}(\xi)) - 1$ has $q(k+1)$ distinct zeros ξ_j, $j = 1, 2, \cdots, q(k+1)$. Then there exist $\delta > 0$, $D_j = \{\xi : |\xi - \xi_j| < \delta\}, j = 1, 2, \cdots, q(k+1)$, such that $\xi_j \in D_j$, where $D_i \bigcap D_j = \emptyset$, for $i \neq j$.

Thus

$$\begin{aligned} &P(z_0, g_n^{(k)}(\xi)) - 1 \\ =&P(z_n + \rho_n\xi, f_n^{(k)}(z_n + \rho_n\xi)) + H(f_n, f_n', \cdots, f_n^{(k)})(z_n + \rho_n\xi)) - 1 \\ &\to P(z_0, g^{(k)}(\xi)) - 1, \end{aligned}$$

as $n \to \infty$, and from Hurwitz's Theorem, we know that there exist points $\xi_{n,j} \in D_j, \xi_{n,j} \to \xi_j, j = 1, 2, \cdots, q(k+1)$, such that

$$P(z_n + \rho_n\xi_{n,j}, f_n^{(k)}(z_n + \rho_n\xi_{n,j})) + H(f_n, f_n', \cdots, f_n^{(k)})(z_n + \rho_n\xi_{n,j})) - 1 = 0.$$

These imply that $P(z_n + \rho_n\xi, f_n^{(k)}(z_n + \rho_n\xi)) + H(f_n, f_n', \cdots, f_n^{(k)})(z_n + \rho_n\xi)) - 1$ has $q(k+1)$ distinct zeros in D, which contradicts with Theorem's hypothesis.

Hence, the claim holds.

Next, we prove that $g(\xi)$ is not a transcendental meromorphic function. From the above we know $d_i \neq 0, i = 1, 2, \cdots, q$ and we can easily get

$$\overline{N}(r, \frac{1}{g^{(k)} - d_i}) = \overline{N}(r, \frac{1}{(\frac{g}{d_i})^k - 1}).$$

By Lemma 2.2 and noting that $g(\xi) \neq 0$, we have

$$T(r, \frac{g}{d_i}) \leq (2+\frac{2}{k})\overline{N}(r, \frac{1}{(\frac{g}{d_i})^k - 1}) + S(r, \frac{g}{d_i}) \leq (2+\frac{2}{k})\overline{N}(r, \frac{1}{g^{(k)} - d_i}) + S(r, \frac{g}{d_i}).$$

Noting that $g^{(k)} - d_i$ is a factor of $P(z_0, g^{(k)}(\xi)) - 1$, we get

$$\overline{N}(r, \frac{1}{g^{(k)} - d_i}) \leq \overline{N}(r, \frac{1}{P(z_0, g^{(k)}(\xi)) - 1}).$$

Hence,

$$T(r, \frac{g}{d_i}) \leq (2+\frac{2}{k})\overline{N}(r, \frac{1}{P(z_0, g^{(k)}(\xi)) - 1}) + S(r, \frac{g}{d_i}).$$

And from the claim, we have

$$\overline{N}(r, \frac{1}{P(z_0, g^{(k)}(\xi)) - 1}) = O(\log r).$$

Therefore,

$$T(r, g) = O(\log r) + S(r, g).$$

This implies that $g(\xi)$ is a rational function.

By Lemma 2.4, we know that $P(z_0, g^{(k)}(\xi)) - 1$ has at least $q(k+1)$ distinct zeros, and this contradicts with the claim. It follows that our assumption is not true, hence $\mathcal{F}$ is normal in D.

This completes the proof of Theorem 1.6.

Proof of Theorem 1.8**:** We show that $\mathcal{F}$ is normal in D. Otherwise, there exists at least one point $z_0 \in D$ such that $\mathcal{F}$ is not normal at z_0. Then by Lemma 2.1, there exist functions $f_n \in \mathcal{F}$, points $z_n \in D$, $z_n \to z_0$, positive numbers $\rho_n \to 0$, such that $g_n(\xi) = \rho_n^{-\alpha} f_n(z_n + \rho_n \xi)$ converges local uniformly with respect to the spherical metric to a nonconstant meromorphic function $g(\xi)$ on $\mathbf{C}$.

By the hypothesis of Theorem 1.8, we can deduce that $g(\xi) \neq 0$.

From the proof of Theorem 1.6, we have

$$\begin{aligned}
&P(z_n + \rho_n \xi, f_n^{(k)}(z_n + \rho_n \xi)) + H(f_n, f_n', \cdots, f_n^{(k)})(z_n + \rho_n \xi) - a \\
=& P(z_n + \rho_n \xi, f_n^{(k)}(z_n + \rho_n \xi)) \\
&+ \sum_{j=1}^{m} b_j(z_n + \rho_n \xi) \rho_n^{(k+1)\gamma_{M_j} - \Gamma_{M_j}} M_j(g_n, g_n', \cdots, g_n^{(k)})(\xi) - a
\end{aligned}$$

converges local uniformly to $P(z_0, g^{(k)}(\xi)) - a$ on $\mathbf{C}$.

By Hurwitz's Theorem, we have $P(z_0, g^{(k)}(\xi)) - a = 0$, or $P(z_0, g^{(k)}(\xi)) - a \neq 0$.

If $P(z_0, g^{(k)}(\xi)) - a = 0$, we consider two cases.

Case 1. $P(z_0, g^{(k)}(\xi)) \equiv a$. Similarly, by using the same argument as in the case 1 of proof of Theorem 1.6, we get a conclusion that $g(\xi)$ is a constant, a contradiction.

Case 2. $P(z_0, g^{(k)}(\xi)) \not\equiv a$. We know $P(z_0, g^{(k)}(\xi)) = a$ and $P(z_0, g^{(k)}(\xi)) \not\equiv a$, therefore, there exist ξ_0, such that $P(z_0, g^{(k)}(\xi_0)) = a$. According to the a-point of analytic function isolation, there exist $D(\xi_0, \delta) = \{\xi : |\xi - \xi_0| < \delta\}$ and $P(z_0, g^{(k)}(\xi))$ has no other points except for ξ_0 in $D(\xi_0, \delta)$ such that $P(z_0, g^{(k)}(\xi)) = a$. Obviously, $g(\xi_0) \neq \infty$.

By Hurwitz's Theorem, for sufficiently large n, there exist $\xi_n \in D_\delta, \xi_n \to \xi_0$, such that

$$P(z_n + \rho_n \xi_n f_n^{(k)}(z_n + \rho_n \xi_n)) + H(f_n, f_n', \cdots, f_n^{(k)})(z_n + \rho_n \xi_n) = a$$

By the hypothesis of Theorem, $P(z, f^{(k)}) + H(f, f', \cdots, f^{(k)}) = a \Rightarrow f(z) = b$, we have $f_n(z_n + \rho_n \xi_n) = b$.

Therefore, $g_n(\xi_n) = \rho_n^{-k} f_n(z_n + \rho_n \xi_n) = \rho_n^{-k} b$.

Noticing that $b \neq 0$, we also obtain $g(\xi_0) = \lim_{n \to \infty} g_n(\xi_n) = \lim_{n \to \infty} \frac{b}{\rho_n^k} = \infty$, however $g(\xi_0) \neq \infty$, a contradiction.

If $P(z_0, g^{(k)}(\xi)) - a \neq 0$, noting that $P(z_0, 0) = 0$, we know that q zeros (counting multiplicity) of $P(z_0, w) - a$ are zero-free and we write

$$P(z_0, g^{(k)}(\xi)) - a = (g^{(k)} - v_1)(g^{(k)} - v_2) \cdots (g^{(k)} - v_q),$$

where $v_i \neq 0, i = 1, 2, \cdots, q$ are q zeros of $P(z_0, g^{(k)}(\xi)) - a$. Therefore, $g^{(k)} - v_1 \neq 0$, by Lemma 2.3 and noting that $g \neq 0$, we deduce $g(\xi)$ is a constant. This contradicts the fact that $g(\xi)$ is a nonconstant meromorphic function.

Hence, $\mathcal{F}$ is normal in D.

The proof of Theorem 1.8 is completed.

Acknowledgements. Thanks to the support by National Natural Science Foundation of China (No. 11271090) and of National Natural Science Foundation Guangdong Province (No. 2016A030310257 and No. 2015A030313346).

The authors would like to express their hearty thanks to Professor Fang Mingliang and Liao Liangwen for their helpful discussions and suggestions. This work was also supported by the Visiting Scholar Program of Chern Institute of Mathematics at Nankai University when the authors worked as visiting scholars.

References

1. Chang, J.M.: Normality and quasinormality of zero-free meromorphic functions. Acta Math. Sin. (Engl. Ser.) **28**(4), 707–716 (2012)
2. Fang, M.L., Hong, W.: Some results on normal family of meromorphic functions. Math. Meth. App. Sci. **23**, 143–151 (2000)
3. Gu, Y.X.: A criterion for normality of families of meromorphic functions. Sci. Sin. Math. Issue **1**, 267–274 (1979)
4. Gu, Y.X., Pang, X.C., Fang, M.L.: Theory of Normal Families and Its Applications. Science Press, Beijing (2007)
5. Hayman, W.K.: Meromorphic Functions. Clarendon Press, Oxford (1964)
6. Hayman, W.K.: Picard values of meromorphic functions and their derivatives. Ann. Math. **70**(2), 9–42 (1959)

7. Hayman, W.K.: Research Problems in Function Theory. Athlone Press, London (1967)
8. Pang, X.C., Zalcman, L.: Normal families and shared values. Bull. Lond. Math Soc. **32**, 325–331 (2000)
9. Shang H.: Normal families of meromorphic function concerning differential polynomials. J. Chongqing Univ. Posts Telecommun. (Nat. Sci. Ed.). **21**(6) (2009)
10. Schiff, J.: Normal Families. Springer, New York (1993)
11. Yang, L.: Value Distribution Theory. Springer, Heidelberg (1993)
12. Yuan, W.J., Lai, J.C., Huang, Z.F., Liu, Z.: Normality of meromorphic functions and differential polynomials share values. Adv. Differ. Equ. **120**(1), 1–9 (2014)
13. Zalcman, L.: Normal families: new perspectives. Bull. Amer. Math. Soc. **35**, 215–230 (1998)

Analysis of Error System Structure_Dynamic Structure and Static Structure

Hai-ou Xiong(✉)

Faculty of Port and Shipping Management, Guangzhou Maritime University, Guangzhou, Guangdong 51000, China
7752833@qq.com

Abstract. In this paper, the state structure equation was studied for general systems. The static structure and dynamic structure of the system error were also studied.

Keywords: Error system · Structure

1 Introduction

View things from a structural perspective, with a focus on the holistic connection between their internal elements. For the same set of elements, the system constituted by them will be poor if the structure is unreasonable, whereas will be favorable if the structure is reasonable. The existence value of an element in the system should be assessed based on whether it can achieve global optimization, rather than by its own merits and demerits. Because structure does not refer to the relation between individual or local elements of system, but is instead a holistic connection between various elements inside the system. Therefore, structure reflects the orderliness and organization of things on the whole, which is the internal basis for coordination and imbalance of things. The interrelationship and interaction between elements within a thing are always achieved through a certain structure, and whether the system structure is reasonable is the basis for the coordinated development of things. Thus, to study the optimization of error system, it is necessary to explore the variation regularity of system errors upon change in system structure; to investigate the causes and mechanisms of system errors, as well as the methods and regularities of transmission and transformation; and to study the temporal or spatial organic connections and interactions between various elements of error system, i.e. the structure of error system [1].

2 State Equation for General System Structures

Definition 1. System $Z(n)$ is defined as an entirety that is constituted by n number of interrelated parts $e(1)$, …, $e(i)$, …, $e(n)$, which is expressed as

B.-Y. Cao and Y.-B. Zhong (Eds.): ICFIE 2017, AISC 872, pp. 334–339, 2019.
https://doi.org/10.1007/978-3-030-02777-3_29

$$Z(n) = \{E(n), R_z^*\}$$

where

$$E(n) = \{e(i)/i = 1, 2, \cdots, n; n \geq 2\}$$

R_z^* represents the set of associations that exist between parts $e(1)$, …, $e(i)$, …, $e(n)$. For a system $Z(n)$:

(1) Its internal state

$$S_{in} = (s_1, \cdots, s_i, \cdots, s_n)^T,\ s_i \in A_i = \{a_1^i, a_2^i, \cdots, a_r^i\},\ (i = 1, 2, \cdots, n)$$

Where s_i and A_i denote the state and state space of part $e(i) \in Z(n)$, respectively.

(2) Its system behavior H_Z refers to certain external activity or performance of system $Z(n)$, so it is a function of the system internal state S_{in} and the system input R,

$$H_Z = F_h(R, S_{in})H_Z$$

(3) Its system state S_Z refers to the state that can characterize the existence of system,

$$S_Z = F_s(S, R, S_{in}).$$

where S represents the state of environment in which the system $Z(n)$ resides [2, 3].

Definition 2. System environment $E(S)$ refers to a set of system's external parts that are associated with the system $Z(n)$, where

$$S \in B = \{b_1, b_2, \cdots, b_m\}$$

S and B denote the state and state space of environment $E(S)$, respectively.

Definition 3. Relation $R_{ij}(t)$ refers to the acting factor of part $e(i)$ on $e(j)$ at time t. The part $e(i)$ acts on part $e(j)$ via this factor, so that a certain association is generated between these two parts, as shown in Fig. 1. Hence, the following relational equation exists:

$$f\left(s_i(t), R_{ij}(t), s_j(t)\right) = 0$$

Where $s_i(t)$ and $s_j(t)$ represent the states of parts $e(i)$ and $e(j)$ at time t, respectively.

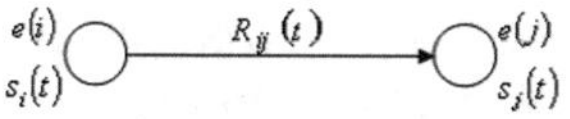

Fig. 1. Relation $R_{ij}(t)$

Definition 4. Suppose that the system *Z*(*n*) has such k number of relations $R_{p+i-1,p+i}(t)$ $(i=1,2,\cdots,k;\ k\geq 2; e(p+k)\equiv e(p))$ at time t, then the set of relations $R_{p+i-1,p+i}(t)$ $(i=1,2,\cdots,k;\ k\geq 2; e(p+k)\equiv e(p))$ is called a relational cycle, which is expressed as

$$Y(t)=\{E_0(k),R_0(t)\}$$

where

$$E_0(k)=\{e(i)/i=1,2,\cdots k; k\geq 2; e(p+k)\equiv e(p)\};$$

$$e(p+k)\equiv e(p)\}$$

Let the system $R_0(t)=\{R_{p+i-1,p+i}(t)/f_{p+i-1,p+i}(s_{p+i-1}(t),R_{p+i-1,p+i}(t),$ $s_{p+i}(t))=0; i=1,2,\cdots k; k\geq 2;$. *Z*(*n*) has θ number of different relational cycles $Y_1(t)$, …, $Y_i(t)$, …, $Y_j(t)$, …, $Y_\theta(t)(Y_i\neq Y_j; i\neq j)$ at time t, then θ is called the relational cycle number of system *Z*(*n*) (Fig. 2).

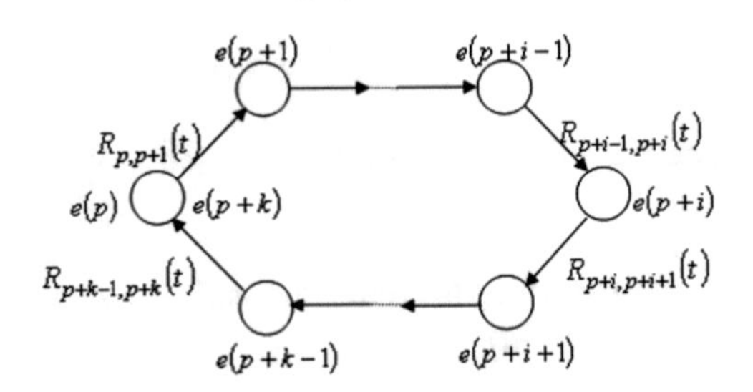

Fig. 2. Relational cycle $Y(t)$

Definition 5. Suppose that the distance between any two parts of system *Z*(*n*) is *r*(*i*,*j*), then $d=max\{r(i,j)\}\ i,j=1,2,\ldots,n.\ i\neq j$ is called the diameter of system [4].

Definition 6. In the system *Z*(*n*), let the transmission speed of relation $R_{ij}(t)$ between parts $e(i)$ and $e(j)(e(i),e(j)\in Z(n))$ be $u_{ij}(m/s)$, then it is called the relation transmission speed.

3 Basic Structure of System Error

3.1 Static Basic Structure of System Error

According to the primary classification of system structure, the basic structure of system error can also be classified into the following six basic types [5].

1. Cascade structure
   For a cascade structure system $CS(s_1, s_2, ..., s_n)$, the error description function is $CSC(s_1, s_2, ..., s_n)$.
   The logic propositional expression for basic structural error:

$$X(S) = X(s_1) \vee X(s_2) \vee \ldots \vee X(s_n).$$

2. Parallel structure
   For a parallel structure system $BS(s_1, s_2, ..., s_n)$, the error description function is $BSC(s_1, s_2, \ldots, s_n)$.
   The logic propositional expression for basic structural error:

$$X(S) = (X(s_{11}) \vee X(s_{12}) \vee \ldots \vee X(s_{1n1})) \wedge (X(s_{21}) \vee (X(s_{22}) \vee \ldots \vee X(s_{2n2}))$$

3. Amplification-reduction structure
   (1) Amplification structure
   The logic propositional expression for basic structural error:

$$X(S) = (X(s_0) \vee (X(s_1) \vee X(s_2) \vee \ldots \vee X(s_n))$$

   (2) Reduction structure
   For an amplification-reduction structure system $KS(s_1, s_2, ..., s_n)$, the error description function is $KSC(s_1, s_2, ..., s_n)$.
   The logic propositional expression for basic structural error:

$$X(S) = (X(s_1) \vee X(s_2) \vee \ldots \vee X(s_n)) \vee X(s_0).$$

4. Implication structure
   For an implication structure system $YS(s_1, s_2, ..., s_n)$, the error description function is $YSC(s_1, s_2, ..., s_n)$.
   The logic propositional expression for basic structural error:

$$X(S) = X(s_1) \vee X(s_2) \vee \ldots \vee X(s_n).$$

5. Feedback structure
   For a feedback structure system $FS(s_1, s_2, \ldots, s_n)$, the error description function is $FSC(s_1, s_2, ..., s_n)$.
   The logic propositional expression for basic structural error:

$$X(S) = X(s_1) \vee X(s_2) \vee \ldots \vee X(s_n).$$

6. Other basic types

3.2 Dynamic Basic Structures of System Error

1. Variable structures of error
   (1) Transformation structure

(2) Gradation structure

The main path of error transformation and gradation and relevant transformation modes are as follows:

Path: (*a*) domain; (*b*) subsystem (element); (*c*) structure ((1) cascade, (2) parallel, (3) amplification-reduction, (4) implication, (5) feedback); (*d*) time.

Transformation:

(1) Similarity transformation $T_x \subseteq \{T_{xly}, T_{xzx}, T_{xys}, T_{xjg}, T_{xsj}\}$ (similarity); corresponds to T_x^{-1} (inverse similarity transformation). Similarity transformation includes: domain similarity, subsystem similarity, element similarity, structure similarity and time similarity.

(2) Replacement transformation $T_z \subseteq \{T_{zly}, T_{zzx}, T_{zys}, T_{zjg}, T_{zsj}\}$ (replacement); corresponds to T_z^{-1} (inverse replacement transformation). Replacement transformation includes: domain replacement, subsystem replacement, element replacement, structure replacement and time replacement.

(3) Increment transformation $T_{zn} \subseteq \{T_{znly}, T_{znzx}, T_{znys}, T_{znjg}, T_{znsj}\}$ (increment); corresponds to T_{zn}^{-1} (decrement transformation). Increment transformation includes: domain increment, subsystem increment, element increment, structure increment and time increment.

(4) Decomposition transformation $T_f \subseteq \{T_{fly}, T_{fzx}, T_{fys}, T_{fjg}, T_{fsj}\}$ (decomposition); corresponds to T_f^{-1} (combinatorial transformation). Decomposition transformation includes: domain decomposition, subsystem decomposition, element decomposition, structure decomposition and time decomposition.

(5) Destruction transformation $T_h \subseteq \{ T_{hly}, T_{hzx}, T_{hys}, T_{hjg}, T_{hsj}\}$ (destruction); corresponds to T_h^{-1} (generation transformation). Destruction transformation includes: domain destruction, subsystem destruction, element destruction, structure destruction and time destruction.

(6) Unitary transformation T_d; corresponds to T_d^{-1} (inverse unitary transformation);

(7) Transformation system (with, or, inverse).

2. Fuzzy structures of error [6]
   (1) Cognitive fuzzy structures;
   (2) Objective fuzzy structures;
   (3) Dealing fuzzy structures.
3. Stability structures of error
   (1) Conditional stability structures. The system is stable under certain conditions.
   (2) Unconditional stability structures. The system is stable under any condition.
   (3) Interval stability structures. The system is stable within certain intervals.

4 Conclusion

The basic structure of system is fundamentally consistent with the static basic system structure of system error. The basic system structures of system error are classified into: 1. Static basic system structures of system error, including six basic types, namely the cascade, parallel, amplification-reduction, implication, feedback and other structures;

and 2. Dynamic basis system structures of system error, including transformation, gradation, stability and fuzzy structures.

Acknowledgements. Thanks to the support by Scientific Research Project of Guangzhou Maritime University.

Recommender:

Guo Kaizhong, Guangdong University of Technology, Professor.

References

1. Lin, F.Y., Wu, J.J.: General system structure theory and its application. J. Syst. Eng. **12**(3), 21–27 (1997)
2. Xu, G.Z.: System Science and Engineering Research, p. 10. Shanghai Scientific & Technological Education Publishing House, Shanghai (2000)
3. Lin, F.Y.: Structural Theory of General Systems. Jinan University Press, Guangzhou (1998)
4. Zhang, Q.R., Lin, F.Y.: A Study on the origin of complexity in the science of complexity. Syst. Eng. **22**(10) (2002)
5. Guo, K.Z., Zhang, S.Q.: The Theory of Error Set. Central South University Press, Changsha (2000)

Adaptive Synchronization Control of the Fractional Order Hyper Chaotic Systems with Unknown Parameters

Xiong Li[1] and Sheng-quan Ma[2(✉)]

[1] Northwest Agriculture and Forestry University, Yangling 712100, Shannxi, China

[2] School of Information Science and Technology, Hainan Normal University, Haikou 571158, Hainan, China

8701@hainnu.edu.cn

Abstract. Based on the stability theory of fractional order systems, the controller and recognizing rules of unknown parameters are designed. By using the square of the Lyapunov function, the controller is designed for synchronizing the fractional order hyper chaotic Chen system and fractional order hyper chaotic Lorenz system. The final results of the numerical simulation showed that the control method proposed is feasible.

Keywords: Hyper chaotic system · Fractional-order chaotic system Adaptive rule · Synchronous control

1 Introduction

The fractional calculus almost has the same long history with the integer order calculus. However, the development of the theoretical study is always lacking in progress for the reason of lack of actual application background, and also limited by the complexity of the theory [1]. The fractional theory not only provided a new mathematical tool for many practical systems, but also was suitable for describing the dynamic behavior of some physical systems. In some systems like diffuse, spectral analysis and dielectric, some mathematicians, physicists and engineers have began to solve problems with the application of fractional calculus. In the complex dynamic system, the model that established the application of fractional calculus tends to be more accurate than the integer order system [2].

In recent years, the project of the adaptive synchronization of chaotic systems with unknown parameters has caused wide public concern. The Ref. [3] have discussed the modified projective synchronization of hyper chaotic system with two unknown parameters; the Ref. [4] have discussed the adaptive generalized function projective lag synchronization of the different structured chaotic systems with uncertain parameters; the Ref. [5] have discussed the adaptive function of modified projective synchronization of the unknown Chaotic System with unequal dimensions. In a certain degree, this kind of reference put forward some adaptive synchronization methods that center on the Systems with uncertain parameters that have the applicability, but due to the

B.-Y. Cao and Y.-B. Zhong (Eds.): ICFIE 2017, AISC 872, pp. 340–348, 2019.
https://doi.org/10.1007/978-3-030-02777-3_30

influence of the external environment, some parameters of the chaotic system will change causing the inconsistencies between the practical application of the chaotic system and the expected Chaotic system, then failed, to achieve its application value. Aiming at this problem, this paper not only studies the problem of synchronization of fractional hyper chaotic system, but also tries to identify some of the key parameters about the hyper chaotic system, then test and verify whether they are identical between the real value of parameters and the expected value.

In this paper, The author use the square Lyapunov function in the stability analysis of the integer order system, with Lyapunov's second method of fractional order system proposed by Ref. [2], the control and the stability analysis of fractional order nonlinear system have gradually become a spotlight recently. However, exactly as the way pointed by Refs. [13, 14], it is difficult to use the square Lyapunov function in the stability analysis of the fractional order system for the reason of the fractional derivative of square functions has complicated forms so that there is nearly no reference that can realize the control of fractional order chaotic systems or synchronize by using the adaptive control so far. On the basis of the research of the related literature above, this paper mainly studies the problem of synchronization of hyper chaotic system with unknown parameters, and then designs the parameter adaptive rule to the estimated parameter. The main conclusions of this paper is as following: (1) The synchronization of fractional order hyper chaotic Chen system and fractional order hyper chaotic Lorenz system. (2) The successful application of the square Lyapunov function in the stability analysis and strictly proved the stability of the system. (3) Especially aiming at the fractional order nonlinear system with a time-varying coefficient matrix, and then put forward a ratiocination that can judge the stability that will certainly bring the convenience to the stability of this kind of system.

2 The Stability Theory of the Fractional Order System

During the process of research, a variety of frequently used definitions of the fractional calculus have been put forward, Here we use Caputo's definition for the reason of the initial value of this system that is corresponding with the value of the integer order system so that it has good physics significance. The definition of the Caputo fractional differential:

$$ {}_0^C D_t^\alpha f(x) = \frac{1}{\Gamma(n-\alpha)} \int_0^t (t-\tau)^{n-\alpha-1} f(\tau)^{(n)} d\tau. $$

Among them, $n-1<\alpha<n$.

When $0<\alpha<1$, The solution of the Caputo Fractional differential is equal to

$$ f(t) = f(0) + \frac{1}{\Gamma(\alpha)} \int_0^t (t-\tau)^{\alpha-1} {}_0^C D_t^\alpha f(\tau) d\tau. $$

Lemma 1. *We suppose that* $x(t) \in R^n$ *and also has continuous first derivative, then*

$$\frac{1}{2}{}_0^C D_t^\alpha x^T(t)Px(t) \leq x^T(t)P{}_0^C D_t^\alpha x(t).$$

Among them, P is an arbitrary n order positive definite matrix.

Lemma 2. *(Fractional order Lyapunov the second method) we suppose that the base point is the balance point of the fractional order nonlinear system as follows:*

$$ {}_0^C D_t^\alpha x(t) = f(t, x(t)). \tag{1}$$

Among them, $x(t) \in R^n$ *is system variables,* $f(t, x(t))$ *is the nonlinear function that accord with part of the Lipschitz condition. If there is a function* $V(t, x(t))$ *and* κ-*function* $\alpha_i (i = 1, 2, 3)$ *making:*

$$\alpha_1 ||x(t)|| \leq V(t, x(t)) \leq \alpha_2 ||x(t)||,$$

$$ {}_0^C D_t^\alpha V(t, x(t)) \leq -\alpha_3 ||x(t)||.$$

Then it showed that the system (1) is asymptotic stability.

Lemma 3. *Supposed that*

$$V(t) = \frac{1}{2}x^T(t)Px(t) + \frac{1}{2}y^T(t)Qy(t),$$

$x(t)$ *and* $y(t) \in R^n$ *have continuous first derivative,* $P, Q \in R^{n \times n}$ *are two positive definite matrices, if there exists a positive definite matrix M and positive constant h making:*

$$ {}_0^C D_t^\alpha V(t) \leq -hx^T(t)Mx(t),$$

Then $||x(t)||$ *and* $||y(t)||$ *are bounded functions and* $\lim_{t \to \infty} ||x(t)|| = 0$.

Corollary 1. *Consider the fractional order system as follows:*

$$ {}_0^C D_t^\alpha x(t) = Ax(t).$$

Among, $0 < \alpha < 1$, *A is coefficient matrix, if there exists a symmetric positive definite matrix P, making the* $\frac{1}{2}{}_0^C D_t^\alpha x^T(t)Px(t) + x^T(t)P{}_0^C D_t^\alpha x(t) + x^T(t)Px(t) = 0$ *true.*

Then it showed the system asymptotic stability.

Proof: Selecting the Lyapunov function as follows:

$$V = \frac{1}{2}x^T(t)Px(t)$$

By calculating the α order derivative to the two sides of V we can get:

$${}_0^C D_t^\alpha V = \frac{1}{2} {}_0^C D_t^\alpha x^T(t)Px(t) \leq x^T(t)P {}_0^C D_t^\alpha x(t)$$

$$\frac{1}{2} {}_0^C D_t^\alpha x^T(t)Px(t) + \frac{1}{2} {}_0^C D_t^\alpha x^T(t)Px(t) + x^T(t)Px(t) \leq 0$$

$$2 {}_0^C D_t^\alpha V \leq -x^T(t)P {}_0^C x(t)$$

$${}_0^C D_t^\alpha V \leq -\frac{1}{2} x^T(t)Px(t)$$

P is a positive real symmetric matrix, we know from Lemma 3 that $x(t)$ is asymptotically to 0, means $\lim_{t \to \infty} ||x(t)|| = 0$, then it showed this system is asymptotic stability. QED.

Remark 1. Fractional order system equation coefficient matrix A can be a constant matrix and also can contain variables within x, this is also the extension of literature [14].

3 The Design of the Controller and Analysis of the Stability

By considering the fractional order chaotic system, here we suppose the drive system as the

$${}_0^C D_t^\alpha x(t) = f(t, x)$$

Take the response system as:

$${}_0^C D_t^\alpha y(t) = g(t, y) + u(t, x, y)$$

Among them, $x(t), y(t) \in R^n, f, g : R^n \times R \to R^n$ is the nonlinear function which is continuously differentiable, $u(t, x, y)$ is the control input of the system.

The synchronous error is $e(t) = x(t) - y(t)$, the purpose of this paper is to design a suitable controller to make the synchronization error asymptotically to 0.

By considering the fractional order hyper chaotic Chen system (Fig. 1):

$$\begin{aligned}
{}_0^C D_t^\alpha x_1 &= a_1(x_2 - x_1) + x_4, \\
{}_0^C D_t^\alpha x_2 &= d_1 x_1 - x_1 x_3 + c_1 x_2, \\
{}_0^C D_t^\alpha x_3 &= x_1 x_2 - b_1 x_3, \\
{}_0^C D_t^\alpha x_4 &= x_2 x_3 + r_1 x_4.
\end{aligned} \tag{2}$$

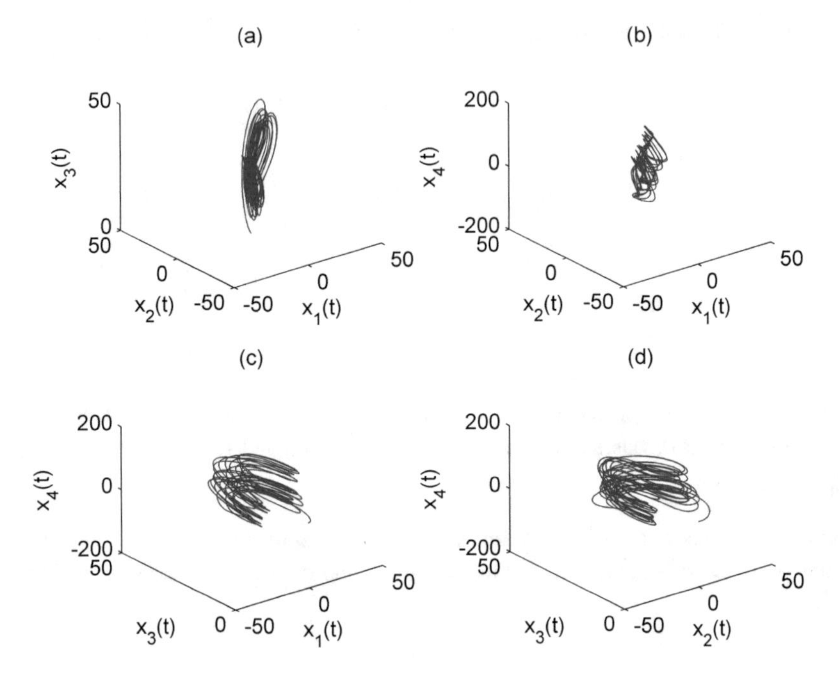

Fig. 1. Fractional order hyper chaotic Chen system.

When $a_1 = 35$, $b_1 = 3$, $c_1 = 12$, $d_1 = 7$, $r_1 = 0.5$, $\alpha = 0.95$, this system is hyper chaos.

The controlled fractional order hyper chaotic Lorenz system

$$
\begin{aligned}
{}_0^C D_t^\alpha y_1 &= a_2(y_2 - y_1) + y_4 + u_1, \\
{}_0^C D_t^\alpha y_2 &= c_2 y_1 - y_1 y_3 - y_2 + u_2, \\
{}_0^C D_t^\alpha y_3 &= y_1 y_2 - b_2 y_3 + u_3, \\
{}_0^C D_t^\alpha y_3 &= -y_2 y_3 + r_2 y_4 + u_4.
\end{aligned} \tag{3}
$$

Without the control variables $u_i (i = 1, 2, 3, 4)$, when $a_2 = 10$, $b_2 = 8/3$, $c_2 = 28$, $r_2 = -1$, $\alpha = 0.95$ then this system is hyper chaos (Fig. 2).

Then use the formula (2) minus formula (3) we get the synchronization error system equation here:

$$
\begin{aligned}
{}_0^C D_t^\alpha e_1 &= a_1(x_2 - x_1) + x_4 - a_2(y_2 - y_1) - y_4 - u_1, \\
{}_0^C D_t^\alpha e_2 &= d_1 x_1 - x_1 x_3 + c_1 x_2 + c_2 y_1 + y_1 y_3 + y_2 - u_2,
\end{aligned}
$$

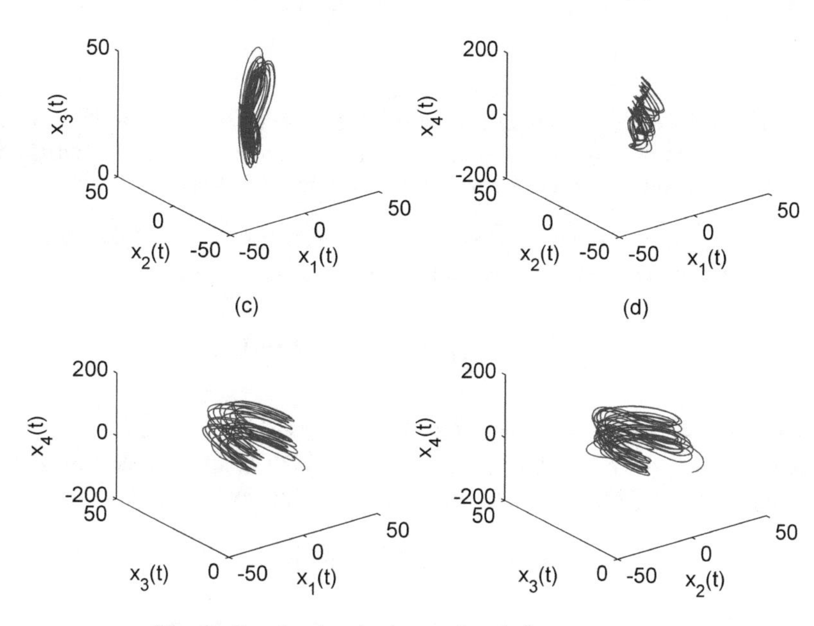

Fig. 2. Fractional order hyper chaotic Lorenz system.

$$
\begin{aligned}
{}_0^C D_t^\alpha e_3 &= x_1x_2 - b_1x_3 - y_1y_2 + b_2y_3 - u_3, \\
{}_0^C D_t^\alpha e_4 &= x_2x_3 + r_1x_4 + y_2y_3 - r_2y_4 - u_4.
\end{aligned} \tag{4}
$$

We suppose that $\Theta = (a_1, b_1, c_1, d_1, r_1, a_2, b_2, c_2, r_2)^T$, its estimated value $\hat{\Theta} = (\hat{a}_1, \hat{b}_1, \hat{c}_1, \hat{d}_1, \hat{r}_1, \hat{a}_2, \hat{b}_2, \hat{c}_2, \hat{r}_2)^T$, Then the system parameter errors

$$\tilde{\Theta} = \Theta - \hat{\Theta} = (\tilde{a}_1, \tilde{b}_1, \tilde{c}_1, \tilde{d}_1, \tilde{r}_1, \tilde{a}_2, \tilde{b}_2, \tilde{c}_2, \tilde{r}_2)^T.$$

According to the above discussion we can design the controller $u(t)$ is:

$$
\begin{aligned}
u_1 &= \hat{a}_1(x_2 - x_1) + \hat{a}_2(y_2 - y_1) - e_4 + k_1e_1, \\
u_2 &= \hat{d}_1x_1 - x_1x_3 + \hat{c}_1x_2 + \hat{c}_2y_1 + y_1y_3 + y_2 + k_2e_2, \\
u_3 &= x_1x_2 - \hat{b}_1x_3 - y_1y_2 + \hat{b}_2y_3 + k_3e_3, \\
u_4 &= x_2x_3 + \hat{r}_1x_4 + y_2y_3 - \hat{r}_2y_4 + k_4e_4.
\end{aligned} \tag{5}
$$

Among them, $G = [k_1, k_2, k_3, k_4]^T$ is control gains, the constant $k_i > 0 (i = 1, 2, 3, 4)$. The fractional parameter adaptive rules designed in this paper are as follows:

$$ {}_0^C D_t^\alpha \hat{\Theta} = (e_1(x_2 - x_1), -x_3e_3, -e_2x_2, e_2x_1, e_4x_4, -e_1(y_2 - y_1), -y_3e_3, -e_1y_1, -e_4y_4)^T \quad (6) $$

Theorem 1. With the given initial conditions and the adaptive controller (4) and also the fractional adaptive rule (6) can realize the hyper chaotic system (2) and (3) synchronization. It showed that error system (4) is asymptotically stable.

Proof: Here we generate the Lyapunov function: $V(t) = \frac{1}{2}e^T e + \frac{1}{2}\tilde{\Theta}^T\tilde{\Theta}$, by calculating the α order derivative to the two sides of V we can get:

$$ {}_0^C D_t^\alpha V(t) = \frac{1}{2} {}_0^C D_t^\alpha e^T e + \frac{1}{2} {}_0^C D_t^\alpha \tilde{\Theta}^T \tilde{\Theta} \quad (7) $$

Because of the constant's α order Caputo derivatives is 0, so that ${}_0^C D_t^\alpha \tilde{\Theta} = -{}_0^C D_t^\alpha \hat{\Theta}$.

According to Lemma 1: ${}_0^C D_t^\alpha V(t) \leq e^T {}_0^C D_t^\alpha e + \tilde{\Theta}^T {}_0^C D_t^\alpha \tilde{\Theta}$, left multiplied by e^T to (3) then bring (4) and (5) into (6), after some simple calculations we get:

$$ {}_0^C D_t^\alpha V(t) \leq -e^T Ge \leq -he^T Me \quad (8) $$

Among $h = \frac{k_{\min}}{\lambda_{\max}}$, $k_{\min} = \min\{k_1, k_2, k_2, k_4\}$, is the largest eigenvalue of positive definite matrix M and from the Lemma 3 we can see that the synchronization error $e(t)$ asymptotically to 0. That is $\lim_{t\to\infty} ||e(t)|| = 0$, QED.

4 Numerical Simulation

In order to verify the validity of this method we proposed in this paper now we do the numerical simulation to the results. Here select the fractional order $\alpha = 0.95$, and the time step $h = 0.01$.

Select the system (1) parameters $a_1 = 35, b_1 = 3,\ c_1 = 12, d_1 = 7, r_1 = 0.5$. Select the system (2) parameters $a_2 = 10, b_2 = 8/3, c_2 = 28, r_2 = -1$.

Select the variable initial value at random $x(0) = [0.21 \quad -0.15 \quad 0.32 \quad 1]^T$, $y(0) = [2.13\, 1.13 \quad 2.34 \quad 1.51]^T$, $\hat{\Theta} = [1 \quad 2 \quad 1 \quad 1 \quad 3 \quad 1 \quad 1 \quad 2 \quad 1]^T$, the control gain $G = [3 \quad 3 \quad 3 \quad 3]^T$.

The simulation results are shown in Figs. 3 and 4, according to the numerical simulation above we can learn that if the implementation of the control is in accordance with the method proposed by this paper, it will make the error of the system quickly reache a steady state, so that the drive system and response system achieve synchronization.

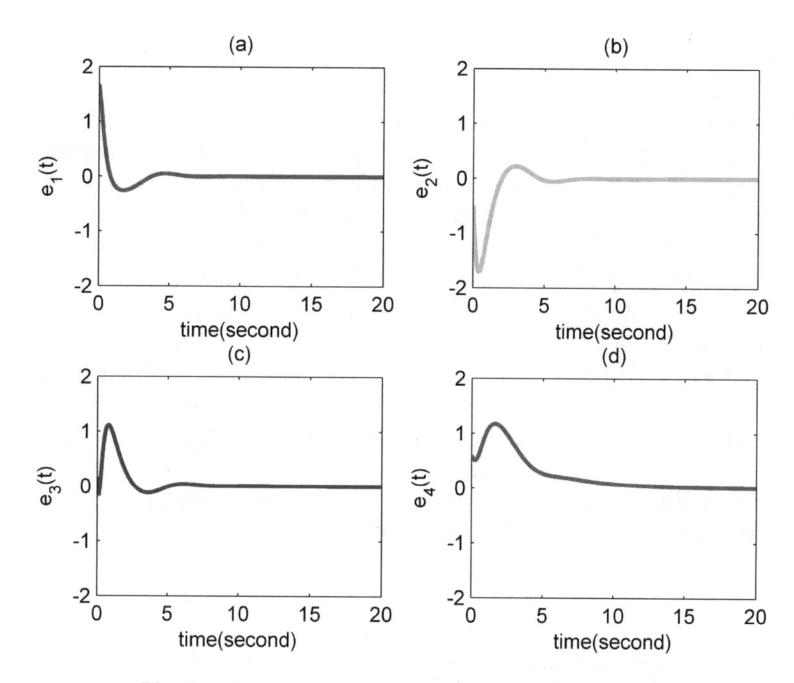

Fig. 3. The error curve under the controlled condition

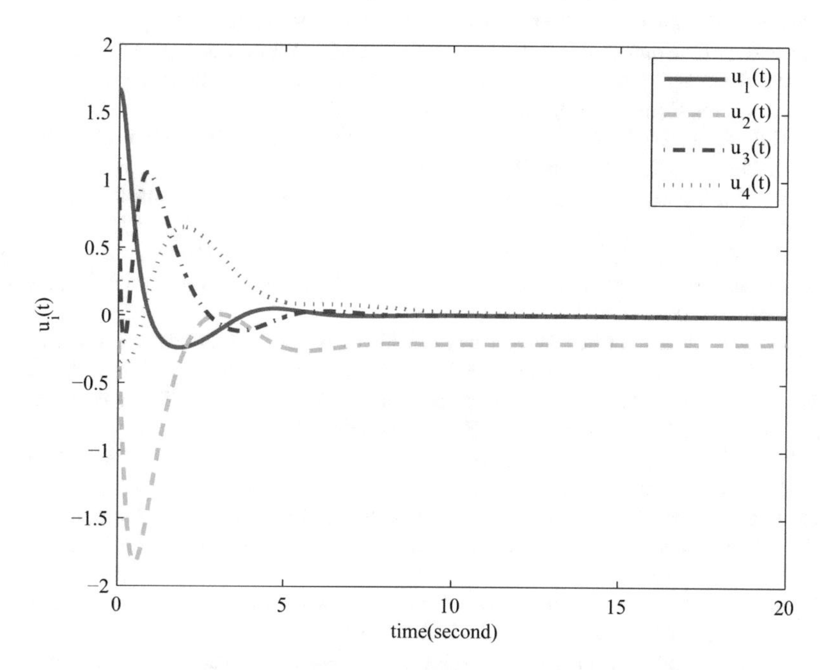

Fig. 4. Time response of the control inputs.

5 Conclusion

The design of the adaptive synchronization controller and the adaptive rule of parameters of fractional order are based on the stability theory of fractional order Lyapunov. It realized the synchronization of the fractional order hyper chaotic systems with unknown parameters by using the fractional derivative of the square. The Lyapunov function discusses the stability of the system, then takes the fractional order hyper chaotic Chen system and fractional order hyper chaotic Lorenz system with different structure with time-varying structure parameters and random initial values as an example to design the active controller and parameter adaptive rule. This new method is simple and effective and also can solve the problem of parameter perturbation at the same time. Additionally, it also has good robustness. The numerical simulation verifies the effectiveness of the method that was proposed in this paper.

Acknowledgements. Thanks to the support by The Ministry of Science and Technology of People's Republic of China programme (no. 2012DFA11270).

References

1. Podlubny, I.: Fractional Differential Equations. Academic Press Publishers, New York (1999)
2. Li, Y., Chen, Y.Q., Podlubny, I.: Mittag-Leffler stability of fractional order nonlinear dynamic systems. Automatica **45**, 1965–1969 (1996)
3. Matignon, D.: Stability results for fractional differential equations with applications to control processing. In: Computational Engineering in Systems Applications, pp. 963–968 (1996)
4. Kuntanapreeda, S.: Robust synchronization of fractional-order unified chaotic systems via linear control. Comput. Math Appl. **63**, 183–190 (2012)
5. Li, C.P., Deng, W.H., Xu, D.: Chaos synchronization of the Chua system with a fractional order. Phys. A **360**, 171–185 (2006)
6. Abd, M.S., Hamri, N., Wang, J.: Chaos control of a fractional-order financial system. Math. Probl. Eng., 270646 (2010)
7. Chen, L.P., Chai, Y., Wu, R.C., Yang, J.: Stability and stabilization of a class of nonlinear fractional-order systems with Caputo derivative. IEEE Trans. Circuits Syst. II Express Briefs **59**, 602–606 (2012)
8. Yin, C., Zhong, S.M., Chen, W.F.: Design of sliding mode controller for a class of fractional-order chaotic systems. Commun. Nonlinear Sci. Numer. Simul. **17**, 356–366 (2012)
9. Wen, X.J., Wu, Z.M., Lu, J.G.: Stability analysis of a class of nonlinear fractional-order systems. IEEE Trans. Circuits Syst. **55**, 1178–1182 (2008)
10. Fallahi, K., Leung, H.: A chaos secure communication scheme based on multiplication modulation. Commun. Nonlinear Sci. Numer. Simul. **15**, 368–383 (2010)
11. Du, H.B., Cheng, Y.Y., He, Y.G., Jia, R.T.: Finite-time output feedback control for a class of second-order nonlinear systems with application to DC-DC buck converters. Nonlinear Dyn. **78**, 2021–2030 (2014)
12. Huang, L.L., Ma, N.: A new method for projective synchronization of different fractional-order chaotic systems. Acta Phys. Sin., 160501 (2012)
13. Rivero, M., Trujillo, J., Vazquez, L., Velasco, M.: Fractional dynamics of populations. Appl. Math. Comput. **218**, 1089–1095 (2011)
14. Li, C., Deng, W.: Remarks on fractional derivatives. Appl. Math. Comput. **187**, 777–784 (2007)

Pricing Problem on the Task of Making Money by Taking Photos

Geng Zhang(✉) and Yuan Duan

Department of Public Foundation, Guangdong University of Science and Technology, Dongguan 523083, China
409251579@qq.com

Abstract. This paper intends to establish a mathematical model to analyze the question B “Pricing problem on the task of making money by taking photos” assigned on the National Mathematical Contest on Modeling in 2017 to find out the pricing rules of the project and also to analyze the reasons responsible for the failure of this task. We have applied the limited multi-arm gambling machine problem model to redesign the pricing scheme and have tested the results. The conclusion of this research has certain reference value to the market.

Keywords: Pricing mechanism · Incentive mechanism · Credit rating Multi-arm gambling machine problem model

1 Background Information of This Question

This article will discuss the pricing problem on the task of making money by taking photos. In nowadays, with the rapid development of information technology, the relationship between information technology, networking and human beings is intimate and inseparable with each other. Mobile Internet has brought great convenience to people’s lives and has made life data possible. Meanwhile, a great number of Internet-based APPs have also developed rapidly. The self-service mode of making money by taking photos is one of the many focuses. The crowdsourcing platform provides enterprises with a large amount of commercial information rapidly, which can save a great amount of costs compared with the traditional methods of investigation and can effectively ensure the authenticity of the survey data. Therefore, it is strongly sought after by many enterprises and APP users. However, there are both advantages and disadvantages of the APP, because the pricing issue is the core element for this APP. Therefore, if it is unreasonable, some tasks will be left neglected, and will lead directly to the failure of commodity inspection. This article will be based on the above-mentioned study to extend to the following issues:

(1) Study the pricing rules of the items in Annex I and analyze the reasons for the unfinished tasks according to the pricing rules obtained from the research.
(2) Re-design the pricing plan for the items in Annex I and compare it with the original plan.
(3) In reality, a number of tasks leading to users’ competition may be attributed to their intensive location, one of the ideas is to unite all of the tasks and publish

B.-Y. Cao and Y.-B. Zhong (Eds.): ICFIE 2017, AISC 872, pp. 349–361, 2019.
https://doi.org/10.1007/978-3-030-02777-3_31

them in a package. With this in mind, we will modify the previous pricing model and answer the question of what influences will the revised model render to the final completion of the task.

(4) According to the new item in Annex III, we will provide our own task pricing plan and evaluate the implementation effect of the plan.

Note: The above attachment can be found in http://special.univs.cn/service/jianmo/sxjmyw/2017/0724/1164059.shtml

2 Model Assumptions

(1) Assuming that cities with a population of over 10 million are densely populated cities.
(2) Assuming that per capita salary above 7,500 is high salary.
(3) Assuming that the radius of city center downtown is 5 km.
(4) Assuming that the target distance within 3 km can be called a short distance.
(5) Assuming that each worker is individual and rational, they will spare no effort to maximize their profits.

Note: The following analysis and discussion are based on the above-mentioned assumptions.

3 Question Analysis

With the rapid development of information technology, the Internet era is constantly changing our lives and the Internet-based self-service crowdsourcing platform rises in response to the proper time and conditions. However, due to the diversity and complexity of the network environment and some other attributes like time-space separation and so on, frequently, there will be a variety of problems awaiting for us to deal with. Therefore, in order to deal with these problems efficiently, we will conduct the following discussion.

3.1 Analysis of the First Question

In order to see more intuitively the location of the task pricing distribution, combined with task gps latitude and task gps longitude provided in Annex I, with the help of the BIGEMAP software, we show the task positions in the form of points on the map.

As shown in the following figure:

It can be seen that the tasks are mainly distributed in four cities: Foshan, Guangzhou, Shenzhen and Dongguan. Therefore, the four cities are taken as examples to study the task pricing rules of the research projects. Due to the huge amount of data and the big difference in prices, the task is divided into three parts according to the difference in task pricing: the low price part, the regular price part and the high price part (Fig. 1).

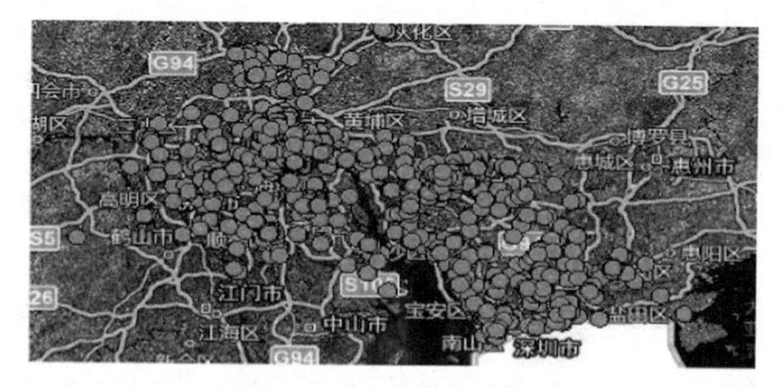

Fig. 1. All tasks location distribution.

We use BIGEMAP to display the location points on the map from low to high according to the price of the task, and analyze the influence of task pricing with the distance from the task center to the nearest center; the living standard in different cities; the resident population and the distribution of the high credit value. The position of the unfinished task is displayed on the map, the unfinished task points and the distribution of low price in each city, the distribution of members and the relationship between the task price are analyzed, and the reasons for the unfinished task are as follows: the price of the task is too low, this location's transportation is not convenient; the travel costs are high; the marked price can not meet the expected income of locals; the distribution is too concentrated; the cheap labor around the location is lacking; the user credibility value around the place is low and so on.

3.2 Analysis of the Second Question

3.2.1 Crowdsourcing, Task Pricing Description

Since the concept of crowdsourcing was proposed, crowdsourcing has become a new business model, the arise of crowdsourcing has caused great changes in production organization. Internet is the key to crowdsourcing in that it can quickly and efficiently access the resourcefulness of crowdsourced participants and break the traditional enterprise's inability to effectively access resource limitations due to geographical separation. However, there are some drawbacks. If the task's pricing is unreasonable, some tasks will be neglected, which will directly lead to the failure of the task. So the pricing for crowdsourcing is very significant. The ideal pricing point for any crowdsourcing APP is in the range of that "the business can make the highest profit" and "the price that crowdsourcing participants can generally afford." It is based on this perspective that the author of this paper decide to place pricing behavior of the crowdsourcing platform publishers as the object of study and design a new task pricing program to conduct a comprehensive comparison between these two ways.

3.2.2 The Basic Process for Crowdsourcing Task Pricing Program

The main body of the value realization of crowdsourcing business model include the employer, the outsourcing, the platform and other stakeholders, collecting intellectual resources around the world, they manage to achieve effective circulation of resources.

In the value network structure model of the crowdsourcing business model, the basic operation process of the value of the realization of crowdsourcing business model is (Fig. 2):

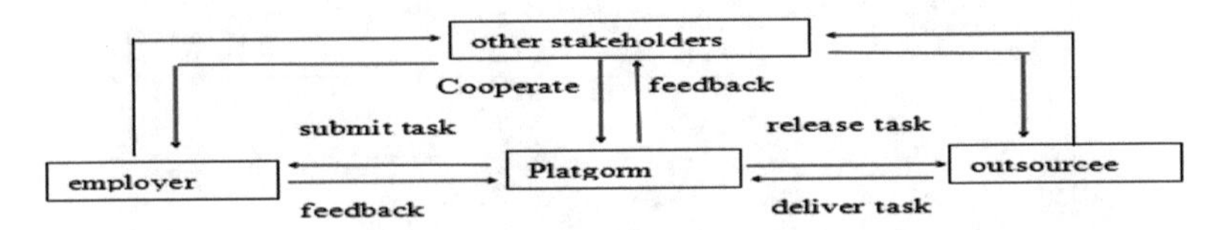

Fig. 2. The operation flow chart of the crowdsourcing business model.

The following questions need constant attention during operation:

1. The task submitted by the employer should be solvable by the general public. Therefore, the enterprises are required to make their tasks more precise and standardized to make the outsourcing parties have their own extra time and ability to match it. But some tasks are too large and complex to be accepted by the general population.
2. The third party can send the information to a specific crowd when attracting potential recipients, and correspondingly raise the remuneration of the task.
3. The crowdsourcing also involves the interests of others. They cooperate with each other under the crowdsourcing system to provide a powerful guarantee for the smooth operation of the crowdsourcing business model. Even though there are four different types of crowdsourcing in theory, depending on whether the target task can be integrated into the operational model, there are broadly two types, they are respectively the open source code tasks and non-open source code tasks [2].

3.2.3 Factors that Need to Be Taken into Account in Setting the Task Pricing

1. The cost price for participants to complete a single task.
2. The maximum number of tasks a participant expects to complete.
3. The total number M of workers participating in the task.
4. Budget L.

3.2.4 Determine the New Task Pricing Program

In reality, the same kind of goods will have a relatively uniform pricing, because it is easier for management. However, for the crowdsourcing platform, there is no uniform price for the similar tasks, this kind of situation is not conducive to unified management. The main content involved in this article is to measure the participants' completion quality of the tasks at the crowdsourcing platform through the gold standard method. Secondly, to give an initial pricing on the tasks according to the common pricing method. Finally, the greedy algorithm is used to maximize the interests of the task publishers and at the same time to mobilize the enthusiasm of the participants from the crowdsourcing platform.

The problem model set up below considers only one task publisher and M workers. Workers are represented respectively by 1, 2, …… M, each worker z_i has two attributes, one is the cost a_i for completing the published single task and the other is the number of tasks b_i the workers are capable of accomplishing. The task publisher does not know the worker's two attributes, but they know the total number of workers involved in the task is M. The purpose of the task publisher is to maximize their own earnings with budget L.

For the worker z_i, suppose he/she is assigned to $\overline{v_i}$ tasks, and the price of each task is j_i. Then, his/her return can be expressed as: $\overline{v_i}(j_i - a_i)$.

Every worker involved in the task must submit the cost of completing a single task: a_i.

The maximum number of tasks expected to be completed: b_i.

Because each worker is individual and rational, therefore, in order to maximize his or her own income, workers may misrepresent their costs and the maximum number of tasks they expect to accomplish. Workers will not be rewarded if they fail to do their job, so they generally will only lie about their costs. This paper seeks to find an incentive-compatible mechanism to ensure that each worker will not lie about his/her own costs. Formally speaking,

For each worker: z_i.

When he submits the real cost: a_i and quantity: b_i, we use f_i to represent the task's unit price given by the publisher d in the absence of a false report, use g_i to represent the number of tasks submitted to the workers in the absence of a false report. Conversely, if he submits a false report on cost of a_i' and number of b_i', we use f_i' to represent the false unit price and g_i' the false assigned number of tasks. Then if you can meet the conditions

$$g_i(j_i - a_i) \geq g_i'(j_i' - a_i').$$

Description is incentive compatible, that is, the dominant strategy here is that workers will not lie about the cost and quantity.

Before setting the unit price, applying the gold standard test method to assign the worker a similar task and explore the quality of the work done. Test results are expressed as:

$$h_1, h_2, h_3 \ldots h_n.$$

Due to the unity of unit price during the pricing stage, therefore, as long as we order it according to the quality of work done, under the restrictions of budgeting and completed tasks, trying as much as possible to choose the workers with the highest quality of tasks, then choose the second-highest quality worker. In this case, the revenue of the task publisher can be maximized.

3.2.5 Pricing Mechanism Design

Intuitively, the threshold price is the price issued by the project publisher and accepted by most workers. The method of calculating the threshold price given below is a variation on the mechanism of proportional sharing. Proportionality sharing mechanism [3] is the basis for designing purchasing mechanisms under budgetary conditions.

1. Input the data $\{(a_1, v_1), (a_2, v_2), \ldots (a_n, v_n)\}$ and budget L submitted by the workers.
2. Initialization: sequence the data pairs by $a_1 \leq a_2 \leq a_3 \leq \ldots\ldots \leq a_n$, set $i = 1$.
3. Repeat the following steps.
4. If $a_i \leq \frac{L}{\sum_{j<i} \bar{v}_j + 1}$ is established.
5. Set $p = a_i$.
6. $\bar{v}_i = \min\left\{v_i, \left|\frac{L}{p}\right| - \sum_{j<i} \bar{v}_j\right\}$.
7. $i = i + 1$.
8. End the cycle.
9. End of the algorithm.

When the unit of the task is $p = a_i$, the remaining budget is

$$L - p\sum_{j-1}^{i-1} \bar{v}_j.$$

Therefore, when it is under the budget and the unit price set is $p = a_i$, the number of tasks that the workers z_i are allotted are shown

$$\bar{v}_i = \min\left\{v_i, \left|\frac{L}{p}\right| - \sum_{j<i} \bar{v}_j\right\}.$$

That is, when the number and pricing needing to be completed submitted by the participants z_i is p, the minimum value among the remaining number of tasks is selected as the number of tasks scheduled by the participants z_i. The reasons for choosing the threshold price are as follows.

The first reason is that we should sort them according to the costs submitted by the workers to reduce the possibility of workers to lie about higher costs.

The second reason is that when the selected unit price is low enough, they can complete more tasks, of course, it should be controlled within the total budget.

The third reason is that for the business, if the task's unit price can be accepted by the majority of participants, it will not result in failure of completing the task.

3.2.6 Algorithm Design

There algorithm presented in this paper, which combines the gold standard method, the pricing mechanism design and the greedy algorithm [4], plays an important role in maximizing the revenue of the task publisher. First of all, according to the gold standard test method, before the participants formally answer a task, select some tasks that the answers are already clear to each participant, and judge the quality of the tasks performed by them according to the results: they can be represented like

$$h_1, h_2, h_3 \ldots h_n.$$

Then, according to the spent costs submitted by the participants with the completed tasks and the number of tasks expected to complete, in accordance with the above-mentioned pricing mechanism to execute operation. You will get the unit price of the task and the number of tasks that the worker are actually assigned. Finally, arrange the quality of the completed tasks in a low to high ranking, choose the person who complete the task with the best quality and assign more tasks to him/her (as long as it does not exceed the budget and the number of tasks can be completed); if it exceeds the number of tasks can be completed, then choose the next-best participants to assign the tasks; and the rest can be done in the same manner until it exceeds the budget. The design of incentive mechanism can not only enable participants to perform their tasks better, but also can produce better works, what's more, it can affect the efficiency of crowdsourcing innovation. Crowdsourcing users can choose what they are interested in and good at through crowdsourcing site. The public from all over the world can undertake the task. The only criterion for judging whether a project is good or bad is to collect and collate the design plan submitted by the public and finally announce the best selected plan and designer to reward them.

The current crowdsourcing innovation incentives are all fixed bonuses. We can start with a simpler models that are used currently (Fig. 3).

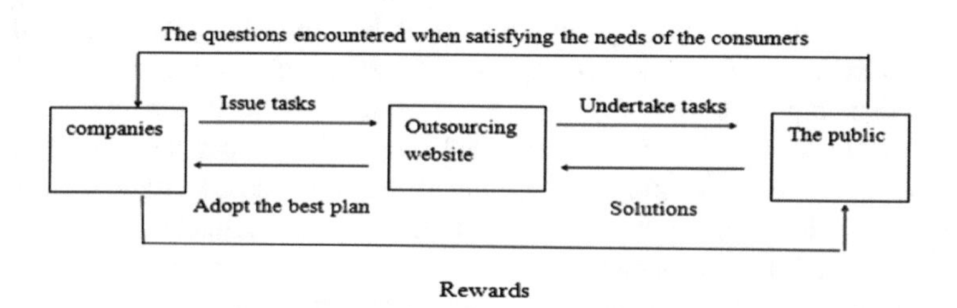

Fig. 3. The basic model for crowdsourcing innovation.

3.2.7 Linear Award Mechanism Can Stimulate the Enthusiasm of Participants

It goes without saying that the linear reward mechanism can stimulate the enthusiasm of participants. Participants can be divided into two types: the hobby-based and the award-winning based.

For the second type of participants, you can set the individual's higher skill and hard working level proportional to the incentive coefficient. There are also some people who participate in purely out of their personal hobbies or expectation to win a prize, which is somewhat similar to Hertzberg's two-factor incentive, which means bonus motivation is still widespread.

With further investigation, we also find that most of the participants who want to win the prize have the highest level of skill in this field or industry. This may be due to that such people believe they have the ability to earn rewards through hard work, so they will be more serious and industrious when completing the tasks, they try to finish their works as perfect as possible. The interpretation in the model is same to this, because:

$$\varphi * (\delta, \gamma) = \frac{\delta\gamma}{2M}.$$

we have:

$$\frac{\partial\varphi(\delta, \gamma)}{\partial\gamma} = \frac{\delta}{2M} > 0.$$

a is k's increasing function, a higher skill level can produce a higher level of effort. In addition, we can also see

$$\frac{\partial\varphi(\delta, \gamma)}{\partial\delta} = \frac{\gamma}{2M} > 0.$$

The level of effort at this time is an increasing function of the incentive coefficient. A higher incentive coefficient motivates participants to work harder to accomplish their task.

To sum up, because of the high-speed development and popularization of the Internet, the crowdsourcing model innovation still shows a much promising future compared with the traditional innovation models. However, it's still well worth exploring whether the efficiency and quality of the crowdsourcing model innovation can be further improved with our joint effort. What is certain, however, is that the incentive mechanism still plays an important role in improving the efficiency of the system. In real life, the employer can improve the operational efficiency of crowdsourcing by carefully examining the innovations of the participants and setting a reasonable linear incentive coefficient.

So how to design a linear incentive coefficient? The contractor should follow the relationship between the incentive coefficient and the participants skill coefficient in the system. Therefore, the outsourcing enterprises should do a good survey job before the analysis of the crowdsourcing platform to have an overall understanding of the level of capacity of the participants and the cost of crowdsourcing about the business operations. Therefore, in order to improve the operating efficiency of crowdsourcing, enterprises should set reasonable incentive mechanism and determine the incentive coefficient through reasonable analysis. Because crowdsourcing innovation is a new and innovative mode and all works of life are constantly exploring it. So the model we proposed also needs to be tested in time, because only time will tell whether it's good or not.

3.2.8 A Comprehensive Comparison Between the New Program and the Original One

The original plan has the following five points as the basis for the task pricing:

1. Population density of the task distribution area.
2. The average salary of the task distribution area.

3. The bustling radius of the city of the task distribution area.
4. The distance from the city center to the task distribution area.
5. The shortest distance between the task distribution area and the members with higher credibility at the surrounding area.

The formula is:

$$T = T_0 + T_{s_1} + T_{s_2} + T_{s_3} + T_{s_4} + T_{s_5}.$$

Obviously, there are too many variables in this formula and their influences to each other is complicated, what's more, the influencing factors of geographical location are too large. Therefore, the task pricing completed based on the same scheme differs greatly in different cities: According to data, The completion of the task is close to 100% in Dongguan, while the completion of the tasks in the three cities Foshan, Guangzhou and Shenzhen is not optimistic and the overall task completion rate is only 62.5%. Obviously, the implementation of the original program is not effective and can not be applied to different cities. And the original program only have the price factor to attract users to complete the task.

In the new task pricing program, we simplify it to minimize the factors influencing the pricing, the calculation formula of completion of a single task is: $j_i - a_i$.

Among them, a_i is the cost of a single task, j_i is the price for each task.

The cost data of a single task of the new program is collected from the user and the relevant scheme is used to eliminate the false data reported by the user. Therefore, the data source is targeted and has high credibility. Considering only one influencing factor can greatly reduce the difficulty of task pricing and can make the price more reasonable.

The new program also introduces a linear reward mechanism, and combined with the task pricing, the two can complement with each other to stimulate the enthusiasm of users to complete the task.

Therefore, by comparing the original program and the new program, we can see that the new program makes pricing much easier and also have wide applicability to different urban areas. The linear reward mechanism of the new program can guide users better than the original program with higher quality and probability in the completion of the tasks.

3.3 Analysis of the Third Question

In view of the actual situation and in order to accomplish the task better, we use a limited model of multi-arm gambling machine to conduct a more comprehensive analysis of the pricing model of the second question. On this basis, we also apply the expert crowdsourcing problem to reflect the current model. Then, referring to the credit rating, the credit ratings of crowdsourcing members are evaluated and we create a rule that applies to credit rating of crowdsourcing members. Finally, we come to the conclusion that the contract-issuing party should follow the positive correlation relationship between the incentive coefficient and the skill coefficient of the participants, to have a full understanding of the overall level of the participants' ability in the crowdsourcing platform, the average value of their reward expectations and the general cost of crowdsourcing. Then apply the limited multi-arm gambling machine problem model as a reference to improve it to come to a more satisfactory results, which has a very positive influence on improving the accomplishment rate of crowdsourcing tasks.

With each successful completion of the task for the members in the crowdsourcing platform, the public, contractors and professional evaluation agencies are entitled to make a quality assessment of the completion situation, the evaluation is also known as membership credit rating. But the random or malicious evaluation should be eliminated, which requires learning from Jingdong's practice on how to deal with complex big data and accurately determine the needs of each potential consumer, of course, the following scoring rules also have a simple elimination.

Evaluation of integral: the evaluation is divided into three categories: "poor", "comment" and "praise", each rating corresponds to an integral evaluation The specific calculation of the evaluation is: "poor" deduct 1 points, "comment" plus 2 points, "praise" plus 3 points.

Credibility: the evaluation of integral is accumulated, the initial credit value of each member is 60, the total point is not capped, but when the point is negative, the member will not be allowed to take crowdsourcing tasks within 30 days. The evaluation points are shown in the member website.

Evaluation period: It is within 15 days after the task is submitted successfully.

3.3.1 Scoring Rules (Including Anonymity)

In each month, the same members or the same public can not evaluate the same member more than 5 times. Evaluation beyond the scope of the scoring rules will not be scored, that is, multiple praise only have 3 points, multiple negative comments only deduct one points. And if there are more than three violations, both of them will be punished by the way of reducing the degree of credibility or prohibiting participation in the evaluation between the two parties, we will not elaborate on this any more (Table 1).

3.3.2 Credit Rating Table

See Table 1.

Table 1. Credit registration form (members with higher credit can be given priority in choosing the task with a higher price)

60t–80 point	1 star	1501–2000 point	1 yellow diamond
81–100 point	2 star	2001–3000 point	2 yellow diamond
101–150 point	3 star	3001–4000 point	3 yellow diamond
91–150 point	4 star	4001–5000 point	4 yellow diamond
151–250 point	5 star	5001–10000 point	5 yellow diamond
251t–300 point	1 white diamond	10001–15000 point	6 yellow diamond
301t–400 point	2 white diamond	15001–20000 point	7 yellow diamond
401–500 point	3 white diamond	20001–25000 point	8 yellow diamond
501–1000 point	4 white diamond	25001–30000 point	9 yellow diamond
1001–1500 point	5 white diamond	Above 30000 point is crown	1 crown

3.3.3 An Overview of the Impact on the Final Task

Based on the new task pricing program we have designed at the second question, it is mainly related to the pricing mechanism of threshold price and innovation incentive mechanism, and applying the gold standard test method for testing, but there are still some shortcomings. In the third question, by taking into account the practical situations, these tasks are packaged together, so we model the pricing model in the second question using a limited multi-arm gambling machine problem model and map the existing model with an expert crowdsourcing problem. In an expert crowdsourcing environment, task publishers can assign tasks to workers and form a collection. This collection is generally determined publicly by the task publisher; finally, the ultimate goal is to maximize the benefits of the budget under the constraints of the budget, it will have a significant impact on the completion of the final task The design of incentive mechanism can not only enable participants to perform their tasks better but also can influence the efficiency of crowdsourcing innovation, and crowdsourcing users can choose what they are interested in and good at through crowdsourcing platforms. The whole public from all over the world can undertake the task. The public submits its own design proposal, and the company collects and organizes them, finally, they will announce the best plan and its designer and reward it.

The impact and enlightenment on our final completion of the task: The employer should follow the relationship between the incentive coefficient and the skill coefficient of participants in the system. Therefore, the outsourcing enterprises should do a good survey job before outsourcing it, including a general understanding of the capacity of participants and the costs of business operations at the crowdsourcing platform. Therefore, in order to improve crowdsourcing completion rate, enterprises should set reasonable incentive mechanism [6], through reasonable analysis to determine the incentive coefficient, meanwhile using the limited multi-arm gambling machine problem model as a certain reference to constantly improve it, to get a model that is the closet to reality, to improve the completion rate of crowdsourcing tasks.

3.4 Analysis of the Fourth Question

The method of subregional research was used to establish the task pricing scheme. The task distribution location was divided into six areas. The average cost price and the average acceptable minimum income of six areas were obtained in the form of questionnaires to get an initial pricing. What's more, according to the scheduled quotas, the task-intensive the member can accomplish and the number of members, we can amend the initial pricing to get a final pricing. Analysis shows that the program's task pricing can meet the user's expectations and the implementation of it produces desirable results.

3.5 An Evaluation of the Implementation Effect

The influence of population density, average salary, the distance from the city center to the task point, the travel cost and the user's expected income on the price formulation are all reflected in the variable of the cost for the user to complete a single task.

The cost data of a single task of the task pricing scheme are collected from users in six regions respectively. The relevant schemes are used to eliminate the user-misrepresented and invalid data, so the data sources are targeted and credible.

The cost of completing a single task by the user can result in a preliminary task price. Then considering the distribution of members, the task center distance from low-cost center, the final price of the task can be drawn.

Considering the subjective and objective factors, the final task price is expected to be accepted by most users, and the task completion rate will be guaranteed.

Therefore, the scheme can not only meet the users' requirements for income but also reduce the enterprise cost to the maximum while improving the task completion rate, making the pricing much easier, and also having good applicability to different urban areas.

4 The Advantages and Disadvantages of the Model

4.1 An Evaluation of the First Question

For the first question. the advantage of it is the using of BIGEMAP to accurately display the distribution of task locations in Annex I and provide an effective reference for finding out the task pricing rules and the reasons for the unfinished tasks. However, the influence of user's different subjective psychology, gender, occupations, ages are neglected when analyzing the reasons for the unfinished tasks.

4.2 An Evaluation of the Second Question

For the second question, the advantage is that the pricing of new tasks through the pricing mechanism and innovation incentive mechanism can not only achieve the desired requirements, but also in a more scientific principle. It can increase the enthusiasm of users to participate in and reduce the task unfinished rate. However, the disadvantages are also obvious. For example, the calculation process is complicated and the data accuracy is highly demanding. Once the data is wrong, the task pricing will be biased.

4.3 An Evaluation of the Third Question

For the third question, first: we can modify it according to the pricing model established in the second question to solve the practical problem (multiple tasks are bundled together). Second, the multi-arm gambling machine problem model can be used to conduct a more comprehensive analysis when modify it. Moreover, the newly built credit rating rules of crowdsourcing members can be better adapted to the future development of crowdsourcing and improve the completion quality and completion rate. However, there are still some drawbacks, for example: the calculation process is too complicated, and more factors need to be considered in the calculation and the quality of the early research is poor.

4.4 An Evaluation of the Fourth Question

To the fourth question, based on the quantitative and qualitative results of our previous three questions, it can scientifically reflect the pricing situation of the areas shown in Annex III, and in combination with the results of the field survey, we can give a more reasonable pricing. However, there are still some shortcomings. The "making money by taking photos" task pricing survey has great limitations. Due to the limited time and school-based personnel constraints, the results of the survey do not have extensive representation.

5 Conclusion

Our scheme mainly adopts pricing mechanism of threshold price and innovation incentive mechanism as the basis, and flexibly determines the task pricing based on factors such as the credit rating of members, the intensity of tasks and living standards in different regions, and can better guide the members to complete the task with high quality and great chance. In this era of rapid Internet development, the conclusions we have come to have some market reference value.

Acknowledgements. Thanks to the support rendered by Specialty Innovation Program (education scientific research) of Guangdong (No. 2016GXJK177) and Quality Engineering Project of Guangdong University of Science and Technology (No. CQ2017020).

References

1. Li, C.N.: Crowdsourcing system multi-dimensional credit rating model design and implementation. Doctoral dissertation, Guangxi University, pp. 8–9 (2016)
2. Xie, Q.N.: Based on the value of the crowdsourcing business model evaluation system. Doctoral dissertation, Xidian University, pp. 7–8 (2015)
3. Liu, Q.: Incentive effect and reasonable range of labor cost in China's producer services. Doctoral dissertation, Beijing Jiaotong University, pp. 22–26 (2014)
4. Hao, L.N., Hou, W.H., Liu, M.: Analysis and prospect of crowdsourcing innovation mode. Nankai University, pp. 156–160 (2014)
5. Shen, H.: Crowdsourcing participatory behavior influencing factors. Doctoral dissertation, Zhejiang University of Finance and Economics, pp. 9–14 (2014)
6. Sun, X.X.: Task distribution technology in crowdsourcing environment. Doctoral dissertation, Yangzhou University, pp. 34–38 (2016)

Others

Application of Fuzzy System Method in Economic Data Analysis

Cheng Zhang(✉), Shuai Wang, and Fu-cong Teng

College of Information Engineering, Dalian University, Dalian 116622, China
zhangcheng426@sina.com

Abstract. In this paper, the fuzzy system modeling method and the probability representation of fuzzy system method are used to establish model based on economic data. Compared their sum of residual squares with that of classical linear regression analysis model, which indicates that the results represent by the fuzzy system model method and the probability representation of fuzzy system model method are better than the linear regression analysis model method. Analogy to the linear regression analysis theory, the deterministic coefficient is introduced for the fuzzy system model and the probability representation of fuzzy system model, and given a new index $\tilde{R}^2$ to measure the effect of fit, then we use this index to compare the degree of fit of each models. The results show that the linear regression analysis model is the best fit, but the probability representation of fuzzy system model fitting the worst. The research of this paper has theoretical significance value to the fuzzy system theory in the analysis of economic data.

Keywords: Fuzzy system · Linear regression
Probability representation of fuzzy system · Residual sum of squares
Goodness of fit

1 Introduction

In econometrics, we usually use linear regression analysis method to deal with economic data. For the collected economic data, determining the type of model according to the economic theory, then to determine the regression coefficients by using the method of least squares, the regression model is also tested by relevant statistical tests and economic significance tests. Finally, the model can be used to predict and make decisions [1]. But when the model is tested by hypothesis, it is usually necessary to assume that the random disturbance is subject to normal distribution, of course, this assumption is reasonable in theory. Economic activity is all about human participation, the regression analysis model pays attention to the data itself, but ignores the factors of human participation. The fuzzy system modeling not only considers the data factors, but also embodies the human factors in the model [2]. Therefore, in theory, it is more reasonable to use the fuzzy system method to model the economic data of the factors that human

B.-Y. Cao and Y.-B. Zhong (Eds.): ICFIE 2017, AISC 872, pp. 365–374, 2019.
https://doi.org/10.1007/978-3-030-02777-3_32

are involved. But human thinking is complicated and how to describe economic phenomena with fuzzy system method is also a problem worthy of further study. For the fuzzy system theory, Professor Li Hongxing first proposed a control system modeling method based on fuzzy reasoning, called fuzzy reasoning modeling method [3]. Then, the probability meaning of fuzzy system is revealed, and the common method of clearness in fuzzy system is pointed out, that is, the center of gravity method is reasonable and the method is optimal in the mean square sense [4]. According to some properties of fuzzy system probability distribution, demonstrates the CRI algorithm of fuzzy system proposed by Zadeh is reasonable and effective [5]. In addition also depicts the special function of uniform probability distribution in fuzzy system, reveals the fuzzy inference significance of random system. This establishes a connection between fuzzy systems and stochastic systems. After this, Professor Yuan Xuehai using fuzzy reasoning to construct the joint probability density function of the input - output data [6], derived the regression function based on the joint probability density function, it is actually the fuzzy system of the center of gravity method based on given fuzzy implication operator, and proved that the fuzzy system with the first-order approximation accuracy, and further enrich the probability representation theory of fuzzy system [7]; Professor Yuan Xuehai also presented a method to construct fuzzy system by fuzzy transformation, and the constructed fuzzy system has very good universal approximation property [8], then the sufficient condition of the center of gravity method fuzzy system based on the input - output data with universal approximation property are given [9]. At the same time, Professor Guo Sicong proposed a fuzzy inference model with self - learning and rule self- generating when the rule is unknown in advance, and put forward a new way for the study of fuzzy rules [10]; Based on Zadeh's fuzzy reasoning rules, he presented a mathematical model of generalized fuzzy feature synthesis reasoning, which gets the information rate of factors on the system by meaning of the method of information theory, and given the characteristic coefficients weighting deterministic method based on statistical information using the information rate of factors on the system [11]. The method laid a theoretical foundation for the application of fuzzy system method in practice, Professor Chen Gang obtained satisfactory results using multiple multidimensional fuzzy reasoning method for student achievements [12]. On the basis of the above work, the application example of comparison between the fuzzy system method and the linear regression method is given in this paper. That is, according to the economic data provided in the paper [1], using the fuzzy system modeling method and the probability representation of fuzzy system modeling method to deal with it, and compared the results with the result obtained by linear regression modeling method. The sum of residual squares and the goodness of fit of the three methods are analyzed, and the obtained results are explained. Finally, the summary and prospect of this work are given.

2 Preliminaries

Linear regression modeling method: The paper [1] have given the weekly income and weekly consumption data of 60 families (see Appendix). In order to analyze the relationship between weekly income and weekly consumption, the paper [1] has used the regression analysis method to get the linear regression equation $\hat{Y}_R = 0.591x + 22.345$ (The estimate obtained by the regression method is expressed in $\hat{Y}_R$). In the regression analysis, in order to verify the prediction fitting effect of model, clear the effect of explanation variables and random errors, and reflect the uncertainty of dependent variable with the total sum of squares (TSS). It can be shown in the least squares sense, TSS is made up of the residuals sum of squares (RSS) which represents for random errors and the explain sum of squares (ESS) caused by the fluctuation of explanatory variables, the proportion of ESS in TSS is also known as the coefficient of determination (written as R^2), it can be used as the index of comprehensive measures the goodness of fit of linear regression model for the sample observation, its value closes to 1, showing that the part explained by model have greater proportion in the total variation, the model fitting degree is better.

Fuzzy system modeling method: To construct a fuzzy system generally requires the following four steps [2]: First, fuzzy the data. It divided into single point fuzzy and non single point fuzzy; The second is to establish the fuzzy inference rules; Then the fuzzy reasoning by choosing appropriate fuzzy implication operators, according to the situation can be the first inference after polymerization, can also be the first polymerization after inference; The last is defuzzification, including the center of gravity method, the maximum defuzzification method, etc. The estimate by the fuzzy system modeling method denoted as $\hat{Y}_S$ in this paper.

Probability representation of fuzzy system modeling method: Based on the fuzzy system modeling method, professor Li Hongxing reveals the probability meaning of fuzzy system [4]. The general approach is using fuzzy inference rules to get a joint probability density, and regards it as a approximate of the distribution of data. The density function of the probability distribution is:

$$f(x,y) = \frac{[\bigvee_{i=1}^{n} \theta(A_i(x), B_i(y))] I_{X\times Y}}{\int_X \int_Y [\bigvee_{i=1}^{n} \theta(A_i(x), B_i(y))] dxdy} \tag{1}$$

where $I_{X\times Y}(X,Y)$ is the indicator function of the set $X \times Y$, $\theta : [0,1]^2 \to [0,1]$ is Fuzzy implication operator.

Calculate the conditional mathematical expectation of the approximate probability distribution, which is denoted as $\hat{Y}_P$:

$$\hat{Y}_P(x) = E(\eta|\xi = x) = \frac{\int_{-\infty}^{+\infty} y f(x,y)dy}{\int_{-\infty}^{+\infty} f(x,y)dy} \tag{2}$$

3 The Application of Fuzzy System Modeling Method

The case presented data can be regarded as a set-valued mapping $f : X \to P(Y)$, i.e.

$$80 \mapsto [60, 68, \cdots 77], 100 \mapsto [68, 78, \cdots 88], \cdots\cdots, 260 \mapsto [150, 152, \cdots 191],$$

then use the fuzzy system method modeling:

Step 1 Fuzzy

We use single point fuzzification in this paper,

$$A(x) = \begin{cases} 1 & x = x^* \\ 0 & x \neq x^* \end{cases},$$

where x^* is input value.

Step 2 Establishment the fuzzy reasoning rules

Firstly, we give the antecedent and the posterior of the fuzzy rules according to the data:

Using the equidistant triangular fuzzy (the spacing is 20) to construct the membership function $A_i (i = 1, 2, \cdots, 10)$:

$$A_i(x) = \begin{cases} \dfrac{x - (40 + 20i)}{20} & x \in [40 + 20i, 60 + 20i] \\ \dfrac{(80 + 20i) - x}{20} & x \in (60 + 20i, 80 + 20i] \end{cases},$$

that is, the antecedent of the fuzzy rules.

To fuzzy the element of $P(Y)$, the specific method is: calculate the mean of each set firstly, i.e.

$$70, 81, 93, 103, 117, 130, 140, 156, 167, 171$$

Written the 10 mean values respectively as $\bar{Y}_i (i = 1, 2, \cdots, 10)$. Then let the mean value as the peak point of each group to triangular fuzzy, the obtained membership functions are:

$$B_1(y) = \begin{cases} \dfrac{y - 60}{10} & y \in [60, 70] \\ \dfrac{77 - y}{7} & y \in (70, 77] \end{cases}, \quad B_2(y) = \begin{cases} \dfrac{y - 68}{13} & y \in [68, 81] \\ \dfrac{88 - y}{7} & y \in (81, 88] \end{cases},$$

$$B_3(y) = \begin{cases} \dfrac{y - 82}{11} & y \in [82, 93] \\ \dfrac{101 - y}{8} & y \in (93, 101] \end{cases}, \quad B_4(y) = \begin{cases} \dfrac{y - 86}{17} & y \in [86, 103] \\ \dfrac{120 - y}{17} & y \in (103, 120] \end{cases},$$

$$B_5(y) = \begin{cases} \dfrac{y - 105}{12} & y \in [105, 117] \\ \dfrac{125 - y}{8} & y \in (117, 125] \end{cases}, \quad B_6(y) = \begin{cases} \dfrac{y - 112}{18} & y \in [112, 130] \\ \dfrac{148 - y}{18} & y \in (130, 148] \end{cases},$$

$$B_7(y) = \begin{cases} \dfrac{y-125}{15} & y \in [125,140] \\ \dfrac{155-y}{15} & y \in (140,155] \end{cases}, \qquad B_8(y) = \begin{cases} \dfrac{y-142}{14} & y \in [142,156] \\ \dfrac{170-y}{14} & y \in (156,170] \end{cases},$$

$$B_9(y) = \begin{cases} \dfrac{y-137}{30} & y \in [137,167] \\ \dfrac{189-y}{22} & y \in (167,189] \end{cases}, \qquad B_{10}(y) = \begin{cases} \dfrac{y-150}{21} & y \in [150,171] \\ \dfrac{191-y}{20} & y \in (171,191] \end{cases}.$$

That is, the posterior part of fuzzy rules.

After that, establish fuzzy rules:

$$If\ x\ is\ A_i\ then\ y\ is B_i (i = 1, 2, \cdots, 10)$$

where the article i inference rules formation the fuzzy relation $R_i \in F(X \times Y)(i = 1, 2, \cdots, 10)$ about this rule,it is determined by a fuzzy implication operator $\theta : [0,1]^2 \to [0,1]$, i.e. $R_i(x,y) = \theta(A_i(x), B_i(y))$. we take Larsen operator, i.e. $\theta(a,b) = ab$ in this paper.

Step 3 Fuzzy reasoning

According to the obtained fuzzy reasoning rules, we use the First polymerization, After reasoning method, that is, these ten reasoning relation $R_i(i = 1, 2, \cdots, 10)$ form general reasoning relation $R = \bigcup_{i=1}^{10} R_i$, have

$$R(x,y) = \bigvee_{i=1}^{10} (A_i(x) \cdot B_i(y))$$

For $A \in F(X)$, getting reasoning result $B \in F(Y)$ through the general reasoning relation R , where

$$B(y) = A(x) \circ R(x,y) = \underset{x \in X}{\vee} (A(x) \wedge (\bigcup_{i=1}^{10} R_i(x,y))) = \underset{x \in X}{\vee} (A(x) \wedge (\bigvee_{i=1}^{10} (A_i(x) \cdot B_i(y))))$$

Step 4 Defuzzy

This paper uses the center of gravity method to defuzzy B, i.e.

$$\hat{Y}_s(x) = \frac{\int_Y yB(y)dy}{\int_Y B(y)dy} = \frac{\int_Y y(\underset{x \in X}{\vee} (A(x) \wedge (\bigvee_{i=1}^{10} (A_i(x) \cdot B_i(y))))) dy}{\int_Y \underset{x \in X}{\vee} (A(x) \wedge (\bigvee_{i=1}^{10} (A_i(x) \cdot B_i(y)))) dy}$$

With the help of fuzzy system control module in Matlab, the estimated value was shown in the following Table 1:

Table 1. Values of fuzzy system modeling (dollars)

Primary data x	80.0	100.0	120.0	140.0	160.0	180.0	200.0	220.0	240.0	260.0
Estimated value $\hat{Y}_{Si}$	69.0	79.0	92.0	103.0	116.0	130.0	140.0	156.0	164.0	171.0

4 The Application of the Probability Representation of Fuzzy System

Take Larsen Implication Operator $\theta(a,b) = ab$ into Formula (1), have

$$f(x,y) = \frac{[\bigvee_{i=1}^{10}(A_i(x)\cdot B_i(y))]I_{X\times Y}}{\int_X\int_Y[\bigvee_{i=1}^{10}(A_i(x)\cdot B_i(y))]d_xd_y} \quad (3)$$

That is, the original distribution is approximated by the Larsen distribution mentioned in literature [4].

Substituting Formula (3) into Formula (2), get $\hat{Y}_p(x)$. This involves the calculation of the double integral and the conditional mathematical expectation. For different x values, we divide the universe $X \times Y$, then calculate the double integral of each block, and seek its sum finally. Get

$$\int_X\int_Y[\bigvee_{i=1}^{10}(A_i(x)\cdot B_i(y))]d_xd_y = 2970$$

So (3) can be transformed into

$$f(x,y) = \frac{[\bigvee_{i=1}^{10}(A_i(x)\cdot B_i(y))]I_{X\times Y}}{2970} \quad (4)$$

The conditional mathematical expectation is calculated according to Formula (2). The estimates based on this method as shown in the following Table 2:

Table 2. Values of the probability representation of fuzzy system modeling (dollars)

Primary data x	80.0	100.0	120.0	140.0	160.0	180.0	200.0	220.0	240.0	260.0
Estimated value $\hat{Y}_{Si}$	69.0	78.8	92.0	103.0	115.7	130.0	140.0	156.0	164.3	170.7

5 Analysis and Comparison of Results

Summarizing the above results into the following table:

The estimated values of the third row are obtained by the one-dimensional linear regression method.

Let $\hat{u}_{Ri} = \bar{Y}_i - \hat{Y}_{Ri}$, $\hat{u}_{Si} = \bar{Y}_i - \hat{Y}_{Si}$, $\hat{u}_{Pi} = \bar{Y}_i - \hat{Y}_{Pi}$

So $RSS_{Ri} = \sum_{i=1}^{10}\hat{u}_{Ri}^2$, $RSS_{Si} = \sum_{i=1}^{10}\hat{u}_{Si}^2$, $RSS_{Pi} = \sum_{i=1}^{10}\hat{u}_{Pi}^2$.

In the regression analysis, the least squares method is based on estimate $\hat{Y}_{Ri}$ which got by minimum RSS_{Ri}. We can calculate through Table 3, get

$$RSS_{Ri} = 53.30,\ RSS_{Si} = 16.00,\ RSS_{Pi} = 15.91,$$

Table 3. Analysis result table (dollars)

Primary data x	80.0	100.0	120.0	140.0	160.0	180.0	200.0	220.0	240.0	260.0
Estimated value $\bar{Y}_i$	70.0	81.0	93.0	103.0	117.0	130.0	140.0	156.0	167.0	171.0
Estimated value $\hat{Y}_{Ri}$	69.6	81.4	93.3	105.1	116.9	128.7	140.5	152.3	164.2	176.0
Estimated value $\hat{Y}_{Si}$	69.0	79.0	92.0	103.0	116.0	130.0	140.0	156.0	164.0	171.0
Estimated value $\hat{Y}_{Pi}$	69.0	78.8	92.0	103.0	115.7	130.0	140.0	156.0	164.3	170.7

Thus visible, $RSS_{Pi} < RSS_{Si} < RSS_{Ri}$. Therefore, the probability representation of fuzzy system method is better than the fuzzy system method from point of view of residuals, and the fuzzy system method is superior to the one-dimensional linear regression analysis method.

In order to compare the goodness of fit, the determinants are introduced in the regression analysis to determine the goodness of fit of the regression line. Similar to the method of regression analysis delimit:

$$R^2 = \frac{ESS}{RSS}$$

Among them, the total sum of squares is $TSS = \sum y_i^2$, the description is the total variation of the true value Y around its mean $\bar{Y}$; The explain sum of squares is $ESS = \sum \hat{y}_i^2$, the description is the variation of the estimated value of Y around its mean $\bar{Y}$.

$$Let\ \bar{Y} = \frac{1}{10}\sum_{i=1}^{10} \bar{Y}_i = 122.8,\ we\ have$$

$$TSS = \sum(\bar{Y}_i - \bar{Y})^2 = 11575.6$$

$$ESS_{Ri} = \sum(\hat{Y}_{Ri} - \bar{Y})^2 = 11525.1$$

$$ESS_{Si} = \sum(\hat{Y}_{Si} - \bar{Y})^2 = 11670.4$$

$$ESS_{Pi} = \sum(\hat{Y}_{Pi} - \bar{Y})^2 = 11688.1$$

So we have

$$R_{Ri}^2 = \frac{ESS_{Ri}}{RSS} = 0.99564$$

$$R_{Si}^2 = \frac{ESS_{Si}}{RSS} = 1.00819$$

$$R_{Pi}^2 = \frac{ESS_{Pi}}{RSS} = 1.00972$$

We know that for linear regression models have $TSS = ESS+RSS$. Through the above results we can found that, $TSS = ESS + RSS$ no longer holds in the fuzzy case. ESS may be greater than TSS. So the coefficient of determination

R^2 may be greater than 1. Hence, we introduce a new index to measure the degree of fit:

$$\tilde{R}^2 = |1 - R^2|.$$

We think that the smaller $\tilde{R}^2$ is, the better the degree of fit. Thus we can have

$$\tilde{R}^2_{Ri} = 0.00436,\ \tilde{R}^2_{Si} = 0.00819,\ \tilde{R}^2_{Pi} = 0.00972.$$

It can clearly be seen $\tilde{R}^2_{Ri} < \tilde{R}^2_{Si} < \tilde{R}^2_{Pi}$, so from the point of view of goodness of fit, the linear regression model has the best fitting effect, the probability representation of fuzzy system model has the worst effect.

6 Conclusion and Prospect

In this paper, we use the fuzzy system method and the probability representation of fuzzy system method to model the existing economic data, and compared the results with the existing results of linear regression model. Conclusions as followed: The residual sum of squares obtained by the probability representation of fuzzy system method is smaller than the residuals sum of squares obtained by the fuzzy system method, and the residuals sum of squares obtained by the fuzzy system method is less than the residual sum of squares obtained by the regression analysis method. This shows that, in this case, the probability representation of fuzzy system modeling method is superior to the fuzzy system modeling method, and the fuzzy system modeling method is superior to the classical linear regression analysis modeling method. In this paper, we introduce the deterministic coefficients for the fuzzy system model and the probability representation of fuzzy system model similar to the linear regression analysis model, and proposed a new measurement index to measure the goodness of fit with the regression line. The results show that the goodness of fit of the linear regression model is better than that of the fuzzy system model, and the fuzzy system model is better than that of the probability representation of fuzzy system model. Therefore, comparing the models obtained by the three methods for the data, the degree of independent variable of the linear regression analysis model explains the total change is highly, but the difference between the estimated value and the actual measured value is large; The degree of independent variable of the probability representation of fuzzy system model explains the total change is poorly, but the difference between the estimated value and the actual measured value is least. This may be that regression analysis in economic data only analyzes the raw data 'bluntly' without taking human factors into account. Although this argument is highly explanatory to the dependent variable, the predicted results are 'unreasonable'.

It should be mentioned that all these three models have their reasonable place. Random interference terms of linear regression analysis model is a substitute for all variables that ignore the dependent variable from the model, but the impact of weekly expenditure is not just weekly income, there may be have other non-measurable factors (for example, family wealth is also a factor in expenditure). The fuzzy system method and the probability representation of fuzzy system method is taken the artificial selection and setting into account in the construction of their model, this is equivalent to combine human will and data to build a 'personal custom' system. This paper only analyzes the application of fuzzy system in economic case, it is just a way to try, and how to get the further theoretical analysis of the results remains to be further studied. In addition, we find that there are many differences between the model obtained by the fuzzy system method and the regression method, which is worth our further study.

Appendix

See Table 4.

Table 4. Economic data (dollars)

Weekly income X	80	100	120	140	160	180	200	220	240	260
Weekly expenditure Y	60	68	82	86	105	112	125	142	137	150
	68	78	89	94	111	118	136	145	151	152
	71	81	95	95	118	127	140	150	165	165
	74	84	98	103	120	135	144	156	175	175
	77	87	101	108	123	140	155	162	185	185
		88		115	125	148		167	189	189
				120				170		191
Total	350	486	465	721	702	780	700	1092	1002	1197
Mean value	70	81	93	103	117	130	140	156	167	171

References

1. Gujarati, D.N., Porter, D.C., Fei, J.: Basic Econometrics, 5th edn. Renmin University of China Press, Beijing (2010)
2. Wang, L.: Fuzzy System and Fuzzy Control. By Inmon, W.H., Beijing (2003)
3. Li, H., Wang, J., Miao, Z.: Modeling of fuzzy control system. Sci. China Ser. A **32**(9), 772–781 (2002)
4. Li, H.: Probabilistic representation of fuzzy system. Sci. China Ser. E Inf. Sci. **36**(4), 373–397 (2006)
5. Li, H.: Unification of uncertain systems. J. Eng. Math. **24**(1), 1–21 (2007)
6. Yuan, X., Li, H.: Probability distribution and digital characteristics based on input output data. Fuzzy Syst. Math. **25**(1), 69–78 (2011)
7. Yuan, X., Lu, W., Li, H.: Cente-of-gravity fuzzy system and its probability representation theory. J. Liaoning Tech. Univ. Nat. Sci. **29**(5), 795–798 (2010)
8. Yuan, X., Li, H., Yang, X.: Fuzzy system and fuzzy inference modeling method based on fuzzy transformation. Acta Electronica Sinica **41**(4), 674–680 (2013)

9. Yuan, X., Li, H.: Center of gravity fuzzy systems and their universal approximations based on a set data of input-output. Fuzzy Syst. Math. **25** (2011)
10. Guo, S., Chen, S.: A fuzzy reasoning model with self-learning. J. Fuxin Min. Inst. Nat. Sci. **3**, 89–95 (1993)
11. Guo, S., Peng, X.: Fuzzy reasoning method based on statistical information. J. Liaoning Tech. Univ. **23**(2), 268–271 (2004)
12. Chen, G.: Knowledge mining of managerial date base on fuzzy reasoning methods. J. Liaoning Tech. Univ. **22**(3), 416–418 (2003)

A Interval Number Ranking Method Based on Interval Quantile and TOPSIS for Decision Problems

Xin Wang[1(✉)], Su Feng[2], and Li-dong Wang[1]

[1] Mathematics Department, Dalian Maritime University, Dalian 116026, Liaoning, China
xenawang@163.com
[2] Statistical Institute, Nankai University, Tianjin 300073, China

Abstract. Most of the decision problems tend to be complex and fuzzy in the real life. So, we have to face many decision problems with interval numbers. This paper presents a new method for decision problems with interval numbers based on interval quantile and TOPSIS, which is called Q-TOPSIS. In this approach, firstly we select the appropriate quantiles of interval numbers, by which we can convert interval numbers into information vector, followed by selecting ideal information vector under a given evaluation standard. Secondly, we calculate the similarity between information vector and ideal information vector and obtain the order of interval numbers according to the similarity. Finally, one illustrative example of interval numbers and multiple criteria decision making is analyzed for the use of demonstrating the feasibility and effectiveness of the proposed method.

Keywords: Interval number · Ranking · Vector similarity · TOPSIS Quantile

1 Introduction

In reality, one often encounters a number of alternatives which need to be evaluated on the basis of a set of criteria in investment decision, universities ranking, road safety risk evaluation, and so on. In these cases, the alternatives and the related criteria are often combined to a table. In decision making, one needs to rank these alternatives by some criteria that are characterized by attributes in the table according to an increasing or a decreasing preference. This kind of decision making tasks are called ranking decision, which is becoming a research focus in decision making analysis. Until now, ranking decision has been widely used in economy, management, engineering and other broad areas [1]. In decision making analysis, we often need to deal with various types of data sets, where objects may be characterized by single value, set value, null value, or interval value. Interval data, which have drawn a lot of attention, are the generalized form of single-valued data. Hence, how to rank objects with interval values has become a very desirable issue.

B.-Y. Cao and Y.-B. Zhong (Eds.): ICFIE 2017, AISC 872, pp. 375–383, 2019.
https://doi.org/10.1007/978-3-030-02777-3_33

There are many different methods for interval number ranking. Xu et al. [2] proposed a ranking approach based on possible degree which has been widely used these years. However, the consistency of the fuzzy complementary judgment matrix established by this method is still need to be improved. Young [3] and Sevastianov et al. [4] took the endpoint or the median as the sorting basis of interval numbers. This method has the advantages of low computational complexity and flexible application, although its results are lack of reasonableness. Qiu et al. [5] proposed a method to rank interval numbers by employing probability reliability, which is suitable for all kinds of interval numbers. The difficulty of this method is the determination of the probability density function. Atanu et al. [6] presented a multi-level measure indicator based on mean and width of interval. The priority of the interval can be quantitatively described using this indicator. However, it cannot rank two same-midpoint intervals. Besides, many researches [7–10] consider the width for interval number ranking, which can not reflect the size of the interval for the decision-makers who hold different attitudes. Considering all reviews mentioned above, in this paper, we propose a interval number ranking method based on interval quantile and TOPSIS.

An interval number can be thought as an extension of the concept of a real number and also as a subset of the real line R. It can be assumed as the closed interval on the real line which contains countless points. If all points of the interval number can be considered for interval number ranking, it seem to lose no information. However, this is not feasible by the denseness of the real numbers. So developing approaches to obtain the necessary information which can reflect the size of interval number is desirable. In this study, first we introduce the concept of interval numbers and vector similarity. Then, through the experimental comparison and analysis, we determine the information vector and define the ideal information vector on the basis of TOPSIS [11]. Finally, we report on a experiment when using the data of the college evaluation [2], which gives the same ranking result with the original method.

2 Backgrounds

2.1 The Basic Concept and Operations of Interval Numbers

Definition 2.1 [12]. Let $\tilde{a} = [a^L, a^U] = \{x | a^L \leq x \leq a^U, a^L, a^U \in R\}$, we call $\tilde{a}$ as an interval number.

Definition 2.2 [12]. Let $\tilde{a} = [a^L, a^U]$, $\tilde{b} = [b^L, b^U]$ be two positive interval numbers, then we have the following operations.

(1) Addition: $\tilde{a} + \tilde{b} = [a^L, a^U] + [b^L, b^U] = [a^L + b^L, a^U + b^U]$,
(2) Subtraction: $\tilde{a} - \tilde{b} = [a^L, a^U] - [b^L, b^U] = [a^L - b^L, a^U - b^U]$,
(3) Multiplication: $\tilde{a} \cdot \tilde{b} = [a^L, a^U] \cdot [b^L, b^U] = [a^L b^L, a^U b^U]$, particularly $\lambda a = [\lambda a^L, \lambda a^U]$, $\lambda \geq 0$,
(4) Division: $\tilde{a} / \tilde{b} = [a^L, a^U]/[b^L, b^U] = [a^L, a^U][1/b^U, 1/b^L] = [a^L b^U, a^U b^L]$, particularly, $a^{-1} = [1/a^U, 1/a^L]$.

2.2 TOPSIS Method

TOPSIS method [11], a widely used approach for MCDM problems, is developed by Hwang and Yoon in 1981 and further developed by Yoon and Hwang et al. In this method the best decision should be the closest to the ideal solution and farthest from the nonideal solution. It assumes that each attribute is monotonically increasing or decreasing, making it easy to locate the best and worst selection visually. The final rank is reached by comparing the Euclidean distances of all alternatives. Topsis is considered as a good choice for MCDM problems because of the following reasons. (1) It is relatively easy and fast. (2) It is useful for both qualitative and quantitative data. (3) The output can be a preferential ranking using both negative and positive criteria. Because of these advantages, TOPSIS has been widely used and developed to deal with MCDM problems. In view of all these advantages and wide application, we choose TOPSIS to solve the MCDM problems with interval numbers.

2.3 The Basic Concept of Vector Similarity

Definition 2.3 [13]. Suppose that $\alpha = (a_1, a_2, \cdots a_n)$ is the reference vector, $\beta = (b_1, b_2, \cdots b_n)$ is compared vector. Vector norm similarity γ of α and β is defined as

$$\gamma = \begin{cases} 1 - \frac{|\|\alpha\| - \|\beta\||}{\|\alpha\|}, & \|\beta\| \leq 2\|\alpha\| \\ 0 & \|\beta\| \succ 2\|\alpha\| \end{cases}; \tag{1}$$

Vector direction similarity η of α and β is defined as

$$\eta = 1 - \frac{\langle \alpha, \beta \rangle}{90^\circ}; \tag{2}$$

Vector similarity S of α and β is defined as

$$S(\alpha, \beta) = \gamma \cdot \eta. \tag{3}$$

It can be seen that the similarity of vectors is the product of vector norm similarity and direction similarity. We know that vectors are quantities with both magnitude and direction. Therefore, both size and direction of the vector are taken into account when we use formula (3) to compute vector similarity. γ increases with the reduction of the difference in the norm between the vectors. Similarly, η also increases with decreasing of vector angle $\langle \alpha, \beta \rangle$. Thus, the more similar the two vectors are, the larger γ is and the smaller the angle $\langle \alpha, \beta \rangle$ is, the greater the η is, hence $S(\alpha, \beta)$ is also greater.

3 Interval Number Information Transformation and Interval Number Ranking

In many sorting methods of interval numbers, the width of the interval number is used as an important indicator for the size of the interval. For instance, in the literature [14], the uncertainty of the interval number is measured by the width of the interval number. The bigger the interval width is, the greater the degree of uncertainty is. Cheng and Zhu [15] believed that in the case of other conditions unchanged, the smaller the radius of the interval number, the greater the interval number is. This is feasible to risk-averse decision-makers. But for decision-makers who prefer risk, this runs counter to the goal. For instance, the intervals $\widetilde{a} = [1, 100]$ and $\widetilde{b} = [35, 36]$ represent the expected benefits obtained from two different methods respectively. Risk-averse decision makers believe that $\widetilde{b}$ is better of income effect. However, the risk preference decision-makers think $\widetilde{a}$ is better. The sorting result is $\widetilde{b} \succ \widetilde{a}$ using the method in [15], which goes against the attitude of the decision-makers who pursue the risk. Therefore, using the interval radius to represent the size of the interval is too limited.

3.1 Information Extraction of Interval Numbers

An interval number can be taken as a closed interval. We know that any interval is composed of innumberable real numbers. If all points of closed intervals are used for the information vector of the interval number to calculate the vector similarity, it does not seem to lose any information of the interval number. However, according to the denseness of real numbers, all numbers cannot be enumerated. Therefore, it is necessary to choose a few important numbers on behalf of interval number.

Definition 3.1. Quantile

Suppose X is a continuous random variable, its distribution function is $F(x)$, the probability density function is $p(x)$. When $F(x_p) = P(X \leq x_p) = p$, $0 \prec p \prec 1$.

Then x_p is called quantile [16] or underside quantiles of this distribution.

According to the knowledge of statistics, description of data distribution used to adopt some indicators which includes the mean, the mode, the median and the quantile, etc. Mode is of no practical significance for interval numbers of uniform distribution. Mean and median can be seen as 1/2 quantiles. Therefore, we can use the upper bound, the quartile, the median, the lower quartile and the lower bound as the representative data of the interval based on the concept of the boxplot. Thus, we choose them as interval number information vector.

Definition 3.2. Suppose $\widetilde{a} = [a^L, a^U]$ is a interval number, we call $\alpha = (a^L, Q_1, Q_2, Q_3, a^U)$ as information vector of $\widetilde{a}$.

Definition 3.3. Let $\widetilde{a}_1, \widetilde{a}_2, \cdots \widetilde{a}_n$ be a set of interval numbers. The corresponding information vectors are $\alpha_i = (a_i^L, Q_{1i}, Q_{2i}, Q_{3i}, a_i^U)$ $(i = 1, 2, \cdots n)$. Then we can define $p = (p_1, p_2, p_3, p_4, p_5)$ as the ideal information vector of $\alpha_i = (a_i^L, Q_{1i}, Q_{2i}, Q_{3i}, a_i^U)$ $(i = 1, 2, \cdots n)$. Where $p_1 = \max\{a_i^L, i = 1, 2, \cdots, n\}$, $p_2 = \max\{Q_{1i}, i = 1, 2, \cdots, n\}$, $p_3 = \max\{Q_{3i}, i = 1, 2, \cdots, n\}$, $p_4 = \max\{Q_{4i}, i = 1,$

$2, \cdots, n\}$, $p_5 = \max\{a_i^U, i = 1, 2, \cdots, n\}$, where Q_1, Q_2, Q_3 are upper quartile, the median and lower quartile for interval numbers respectively.

Definition 3.4 Let $\widetilde{a}_1, \widetilde{a}_2, \cdots \widetilde{a}_n$ be a set of interval numbers. The information vectors are $\alpha_i = (a_i^L, Q_{1i}, Q_{2i}, Q_{3i}, a_i^U)$ $(i = 1, 2, \cdots n)$. The ideal information vector is $p = (p_1, p_2, p_3, p_4, p_5)$. If $S(\alpha_i, p) \succ S(\alpha_j, p)$, we call $\widetilde{a}_i$ is better than $\widetilde{a}_j$, which is denoted as $\widetilde{a}_i \succ \widetilde{a}_j$.

3.2 Interval Number Ranking

According to the above analysis, we extract the five important points of interval number as the components of information vector, and then we can compare and sort a set of interval numbers $\widetilde{a}_1, \widetilde{a}_2, \cdots \widetilde{a}_n$ by vector similarity. Concrete steps are as follows:

(1) Information transformation of interval number.
The information of interval numbers $\widetilde{a}_i = [a_i^L, a_i^U]$ are extracted. We translate them into information vectors $\alpha_i = (a_i^L, Q_{1i}, Q_{2i}, Q_{3i}, a_i^U)$ $(i = 1, 2, \cdots n)$, where, Q_1 is the upper quartile, Q_2 is the median and Q_3 is the lower quartile for interval numbers.
(2) Determine the ideal information vector $p = (p_1, p_2, p_3, p_4, p_5)$, where $p_1 = \max\{a_i^L, i = 1, 2, \cdots, n\}$, $p_2 = \max\{Q_{1i}, i = 1, 2, \cdots, n\}$, $p_3 = \max\{Q_{3i}, i = 1, 2, \cdots, n\}$, $p_4 = \max\{Q_{4i}, i = 1, 2, \cdots, n\}$, $p_5 = \max\{a_i^U, i = 1, 2, \cdots, n\}$.
(3) Calculate the vector similarity S_i between the interval number information vector and the ideal information vector by the formula (1), (2) and (3).
(4) Rank interval numbers according to the vector similarity. The greater the S_i is, the more optimal the corresponding interval number is.

Example 1: Let $\widetilde{a} = [0.4, 0.6]$ and $\widetilde{b} = [0.2, 0.8]$, give the rank of $\widetilde{a}$ and $\widetilde{b}$.
According to the above steps, we can obtain: $S_1 = 0.7859 \prec S_2 = 0.8351$, thus $\widetilde{a} \prec \widetilde{b}$. It is the opposite of the result by using method in the [15].

Example 2: Let $\widetilde{a} = [1, 100]$ and $\widetilde{b} = [35, 36]$, give the rank of $\widetilde{a}$ and $\widetilde{b}$.
In the same way, we can get: $S_1 = 0.8030 \succ S_2 = 0.1415$, thus $\widetilde{a} \succ \widetilde{b}$, which is contrary to the result of the original method and it is applicable to the decision-makers who pursue the risk.

From the intuition that the more points are selected in the interval, the more information can be used for ranking. However, the experiments exhibited that selecting 5 points give the most suitable results comparing with 100, 50, 10, 5, 3 and 2 points. We can see from the Table 1 which showed the results, the vector similarity between information vector and ideal information vector changes from $S_1 \prec S_2$ to $S_1 \succ S_2$ of the interval numbers $[1, 100]$ and $[53, 60]$ as well as $[1, 100]$ and $[55, 61]$ when we executed the experiment for 3 points. We selected the number of integer points between 1–100 in the experiments, the results are shown in Fig. 1.

Table 1. Similarity between the information vector and ideal information vector for selecting different points

S	$[1,100]$ $[35,36]$	$[1,100]$ $[2,99]$	$[1,100]$ $[53,60]$	$[1,100]$ $[55,61]$	$[1,100]$ $[80,110]$	$[1,100]$ $[50,150]$
100	S_1:0.8403 f S_2:0.2268	S_1:0.9943 f S_2:0.9916	S_1:0.6668 p S_2:0.7202	S_1:0.6488 p S_2:0.7387	S_1:0.2620 p S_2:1.0000	S_1:0.1779 p S_2:1.0000
50	S_1:0.8384 f S_2:0.2229	S_1:0.9942 f S_2:0.9915	S_1:0.6657 p S_2:0.7170	S_1:0.6478 p S_2:0.7355	S_1:0.2640 p S_2:1.0000	S_1:0.1802 p S_2:1.0000
10	S_1:0.8233 f S_2:0.1899	S_1:0.9941 f S_2:0.9910	S_1:0.6572 p S_2:0.6903	S_1:0.6394 p S_2:0.7088	S_1:0.2812 p S_2:1.0000	S_1:0.2000 p S_2:1.0000
5	S_1:0.8030 f S_2:0.1415	S_1:0.9941 f S_2:0.9904	S_1:0.6457 p S_2:0.6522	S_1:0.6285 p S_2:0.6711	S_1:0.3043 p S_2:1.0000	S_1:0.2274 p S_2:1.0000
3	S_1:0.7742 f S_2:0.0675	S_1:0.9942 f S_2:0.9894	S_1:0.6348 f S_2:0.5950	S_1:0.6719 f S_2:0.6136	S_1:0.3384 p S_2:1.0000	S_1:0.2697 p S_2:1.0000
2	S_1:0.7450 f S_2:0.0000	S_1:0.9935 f S_2:0.9898	S_1:0.6044 f S_2:0.4981	S_1:0.5893 f S_2:0.5207	S_1:0.3878 p S_2:1.0000	S_1:0.3358 p S_2:1.0000

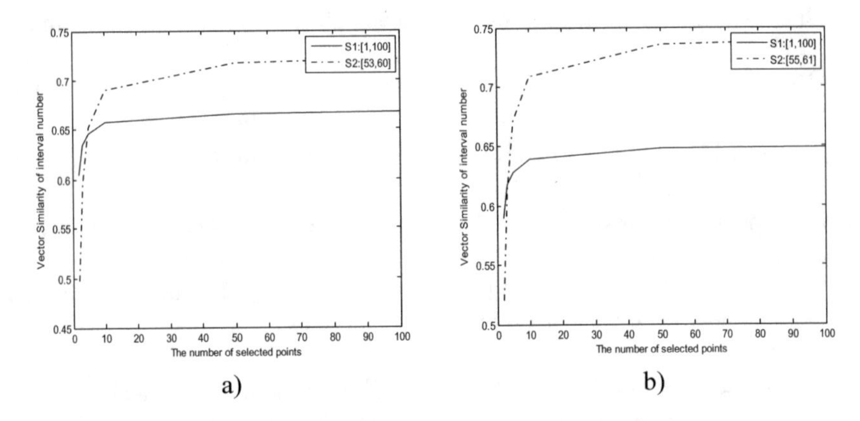

Fig. 1. The vector similarity corresponding different points

The result is $S_1 \prec S_2$ for selecting 5–100 points (including 5 points). However, the result becomes $S_1 \succ S_2$ when the number of selected points is less than 5. Therefore, it is reasonable that 5 points can catch enough information contained in the interval number and is easy for computing.

4 Q-TOPSIS Ranking Method of Interval Number

MCDM is a sub-discipline of operations research that explicitly evaluates multiple conflicting criteria in decision making (both in daily life and in professional settings). Due to the complexity and uncertainty of decision-making problems, we will often encounter the problem of decision-making, which is not expressed in numerical points, but in the form of interval numbers. Therefore, it is necessary to study the MCDM problem with interval number.

4.1 Detailed Steps of the MCDM Using Q-TOPSIS Method

Let $A=\{A_1, A_2, \cdots, A_n\}$ be the solution set of n schemes, $Q=\{q_1, q_2, \cdots, q_m\}$ is the set of attributes with m properties. Weight vector of attributes is $\omega = (\omega_1, \omega_2, \cdots, \omega_m)$, $\omega_i \in R$, $\sum_{i=1}^{m} \omega_i = 1$. The attribute values of scheme A_i for attribute q_i are $\widetilde{\alpha}_{ij} = \left[a_{ij}^L, a_{ij}^U\right]$, $(i = 1, 2\cdots, n, j = 1, 2\cdots, m)$, which form decision matrix $\widetilde{A} = \left(\widetilde{\alpha}_{ij}\right)_{n\times m}$. We calculate the optimal solution by the following steps.

Step 1. Standardize the decision matrix with interval numbers. In order to eliminate the influence of different physical dimension on the decision results, we need normalize decision matrix with interval numbers by the method in the literature [17].

The normalization matrix is denoted by $R = \left(\widetilde{r}_{ij}\right)_{n\times m}$, where $\widetilde{r}_{ij} = \left[a_{ij}^L, a_{ij}^U\right]$, $(i = 1, 2\cdots, n; j = 1, 2\cdots, m)$ is the normalized interval number.

For benefit attributes For cost attributes

$$\begin{cases} r_{ij}^L = a_{ij}^L \Big/ \sqrt{\sum_{i=1}^{n}\left(1/a_{ij}^U\right)^2} \\ r_{ij}^U = a_{ij}^U \Big/ \sqrt{\sum_{i=1}^{n}\left(1/a_{ij}^L\right)^2} \end{cases}, \quad \begin{cases} r_{ij}^L = \left(1/a_{ij}^U\right) \Big/ \sqrt{\sum_{i=1}^{n}\left(1/a_{ij}^L\right)^2} \\ r_{ij}^U = \left(1/a_{ij}^L\right) \Big/ \sqrt{\sum_{i=1}^{n}\left(1/a_{ij}^U\right)^2} \end{cases}.$$

Step 2. Calculate the comprehensive attribute value $\widetilde{z}_i(\omega) = \left[z_i^L, z_i^U\right]$ of scheme A_i using the formula $\widetilde{z}_i(\omega) = \sum_{j=1}^{m} \omega_j \widetilde{r}_{ij} (i = 1, 2\cdots, n)$, where $z_i^L = \sum_{j=1}^{m} \omega_j r_{ij}^L$, $z_i^U = \sum_{j=1}^{m} \omega_j r_{ij}^U$.

Step 3. Convert the comprehensive attribute values of the schemes into information vectors $\alpha_i = \left(a_i^L, Q_{1i}, Q_{2i}, Q_{3i}, a_i^U\right)$.

Step 4. Select the ideal information vector $p = (p_1, p_2, p_3, p_4, p_5)$, where $p_1 = \max\left\{a_i^L, i = 1, 2\cdots, n\right\}$, $p_2 = \max\{Q_{1i}, i = 1, 2\cdots, n\}$, $p_3 = \max\{Q_{2i}, i = 1, 2\cdots, n\}$, $p_4 = \max\{Q_{3i}, i = 1, 2\cdots, n\}$, $p_5 = \max\left\{a_i^U, i = 1, 2\cdots, n\right\}$.

Step 5. Calculate the norm similarity γ_i and direction similarity η_i of the information vector and the ideal information vector by the formula (1) and (2).

Step 6. Calculate the similarity S_i between the information vector and the ideal information vector according to the formula (3).

Step 7. Rank S_i and sort the schemes. Give the optimal scheme.

4.2 MCDM Problem of College Evaluation

In this section, a university college evaluation problem [2] is considered. In general, some universities use three attributes (teaching, research and service) as the evaluation indicators. Suppose that five colleges A, B, C, D, E will be evaluated, and the weights of the attribute are assumed to be the component of the vector $\omega = (0.4, 0.4, 0.2)^T$. The decision makers give the attribute values of each scheme in the form of interval numbers. Its normalized decision matrix is shown in the Table 2.

Table 2. Standardized decision matrix of attribute values of each scheme

	A	B	C	D	E
μ_1	[0.214, 0.220]	[0.206, 0.225]	[0.195, 0.204]	[0.181, 0.190]	[0.175, 0.184]
μ_2	[0.166, 0.178]	[0.220, 0.229]	[0.192, 0.198]	[0.195, 0.205]	[0.193, 0.201]
μ_3	[0.184, 0.190]	[0.182, 0.191]	[0.220, 0.231]	[0.185, 0.195]	[0.201, 0.211]

(1) The comprehensive attribute values of the colleges A, B and C are calculated as step 2 in the form of interval number: $z_1 = [0.1888, 0.1972]$, $z_2 = [0.2068, 0.2198]$, $z_3 = [0.1988, 0.2070]$, $z_4 = [0.1874, 0.1970]$, $z_5 = [0.1874, 0.1962]$.
(2) We translate the comprehensive attribute values of all solutions with interval number into information vector: $\alpha_1 = (0.1888, 0.1909, 0.1930, 0.1951, 0.1972)$, $\alpha_2 = (0.2068, 0.2101, 0.2133, 0.2166, 0.2198)$, $\alpha_3 = (0.1988, 0.2009, 0.2029, 0.2049, 0.2070)$, $\alpha_4 = (0.1874, 0.1898, 0.1922, 0.1946, 0.1970)$, $\alpha_5 = (0.1874, 0.1896, 0.1918, 0.1940, 0.1962)$.
(3) We select ideal information vector based on the step 3:

$$p = (0.2068, 0.2101, 0.2133, 0.2166, 0.2198)$$

(4) According to the step 5 and step 6, the similarities between the information vector and the ideal information vector are obtained: $S_1 = 0.8911$, $S_2 = 1.0000$, $S_3 = 0.9441$, $S_4 = 0.8878$, $S_5 = 0.8847$.
(5) It is easy to get $S_2 \succ S_3 \succ S_1 \succ S_4 \succ S_5$, hence the corresponding 5 colleges are sorted as: $B \succ C \succ A \succ D \succ E$. Thus the comprehensive assessment of college B is the best. This is the same with the result using the method of the possibility [11].

5 Conclusion

In this paper, a hybrid method called Q-TOPSIS is proposed to solve MCDM problems with interval data. Both quantile and TOPSIS are applied within the proposed framework. Some experiments exhibited the effectiveness of the selection of information vector. We choose TOPSIS to determine the ideal information vector in view of TOPSIS s simpleness and wide applications. Finally, one example is used to verify the feasibility and effectiveness of the method. The proposed method can be extended to other distribution of the interval number, such as uniform distribution, normal distribution and so on. Their quantile can be calculated as information vector of interval number according to the different distribution function.

Acknowledgements. This research was supported by the National Natural Science Foundation of China (No. 61803065 and 61773352) and the Fundamental Research Funds for the Central Universities of China, Dalian Maritime University (3132017047 and 3132018228).

References

1. Song, P., Liang, J., Qian, Y.: A two-grade approach to ranking interval data. Knowl.-Based Syst. **27**(10), 234–244 (2012)
2. Xu, Z.S., Da, Q.L.: Possibility degree method for ranking interval numbers and its application. J. Syst. Eng. **18**(01), 67–70 (2003)
3. Young, R.C.: The algebra of many-valued quantities. Ann. Math. **31**(2), 260–290 (1931)
4. Sevastianov, P.: Numerical methods for interval and fuzzy number comparison based on the probabilistic approach and Dempster-Shafer theory. Inf. Sci. **177**(21), 4645–4661 (2007)
5. Qiu, D.S., He, C., Zhu, X.M.: Ranking method research of interval numbers based on probability reliability distribution. Control Decis. **27**(12), 1894–1898 (2012)
6. Atanu, S., Tapan, K.P.: On comparing interval number. Eur. J. Oper. Res. **127**(1), 28–43 (2000)
7. Morse, W.H., Mcmillan, D.E.: Schedules using noxious stimuli. I. Multiple fixed-ratio and fixed-interval termination of schedule complexes. J. Exp. Anal. Behav. **9**(03), 267–290 (1966)
8. Moore, R.E.: Methods and applications of interval analysis. Soc. Ind. Math. **2**(04), 59–86 (1978)
9. Zeng, W.Y., Luo, C.Z., Haji, R.: Comprehensive decision model of interval number. Syst. Eng. Theory Pract. **34**(11), 49–51 (1997)
10. Liu, J.S., Wang, X.Z., Zhang, B.Y.: The ranking of interval numbers. J. Eng. Math. **18**(04), 103–109 (2001)
11. Wang, P., et al.: A new method based on TOPSIS and response surface method for MCDM problems with interval numbers. Math. Probl. Eng. **2015**(938535), 1–11 (2015)
12. Senguta, A., Pal, T.K.: On comparing interval numbers. Eur. J. Oper. Res. **127**(01), 28–43 (2000)
13. Zeng, J., Wang, J., Guo, J.Y.: Local multi-model method based on similarity of vector. Appl. Res. Comput. **29**(05), 1631–1633 + 1640 (2012)
14. Lan, J.B., Hu, M.M., Ye, X.M.: Ranking interval numbers based on similarity. Comput. Eng. Des. **32**(04), 1419–1421 (2011)
15. Chen, C.H., Zhu, C.X.: Interval number sorting method based on vector similarity and its application. Control Decis. **76**(03), 76–78 (2014)
16. Li, Z., et al.: Analysis of wind power fluctuation interval based on quantile regression. Autom. Electr. Power Syst. **35**(03), 83–87 (2011)
17. Gorzalczany, M.B.: A method of inference in approximate reasoning based on interval-valued fuzzy sets. Fuzzy Sets Syst. **21**(1), 1–17 (1987)

Study on Rapidity Decision of Highway Lane Changing Based on OS-ELM

Senlin Cheng(✉), Yang Xu, Ruixue Zong, and Renming Deng

School of Automation, Chongqing University, Chongqing, China
csl@cqu.edu.cn

Abstract. When drivers make a lane changing maneuver in highway merging area, the personalized lane changing models of traditional ADAS usually combine several algorithms, which make the time of running models become longer. Therefore, this paper presented a highway lane changing model based on OS-ELM algorithm. By continuously adding data from the lane changing process to the model one-by-one, the lane changing model of the driver will be online updated constantly. It can be used as a kernel component of ADAS to provide personalized recommendation for the driver. The simulation result demonstrated that the running time of OS-ELM algorithm is about 3 times faster than the SVM algorithm, about 41 times faster than the BP algorithm, under the condition of assuming the predictive accuracy is 86.8% for merging events and 87% for non-merging events. The simulation results show that OS-ELM algorithm can better meet the rapidity demand of running personalized lane changing model.

Keywords: Personalized lane change model · OS-ELM algorithm
Mandatory lane change · Highway merging area

1 Introduction

When drivers make a lane changing maneuver in highway merging area, the vehicle speed of the target lane is very fast, and the slope or the curvature of the road change greatly, which make the traffic flow instable in this area. If the driver make a wrong lane changing decision, it may cause traffic congestion or accidents. Now many scholars have established many models for the lane changing auxiliary of ADAS. For example, they divide lane changing decision models of traffic characteristics into two parts, one part is rigid mechanistic models. Gipps proposed a framework for the structure of lane changing decisions and the execution of lane changing [1]. Wiedemann and Reiter developed a theoretical lane changing decision model to explain the human decision process during the lane changing maneuver which is influenced by the driver's perception of surrounding vehicles [2]. Ahmed defined three categories of lane changing maneuvers, Mandatory Lane Changing (MLC), Discretionary Lane Changing (DLC) and forced merging [3]. Toledo developed an integrated probabilistic lane changing decision model which allows drivers to consider both MLC and DLC at the same time [4]. Due to the complexity and uncertainty of the lane changing maneuver, it is difficult to describe it using traditional models. However, artificial intelligence

B.-Y. Cao and Y.-B. Zhong (Eds.): ICFIE 2017, AISC 872, pp. 384–393, 2019.
https://doi.org/10.1007/978-3-030-02777-3_34

models can establish model for complex traffic flow system based on data in spite of unknown mechanism and incomplete information, So the other part is artificial intelligence model. Das et al. proposed a new microscopic simulation methodology based on fuzzy IF-THEN rules and called the software package as Autonomous Agent SIMulation Package (AASIM) [5]. Wu et al. developed a fuzzy logic motorway simulation model (FLOWSIM) and established fuzzy sets and systems for the model [6, 7]. Kumar proposed a solution to lane change intention prediction based on a combination of a multiclass SVM classifier and Bayesian filtering [8]. W. Yao and etc. developed a model to predict the lane change trajectory by the human driving data according to the traffic environment [9]. Hou has made some progress with Bayes and Decision Trees algorithm recently [10, 11]. Dou et al. have developed a Support Vector Machine and Artificial Neural Network Classifiers model to predict merging behavior of drivers at lane drops [12].

However, current ADAS provide personalized lane changing models usually combine several algorithms, which make the time for running personalized lane changing models become longer. This paper presented a lane changing model based on online sequential extreme learning machine (OS-ELM) algorithm, which has very fast learning speed and good generalization ability. By continuously adding data from the lane changing process to the model one-by-one, the lane changing model of the driver will be online updated constantly, further, the running time of OS-ELM algorithm is very fast. When collecting data from a individual driver, and then retraining the model, the resulting model is suitable for the driver's lane changing maneuver. While drivers make a lane changing maneuver, the personalized model can be used to provide recommendation to the driver. Further, personalized recommendation are real-time with very high predictive accuracy.

2 Brief Introduction of OS-ELM Algorithm

Extreme learning machine (ELM) is for single hidden layer feed forward neural networks (SLFNs) [13], which needn't to adjust the input weights and the bias of the hidden layer nodes [14], it only needs to set the number of hidden layer nodes of the network, further, the output weights are determined analytically using least-squares method. Given N arbitrary distinct data samples $(x_j, t_j) \in R^d \times R^m$, the mathematical model of SLFNs with n hidden layer nodes is defined as Eq. (1).

$$\sum_{i=1}^{n} \beta_i g_i(x_j) = \sum_{i=1}^{n} \beta_i G(a_i, b_i, x_j), j = 1, 2, \ldots, N \tag{1}$$

Where $a_i = [a_{i1}, \ldots, a_{in}]$ is the weight vector connecting ith hidden neuron and input neuron, $\beta_i = [\beta_{i1}, \ldots, \beta_{in}]$ is the weight vector connecting ith hidden node to output neuron, b_i is the threshold of ith hidden neuron, $g_i(x_j) = G(a_i, b_i, x_j)$ is the output of the ith hidden layer node about the input vector x_j. SLFNs can universally approximate the N sample data, so the relationship between (a_i, b_i) and β_i can be obtained, as shown in Eq. (2).

$$\sum_{i=1}^{n} \beta_i G(a_i, b_i, x_j) = t_j \,, j = 1, 2, \ldots, N \tag{2}$$

The above N equations can be written in the matrix form, as shown in Eq. (3).

$$H\beta = T \tag{3}$$

Where, $H = \left[h_1^T \, h_2^T \cdots h_N^T\right]^T = \begin{bmatrix} G(a_1, b_1, x_1) & \cdots & G(a_n, b_n, x_1) \\ \vdots & \ddots & \vdots \\ G(a_1, b_1, x_N) & \cdots & G(a_n, b_n, x_N) \end{bmatrix}_{N \times n}$,

$$\beta = \begin{bmatrix} \beta_1^T \\ \beta_2^T \\ \vdots \\ \beta_N^T \end{bmatrix}_{n \times m}, T = \begin{bmatrix} t_1^T \\ t_2^T \\ \vdots \\ t_N^T \end{bmatrix}_{N \times m}$$

According to the interpolation theory in [13], as long as the hidden layer activation function is infinitely differentiable, (a_i, b_i) can be assigned randomly and remained unchanged, and now H is a constant matrix, The problem of SLFNs is then converted to solve the least squares $\hat{\beta}$ of linear systems $H\beta = T$, as shown in Eq. (4).

$$\left\| H\hat{\beta} - T \right\| = \min_{\beta} \left\| H\,\beta - T \right\| \tag{4}$$

Therefore, if it knows the values of hidden layer output matrix H and output matrix T, then the solution of output parameter $\hat{\beta}$ can be obtained, as shown in Eq. (5).

$$\hat{\beta} = H^{\dagger} T \tag{5}$$

where $H^{\dagger}$ is the Moore–Penrose generalized pseudo inverse of the hidden layer output [15], while $H^T H$ is non-singular matrix, the Eq. (5) can be written as Eq. (6).

$$\hat{\beta} = H^{\dagger} T = \left(H^T H\right)^{-1} H^T T \tag{6}$$

The [16] proposed an OS-ELM algorithm, and it can learn data one-by-one or chunk-by-chunk (a block of data) with fixed or varying chunk size. OS-ELM algorithm learning process is as follows.

In the initialization phase, giving a initial training subset $\Omega_{k-1} = \left\{\left(x_j, t_j\right) | j = 1, 2, \ldots, k-1.\right\}$. From the Eq. (6), it can calculate the initial output weight, as shown in Eq. (7).

$$\beta_{k-1} = P_{k-1} H_{k-1}^T T_{k-1} \quad (7)$$

Where, $P_{k-1} = \left(H_{k-1}^T H_{k-1}\right)^{-1}$,

$$T_{k-1} = [t_1\, t_2\, \cdots\, t_{k-1}]^T, H_{k-1} = \left[h_1^T\, h_2^T\, \cdots\, h_{k-1}^T\right]^T.$$

In the sequential learning phase, a new training sample (x_k, t_k) is obtained each time, the output weights are recursively calculated, as shown in Eqs. (8) and (9).

$$P_k = P_{k-1} - \frac{P_{k-1} h_k^T h_k P_{k-1}}{1 + h_k P_{k-1} h_k^T} \quad (8)$$

$$\beta_k = \beta_{k-1} + P_k h_k^T (t_k - h_k \beta_{k-1}) \quad (9)$$

Where, $h_k = [G(a_1, b_1, x_k) \cdots G(a_n, b_n, x_k)]$.

3 Selection of Lane Changing Feature Variable

In this paper, the data come from Next Generation Simulation (NGSIM) project in the U.S, it include vehicle trajectories on a segment of southbound U.S. Highway 101 in Los Angeles, and a segment of Interstate 80 in San Francisco. Given the focus of this paper on MLC, so only trajectory data of vehicles in the auxiliary lane and the adjacent lane are used for model development. Trajectory data include the longitudinal and lateral coordinates, acceleration, speed and headway for each vehicle at a resolution of 10 frames/s.

This paper defines the acceleration lane and the emergency lane as the merge lane, and defines the adjacent lane of the merge lane as the target lane. The merge event defines when a vehicle's lateral coordinate began to shift toward to the adjacent target lane direction without oscillations. Otherwise, it is deemed as a non-merge event. The MLC process in highway merging area is shown in Fig. 1.

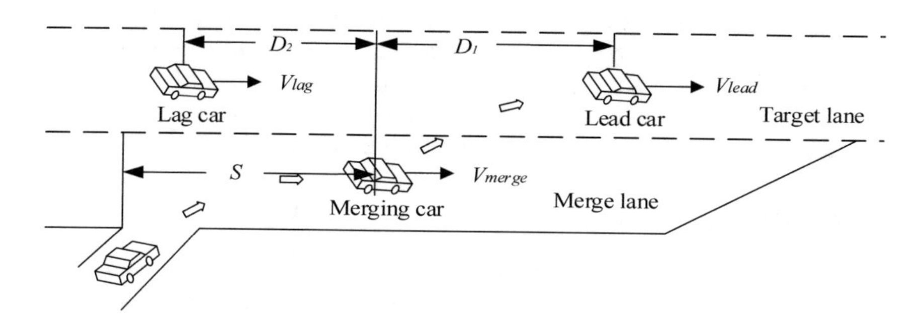

Fig. 1. Sketch map of vehicle merge and related parameters

During the lane changing process, there are many factors to affect the driver's lane changing maneuver. For example, the merging vehicle's speed and acceleration, the distance from the merging vehicle to the beginning of the merge lane, the speed or gap distance difference between the vehicle in the merging lane and the merging vehicle, the speed or gap distance difference between the vehicle in the target lane and the merging vehicle, and the driver's own state. But too many factors will affect the learning speed of the algorithm and increase the number of hardware devices such as sensors. Little factors impact prediction accuracy of the algorithm. Therefore, selecting the appropriate input factors is very important. In this paper, selecting the five factors as the feature variables [17].

1. S (feet) is the distance from the merging vehicle to the beginning of the merge lane, and it can reflect the urgency of the lane changing process, as shown in Fig. 1.

2. V_1 (feet/s) is the speed difference between the lead vehicle in the target lane and the merging vehicle, V_2 (feet/s) is the speed difference between the lag vehicle in the target lane and the merging vehicle, V_1 and V_2 can reflect the latent danger of lane changing process, which is defined respectively as Eqs. (10) and (11).

$$V_1 = V_{merge} - V_{lead} \tag{10}$$

$$V_2 = V_{merge} - V_{lag} \tag{11}$$

Where V_{lead} is the speed of the lead vehicle, V_{merge} is the speed of the merging vehicle. V_{lag} is the speed of the lag vehicle. D_1 (feet) is the gap distance between the lead vehicle in the target lane and the merging vehicle, D_2 (feet) is the gap distance between the lag vehicle in the target lane and the merging vehicle D_1 and D_2 can reflect whether it has a suitable acceptable gap, as shown in Fig. 1.

4 Experimental Simulation and Result Analysis

4.1 Experimental Data Selection

Based on the analysis of previous feature selection, the five features can be determined as input variables of OS-ELM algorithm, namely $x_i = [S_i, \mathrm{V}_{1i}, \mathrm{V}_{2i}, \mathrm{D}_{1i}, \mathrm{D}_{2i}]$, $i = 1, \ldots, n$. The output variable is non-merge or merge, here, the symbol "0" represents the non-merge event, symbol "1" represents the merge event, namely $t_i = 0$ or 1, $i = 1, \ldots, n$. Vectors in matrix x_i have different dimensions, and therefore, in order to eliminate the dimension influence among them, and it can make the model be more robust, the input variables are normalized to the interval on [−1, 1].

4.2 Experimental Simulation

In all the experiments of simulation, the running environment of the algorithms is as following. (1) Operating system, Windows 7 ultimate version. (2) CPU, Intel core i5. (3) Memory, 2 GB. (4) Software, comparison experiments are made in Matlab R2016 (a). In this paper, it adopts the OS-ELM, SVM and BP algorithm for convenience to conduct the comparison of experimental effect. SVM algorithm package adopts by LIBSVM [18], BP algorithm adopts by Matlab Neural Network ToolboxTM. Sigmoidal function is chosen as the activation function of OS-ELM algorithm, SVM algorithm adopts RBF as the kernel function, The activation function of BP network is default tangent Sigmoidal function. There are also some main parameters needed to determine, such as the number of hidden layer nodes in OS-ELM and BP algorithm, the value of the penalty factor c and the kernel function radius g in SVM algorithm.

4.3 Simulation Based on U.S. Highway 101 Data

In case 1, a total of 686 sample data are obtained from U.S. Highway 101. In simulation, 80% of the sample data is used for training and 20% for testing [19, 20], so in which, 268 sample data for merge events and 281 sample data for non-merge events as training data, 68 sample data for merge events and 69 sample data for non-merge events as testing data. First, when the number of hidden layer nodes increases to a certain value, the test accuracy appears to stable or decrease status, the number of hidden layer nodes is 169 that makes test accuracy reach to maximum for the first time, as shown in Fig. 2. The experimental simulation result of OS-ELM algorithm is shown in Table 1.

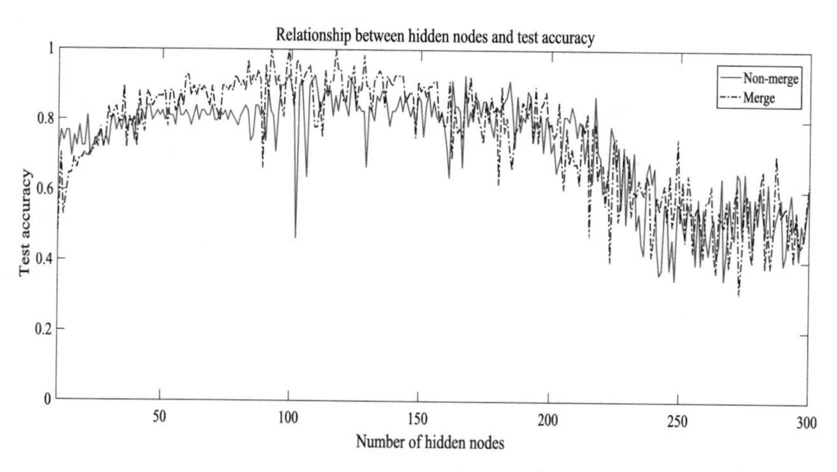

Fig. 2. Relationship between hidden layer nodes and test accuracy

Table 1. Experimental results of OS-ELM, SVM and BP algorithm

Algorithms	Decision	Number of training data	Number of testing data	Test accuracy (%)	Time (s)
OS-ELM	Merge	268	68	92.6	0.074
	Non-merge	281	69	94.2	
SVM	Merge	268	68	95.6	0.192
	Non-merge	281	69	94.2	
BP	Merge	268	68	94.1	1.856
	Non-merge	281	69	88.4	

Second, the method of determining the best parameters c and g in SVM algorithm. In the range of c and g taking value, the both are from 2–10 to 2–10, by means of using the grid search and the 3-fold cross validation method, it can obtain that the best c is 111.4305, g is 64 in SVM algorithm, as shown in Fig. 3. Third, the method of determining the number of hidden layer nodes of BP algorithm is the same as OS-ELM algorithm. So the number of hidden layer nodes is 422 that the BP algorithm test accuracy reach the maximum, and the experimental result of SVM and BP algorithm is shown in Table 1.

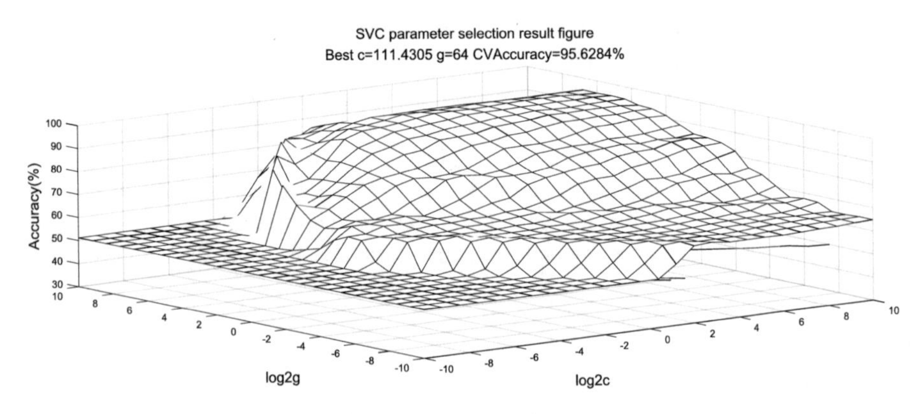

Fig. 3. SVM algorithm parameters optimization

Taking into account of the result of each running time, there would be inevitably small differences, so the running result of all the three algorithms takes the mean of 10 running results.

4.4 Simulation Based on U.S. Highway 101 and Interstate 80 Data

In case 2, a total of 263 sample data are obtained from Interstate 80, in which, 109 sample data of them are used for merge events, and another 154 sample data of them are used for non-merge events. Now the training data obtain from U.S. Highway 101 and Interstate 80 together, and testing data is the same as case 1. Determining the main parameters is just like case 1. First, the number of hidden layer nodes of OS-ELM algorithm is 131, as shown in Fig. 4. Second, the best c is 1024, and best g is 36.7583 in SVM algorithm, as shown in Fig. 5. Third, the number of hidden layer nodes is 517. Experimental results of OS-ELM, SVM and BP algorithm are shown in Table 2.

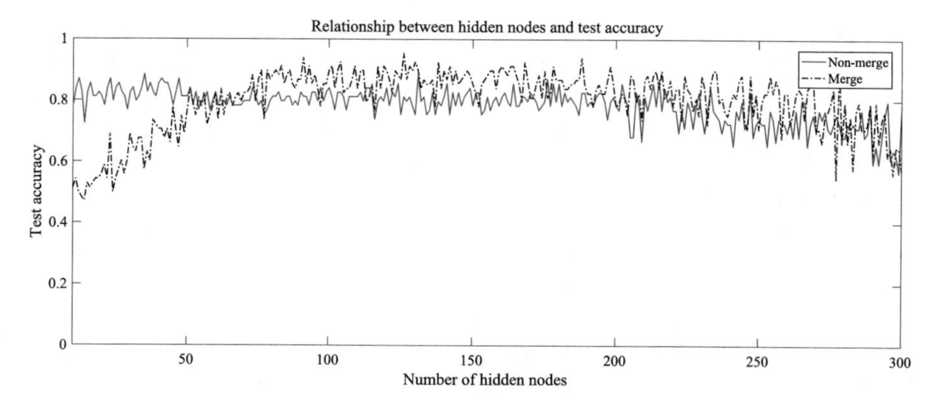

Fig. 4. Relationship between hidden layer nodes and test accuracy

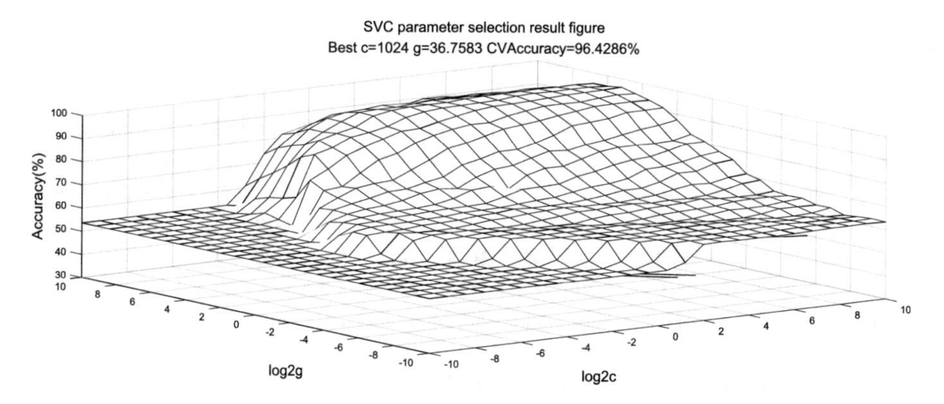

Fig. 5. SVM algorithm parameters optimization

Table 2. Experimental results of OS-ELM, SVM and BP algorithm

Algorithms	Decision	Number of training data	Number of testing data	Test accuracy (%)	Time (s)
OS-ELM	Merge	377	68	86.8	0.078
	Non-merge	435	69	87.0	
SVM	Merge	377	68	86.8	0.238
	Non-merge	435	69	88.4	
BP	Merge	377	68	88.2	3.19
	Non-merge	435	69	85.5	

4.5 Analysis of Experiment Result

From Table 1, it can be found that all the three algorithms have reached up to 88.4% or the above for the merge and non-merge predictive accuracy in highway merging area. In a lane changing event, not only the high predictive accuracy is important, but the running time of model is also very important. The predictive accuracy of SVM algorithm is slightly higher than OS-ELM algorithm, but the running time of OS-ELM algorithm is about 3 times faster than the SVM algorithm. Further, the running time of OS-ELM algorithm is about 25 times faster than the BP algorithm. The reason that why the mathematical model of OS-ELM algorithm is chose is quite simple, and it needs not to make the iteration operation.

From Table 2, U.S. Highway 101 and Interstate 80 data are used together to train the model, some of the U.S. Highway 101 data as testing data. OS-ELM algorithm predictive accuracy is 86.8% for merge events, and 87% for non-merge events, the simulation results demonstrated that the OS-ELM algorithm has stronger generalization ability. Further, the running time of OS-ELM algorithm is about 3 times faster than the SVM algorithm, and about 41 times faster than the BP algorithm, respectively. So it can better meet the rapid demand of running personalized lane changing model, this would prevent many traffic accidents.

5 Conclusions

Current ADAS provides personalized lane changing models usually combine several algorithms, which make the time for running personalized lane change models become longer. However, the fast speed of running the model is very important in highway lane changing. This paper presented a lane changing model based on OS-ELM algorithm, which has very fast learning speed and good generalization ability. By continuously adding data from the lane changing process to the model one-by-one, the lane changing model of the driver will be online updated constantly, further. When collecting data from a individual driver, and then retraining the model, the resulting model is suitable for the driver's lane changing maneuver. While the driver wants to make a lane changing decision, the personalized model can provide recommendation service to the driver. The research results show that the presented model can better provide recommendation in real-time with very high predictive accuracy.

Acknowledgment. Thanks to the support by the National Natural Science Foundation of China (Grant No. 61573075).

References

1. Gipps, P.G.: A model for the structure of lane-changing decisions. Transp. Res. Part B: Methodol. **20**(5), 403–414 (1986)
2. Wiedemann, R., Reiter, U.: Microscopic traffic simulation: the simulation system MISSION, background and actual state. Proj. ICARUS (V1052) Final. Rep. **2**, 1–53 (1992)
3. Ahmed, K.I.: Modeling drivers' acceleration and lane changing behavior. Massachusetts Institute of Technology (1999)
4. Toledo, T., Koutsopoulos, H.N., Ben-Akiva, M.: Estimation of an integrated driving behavior model. Transp. Res. Part C: Emerg. Technol. **17**(4), 365–380 (2009)

5. Das, S., Bowles, B.A.: Simulations of highway chaos using fuzzy logic. In: NAFIPS. 18th International Conference of the North American Fuzzy Information Processing Society, pp. 130–133. IEEE (1999)
6. Wu, J., Brackstone, M., McDonald, M.: Fuzzy sets and systems for a motorway microscopic simulation model. Fuzzy Sets Syst. **116**(1), 65–76 (2000)
7. Wu, J., Brackstone, M., McDonald, M.: The validation of a microscopic simulation model: a methodological case study. Transp. Res. Part C: Emerg. Technol. **11**(6), 463–479 (2003)
8. Kumar, P., Perrollaz, M., Lefevre, S., et al.: Learning-based approach for online lane change intention prediction. In: 2013 IEEE Intelligent Vehicles Symposium (IV), pp. 797–802. IEEE (2013)
9. Yao, W., Zhao, H., Bonnifait, P., et al.: Lane change trajectory prediction by using recorded human driving data. In: 2013 IEEE Intelligent Vehicles Symposium (IV), pp. 430–436. IEEE (2013)
10. Hou, Y., Edara, P., Sun, C.: A genetic fuzzy system for modeling mandatory lane changing. In: 2012 15th International IEEE Conference on Intelligent Transportation Systems (ITSC), pp. 1044–1048. IEEE (2012)
11. Hou, Y., Edara, P., Sun, C.: Modeling mandatory lane changing using Bayes classifier and decision trees [J]. IEEE Trans. Intell. Transp. Syst. **15**(2), 647–655 (2014)
12. Dou, Y., Yan, F., Feng, D.: Lane changing prediction at highway lane drops using support vector machine and artificial neural network classifiers. In: 2016 IEEE International Conference on Advanced Intelligent Mechatronics (AIM), pp. 901–906. IEEE (2016)
13. Huang, G.B., Zhu, Q.Y., Siew, C.K.: Extreme learning machine: theory and applications. Neurocomputing **70**(1), 489–501 (2006)
14. Ojala, T., Pietikainen, M., Maenpaa, T.: Multiresolution gray-scale and rotation invariant texture classification with local binary patterns. IEEE Trans. Pattern Anal. Mach. Intell. **24** (7), 971–987 (2002)
15. Banerjee, K.S.: Generalized Inverse of Matrices and Its Applications. Wiley, Hoboken (1971)
16. Liang, N.Y., Huang, G.B., Saratchandran, P., et al.: A fast and accurate online sequential learning algorithm for feedforward networks. IEEE Trans. Neural Netw. **17**(6), 1411–1423 (2006)
17. Ahmed, K.L., Ben-Akiva, M., Koutsopoulos, H., et al.: Models of freeway lane changing and gap acceptance behavior. Transp. Traffic Theory **13**, 501–515 (1996)
18. http://www.csie.ntu.edu.tw/~cjlin/libsvm/
19. Martin, T.M., Harten, P., Young, D.M., et al.: Does rational selection of training and test sets improve the outcome of QSAR modeling? J. Chem. Inf. Model. **52**(10), 2570–2578 (2012)
20. Edara, P., Teodorovi, D., BAIK, H.: Using neural networks to model intercity mode choice. In: Intelligent Engineering Systems Through Artificial Neural Networks, vol. 17. ASME Press (2007)

Study of the Impact on Traffic Capacity Made by Opening Block Type Housing Estates

Yuan Duan(✉), Chen Juan, Geng Zhang, and Hu Guo-yi

Department of Public Foundation, Guangdong University of Science and Technology, Dongguan 523083, China
dy_01155@126.com

Abstract. The government is putting forward to promote blocks and open some closed housing estates recently. It brings a heated argument about the influence on traffic capacity of the city. In this thesis, an appropriate evaluation index system is based on Analytic Hierarchy Process (AHP). With the help of quantitative comparison and programming in C, a mathematical model about traffic capacity influenced by open housing estates of block is well built. It comes out that traffic capacity would be greatly increased by opening more closed housing estates. According to this conclusion, we provide some suggestions of optimizing the traffic to traffic department in the end.

Keywords: Traffic capacity · Wardrop equation · Poisson distribution Quantitative comparison

1 Introduction

According to the real traffic situation, the model is evaluated by Analytic Hierarchy Process (AHP), which includes intersection index, road section index and district index. We collect data on traffic volume and car distance of the intersection, as well as data from traffic department on vehicle saturation and delay time. Combining with comprehensive analysis on various types of housing estates in different district, an objective and scientific index system that evaluates opening housing estates' impact on roads around is well built.

We choose a no signal intersection to build the model. Traffic conflict volume, critical crossing gap and follow-up time are used to figure out total traffic saturation and delay time with the help of formula method. We take an open house holding as an example, studying car volume and traffic conflict volume around it and figuring out total traffic saturation and delay time on the basis of the model. Through quantitative comparison, the study shows that opening housing estates could greatly improve traffic capacity of the roads around. Therefore, suggestions of improving city planning and traffic are given in the end of the paper.

B.-Y. Cao and Y.-B. Zhong (Eds.): ICFIE 2017, AISC 872, pp. 394–405, 2019.
https://doi.org/10.1007/978-3-030-02777-3_35

2 Determine the Evaluation Index System

2.1 Basic Conditions Assumed

(1) Traffic flow fluctuation is stable as it's not a holiday.
(2) Drivers are familiar with the road.
(3) Model conform to Wardrop Equation.
(4) Traffic condition is normal and no trouble because of the traffic accident.
(5) Different types of vehicles are converted to ordinary cars. Mid-size, large car, and giant car are written with 1, 1.2, 2.0 and 4.0 respectively.

2.2 Mark of Each Index

The indicators are listed in the following Table 1.

Table 1. The indicators

v_0	Maximum speed limit
L	Length of the vehicle
v_p	Traffic conflict by different driving types
K	The critical crossing gap
t	Follow-up time
C_n	Traffic capacity
V/C	Saturation
d	Delay time
C	Design capacity

2.3 Determine the Capacity Evaluation Index System

At present, each nation has different standards of the problem of traffic congestion. In order to reflect the degree of congestion quantitatively, we set up an index system under the principle of completeness, objectivity, operability and scientific nature [1, 2], to evaluates road traffic capacity in the chart below.

Considering real traffic situation, this index system includes intersection, road section and regional indicator. Intersection indicators contain car volume, saturation, delay time, length of queue and re-queue by traffic signal. Road indicators contain car volume and length of the road. In this thesis, we choose car volume of intersection, saturation and delay time to evaluate the road capacity (Table 2).

Traffic capacity refers to car volume that a road can handle, which means in certain period of time (usually 15 min or 1 h) and ordinary road state, the amount of cars that travel through a point of the road [3, 4]. It's generally indicated by veh/h, pcu/h, with a basic unit of pcu/h.ln [5, 6]. Car volume of the intersection refers to the amount of car traveling on the road. Afterwards, saturation can be gained from dividing car volume

Table 2. The evaluation index system

	First indicators	Secondary indicators
The evaluation index system	The intersection indicators	Car distance
		Extension of the time
		Road speed limit
		Driving speed
		Car volume
		Traffic capacity
		Car length
	Road indicators	Length
		Width
	Regional indicators	Suburb
		Urban area

by traffic capacity, which is an indicator showing the burden of the road. It is usually called V/C. The total traffic saturation is like

$$x\frac{\sum v}{c} = \mathrm{E},$$

where E equals to the summation of car saturation from three priorities [7, 8].

General delay time is an indicator that describes extra traveling time, which caused by car deceleration as the road condition is bad. We would compare data before opening the housing estates to that of afters. If car volume rises to a high level, meanwhile saturation and delay time stay in low level after opening the housing estates, it means road capacity improves.

3 No Signal Intersection Model and the Analysis of Solution

3.1 No Signal Intersection Traffic Model

Traffic congestion takes place frequently in intersections. Lantianti Road is a 600 m long road near a housing estate Lijingyuan and the national highway. It collects Qingyun Road and Green Way with only two lanes. Traffic on the national highway is heavy, while Lantianti Road is the only road through three large housing estates. Therefore, it is always congested in Lantianti Road, but there's no way to extend it (Fig. 1).

It is known that congestion often happens in the evening. To the left side of the road, it's two large housing estates and the Green Way. To the right, there's a park in which lots citizens jog after dinner.

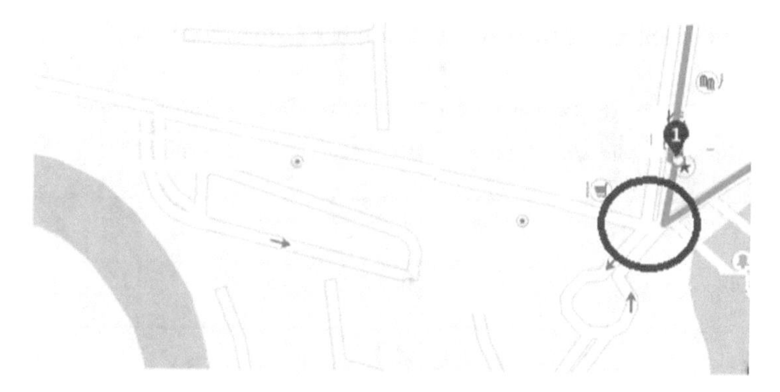

Fig. 1. The two-lane no signal intersection of Qingyun Road

Analysis on the running state of no signal intersection and the traffic (Fig. 2):

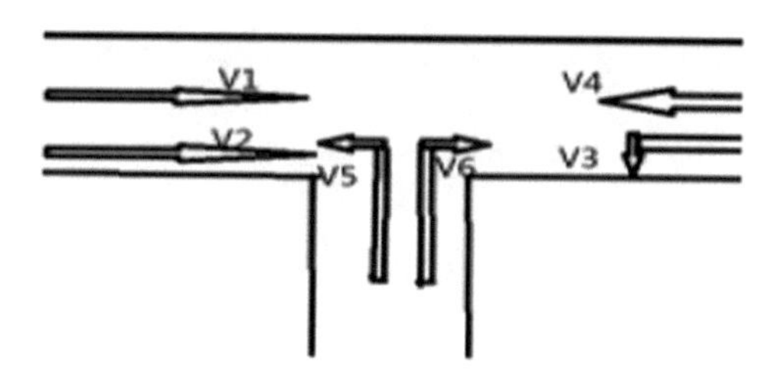

Fig. 2. No signal intersection

No signal intersection, all kinds of traffic priority is as followes [9, 10]:

Priority level 1: the main road go straight and turn right, direction of 1, 2, 4.
Priority level 2: turn left in the main road and turn right at branch, direction of 3, 6.
Priority level 3: turn left at branch, direction of 5.

Mathematical model is set up [11]:

$$C_n = \frac{v_p e^{-v_p \cdot \kappa}}{1 - e^{-v_p \cdot t}}.$$

where v_p is different ways of driving traffic conflict. κ is the critical crossing gap. t is the follow-up time.

3.2 No Signal Intersection Model Data Preprocessing

(1) The conversion coefficient: in order to add up equivalent traffic flow, we need to convert different types of car (Table 3):

Table 3. Convert coefficient method

Small car	Midsize car	Large car	Container	Motorcycle	Bike
1.0	1.2	2.0	4.0	0.4	0.2

(2) The equivalent traffic summary is as follows (Tables 4 and 5):

Table 4. The two-lane traffic flow (m/h) outside the housing estates

	Small car	Midsize car	Large car	Container	Motorcycle	Bike	Equivalent car
East import	83	23	4	0	8	8	124
West import	44	12	3	0	6	15	70

Table 5. Qingyun Road traffic (m/h)

	Small car	Midsize car	Large car	Container	Motorcycle	Bike	Equivalent car
Turn left	21	12	3	0	13	3	48
Turn right	25	5	0	0	6	5	35

(3) The determination of critical crossing gap and follow-up time (Tables 6 and 7):

Table 6. κ Recommended value (2/4) (unit: s)

Vehicle type	Small car	Midsize car	Large cars	Articulated vehicle
Main road left	6.0	7.0	8.0	8.0
Branch left	6.5	7.5	8.5	9.0
Branch right	4.0	4.5	5.0	5.5

Table 7. t Recommended value (2/4) (unit: s)

Vehicle type	Small car	Midsize car	Large car	Articulated vehicle
Main road left	2.0	2.5	3.0	4.0
Branch left	2.5	3.0	3.5	4.0
Branch right	1.6	2.5	2.5	3.0

(4) The conflict amount (Table 8):

Table 8. The conflict of traffic flow in different driving ways

Driving way	Label	Conflict amount	In total
Main road left	3	V1+V2	70
Branch left	6	V1+0.5V2	53
Branch right	5	V1+0.5V2	53

3.3 Solution of no Signal Intersection Model

(1) Formula of capacity calculation [11]

$$C_n = \frac{v_p e^{-v_p \cdot \kappa}}{1 - e^{-v_p \cdot t}}.$$

where v_p is conflict amount of different driving ways, κ is critical crossing gap, t is follow-up time.

The first priority traffic maximum capacity

$$C_0 = 2500\ (\text{pcu/h}).$$

and the second priority traffic maximum capacity:

$$C_{c,3} = C_{n,3} = \frac{70 \times e^{-70 \times 6/3600}}{1 - e^{-70 \times 2/3600}} = 1438\ (\text{pcu/h}).$$

$$C_{n,6} = \frac{53 \times e^{-53 \times 4/3600}}{1 - e^{-53 \times 1.6/3600}} = 1680\ (\text{pcu/h}).$$

The third priority traffic maximum capacity:

$$P_{0,3} = \left(1 - \frac{48}{1638}\right) \times \left(1 - \frac{35}{2080}\right) = 0.954.$$

$$C_{n,5} = \frac{53 \times e^{-53 \times 6.5/3600}}{1 - e^{-53 \times 2.5/3600}} = 1205\ (\text{pcu/h}).$$

$$C_{c,5} = P_{0,3} \cdot C_{n,5} = 0.954 \times 1205 = 1149\ (\text{pcu/h}).$$

Total capacity before opening of housing estates:

$$C_n = 1438 + 1680 + 1205 = 4323.$$

(2) Saturation calculation and service level evaluation.
The first priority traffic saturation:

$$\frac{V_1}{C_0} = \frac{124 + 70}{2500} = 0.077.$$

The second priority traffic saturation:

$$\frac{V_2}{C_{c,3}} = \frac{48}{1638} = 0.029.$$

$$\frac{V_2}{C_{c,6}} = \frac{35}{2080} = 0.017.$$

The third priority traffic saturation:

$$\frac{V_3}{C_{c,5}} = \frac{48}{1149} = 0.042.$$

The total traffic saturation:

$$x\frac{\sum V}{C} = 0.077 + 0.029 + 0.017 + 0.042 = 0.165.$$

Saturation of x is less than 0.25, the calculation formula for the delay:

$$d = 0.13e^{4.28x} = 0.13 \times e^{4.28\times 0.165} = 2.6\ (\mathrm{s}).$$

After opening, the capacity calculation formula [11]:

$$C_n' = x\frac{v_p e^{-v_p \kappa}}{1 - e^{-v_p t}}$$

As car flow gets separated, traffic volume change as follows (Tables 9, 10 and 11):

Table 9. After opening, outer two-lane traffic volume (m/h)

	Small car	Midsize car	Large car	Container truck	Motorcycle	Bike	Equivalent car
East import	42	11	4	0	3	2	62
West import	15	10	3	0	2	5	35

Table 10. After opening, Qingyun Road traffic volume (m/h)

	Small car	Midsize car	Large car	Container	Motorcycle	Bike	Equivalent car
Turn left	14	7	3	0	5	1	30
Turn right	16	5	0	0	1	0	22

Table 11.

Driving way	Label	Conflict in circulation	In total
Main road left	3	V1+V2	35
Branch left	6	V1+0.5V2	29
Branch right	5	V1+0.5V2	29

With the calculation above, we work out:

$$C_{c,3} = C_{n,3} = \frac{35 \times e^{-35\times6/3600}}{1 - e^{-35\times2/3600}} = 980 \text{ (pcu/h)}.$$

$$C_{n,6} = \frac{29 \times e^{-29\times4/3600}}{1 - e^{-29\times1.6/3600}} = 2160 \text{ (pcu/h)}.$$

$$P_{0,3} = \left(1 - \frac{30}{980}\right) \times \left(1 - \frac{22}{2160}\right) = 0.959.$$

$$C_{n,5} = \frac{29 \times e^{-29\times6.5/3600}}{1 - e^{-29\times2.5/3600}} = 1376 \text{ (pcu/h)}.$$

$$C_{c,5} = P_{0,3} \cdot C_{n,5} = 0.959 \times 1376 = 1319 \text{ (pcu/h)}.$$

Capacity after opening:

$$C'_n = 980 + 2160 + 1376 = 4516 \text{ (pcu/h)}.$$

$$\frac{V_1}{C_0} = \frac{62+35}{2500} = 0.039.$$

$$\frac{V_2}{C_{c,3}} = \frac{30}{980} = 0.031.$$

$$\frac{V_2}{C_{c,6}} = \frac{22}{2160} = 0.010.$$

$$\frac{V_3}{C_{c,5}} = \frac{30}{1319} = 0.023.$$

The total traffic saturation after opening:

$$x\frac{\sum V}{C} = 0.039 + 0.031 + 0.010 + 0.023 = 0.103.$$

The delay time after village opening:

$$d = 0.13e^{4.28x} = 0.13 \times e^{4.28 \times 0.103} = 2.1\ (\mathrm{s}).$$

The ratio of the capacity before and after opening:

$$x\frac{v_p e^{-v_p \cdot K}}{1 - e^{-v_p \cdot t}} \div \frac{v_p e^{-v_p \cdot K}}{1 - e^{-v_p \cdot t}} = \frac{4516}{4313} = 1.045 > 1$$

The traffic capacity of community opened:

$$C_n' = \frac{1.045 v_p e^{-v_p K}}{1 - e^{-v_p t}}.$$

So, capacity of the road after opening the housing estate is 1.045 times of before, which means road traffic conditions are effectively improved.

3.4 No Signal Intersection Model Results

According to service standard of no signal intersection, delay time of 2.1 < 15 indicates that Qingyun Road and the intersection outside the housing estate are in grade one standard, which means traffic flow is smooth with a bit congestion.

4 Analysis of Traffic Impact Before and After Opening of Housing Estates of Block

Residential areas are divided into the following categories: block layout, layout of centripetal type, axis type layout, enclosure layout, intensive layout and metaphor type layout [6]. We now analyze capacity of open block housing estates due to the no signal intersection model:

For the plot after opening, the capacity calculation formula is (Fig. 3):

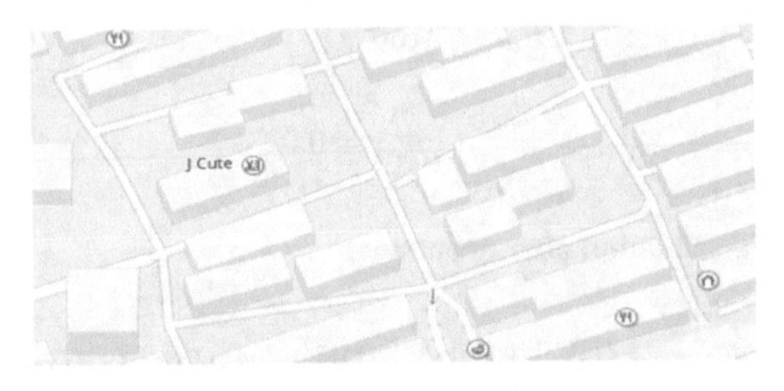

Fig. 3. Block type housing estate

$$C_n = x\frac{v_p e^{-v_p K}}{1 - e^{-v_p t}}.$$

Because car flow is separated after the opening, traffic volume of the road would be changed (Tables 12, 13 and 14).

Table 12. Traffic volume of Dongxing Road after opening (m/h)

	Small car	Midsize car	Large car	Container	Motorcycle	Bike	Equivalent car
East import	122	38	61	7	5	0	320
West import	130	36	57	5	7	2	309

Table 13. Traffic volume of Swan Lake Road after opening (m/h)

	Small car	Midsize car	Large car	Container	Motorcycle	Bike	Equivalent car
Turn left	47	21	3	0	0	3	79
Turn right	39	14	1	0	7	1	61

Table 14. Conflict of traffic flow after opening

Drive way	Conflict combined circulation
South Lane road left	128
North Lane road left	114
North Lane road right	114

Using formula above:

$$C_{c,3} = C_{n,3} = \frac{128 \times e^{-128 \times 6/3600}}{1 - e^{-128 \times 2/3600}} = 1493 \text{ (pcu/h)}.$$

$$C_{n,6} = \frac{114 \times e^{-114 \times 4/3600}}{1 - e^{-114 \times 1.6/3600}} = 2009 \text{ (pcu/h)}.$$

$$P_{0,3} = \left(1 - \frac{79}{1493}\right) \times \left(1 - \frac{61}{2009}\right) = 0.919.$$

$$C_{n,5} = \frac{114 \times e^{-114 \times 6.5/3600}}{1 - e^{-114 \times 2.5/3600}} = 1289 \text{ (pcu/h)}.$$

$$C_{c,5} = P_{0,3} \cdot C_{n,5} = 0.919 \times 1289 = 1184\ (\text{pcu/h}).$$

$$\frac{V_1}{C_0} = \frac{320 + 309}{2500} = 0.252.$$

$$\frac{V_2}{C_{c,3}} = \frac{79}{1493} = 0.053.$$

$$\frac{V_2}{C_{c,6}} = \frac{61}{2009} = 0.030.$$

$$\frac{V_3}{C_{c,5}} = \frac{79}{1319} = 0.060.$$

Total traffic saturation after opening:

$$x\frac{\sum V}{C} = 0.252 + 0.053 + 0.030 + 0.060 = 0.395.$$

Formula of delay after opening:

$$d = 0.13e^{4.28x} = 0.13 \times e^{4.28 \times 0.395} = 7.05\ (\text{s}).$$

Therefore, capacity of block type housing estate is greatly improved after opening.

5 Conclusion and Advice

Suggestions of improving city planning and traffic are as follows:

(1) To the construction of residential inside. Inner structure of newly built housing estates should follow more reasonable rules. Dead end road and sealed tree form community should be avoided. The study recommends block type as the typical layout. To those existing community, connecting inner path and opening some appropriate residential are good ways of improving traffic capacity.
(2) To urban construction outside residential. By using the modern advanced science and technology: GIS software, traffic volume's data of all roads can be collected, stored and analyzed, giving out an optimal way to avoid traffic congestion. All communities should link to the main highway, which combines inner paths of housing estates and traffic network of the city together. This requires a deep popularization of public transport network.
(3) Traffic management advice after opening blocks. On the basis of traffic regulation, appropriate rules of using paths in the open blocks should be formulated. For example, these inner paths open only in rush hour. No cars should be allowed to travel during night time. Disorderly parking is also forbidden.

References

1. Duan,Y., Gao, T.: Based on the fuzzy comprehensive evaluation method of urban road congestion evaluation method research. In: Logistics Technology, 15–33 October (2013)
2. Feng, C., Liang, J.: Solve the more general travelling salesman problem. AMSE J. Ser.: Model. D **35**(1), 9–23 (2014)
3. Cong, X.: Besearch to the influential factors of urban road traffic capacity. [DB/OL] college students paper joint than in the library, May (2016)
4. Feng, C., Liang, J.: The solution of the more general traveling salesman. AMSE J. Ser.: Adv. A **51**(1), 27–40 (2014)
5. Li, J.: Based on the influence of road network logistics cost research. Master thesis, Hefei University of Technology, pp. 6–48, April (2009)
6. Yu, C.: Thinking on some issues of traffic management. Digital communication world, pp. 1–38 October (2016)
7. Wang, Y.: Rail transit network changes and the response of the urban space development strategy research. Tianjin University Doctoral Dissertation, pp. 1–117, December (2013)
8. Duan, Y.: Optimization design of the single processor scheduling algorithm in real-time system research. J. Oper. Res. **17**(1), 27–34 (2013)
9. Lin, L., Cheung, W.: Petri network application in uncontrolled intersection conflict analysis. Forest engineering, pp. 60–63 November (2009)
10. Duan, Y.: Comparative study of different genetic operator combination to solve TSP problem. Sci. Technol. **28**(5), 27–31 (2012)
11. Wang, Z.: Road traffic capacity analysis (new). [DB/OL] baidu library, September (2013). https://wenku.baidu.com/view/0cc8620dbd64783e09122b8b.html

Discussion on Spread of Fuzzy Mathematics from Publication of *Fuzzy Sets Theory Preliminary*

Liao Yun-zhang[1], Wang Pei-hua[1,2], Bing-yuan Cao[1,2,3(✉)], and Hao-ran Lin[4]

[1] Guangzhou Vocational College of Science and Technology, Guangzhou 510550, China
j_ifiecao@163.com
[2] School of Mathematics and Information Science, Guangzhou University, Guangzhou 510006, China
[3] Foshan University, Foshan 528000, China
[4] Shanghai China High School, Shanghai, China

Abstract. The first textbook in the world for high school fuzzy mathematics on Fuzzy Sets Theory Preliminary was published and issued in February 2018. Based on the necessity, feasibility, and practicality of high school mathematics courses to disseminate fuzzy mathematics, the authors, who combined high school mathematics at home and abroad, put forward for the first time a proposal for curriculum standards for elective courses in General High School in the development trend of curriculum modernization, which offers a preliminary course on "Fuzzy Set Theory—High School Mathematical Expansion Course", hoping that the proposal will provide inspiration for mathematics education in high schools in the future.

Keywords: Fuzzy Sets Theory · High school mathematics textbooks
Curriculum standards · Innovation · Spread

1 Preface

In 2015, invited to visit China, Professor Zade participated in the founding meeting of the Fuzzy Information and Engineering Branch of the Operations Research Society of China, hosted by Guangzhou University, and delivered an enthusiastic speech. During the meeting, he read the preliminary "Introduction to Fuzzy Mathematics (High School Mathematics Development Course)", a Chinese high school mathematics textbook sent by a senior Chinese teacher named Lin Haoran from Shanghai Middle School, China. He was very happy to ask if he had an English version. This question has injected great momentum for us to publish the world's first high school mathematics textbook for students. The world-class publisher Springer and China Science and Education Publishing House jointly published in September 2015 a Chinese version of the book

B.-Y. Cao and Y.-B. Zhong (Eds.): ICFIE 2017, AISC 872, pp. 406–417, 2019.
https://doi.org/10.1007/978-3-030-02777-3_36

"Introduction to Fuzzy Mathematics (High School Mathematics Extension Course) - Will Washing Machines Think? (ISBN 978-988-15759-1-3). After arduous efforts, the English version of A Further-Studied Textbook for Senior Middle School Mathematics, Fuzzy Sets Theory Preliminary—Can a Washing Machine Think? (ISBN 978-3-319-70749-5) was also officially published in February 2018 in the world. Up to now, Professor David Zade has been in paradise for half a year. He can not see the publication of this book, but his lifelong history of creating, popularizing, and perfecting Fuzzy Sets will go down in history. By his inspirit, we popularize Fuzzy Sets, presenting the textbook to Mr. Zade to complete his ideal.

From October 25 to 26, 2016, the Xiangshan Science Conference, a high-level academic event of the Chinese scientific community, was held in Xiangshan Hotel, Beijing, China. The meeting recommended that the Academic Degrees Committee of the State Council upgrade our current "secondary disciplines of intelligent science and technology" to the first-level disciplines in order to form a systematic and complete an intelligent science and technology personnel training system. At the same time, it suggests that the Ministry of Education open "intelligent science and technology in primary and secondary schools". "Basic knowledge" courses to develop extra-curricular interest-building activities. As for fuzzy sets, the theory and application of fuzzy sets is playing an important part to the basic knowledge of intelligent science and technology. Therefore, Chinese scholars are suppose to use this strong force to work hard to incorporate the book, Fuzzy Sets Theory Preliminary—Can A Washing Machine Think?, into the textbooks on the basic knowledge of intelligent science and technology in primary and secondary schools. Scholars from all over the world are also suggested to use the book as a new starting point to speed up the popularization of fuzzy mathematics in secondary schools.

2 Contents of *Fuzzy Sets Theory Preliminary*

The book includes six chapters with its preface. The first chapter introduces Human Brain, Computer and Fuzzy Mathematics; the second chapter brings in Fuzzy Relation and Fuzzy Matrix, while reviewing Matrix; the third chapter describes the most successful application of fuzzy set: Fuzzy Control; Chapter 4 demonstrates Fuzzy Statistics and Fuzzy Probability and the last chapter involves the preliminary introduction of Fuzzy Linear Programming. At the end of the chapter, there appear exercises with Reading and Reflecting, and Summary. Its writing presents ideas in a simple and easy way and all of the explanation is completed within 20 or 40 periods.

Its cover and contents are presented as follows:

It contains 6 chapters with totally 176 pages covering the whole book, including 1. Introduction, 2 Matrix, Fuzzy Relation and Fuzzy Matrix, 3 Fuzzy Control, 4 Fuzzy Statistics and Fuzzy Probability, and 5 Fuzzy Linear Programming. In addition, the book is also equipped with its learning guides, dealing with its focus and difficulty materials, answering or suggesting exercises. Learning guide materials is to be published late.

3 Basic Mathematics Literacy of Fuzzy Collection for People and High School Students to Learn

3.1 Fuzzy Sets—A Modern Branch of Mathematics with Wide Applications

Mathematics is a science used to study real-world quantitative relations and spatial forms (Engels), which experiences four stages of development, including constant mathematics, variable mathematics, stochastic mathematics and fuzzy mathematics, each marking humanity changes in mathematical to understand objects and methods. Constant and variable mathematics are used for study of deterministic phenomena, random for random phenomena. Since the objective world has a large number of ambiguous phenomena (such as the computer's poor ability to recognize fuzzy phenomena), the generation of fuzzy mathematics is the historical necessity of the development of science, technology and mathematics.

Zadeh's "Fuzzy Sets" marks the birth of the discipline of [1]. The merit of fuzzy mathematics lies in fuzzy theory, which could be used for people to find effective ways and methods to ambiguous objects, so as to connect a gap between mathematics and uncertainty of deterministic objects, compensating the lack of deterministic and random mathematics, expanding the content of classic mathematics, and forming mathematics branches of fuzzy topology, fuzzy group theory, fuzzy graph theory, fuzzy probability, fuzzy programming, fuzzy logic and so on. At the same time, fuzzy mathematics has been widely used in almost all areas of natural sciences, social sciences, and engineering technology, forming an emerging technology based on fuzzy mathematics and fuzzy technology. Successful application examples have appeared in industrial and agricultural production, engineering, technology, information, medical, meteorological, economic and regulatory soft sciences, earthquake disaster prevention and reduction, artificial intelligence and computer, aerospace, military and the like.

3.2 Fuzzy Mathematics—Possess of Basic Mathematics Literacy for People

In contemporary information society, people face with increasingly complex and variety. In a complicated and large system, the complexity always accompanies with ambiguities. For students, who have entered the information age, no matter what careers they will engage in the future, they need to have the ability to process and interpret fuzzy information, in order to make judgments and decisions on a basis of information, including workers, farmers, businessmen, teachers, lawyers, administrators, service personnel, and even ordinary consumers and entertainers, who will encounter a large amount of unorganized forms of information in their daily lives. Therefore, the full use of information will undoubtedly bring incalculable benefits to the society. In other words, fuzzy mathematics is a basic mathematical literacy that contemporary people should possess.

As informatization drives the impact of industrialization on education, rapid development appears in various intelligent technology applications, containing fuzzy technology, experts expect that the integration of information technology and mathematics teaching, including fuzzy technology, will be a major trend of development in mathematical education. Besides, the inclusion of fuzzy mathematics in traditional mathematics could enable students to fully recognize the limitations of precision, understand the deficiencies of the law of exclusion, and have knowledge of the fact that not only ambiguity and accuracy are both inconsistent and mutually dependent, but also the mutual transformation of their internal relationship, in line with the students' cognitive development rules. Therefore, it is necessary to learn some preliminary fuzzy mathematics knowledge, ideas and methods as early as they could in high schools.

In teaching, a teacher does not go from concept to concept in preliminary elective course of fuzzy mathematics, but turns students' perspective to the side of real life. Through a piece of audio-visual teaching material firstly, he/she makes them know dissection of an automatic fuzzy washing machine, and from its operation process and working principle, introduces fuzzy control, fuzzy logic and fuzzy reasoning, pointing out that its computing tool is a fuzzy matrix, and leading students to the systematic study of initial courses of fuzzy mathematics.

3.3 China's Practice—New Attempts for High School Students to Learn Fuzzy Mathematics

Can high school students learn fuzzy mathematics? The first author, Lin Haoran, a senior mathematics teacher from China's middle school in Xuhui District, Shanghai, China, showed that if only a suitable entry topic is found (such as fuzzy sets, fuzzy matrix, fuzzy control, etc.), with meticulous selection of materials and materials, fuzzy mathematics is fully understandable for high school students. It is feasible for them to get knowledge of preliminary fuzzy mathematics.

From 1997 to 2007, Lin Hao-ran began with lecture on fuzzy mathematics for students in Senior One and Two in middle schools. From then on, he compiled the textbook "Introduction to fuzzy mathematics" (including teaching reference books) and conducted preliminary teaching of fuzzy mathematics elective courses. The course is welcomed by the students and for this course he got praises from experts and scholars at home and abroad. The paper "Can Computer Think? - Artificial Intelligence and Fuzzy Mathematics" was completed in 2003, which was awarded the first prize in the scientific paper of Xuhui District Youth Science and Technology Innovation Contest in Shanghai; in 2004 Design and manufacture of fuzzy control smart car won the First Prize of the Shanghai Computer Robot Contest; the "hydropower coal intelligent meter reading and billing system" was designed and produced in 2005, which won the 19th Shanghai Young People's Science and Technology Innovation Contest etc. One member of the research group of "Can a Computer think - artificial intelligence and fuzzy mathematics", Ding Shengxiong had an excellent entrance examination that year, went directly to study abroad.

In teaching, the initial elective course of fuzzy collection does not go from concept to concept, but turns the perspective to the side and turns to real life. First, through a piece of audio-visual teaching material, an automatic fuzzy washing machine is dissected, and from its operation process and working principle, it is blurred. Control, fuzzy logic and fuzzy reasoning, pointed out that its computing tool is a fuzzy matrix. Then start with a simple fuzzy statistics and introduce the fuzzy probability, so that students can learn from the middle school that fuzzy probability statistics and probability statistics are different. Finally, the introduction of fuzzy linear programming enables middle school students to understand that there is no absolute optimum in the world, and there may be suboptimal and satisfactory solutions. This leads to the introduction of a systematic study of initial courses in fuzzy mathematics.

(1) From {0, 1} to [0, 1], the fuzzy system is an innovation

The "Preliminary fuzzy system" dose not simply extend university courses to secondary schools. It popularizes not only the ready-made knowledge that, but the scientific spirit of searches and innovations.

We have positioned the enlightenment and popularization of fuzzy mathematics as follows:

(1) Introducing new ideas and perspectives so that high school students can broaden their horizons with new modes of thinking formed.
(2) Arousing their strong interest.

(3) Informing them of the fact that fuzzy mathematics remains to be explored and improved because it is not perfect yet.
(4) Not sitting idle and enjoying it - instead of simply applying the results of others (or foreigners), they should invest them and enjoy the joy of success after unrelenting efforts.

As in the previous year, the fifth gong that broke the plane geometry, "two parallel lines can intersect at infinity", opened up a new world of non-Euclidean geometry; breaking out of $x^2 > -1$ has an imaginary number, bringing the flourishing development of modern physics, so is fuzzy mathematics, which removes the worship of elemental determinism in ordinary collections, introduces uncertainty, and extends the ordinary set of boundary determinations to the fuzzy set of uncertain borders; The two outcomes of the set are either one or the other, corresponding to two values $\{0, 1\}$, and the multiple results of the fuzzy set also correspond to an infinite number of values $[0, 1]$, creating a fuzzy mathematics a new era, which is another milestone in the development of mathematics.

Traditional mathematics in middle school pursue only correct solutions, while fuzzy mathematics seeks the most satisfactory solutions. It is not necessarily unique, but it is more adaptable to the demands of today's information age. If you ride a bicycle, you must master the balance. It is related to speed, direction, center of gravity, posture and other factors. It is not only a state that can achieve balance, but there are infinite combinations that can easily meet the standard. As long as safety is guaranteed, there exist different and best combinations in different road conditions. In life, this is not only the right thing to do, but only the most satisfying situations abound. In production and business activities, using this mode of thinking can save resources, costs, and achieve an optimization goal. Allowing middle school students to engage early in modern mathematics thinking is not passively accepting ready-made knowledge, but actively "learning" how to use a new perspective to observe the world, analyze problems, and help foster their innovation consciousness and exploration spirit.

(2) Industry 4.0 Intelligent - the Running of the Future

There is a view that Industry 1.0 is marked by a steam engine; Industry 2.0 is marked by an electric motor; Industry 3.0 is informatized; and Industry 4.0 denotes intelligent.

In January 2010, a computer program called "Nell" was launched at the Computer Laboratory of Carnegie Mellon University in the United States. "Nell" is a self-learning program written to self-learn and grow in Internet, learn to judge and analyze. Scientists are patient with Nell. Perhaps a few decades later, "Nell" will suddenly "awaken" in the invisible and inaccessible Internet, creating consciousness and being as intelligent as human beings.

At present, the development momentum of AI is sweeping across the globe, and fuzzy sets plays a decisive role in AI. In the "Elementary Collection of Fuzzy Sets" in the textbook, we simulated the thinking through three control rules such as IF ATHEN B; IF A THEN B, ELSE C; IF A AND B THEN C and their fuzzy mathematics models, making students feel that the washing machine seems to be thinking: "What kind of clothing does the owner wash for me? What kind of washing method should I use to wash it and save water?" How does the washing machine do this? Does the machine really think? Students are very interested in these questions.

Robots controlled by artificial intelligence have been able to do things that many people cannot. Combining teaching with fuzzy control, Teacher Lin Haoran once led students to participate in a Robot World Cup soccer game. Robots now have more judgment and agility than robots before. Both the ball control and the ball-handling ability have made great progress. The international organizers confidently announced that the champion of the World Cup robot soccer will compete with a real-life team in 50 years! Who is the winner? Can the robot's intelligence reach or even exceed humanity? Students were inspired with great suspense and curiosity inquiring… Many students wrote in the feedback after they finished this course that I see the future and the future is running!

(3) Innovation Education Initiated at Children Age—"Internet +"

On January 9, 2006, the China Science and Technology Conference announced the goal of China's science and technology development next 15 years: an innovative country will be built in 2020. Can people become competent, who have a mode of thinking under the influence of exam-oriented education for a long time?

Ten years ago, a newsletter "25 years: two wrong predictions" of the "China's Education News" was very illustrative. In June 1979, the education circles of China and the United States sent delegations to each other for mutual visits. After returning to China, the Chinese side reported that pupils in the second grade primary school in the United States did not know a word, and that the addition and subtraction was still pinching their heads, talking about inventions and creations all day long… Conclusion appeared that, in 20 years, China's science and technology and culture will surely catch up with a superior country. The Americans reported that Chinese pupils in primary schools liked to hold their hands on their chests during class, they liked to get up early… Conclusion is that Chinese students are the most diligent in the world. Their academic performance is comparable to any other country in the same grade. It can be predicted that in 20 years, China will surely leave the United States far behind in terms of technology and culture. Now, 35 years has passed. In the 20 years from 1979 to 1999, the United States has nurtured 43 Nobel Laureates and 197 knowledgeable billionaires. In 2003, there were 11 Nobel Prize winners, and the United States accounted for about half. Academician Qian Xuesen always thought when he was alive: "Why does our country's education train no outstanding talents?" If our young students of one session or another only sink up and down in the sea to cope with examinations, they will only master ways to questions instead of being innovative. Then rejuvenation of our country through science and education became an empty talk. There are few winners of the Nobel Prize in China. The 2015 Nobel winner, Professor Tu You-you, the experience of this three-no- scientist, showed that it is right for us to combine the road of teaching with practice by reforming the education of imperial examinations.

The college entrance examination with justice and equity as a bottom line can select high score students who will solve problems, but they cannot select potential creative talents. Putting university courses on the Internet is what Harvard and other famous foreign universities are doing. In the "Internet Plus" era, this new teaching model is not a problem for China in both technical and financial aspects. With strict entry and strict admission, no college entrance examination is required, so that children who want to enter university can study university courses online and only qualified students can graduate. This will be the fairest and the most justice cause. Dismantling the single-plank bridge of the sole entrance of the college can indeed results in many roads leading to Beijing. By liberating them from heavy examinations, the students could spend time satisfying curiosity and take time to explore and innovate. According to the feedback of the two surveys, students have expressed strong appeal to the college entrance examination reform.

4 Enlightenment and Popularity of Fuzzy Mathematics

4.1 Getting Wild of Enlightenment and Popularization of Fuzzy Mathematics

In the PISA test conducted by two global 65 countries and regions in 2009 and 2012, Shanghai students ranked first in mathematics, reading, and science literacy, which attracted wide attention from abroad. In February 2014, the Minister of Education of the United Kingdom led a delegation to Shanghai for a visit. Three schools were specially selected, three math classes were heard, and much appreciated. Then the Minister of Basic Education of South Africa also came to Shanghai for study.... On March 12th, Britain invited 60 mathematics teachers from Shanghai to attend 30 new mathematics centers in the UK to train English teachers for at least one month. In the summer of 2015, the most used tutoring book for students in Shanghai middle school was "a lesson and practice". It will be translated into English by East China Normal University Press. students in Shanghai have an excellent reputation for mathematics. Under such a background, the English version of "The initial set of fuzzy set theory—would washing machines think?" selected by Shanghai middle school students may be quite successful. And following the basic mathematics course of them, the other side of the Shanghai- middle- school students—the climax of the mathematics innovation curriculum that will be produced abroad - may attract high-level attention from the domestic levels. As a result, the enlightenment and popularization of fuzzy mathematics may get wild.

4.2 Teams from Popularization, Generating Talents, Promoting Future Development

We plan to complete three versions of the fuzzy mathematics textbook. The current popular version is written for ordinary high school students. Its goal is to make those in ordinary middle schools able to know a little fuzzy mathematics, so that it can gradually become a required course from the elective course of university mathematics and become a public course of science and engineering. Classes and basic courses will enable our country to have a large number of engineers, who will master capabilities in fuzzy control design in the future. After the public version got a firm foothold in the middle schools, the second version, an improved version, was suitable for students of key middle- school science classes. It systematically elaborated the basic theory of fuzzy mathematics. The goal is to enable these students to directly study after entering a bachelor's degree, so that they can directly participate in practice to perfect the basic theories of fuzzy mathematics. The third step, we plan to engage in a comprehensive version. The advantage of fuzzy mathematics lies in the fact that it is suitable for the soft science that precision mathematics cannot enter before, such as economics, humanities and other disciplines, doing some research, they could penetrate the basic ideas of fuzzy mathematics to other disciplines, so that they can learn to use mathematical vision to observe the world, analyze the surrounding things, find the right mathematical model, and solve some practical problems.

When reviewing our publication application, the Germans appreciated Lin's point of view on popularizing and creating teams, teams producing talents, and talents driving future development. It is considered that this is also an impetus for the modernization of university courses.

Fuzzy mathematics is a new emerging and still developing discipline that by using mathematical methods, we study and deal with vague phenomena. Since the publication of "Fuzzy Sets" by Professor Zadeh of the University of California in the United States in 1965, it has developed rapidly over the past 40 years and has been widely used in science, technology, economics, and social fields. Many universities use it for a graduate student as a compulsory subject for undergraduates, or Elective courses, and Lin Haoran, a maths teacher at China Middle School in Xuhui District, Shanghai, has also taught fuzzy mathematics in Senior One and Two in high school from 1997 to 2008. Practice shows that fuzzy mathematics has a good educational function in cultivating students' mathematics literacy, mathematics application awareness, and mathematics interest. Based on the current situation of emphasizing science and education innovation, it is necessary to expand the dissemination of fuzzy mathematics in mathematics curriculum in the high schools.

The theory of fuzzy sets was born in the United States (presented by Zadeh-1965) and grew up in Japan (An Hsueh-cheng et al. Here come some development, Fuzzy Control Device for Automatic Stopping of Trains - 1983, Fuzzy Control System for Subway Automatic Driving - 1984. Danji Junichi Et al. studied the pressure control system of an atomic energy reactor based on the fuzzy controller of the self-learning function - 1987; the simulation of the boiling water reactor of the 1.3 million kilowatt atomic power station - 1990, and the household appliances such as fuzzy washing machines and fuzzy refrigerators. The birth, etc.) This is an indisputable fact. Why does the United States fall behind Japan in this area? The reason is that a considerable number of people in the United States (including some academic authorities) objected the theory. Some people did not talk about in their mouths, did not write on their hands either, but they were opposed to the theory in their hearts. Among them, quite a few people held the right to examine and approve various funds of the country and right to publish papers, which influences creating an atmosphere of "fuzzy" researches in the country. Because some people from U.S. and large companies are reluctant to risking ventures in this area, they do not taste the joy like the sweetness of Japan's astonishing profits earned from "fuzzy" technology products in its own country and in the Eurasian and American markets. Fortunately, Professor Zadeh long used the Berkeley Initiative in Soft Computing (BISC) Group to liaise with scholars from all over the world. He adhered to the principle of academic supremacy and a hundred schools of thought contending, and enabled the development of ambiguous research. The "fuzzy fever" which he fervently hoped for has long been formed in Japan. Today, this textbook on fuzzy mathematics has been published and published by Springer, one of the world's top publishers. This is not only a gospel of the fuzzy world of China and the world, but also provides a good platform for the development of fuzzy sets. The reviewer's comments suggested that the title should be changed to "Preliminary Theory of Fuzzy Sets". We expect that our government and relevant departments, as well as experts and scholars, will further support and care for fuzzy mathematics. This will make the birth of fuzzy sets in the United States and growing up in Japan popularized in China, a country with a large area and a large population.

5 Inevitable Trend

In the past of century, people have continuously worked hard to modernize mathematics in middle schools. The modernization of middle school mathematics has also accumulated rich experience in many reforms and experiments.

It is not difficult to see from the modernization process of international middle school mathematics that it is not easy to infiltrate important and basic modern mathematics into middle school mathematics. It can be said that "the road is tortuous but the future is bright". Nevertheless, efforts to reform modern mathematics in mathematics for students in middle schools have never changed, which has always been the ideal of mathematics education.

Over the past 40 years of reforming and opening up, China's high school mathematics curriculum has undergone a series of reforms, including major steps in the modernization of its mathematical content.

It can be foreseen that with the continuous improvement, supplementation, and revision of "mathematics curriculum standards" (experimental) in high schools, fuzzy mathematics, a branch of modern mathematics dealing with blurry phenomena, is widely applied to high school mathematics elective courses. Therefore inclusion of fuzzy mathematics into the high school mathematics curriculum is an inevitable trend in the development of modern mathematics education.

According to the conditions of senior-high-school students' age and their mathematics foundation, scientific and reasonable selection of suitable fuzzy mathematics content is the key to communicate with them. We believe that the basic concept of ordinary high school dissemination of fuzzy mathematics is the following:

(1) Popularizing high school mathematics into an ordinary high school cannot and should not simply decentralize fuzzy mathematics curriculum from universities to secondary schools;
(2) Popularization should not be the prevalence of ready-made knowledge. It should be a popular science popularizing and seeking innovation;
(3) we should introduce new concepts and new perspectives on fuzzy mathematics so that high school students can broaden their horizons, forming a new thinking model;
(4) Activities are implemented through the development of scientific and technological innovations and scientific inquiries related to the subject of fuzzy

mathematics, so that students are initially aware of the course and develop their ability to use fuzzy mathematics to solve problems.

Considering that high school students have already learned basic knowledge such as assembly, simple logic, probability statistics, and linear programming, the following is a draft curriculum standard for the elective course "Introduction to Fuzzy Set Theory —High School Mathematics Extension" which is offered in ordinary high schools.

Fuzzy mathematics is a product of the information age, and it is a new branch of mathematics that studies can use it to deal with fuzzy phenomena and fuzzy concepts. On the one hand, fuzzy mathematics expands classical mathematics theory, on the other hand, it is widely used, especially in the industrialization of fuzzy technology, embodying strong vitality of fuzzy mathematics in the era of knowledge economy. The content and requirements of the book are as follows:

(1) By rich practical problems (such as fuzzy household appliances, Fuzzy technology and spacecraft manned spacecraft, Fuzzy technology and earthquake disaster reduction, fuzzy technology in information processing, fuzzy mathematics and artificial intelligence, fuzzy diagnostic methods, fuzzy geometric planning of power plant waste-water disposal, and the model $^{[6]}$, fuzzy comprehensive evaluation, etc.), students are required to experience the necessity of introducing fuzzy mathematics knowledge.
(2) By reviewing the general set and its operations, and introducing the basic concepts of fuzzy sets by analogy, a simple fuzzy set operation will be performed for students to understand fuzzy operators. The fuzzy sets will be aggregated into normal sets with cut-sets for students to recognize its significance of ambiguities.
(3) By understanding the concept and properties of ordinary relationships, and getting knowledge of the concepts, properties, operations and composition of fuzzy relationships, students will be required to gain access to the definition, operation and properties of fuzzy matrices, and solve simple fuzzy relation equations.
(4) Through the investigation of typical cases (such as fuzzy automatic washing machines, intelligent robots, etc.), students are needed to understand that the concept, basic principles, implementation methods and preliminary application examples of fuzzy control, experience fuzzy control are the most simple and most effective intelligent control.
(5) In the specific context, experiencing the connection and difference between ambiguity and randomness, through the fuzzy statistical test method, fuzzy data and its operation, the descriptive statistics of fuzzy data - fuzzy sample mean, fuzzy sample mode, fuzzy sample median, students will understand the basic method to fuzzy statistics, seeking the probability of some simple fuzzy events.
(6) By reviewing the general linear programming and its solution and through concrete examples, students can understand the binary fuzzy linear programming and its solution, and can use the binary fuzzy linear programming to solve some simple practical problems.
(7) Complete a study summary report.

References

1. Zadeh, L.A.: Fuzzy sets, linguistic variables and fuzzy logic. Science Press (1982)
2. Lin, H., Cao, B., Liao, Y.: A Preliminary in Fuzzy Sets Theory: The Development of Mathematics in Senior High School. Springer, China Science and Education Press (2015)
3. Haoran, L.: Fuzzy mathematics elementary: high school mathematics development course. Shanghai Chinese Secondary School (1997)
4. Haoran, L.: The practice of fuzzy math entering into the extended course in senior high school. J. Hainan Norm. Univ. (Nat. Sci.) **4**, 496–497 (2008)
5. Ministry of Education of the People's Republic of China. Mathematics Curriculum Standards of Senior High School, pp. 5–10. People's Education Press, Beijing (2003)
6. Bingyuan, C.: Application of fuzzy mathematics and system, pp. 34–81. Science Press, Beijing (2005)
7. Haoran, L.: Preliminary Fuzzy Mathematics–High School Mathematics Curriculum Development, pp. 1–129. Shanghai China Middle School (1997)
8. Yunzhang, L.: The spread of fuzzy mathematics in senior high school mathematics new curriculum. Oper. Res. Manag. Fuzzy Math. **1**(2), 43–52 (2014)

Author Index

B.-Y. Cao and Y.-B. Zhong (Eds.): ICFIE 2017, AISC 872, pp. 419–420, 2019.
https://doi.org/10.1007/978-3-030-02777-3

Printed in the United States
By Bookmasters